Supercritical Fluid Extraction of Nutraceuticals and Bioactive Compounds

Supercritical Fluid Extraction of Nutraceuticals and Bioactive Compounds

Edited by

Jose L. Martínez

CRC Press
Taylor & Francis Group
Boca Raton London New York

CRC Press is an imprint of the
Taylor & Francis Group, an **informa** business

CRC Press
Taylor & Francis Group
6000 Broken Sound Parkway NW, Suite 300
Boca Raton, FL 33487-2742

First issued in paperback 2020

ISBN-13: 978-0-367-57762-9 (pbk)
ISBN-13: 978-0-8493-7089-2 (hbk)

Library of Congress Cataloging-in-Publication Data

Supercritical fluid extraction of nutraceuticals and bioactive compounds / [edited by] Jose L. Martinez.
p. cm.
Includes bibliographical references and index.
ISBN 978-0-8493-7089-2 (alk. paper)
1. Supercritical fluid extraction. 2. Functional foods. 3. Bioactive compounds. I. Martinez, José L. (José Luis), 1966-

TP156.E8S835 2007
660.6'3--dc22 2007025441

Visit the Taylor & Francis Web site at
http://www.taylorandfrancis.com

and the CRC Press Web site at
http://www.crcpress.com

Dedication

To Marlene and Alejandro

Contents

Preface ix
Acknowledgments xi
Contributors xiii
Editor xv

Chapter 1 Fundamentals of Supercritical Fluid Technology 1
Selva Pereda, Susana B. Bottini, and Esteban A. Brignole

Chapter 2 Supercritical Extraction Plants: Equipment, Process, and Costs 25
Jose L. Martínez and Samuel W. Vance

Chapter 3 Supercritical Fluid Extraction of Specialty Oils 51
Feral Temelli, Marleny D. A. Saldaña, Paul H. L. Moquin, and Mei Sun

Chapter 4 Extraction and Purification of Natural Tocopherols by Supercritical CO_2 103
Tao Fang, Motonobu Goto, Mitsuru Sasaki, and Dalang Yang

Chapter 5 Processing of Fish Oils by Supercritical Fluids 141
Wayne Eltringham and Owen Catchpole

Chapter 6 Supercritical Fluid Extraction of Active Compounds from Algae 189
Rui L. Mendes

Chapter 7 Application of Supercritical Fluids in Traditional Chinese Medicines and Natural Products 215
Shufen Li

Chapter 8 Extraction of Bioactive Compounds from Latin American Plants 243
M. Angela A. Meireles

Chapter 9 Antioxidant Extraction by Supercritical Fluids 275
Beatriz Díaz-Reinoso, Andrés Moure, Herminia Domínguez, and Juan Carlos Parajó

Chapter 10 Essential Oils Extraction and Fractionation Using Supercritical Fluids 305

Ernesto Reverchon and Iolanda De Marco

Chapter 11 Processing of Spices Using Supercritical Fluids 337

Mamata Mukhopadhyay

Chapter 12 Preparation and Processing of Micro- and Nano-Scale Materials by Supercritical Fluid Technology 367

Eckhard Weidner and Marcus Petermann

Index 391

Preface

In the last decade new trends in the food industry have emerged, enhanced concern over the quality and safety of food products, increased preference for natural products, and stricter regulations related to the residual levels of solvents. Additionally, the nutraceutical and functional food sector represents one of the fastest growing areas in a consumer-driven trend market. These trends have driven supercritical fluid (SCF) technology to be a primary alternative to traditional solvents for extraction, fractionation, and isolation of active ingredients. The aim of this book is to present the current state of the art in extracting and fractionating bioactive ingredients by SCFs.

This book contains twelve chapters that primarily focus on implemented industrial processes and trends of the technology. The content of the chapters includes a review of the major active components in the target material, including chemical, physical, nutritional, and pharmaceutical properties; an analysis of the specific SCF process used; a comparison of traditional processing methods versus SCF technology; and a set of conclusions with supporting data and insight. A review of the fundamentals of the technology and an examination of SCF extraction systems and process economics are also included.

The contributing authors are international experts on the topics covered, and I would like to thank them for their thoughtful and well-written contributions. This book is addressed to food scientists, technologists, and engineers as well as other professionals interested in the nutraceutical and functional food sector. Additionally, I hope that this book will serve to stimulate academia and industry to search for new process and product developments as well as their industrial implementation.

Acknowledgments

The authors of the chapters of *Supercritical Fluid Extraction of Nutraceuticals and Bioactive Compounds* wish to acknowledge the following funding agencies for their support and assistance.

Dr. Feral Temelli would like to acknowledge the financial support from the Natural Sciences and Engineering Research Council of Canada (NSERC).

Dr. Fang et al. gratefully acknowledge the 21st COE program "Pulsed Power Sicence" and Wuhan Kaidi Fine Chemical Industries Co., Ltd., for their support.

Dr. Shufen Li would like to thank Dr. Can Quan, Dr. Shaokun Tang, Dr. Wenqiang Guan, Dr. Yongyue Sun, Ms. Luan Xiao, and Ms. Ying Zhang for their contributions to the research work as well as their assistance in the preparation of Chapter 7.

Dr. Maria Angela Meireles thanks CNPq, CAPES, and FAPESP for supporting the research done at LASEFI – DEA/ FEA – UNICAMP.

Dr. Eckhard Weidner and Dr. Marcus Petermann would like to thank their coworkers and students from the University Bochum as well as Prof. Knez and his coworkers from the University of Maribor and Adalbert-Raps Research Center. They would also like to thank Adalbert-Raps Stiftung, the European Union, the Ewald Doerken AG, and Yara Industrial GmbH for their support.

Contributors

Susana B. Bottini, Ph.D.
Planta Piloto de Ingeniería Química
Universidad Nacional del Sur
Bahía Blanca, Argentina

Esteban A. Brignole, Ph.D.
Planta Piloto de Ingeniería Química
Universidad Nacional del Sur
Bahía Blanca, Argentina

Owen J. Catchpole, Ph.D.
Industrial Research Limited
Lower Hutt, New Zealand

Iolanda De Marco, Ph.D.
Dipartimento di Ingegneria Chimica ed Alimentare
Universita di Salerno
Salerno, Italy

Beatriz Díaz-Reinoso, M.Sc.
Department of Chemical Engineering
Facultade de Ciencias de Ourense
Universidade de Vigo
Ourense, Spain

Herminia Domínguez, Ph.D.
Department of Chemical Engineering
Facultade de Ciencias de Ourense
Universidade de Vigo
Ourense, Spain

Wayne Eltringham, Ph.D.
Industrial Research Limited
Lower Hutt, New Zealand

Tao Fang, Ph.D.
Department of Applied Chemistry and Biochemistry
Kumamoto University
Kumamoto, Japan

Motonobu Goto, Ph.D.
Department of Applied Chemistry and Biochemistry
Kumamoto University
Kumamoto, Japan

Shufen Li, Ph.D.
School of Chemical Engineering & Technology
Tianjin University
Tianjin, China

Jose L. Martínez, Ph.D.
Thar Technologies, Inc.
Pittsburgh, Pennsylvania

M. Angela A. Meireles, Ph.D.
LASEFI-DEAFEA – UNICAMP
Sao Paulo, Brazil

Rui L. Mendes, Ph.D.
Departamento de Energias Renovaveis
INETI
Lisboa, Portugal

Paul H.L. Moquin, B.Sc.
Department of Agricultural, Food, and Nutritional Science
University of Alberta
Edmonton, Canada

Andrés Moure, Ph.D.
Department of Chemical Engineering
Facultade de Ciencias de Ourense
Universidade de Vigo
Ourense, Spain

Mamata Mukhopadhyay, Ph.D.
Chemical Engineering Department
Indian Institute of Technology
Bombay, India

Juan Carlos Parajó, Ph.D.
Department of Chemical Engineering
Facultade de Ciencias de Ourense
Universidade de Vigo
Ourense, Spain

Selva Pereda, Ph.D.
Planta Piloto de Ingeniería Química
Universidad Nacional del Sur
Bahía Blanca, Argentina

Marcus Petermann, Ph.D.
University Bochum
Particle Technology
Bochum, Germany

Ernesto Reverchon, Ph.D.
Dipartimento di Ingegneria
Chimica ed Alimentare
Universita di Salerno
Salerno, Italy

Marleny D.A. Saldana, Ph.D.
Department of Agricultural, Food, and
Nutritional Science
University of Alberta
Edmonton, Canada

Mitsuru Sasaki, Ph.D.
Department of Applied Chemistry and
Biochemistry
Kumamoto University
Kumamoto, Japan

Mei Sun, M.Sc.
Department of Agricultural, Food, and
Nutritional Science
University of Alberta
Edmonton, Canada

Feral Temelli, Ph.D.
Department of Agricultural, Food, and
Nutritional Science
University of Alberta
Edmonton, Canada

Samuel W. Vance, P.E.
Thar Technologies, Inc.
Pittsburgh, Pennsylvania

Eckhard Weidner, Ph.D.
University Bochum
Process Technology
Bochum, Germany

Dalang Yang, M.Sc.
Wuhan Kaidi Fine Chemical Industries
Co. Ltd.
Wuhan, Hubei, China

Editor

Dr. Jose L. Martínez, a native of León, Spain, received his B.S. and Ph.D. degrees from the University of Oviedo (Spain). He is currently General Manager of Thar Technologies, Inc., Process Division (Pittsburgh, USA), a company dedicated exclusively to supercritical fluid technology. He has nearly two decades of experience in conducting R&D and implementing industrial processes in supercritical fluid technology, including applications in extraction, fractionation, chromatography, particle formation, coating, and impregnation for the food, nutraceutical, and pharmaceutical industries.

1 Fundamentals of Supercritical Fluid Technology

Selva Pereda, Susana B. Bottini, and Esteban A. Brignole

CONTENTS

1.1 Introduction 1
1.2 Supercritical Fluids 2
1.2.1 Physical Properties of Supercritical Fluids 4
1.3 Phase Equilibrium with Supercritical Fluids 4
1.3.1 Solid Solubilities 4
1.3.2 Multiple Fluid Phase Equilibrium 6
1.4 Phase Equilibrium Engineering of Supercritical Processes 8
1.4.1 Understanding Phase Behavior 9
1.5 Conceptual Supercritical Process Design 11
1.5.1 Oxychemical Extraction and Dehydration 11
1.5.2 Particle Micronization with Supercritical Fluids 15
1.5.3 Extraction, Purification, or Fractionation of Natural Products with Supercritical Fluids 17
1.5.3.1 Fractionation of Oils 17
1.5.3.2 Extraction from Vegetable Matrices 18
1.5.4 Supercritical Reactions 19
References 21

1.1 INTRODUCTION

Solvents are used in large amounts in the chemical, pharmaceutical, food, and natural-product industries. In the search for environmentally friendly solvents, increasing attention is being paid to supercritical fluids (SCFs) for a wide variety of applications. For instance, supercritical solvents are used in extractions, material processing, micronization, chemical reactions, cleaning, and drying, among other applications. SCFs and near-critical fluids add a new dimension to conventional (liquid) solvents: *their density-dependent solvent power.* The density of SCFs can be easily tuned to the process needs, with changes in temperature, pressure, and/or composition. Other important properties of SCFs are their very low surface tensions, low viscosities, and moderately high diffusion coefficients.

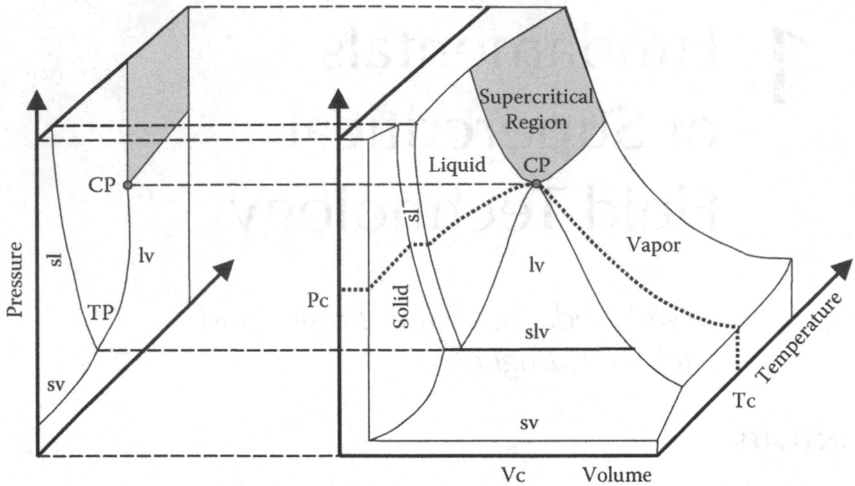

FIGURE 1.1 PVT diagram of a pure substance and its projection on the PT plane.

The design of processes using supercritical solvents is strongly dependent on the phase equilibrium scenario, which is highly sensitive to changes in operating conditions. Therefore, phase equilibrium engineering plays a key role in the synthesis and design of these processes.

1.2 SUPERCRITICAL FLUIDS

The different physical states of a pure substance can be visualized in a three-dimensional pressure–volume–temperature (PVT) diagram, as shown in Figure 1.1. The surfaces represent the different states—solid, liquid, or vapor—that correspond to particular values of pressure and temperature. According to the phase rule, the two-phase (solid–liquid, solid–vapor, and liquid–vapor) regions of a pure substance have only one degree of freedom. Therefore, the equilibrium pressure in each case is a function of temperature. The PT projections of the solid–liquid, solid–vapor, and liquid–vapor equilibrium lines are shown on the left of Figure 1.1. In particular, the vapor–liquid line represents the vapor pressure curve that starts at the triple point (TP) of solid–liquid–vapor coexistence and ends at the critical point (CP). The nature of the CP can be understood following the changes of the fluid properties along the vapor pressure curve. With increasing values of temperature, the density of the liquid phase diminishes and the vapor density increases due to the higher vapor pressure. Eventually, both densities converge at the CP and differentiating the liquid or the vapor state is no longer possible above the critical temperature. When both temperature and pressure are above the critical values (Figure 1.1), the system is considered to be in the supercritical region.

Within a region close to the critical conditions, the system properties are highly sensitive to pressure and temperature; this region is considered near-critical. Usually, the SCF solvent is applied at a temperature close to its critical value and at a pressure high enough for its density to become greater than the fluid critical density. A

TABLE 1.1
Critical Properties of Fluids of Interest in Supercritical Processes

Fluid	Critical Temperature Tc/K	Critical Pressure Pc/bar	Critical Volume Vc/cm³·mol⁻¹
CO_2	304.12	73.7	94.07
Ethane	305.3	48.7	145.5
Propane	369.8	42.5	200.0
Water	647.1	220.6	55.95
Ammonia	405.4	113.5	72.47
n-Hexane	507.5	30.2	368.0
Methanol	512.6	80.9	118.0

list of fluids that have been proposed as SCF solvents is shown in Table 1.1. These fluids can be classified as a) low-critical temperature (low-Tc) and b) high-critical temperature (high-Tc) solvents. Some condensable gases, like carbon dioxide (CO_2), ethane, and propane, are considered low-Tc solvents, whereas the higher alkanes, methanol, and water can be considered high-Tc solvents. Strong differences in solvent power and selectivity characterize the low-Tc and high-Tc solvents.

Francis [1] made a significant contribution on the subject of CO_2 solvent properties by studying its behavior with a large number of solutes. Liquid CO_2 is miscible with alkanes up to approximately carbon number 10, while the range of miscibility increases for ethane up to 20, and propane up to 35. Therefore, these solvents show selectivity for relatively low molecular-weight material. Stahl and Quirin [2] have reported the extractability of a wide range of natural products using CO_2; they showed that: "1) hydrocarbons and other lipophilic organic compounds of relatively low molecular mass and polarity are easily extractable; 2) the introduction of polar functional groups, hydroxyl or carboxyl groups render the extraction more difficult or impossible; 3) sugars and amino acids cannot be extracted; 4) fractionation effects are possible if there are marked differences in mass, vapor pressure or polarity of the constituents of a mixture."

Regarding the use of high-Tc solvents, such as toluene or water, the extraction is carried out at temperatures from 500 to 700 K, where even a mild pyrolysis of high-molecular-weight material takes place. The solvent power of high-Tc fluids is much higher than that of low-Tc solvents, and high-Tc solvents are proper solvents for high molecular weight materials. However, they have low selectivity and the severe operating conditions, on the other hand, degrade thermally labile materials. A good feature of low-Tc solvents, as compared with conventional liquid solvents, is that they operate at moderate temperature and have low solvent power. Therefore, by carefully choosing the pressure and temperature of operation, selective fractions can be extracted from vegetable matrices, such as essential oils, alkaloids, lipids, or oleoresins. These are the preferred solvents for the pharmaceutical and natural-product industries. A key advantage of low-Tc solvents is that they are easily separated from the extract.

TABLE 1.2
Comparison of the Physical Properties of Gas, Liquid, and Supercritical Fluids

Physical Property	Gas ($T_{ambient}$)	SCF (Tc, Pc)	Liquid ($T_{ambient}$)
Density ρ (kg m^{-3})	0.6–2	200–500	600–1600
Dynamic viscosity μ (mPa.s)	0.01–0.3	0.01–0.03	0.2–3
Kinematic viscosity η[a] (10^6 m^2s^{-1})	5–500	0.2–0.1	0.1–5
Thermal conductivity λ (W/mK)	0.01–0.025	Maximum[b]	0.1–0.2
Diffusion coefficient *D* (10^6 m^2s^{-1})	10–40	0.07	0.0002–0.002
Surface tension σ (dyn/cm^2)	—	—	20–40

[a] Kinematic viscosity defined as $\eta = \mu/\rho$

[b] Thermal conductivity presents maximum values in the near-critical region, highly dependent on temperature

SCF-solute interactions in the liquid phase may originate a second liquid phase (gas salting out effect), improving process selectivity and making it possible, for instance, to separate chemical reaction products in situ [3]. A better understanding of supercritical solvent properties will be obtained after considering the phase equilibrium behavior of binary systems that show a different degree of asymmetry in size or intermolecular interactions.

1.2.1 Physical Properties of Supercritical Fluids

The physical properties of SCFs are in-between those of a gaseous and liquid states. Typical values of different physical properties for each fluid state are listed in Table 1.2.

Density and viscosity of SCFs are lower than those of liquids; however, diffusivities are higher. Thermal conductivities are relatively high in the supercritical state and have very large values near the CP because, in principle, the heat capacity of a fluid tends to infinity at the CP. Interfacial tension is close to zero in the critical region. In general, the physical properties in the critical region enhance mass and heat transfer processes.

1.3 PHASE EQUILIBRIUM WITH SUPERCRITICAL FLUIDS

1.3.1 Solid Solubilities

The conditions of phase equilibrium between a SCF (1) and a solid component (2) are formulated on the basis of the isofugacity criterion. If the solid phase is assumed to be a pure component (2), the solubility in the gas phase can be directly obtained as:

$$y_2 = E\frac{p_2^s}{P} \tag{1.1}$$

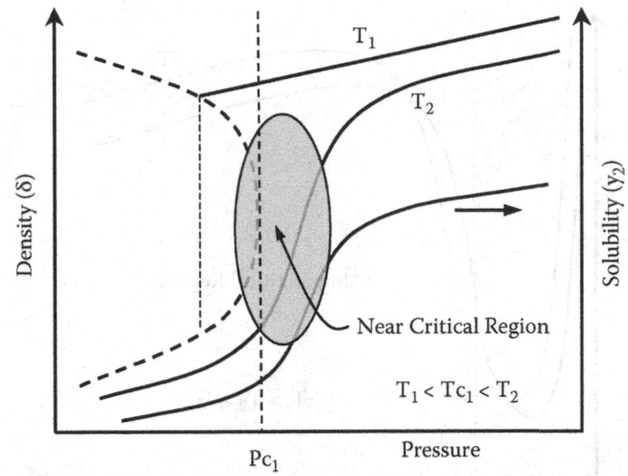

FIGURE 1.2 Density (δ) and solid solubility in fluid phase (y_2) as a function of pressure.

where E is the enhancement factor over the ideal solubility and p_2^s is the sublimation pressure of the solute (2). For a low-volatility, incompressible solid solute, the enhancement factor can be calculated as follows:

$$E = \frac{\exp\left(\frac{(P - p_2^S)v_2^{sol}}{RT}\right)}{\Phi_2} \tag{1.2}$$

where Φ_2 is the fugacity coefficient of the solid solute in the gas phase and v_2^{sol} is the solid molar volume. Φ_2 is strongly dependent on the SCF density. Figure 1.2 shows the region of SCF extraction. This region is characterized by a strong variation of fluid density with pressure, at temperatures close to the SCF critical temperature. For a given isotherm, the increase in solubility closely follows the increase in density, as indicated in Figure 1.2. The drastic increase in solubility in the vicinity of the critical region can be of several orders of magnitude and is mainly due to a sharp decrease of the solute fugacity coefficient Φ_2 in the fluid phase. This is the classical enhancement effect at the near-critical region.

The influence of temperature on the solid solubility is the result of two competing effects: the increase of solid volatility and the decrease of solvent density with temperature rise. Near the critical pressure, the effect of fluid density is predominant. Therefore, a moderate increase in temperature leads to a large decrease in fluid density and a consequent reduction in solute solubility. However, at higher pressures, the increase of solid sublimation pressure with temperature exceeds the density reduction effect, and the solubility increases with temperature. This behavior leads to a region of retrograde behavior of the solid solubility, as illustrated in Figure 1.3. At pressures well above the SCF critical pressure, the isotherms exhibit a maximum in solubility. This maximum is usually observed in the range of 30 to 100 MPa.

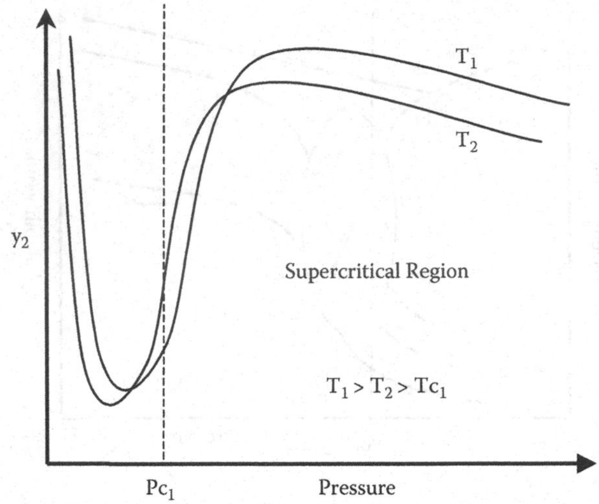

FIGURE 1.3 Typical isotherms of solid solubility in SCF.

Kurnik and Reid [4] have shown that the maximum is achieved when the partial molar volume of the solute in the fluid phase is equal to its solid molar volume.

A quantitative correlation and prediction of the solubility of a pure solid in a SCF is possible if the fugacity coefficient of the solid in the fluid phase is computed using an equation of state. Cubic equations of state, with conventional mixing rules and adjustable binary interaction parameters, have been widely used since the early works of Deiters and Swaid [5] and Kurnik and Reid [4]. However, equations of state that use classical mixing rules, even with energy and size binary interaction parameters, may fail to predict or correlate the solubility of solids with polar or hydrogen-bonding interactions. For instance, Kurnik and Reid [4] found that this approach is not able to model the solubility of stearic acid or n-octanol in CO_2. The limitations of cubic equations of state to model the solubility of polar solids can be tackled by using cubic equations of state with local composition mixing rules [6].

When a nonpolar supercritical solvent is used, the separation process does not present specific selectivities; in this case, the addition of a proper cosolvent can enhance solubility and selectivity. Nonpolar cosolvents increase the solubility of solid aromatics several times, whereas polar cosolvents enhance the solubility of solutes that present specific interactions with the cosolvent. For example, Brenecke and Eckert [7] showed a dramatic effect of the cosolvent tributilphosphate on the solubility of hydroquinone in CO_2. The cosolvent selection follows the general rules applied for classic solvent selection in solid or liquid-liquid extraction. Brunner [8] studied the effects of cosolvents on the extraction of low-volatility liquids and showed that the use of acetone or methanol, for instance, improves selectivity and solvent power in the extraction of hexadecanol from octadecane.

1.3.2 Multiple Fluid Phase Equilibrium

Equilibrium predictions in systems having two or more fluid phases are more complex than those in cases of solid solubilities due to the need to compute fugacities in several phases of different compositions. The use of the same equation of state to compute fugacity coefficients in all phases gives the required continuity in the prediction of phase equilibrium at the critical region. Cubic equations of state of the van der Waals family have been successfully applied in the correlation and prediction of phase equilibria in mixtures of subcritical and supercritical nonpolar components in the natural gas and petrochemical industries. However, their application to size and energy asymmetric systems, typical of the supercritical extraction of natural subtracts, has found little success. De la Fuente et al. [9] tried to correlate both vapor-liquid equilibrium (VLE) and liquid-liquid equilibrium (LLE) of the system sunflower oil + propane using the Soave [10] equation of state with quadratic mixing rules and binary interaction parameters for both, the attractive energy parameter and the covolume. It was not possible to quantitatively describe both VLE or LLE using only one set of parameters for the attractive energy parameter and the covolume. This indicates the limitations of the van der Waals repulsive term to describe these asymmetric mixtures. The failure of cubic equations of state to model phase equilibria in size asymmetric mixtures can be attributed to the large differences in the pure-component covolumes [11].

Espinosa et al. [12] and Ferreira et al. [13] extensively discussed the application of equations of state to model the supercritical processing of natural products. A group contribution approach is particularly useful when dealing with natural products because a large number of compounds, such as triglycerides, fatty acids, esters, and alcohols, can be represented with a small number of functional groups. Group contribution equations of state, such as Modified Huron-Vidal 2 (MHV2) [14, 15] and group contribution equation of state (GC-EOS) [16, 17], are particularly useful to model the complex phase behavior observed in asymmetric mixtures at near-critical conditions. Bottini et al. [18] extended the GC-EOS model to describe both VLE and LLE in mixtures of supercritical gases + vegetable oil mixtures using the same set of parameters. Gros et al. [19] and Ferreira et al. [20] extended this model to represent associating mixtures (GCA-EOS), using a group contribution approach for dealing with self- and cross-associations. The GCA-EOS equation can be derived from a three-term (repulsive, attractive, and associating) Helmholtz residual energy:

$$A = A^{rep} + A^{att} + A^{assoc} \tag{1.3}$$

The repulsive (rep) term is given by the Carnahan-Starling equation for hard spheres, the attractive (att) term is a group contribution version of a density-dependent local composition Non-Random Two Liquids (NRTL) model, and the association (assoc) term is a group contribution expression based on Wertheim's statistical association fluid theory [21]. The hard sphere term performs better than the van der Waals repulsive term when dealing with highly size-asymmetric systems and the other two terms are able to handle strong nonideal specific interactions. The GC-EOS model was compared to MHV2 and PSRK [22] by Espinosa et al. [23]. All three models perform similarly for moderately polar systems of low molecular weight compounds. However, the MHV2 and PSRK models present some limitations when they are applied to very asymmetric systems.

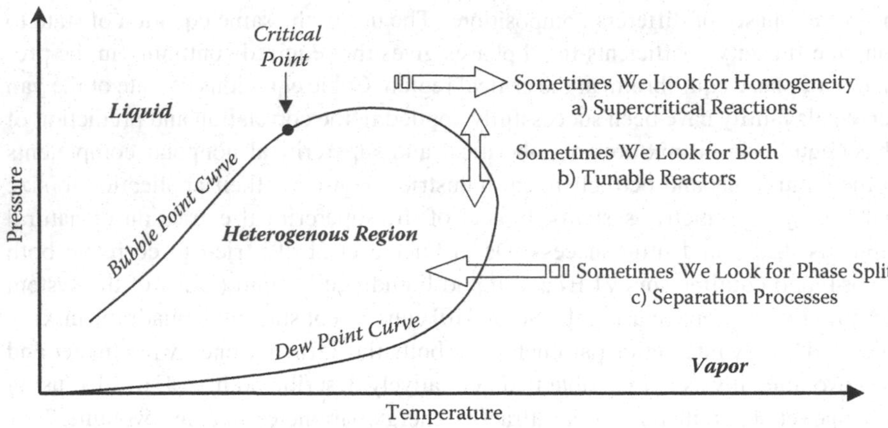

FIGURE 1.4 Possible process phase scenarios.

1.4 PHASE EQUILIBRIUM ENGINEERING OF SUPERCRITICAL PROCESSES

Phase equilibrium engineering is the systematic application of phase equilibrium knowledge to process development. This knowledge comprises data banks, experimental data, phenomenological phase behavior, thermodynamic analysis, and mathematical modeling procedures for phase equilibrium process calculations. Each SCF application has a set of specifications and physical restrictions. In supercritical reactions, for instance, homogeneous phase conditions may be required at the reaction temperature. The solution to this problem is given by the selection of the proper solvent and the determination of a feasible operating pressure range and feed composition to achieve homogeneity in the reaction mixture. On the other hand, a heterogeneous two-phase system may be required to develop supercritical extraction or fractionation processes. Additional phase equilibrium restrictions may include no solid phase precipitation, azeotrope formation, specific solvent solubilities, or saturation conditions.

A multicomponent fluid can be a supercritical mixture, a subcooled liquid, a superheated vapor, or a heterogeneous liquid-liquid, liquid-vapor, or liquid-liquid-vapor mixture. A useful plot to identify each region is a pressure vs. temperature diagram showing the bubble and dew point phase transitions curves, as well as the CP of a given global composition. These lines determine the mixture phase envelope. Different phase scenarios can be selected from this phase envelope (Figure 1.4): a) homogeneous conditions for a supercritical reaction, b) homogenous and heterogeneous conditions for a tunable phase split reactor, or c) phase split for a separation process. Certainly, different phase envelopes are obtained during the course of the reaction or separation process. However, the process trajectory should always remain at the required phase scenario. General conditions can also be set from this plot; for instance, above the maximum pressure of the phase envelope there will be a single phase at any temperature.

Rigorous simulations of equilibrium stage separations at near-critical conditions are needed for the design and optimization of supercritical processes. However, equilibrium calculations in the near-critical region can present serious convergence difficulties. In that respect, Michelsen's [24] phase stability criterion, multiple-phase flash algorithms, and global phase computations are of particular interest for supercritical extraction applications.

Solvent recycle is a major issue in the economic optimization of these processes, because it is the main factor in determining capital and operating costs. Design and synthesis problems have been increasingly solved by formulating mathematical models, which involve continuous and integer variables to represent operating conditions and alternative process topologies [25]. With regard to supercritical extractions, Gros et al. [26] have addressed the synthesis of optimum processes for the extraction and dehydration of oxychemicals as a mixed integer nonlinear programming problem. Espinosa et al. [23] and Diaz et al. [27] have applied these procedures for the synthesis and optimization of citrus oil deterpenation processes.

1.4.1 Understanding Phase Behavior

Van Konynenburg and Scott [28] have shown that the fluid phase behavior observed in binary mixtures can be classified into five main types. In type I phase behavior, complete liquid miscibility is observed at all temperatures. When partial liquid miscibility occurs at low temperatures, the system is of type II. Type I phase behavior is usually found in systems with components of similar chemical nature and molecular size, like mixtures of hydrocarbons, noble gases, or systems that do not deviate greatly from ideal behavior. Type II is typical of nonideal mixtures of similar size compounds, in which nonideality leads to liquid phase split at subcritical conditions. When the liquid immiscibility persists even at high pressures and temperatures, the systems are of type III. This behavior is characteristic, for example, of mixtures of CO_2 with high-molecular-weight alkanes or vegetable oils. When the difference in molecular size becomes significant, in almost ideal systems, liquid-liquid immiscibility is observed near the light-component critical temperature (solvent Tc in supercritical processes). However, complete miscibility is recovered at lower temperatures; this corresponds to type V phase behavior. Type IV, on the other hand, shows discontinued liquid-liquid immiscibility (i.e., liquid immiscibility occurs at low and high temperatures but not at intermediate temperatures). Figure 1.5 is a master chart of the different types of binary fluid phase diagrams [29]. The arrows in Figure 1.5 qualitatively indicate the type of fluid phase behavior that can be expected when the system components exhibit greater molecular interactions, size differences, or both.

Figure 1.6 illustrates, in more detail, a Type V phase diagram. The lines in this diagram indicate the boundaries of phase transitions and the critical locus. The three-phase equilibrium line (l_1l_2g) starts at the lower critical end point (LCEP) and finishes at the upper critical end point (UCEP). This behavior is typical of mixtures of propane with triglycerides, such as sunflower oil or tripalmitin [30]. When the process operating temperatures are above the critical temperature of the solvent, pressures should be higher than the critical pressure of the mixture in order to ensure complete miscibility (i.e., the pressure should be above the $l_1 = l_2$ line).

In the search for an adequate supercritical solvent to achieve homogenous or heterogeneous conditions, two different approaches can be followed: 1) to compare the phase behavior of a given substrate with different solvents or 2) to follow the

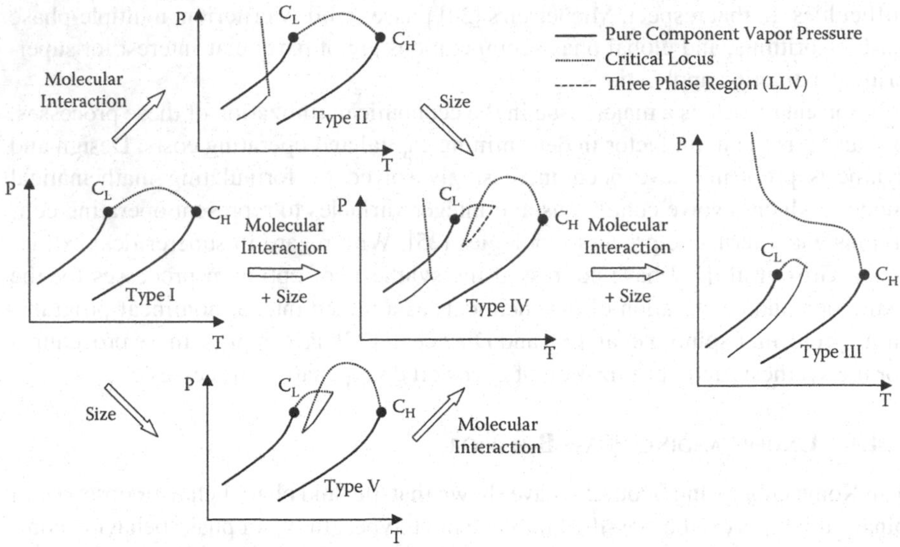

FIGURE 1.5 Modifications of binary phase behavior with size and energy asymmetries. C_L and C_H are the critical points of the light and heavy compounds, respectively.

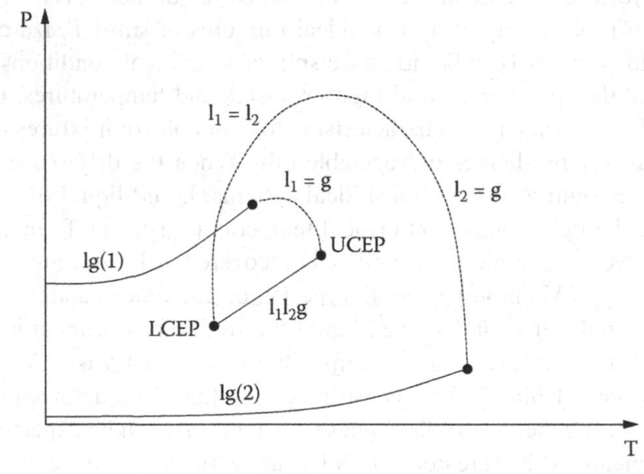

FIGURE 1.6 Type V phase behavior according to Van Konynenburg and Scott classification.

change in the phase behavior of a given solvent with different families of chemical compounds. In the more general case, when the components of the mixture are of different chemical nature, the second approach should be followed to take into account any possible change in the phase behavior during process evolution.

The liquid-liquid immiscibility of type V phase behavior appears in many binary mixtures between supercritical solvents and organic substrates beyond a certain carbon number. Figure 1.7 shows the regions of liquid-liquid immiscibility for binary mixtures of supercritical solvents (ethane and propane) with hydrocarbons of different chain length [31]. Peters [31] also presented similar data on the liquid-liquid

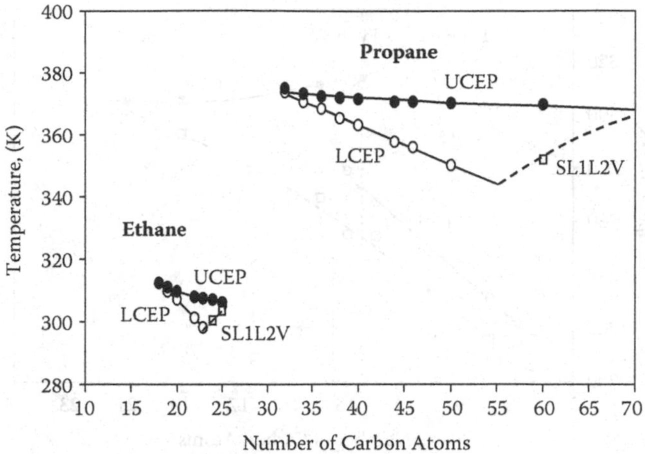

FIGURE 1.7 Phase transitions for the binaries of ethane and propane with paraffins of different chain length. UCEP and LCEP points are upper and lower critical end points, respectively. SL1L2V stands for solid–liquid–liquid–vapor equilibria.

immiscibility domains of the systems ethane + alcohols, ethane + aromatic hydrocarbons, and ethane + alkanes. It becomes clear from these data that ethane is not an adequate supercritical solvent for normal alcohols because it presents liquid-liquid immiscibility even with methanol. However, ethane seems to be a better solvent for aromatic hydrocarbons or paraffins because the liquid-liquid immiscibility appears at carbon numbers greater than 15 or 18, respectively.

CO_2 has been the most studied solvent for supercritical processes. However, it exhibits strong liquid-liquid and gas-liquid immiscibility for hydrocarbons with carbon numbers greater than 13. In addition, CO_2 presents a rather low critical temperature to be used as a solvent for reactions carried out at moderate or high temperatures. Figure 1.8 shows data on the type of phase transition for the families of CO_2 + alkanes compiled by Peters [32], who also showed the behavior of CO_2 + alkanol systems.

Unfortunately, the type of data shown in Figures 1.7 and 1.8 is only known for a limited number of families of organic compounds with some supercritical solvents. Therefore, reliable thermodynamic models are needed to explore the possible phase scenarios found in mixtures between process components and supercritical solvents.

The phase equilibrium engineering approach will be illustrated with several examples, where thermodynamic and modeling tools are applied for supercritical process development. The examples to be covered are alcohol extraction and dehydration, gas antisolvent crystallization, purification of vegetable oils, supercritical fractionation, extraction with near critical fluids, and supercritical reactions.

1.5 CONCEPTUAL SUPERCRITICAL PROCESS DESIGN

1.5.1 Oxychemical Extraction and Dehydration

The supercritical extraction of organic oxygenated compounds from aqueous solutions is of great interest in biotechnological processes. Oxygenated compounds and

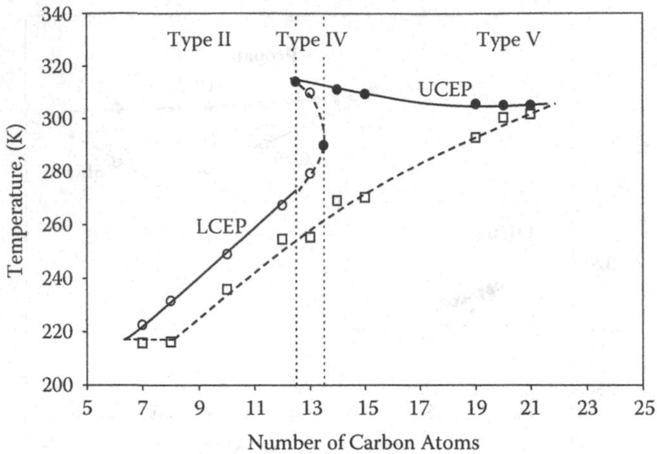

FIGURE 1.8 Phase transitions for the binaries of CO_2 with paraffins of different chain length. UCEP and LCEP points are upper and lower critical end points, respectively. Dashed line (open squares): SL1L2V equilibria.

water have strong hydrogen bonding interactions that complicate their separation with conventional solvents. Moreover, an oxygenated compound dissolved in a nonpolar near-critical solvent will have a rather high activity coefficient (γ_{oxy}^{SCF}), leading to a low value of the distribution coefficient:

$$m_{oxy} = \frac{\gamma_{oxy}^{H2O}}{\gamma_{oxy}^{SCF}} \tag{1.4}$$

This is even more pronounced in the case of alcohols or acids that exhibit self-association. A strategy to overcome this problem may be based on the Koenen and Gaube [33] diagram that classifies binary mixtures in an excess Gibbs function (G^E) versus excess enthalpy (H^E) diagram (Figure 1.9). We can derive the effect of temperature on the activity coefficients directly from this diagram. The aqueous solutions of organic oxygenated compounds are located on the second quadrant of the diagram with negative H^E values, whereas the supercritical solutions that correspond to positive H^E values are located on the first quadrant. In both cases, there are positive deviations to nonideality (positive G^E). From this diagram, we can see that the activity coefficients in the aqueous phase increase with temperature; however, the reverse occurs with the activity coefficients in the SCF phase. Therefore, extracting at high temperatures leads to more attractive values of the distribution coefficients. This behavior is found in the extraction of isopropanol or ethanol from aqueous solutions using CO_2, ethane, or propane as near-critical solvents. However, we should consider another fact to make a proper solvent selection: At optimum extraction temperatures (around 380 to 400K), the solvent power of CO_2 or ethane is drastically reduced due to fluid density decrease at temperatures well above the critical temperature of both fluids (around 304 K). To recover the solvent power, relatively high pressures should be used for the extraction process. This makes propane a better candidate as

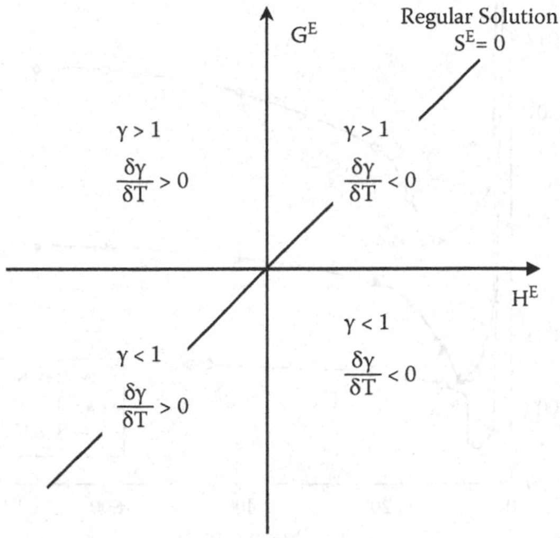

FIGURE 1.9 Value and temperature derivative of activity coefficients, according to the relative values of G^E and H^E.

an extraction solvent because its critical temperature is close to 370 K and it has a lower critical pressure than CO_2 or ethane. Horizoe et al. [34] and Brignole et al. [35] verified the potential of propane as an extracting supercritical solvent.

Dehydration by near-critical solvents finds important applications, for example, in the extraction of solutes from aqueous solutions and in the drying of solid particles after micronization. We will consider first the dehydration of extracted solutes. In low-pressure separations, entrainment agents like cyclohexane or solvents like ethylene glycol have been used to separate water by azeotropic or extractive distillations. In connection with supercritical processes, it is of interest to study the equilibrium between water and a near-critical fluid as a function of temperature and pressure. In the case of CO_2, the data of Wiebbe [36] and Coan et al. [37] show the solubility of water in CO_2 as a function of pressure at subcritical and supercritical temperatures. These data indicate that water follows the classical supercritical effect: the concentration of water in the CO_2 phase increases once the supercritical pressure is exceeded (Figure 1.10).

At the CO_2 saturation pressure, at subcritical conditions, we would have a three-phase VLL equilibrium condition, where the concentration of water in the condensed CO_2 phase exceeds the concentration of water in the vapor phase. Hence, in a CO_2–water separation process, the relative volatility of water with respect to CO_2 is lower than one. This behavior has important consequences for the separation of water from CO_2 extracts. Water, as expected, is less volatile than CO_2; therefore, the extract cannot be obtained free from water in the solvent recovery operation.

When the same phase equilibria analysis is made for water and light alkanes, such as ethane and propane, a different picture is obtained. The data of Kobayashi and Katz [38] for the solubility of water in propane are plotted against pressure at different temperatures (Figure 1.11). For near-critical propane, the solubility of water decreases when the critical pressure is exceeded (see Table 1.1 for the critical properties of propane). This phenomenon can be called a *nonclassical supercritical effect.*

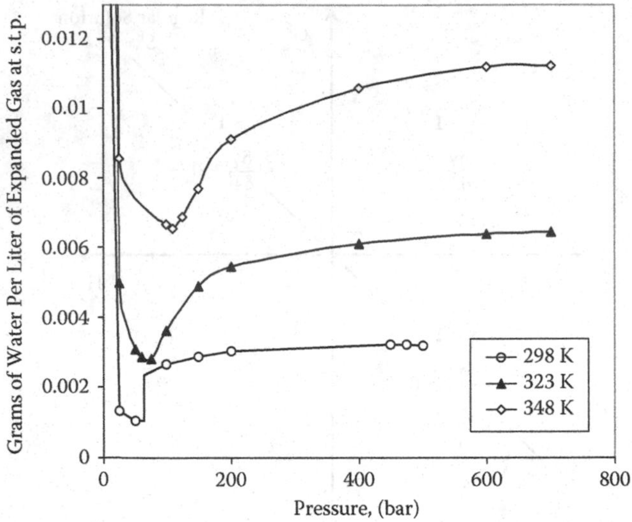

FIGURE 1.10 Composition of the CO_2-rich phase as a function of pressure and temperature. Experimental data from Wiebbe [34].

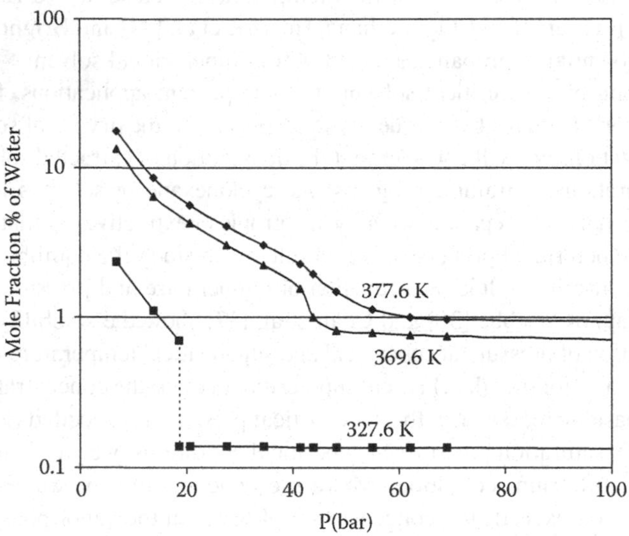

FIGURE 1.11 Experimental water composition in liquid and vapor propane. Data from Kobayashi and Katz [38].

A very attractive property can be derived from this effect. When working at subcritical temperatures, at the propane saturation pressure, we again have VLL equilibria. In this case, the composition of water in the vapor phase is greater than that in the condensed propane phase, leading to a water-propane relative volatility greater than one. This makes it possible to obtain dehydrated organic oxygenated products during the process of solvent recovery from the extract [39].

On the basis of the phase equilibrium engineering concepts presented above, a process for the production of bioethanol or for the dehydration of isopropanol with a near-critical solvent (propane) can be developed. The key features of these processes are:

a) High temperatures and pressures of extraction favor the solubility of alcohol in propane.
b) Liquid-liquid equilibrium at low temperatures is beneficial for reducing the water content in the extract.
c) The alcohol product is obtained dehydrated because the relative volatility of water with respect to propane is greater than one over a certain concentration range of ethanol in the extract mixture.

All these properties were first predicted by group-contribution thermodynamic modeling and thereafter verified by experimental and pilot plant information.

1.5.2 Particle Micronization with Supercritical Fluids

Supercritical micronization processes are based on creating a high degree of solution supersaturation that leads to the formation of a great number of nucleation sites and very small crystals. These processes have found many applications in the last decade [40, 41], mainly in the micronization of pharmaceutical solid compounds. Usually, several components may participate in the process: the solute to be crystallized, the solvent, a supercritical fluid, and a cosolvent. The phase equilibrium between these components plays a key role in the selection of the proper technology for the micronization processes. A better understanding of process selection can be made on the basis of the binaries behavior. First, we shall consider the solute + supercritical fluid binary. If the solute solubility under supercritical conditions is high, then only these components participate in the process and micronization is obtained directly by a drastic reduction in the solute solubility by the rapid expansion of the supercritical solution (RESS process) through a nozzle or other convenient device. The main limitation of this RESS process is that it can only be applied to solutes with high solubilities in the supercritical fluid. The low solvent power of supercritical CO_2 for polar or medium- to high-molecular-weight material makes this approach uneconomical for these mixtures.

When the solute cannot be dissolved in significant amounts in the supercritical fluid, we can look for a good liquid solvent for both the solute and the supercritical gas. In this case, a concentrated solution of the solute in the solvent is prepared and a high degree of supersaturation is obtained by dissolving the supercritical fluid in the liquid phase at high pressure. This technology is called the gas antisolvent (GAS) process and it can be carried out in a batch or semicontinuous process. These processes can be applied to a variety of solutes, but in this case, the ternary phase equilibria should also be evaluated to assure a high degree of supersaturation at the operating pressure and temperature. In the semicontinuous process, both the solution and the supercritical fluid enter together in the precipitation vessel through

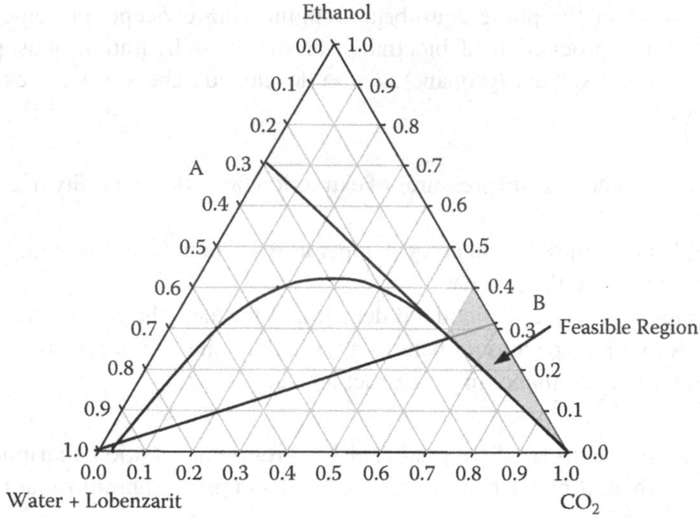

FIGURE 1.12 Feasible operating region for Lobenzarit precipitation using supercritical CO_2 and ethanol as cosolvents.

a mixing device. Very good precipitation conditions are achieved if the operating conditions are above the CP of the solvent + supercritical fluid mixture. Under these conditions, both feeds are completely miscible and no interfacial resistance is offered to mass transfer [42].

Another possible phase scenario appears when the solid solute is not soluble in the SCF, but the solubility of the SCF in the melted solid is high at elevated pressures. Therefore, if the solution is expanded to atmospheric pressure, a large cooling effect occurs that gives rise to the precipitation of micronized solute particles.

A different situation arises with solutes that are only soluble in water, such as some organic salts and proteins [41]. Typical nonpolar supercritical fluids like CO_2 and ethane are not soluble in aqueous solutions, even at high pressures. Therefore, no antisolvent effect can be obtained in a typical GAS process [43]. In this case, a cosolvent that shows complete miscibility with both the SCF and water can be introduced. For example, ethanol was used as a proper cosolvent for the precipitation of an organic salt from aqueous solution [43]. In this application, the aqueous solution is fed as a spray or mist into a precipitation vessel already filled up with a mixture of ethanol + CO_2 at the required composition. To obtain a feasible process, the operating conditions of the precipitation chamber should lie inside the homogeneous region of the triangular phase diagram for water + CO_2 + ethanol at a given pressure and temperature, as shown in Figure 1.12. In this way, the fine water droplets become quickly supersaturated by the ethanol + CO_2 dissolution in the drops and the simultaneous fast evaporation of water. As a result of this process, highly micronized dried salt particles are obtained [43]. All these examples illustrate that a phase equilibrium engineering analysis is a prerequisite for proper technology selection and successful adequate choice of micronization operating conditions.

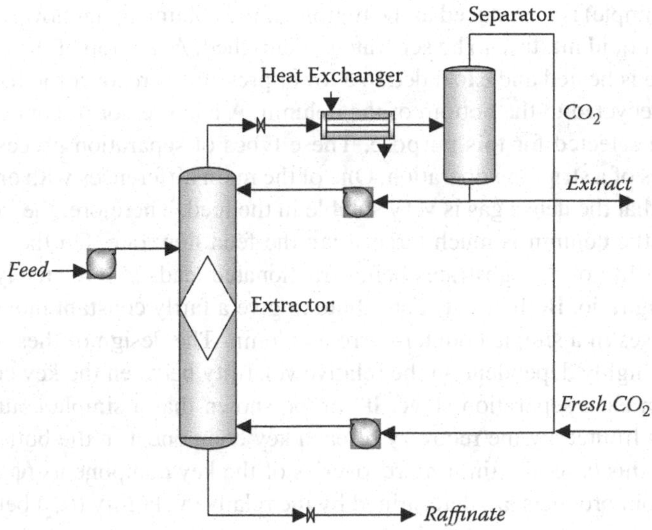

FIGURE 1.13 Dense-gas fractionation scheme process.

1.5.3 Extraction, Purification, or Fractionation of Natural Products with Supercritical Fluids

1.5.3.1 Fractionation of Oils

In the processing of vegetable oils, it is possible to take advantage of the low solubility of triglycerides in CO_2. For instance, both palm oil and sunflower oil give liquid-liquid or liquid SCF immiscibility with CO_2, even at high pressures, typical of type III systems. In these cases, either supercritical or liquid CO_2 can be used as a solvent to remove undesirable components from the oil—for instance, removal of oleic acid from olive oil [44]. Likewise, liquid or near-critical CO_2 can be applied to recover valuable components like tocopherols or squalene from fish oil [45]. When dealing with these separation processes, it is possible to find optimum extraction operating conditions that minimize the solvent-feed ratio and, at the same time, keep the coextraction of oil at a low value. Other solvents that have regions of liquid-liquid immiscibility with fatty oils, such as ethane and propane, may be used as alternative solvents.

SCF solvents can also be used as fractionating agents. This is of interest in the separation of low-volatile substances of close relative volatility. For instance, CO_2 and ethane have been proposed as dense gas extractants to remove the terpene fraction from citrus essential oils[27] and also for the fractionation of highly unsaturated fish oil methyl esters to obtain rich eicosapentaenoic acid and docosahexaenoic acid fractions [46]. The binary systems between CO_2 and these families of compounds are generally of type II, so complete miscibility for all compositions is obtained above the maximum pressure of the vapor-liquid critical locus.

A single dense-gas fractionation column scheme is shown in Figure 1.13. The mixture to be fractionated is fed at an intermediate point in the column. A dense gas

(CO_2, for example) is introduced at the bottom of the column and it flows countercurrently to the liquid mixture to be separated or enriched. At the top of the column, the extract phase is heated and expanded to a lower pressure to recover the light fraction and CO_2 is recycled to the bottom of the column. A compressor or condenser-pump cycle can be selected for this purpose. These types of separation processes follow the principles of a stripping operation. One of the main differences with ordinary gas stripping is that the dense gas is very soluble in the feed. Therefore, the liquid phase flow rate in the column is much larger than the feed flow rate. On the other hand, the low volatility of the substrates being fractionated leads to a relatively high gas-feed stripping ratio. Both effects contribute to give a fairly constant molar overflow for both phases in a simple counter-current column. The design of these separation processes is highly dependent on the relative volatility between the key components of the oils in each separation stage. It can be shown that a simple countercurrent separation is limited by the recovery of each key component in the bottom and top products. In this case, the limiting recoveries of the key components (φ_1, φ_2) in the top and bottom products are determined by the relative volatility (α_{12}) between both components under process conditions:

$$\alpha_{12} = \phi_1 / (1 - \phi_2) \tag{1.5}$$

In most simple countercurrent extraction columns, this constraint limits the recovery and purity of the products in the separation of components of close relative volatility. Therefore, the use of recycle (reflux) of the top product is required: 1) to increase recovery and purity and 2) to assure that the trajectory of the separation process lies inside the two-phase region. Thus, the column and separator operating conditions (pressure, temperature, and compositions) should always be checked in order to verify a heterogeneous operation.

1.5.3.2 Extraction from Vegetable Matrices

The extraction of lipids and oils from vegetable matrices has been extensively covered in the monograph edited by King and List [47]. In the extraction of fatty oils from grounded seeds, it is advantageous to select a solvent that presents complete miscibility with the oil. CO_2 is a cheap, nontoxic solvent; however, the oil solubility in this SCF is very low even at pressures of the order of 30 MPa (type III binary). On the other hand, liquid propane is completely miscible with vegetable oils below the LCEP of this binary. Propane has type II or type IV global phase diagrams with vegetable oils. The main drawback of using propane as a solvent for the extraction of oils from grounded seeds is that it is flammable. Recently, Hegel et al. [48] studied the use of propane + CO_2 solvent mixtures for oil extraction, looking for efficient and safe solvent mixtures. Peter [45] has studied these types of mixtures to improve selectivities in the separation of lecithin from vegetable oils. In the work of Hegel et al. [48], the selected phase scenario was to operate in a region of complete liquid miscibility of the oil + solvent mixture, with a nonflammable vapor phase. The selection of operating conditions was based on experimental data on the LLV region at constant temperature, for the system sunflower oil + propane + CO_2. At constant

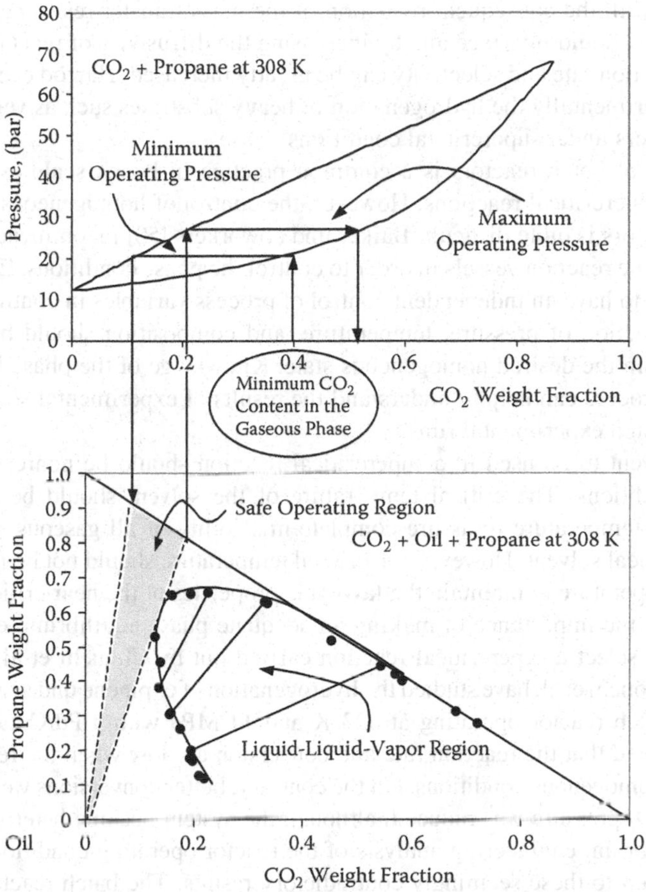

FIGURE 1.14 Safe operating extraction region at 308 K. Experimental data from Hegel et al. [48].

temperature, for a three-component system, the LLV equilibrium is only a function of pressure. Therefore, a binodal curve can be drawn on a triangular diagram, with tie lines linking the two liquid phase compositions at specified pressures (see triangular diagram on Figure 1.14). The binodal curve gives the boundary of the LLV region. The diagram also shows the minimum pressure for which LL immiscibility arises at a given temperature. At lower pressures (i.e., lower CO_2 composition), the solvent has complete miscibility with the oil. However, there is also a minimum operating pressure to avoid vapor phase flammability because, at pressures lower than this, the propane content of the vapor phase is too high. The feasible operating region can be easily determined with the help of Figure 1.14.

1.5.4 Supercritical Reactions

In general, gas-liquid catalyzed reactions are diffusion controlled. The use of an adequate supercritical SCF can bring the reactive mixture into homogeneous conditions, with the consequent reduction of the mass transfer resistance by eliminating the gas-liquid interface and by increasing the diffusivity of reactants. Therefore, the reaction rate and selectivity can be greatly increased. Härröd et al. [49] have studied experimentally the hydrogenation of heavy substrates such as vegetable oils and fatty esters under supercritical conditions.

The use of batch reactors is a common practice in bench scale experimental studies on supercritical reactions. However, the control of homogeneous conditions in these reactors is quite difficult. Baiker and coworkers [50] recommend the use of windows in the reaction vessels in order to control the phase conditions. Even though it is possible to have an independent control of process variables in continuous reactors, the selection of pressure, temperature, and composition should be carefully done to obtain the desired homogeneous state. Knowledge of the phase behavior of a reaction process can help to understand the results of experimental studies and to plan and design experimental runs.

The solvent to be used in a supercritical reaction should be nonreactive under process conditions. The critical temperature of the solvent should be lower than the reaction temperature to assure complete miscibility of all gaseous reactants in the supercritical solvent. However, the critical temperature should not be far from the reaction temperature to maintain the favorable properties of the near-critical state.

To show the importance of making an adequate phase equilibrium engineering analysis, we select a supercritical reaction carried out by Chouchi et al. [51] as an example. Chouchi et al. have studied the hydrogenation of α-pinene under supercritical CO_2 in a batch reactor operating at 323 K and 14 MPa with a Pd/C catalyst. The authors showed that the reaction rate and conversion are low when the reactor operates under homogenous conditions. On the contrary, better conversions were achieved when the CO_2 pressure was reduced, although the system became heterogeneous. A phase equilibrium engineering analysis of the reactor operating conditions can give an explanation to these seemingly contradictory results. The batch reactor was first fed with the catalyst, together with a known amount of α-pinene. Then, the system was pressurized with CO_2 up to the desired pressure (8, 9, 10, or 12 MPa), and, finally, H_2 was fed until a total pressure of 14 MPa was reached. The actual molar composition of the reacting mixture was unknown. This composition may be obtained by using an equation of state suitable for density predictions under the reaction conditions. One possibility is to use the MHV2 [15, 48] equation of state. The computation of the actual mixture compositions requires an iterative procedure for estimating the system compressibility factor, the amounts of each component charged into the cell, and the evolution of the reactor composition with conversion. This analysis indicates that, at the higher CO_2 partial pressure, an important reduction in hydrogen concentration occurs, which is likely the reason for the observed decrease in the reaction rates.

Phase equilibrium engineering analysis of supercritical processes is of the utmost importance in developing new technologies that replace conventional solvents by high-pressure gases to obtain environmentally friendly chemical processes. Several examples of process development clearly demonstrate that a good understanding of phase behavior and application of rigorous modeling tools are essential to process syntheses in which the fluid properties are extremely dependent on pressure, temperature, and composition.

REFERENCES

1. Francis, A.W., Ternary systems of liquid carbon dioxide, *J. Phys. Chem*, 58, 1099, 1954.
2. Stahl, E. and Quirin, K.W., Dense gas extraction on a laboratory scale: A survey of recent results, *Fluid Phase Equilibria*, 8, 93–105, 1983.
3. Eckert, C.A. and Chandler, K., Tuning fluid solvents for chemical reaction, *J. Supercrit. Fluids*, 13, 187–195, 1998.
4. Kurnik, R.T. and Reid, R.C., Solubility extreme in solid-fluid equilibria, *AIChE J.*, 27, 861–863, 1981.
5. Deiters, U.K. and Swaid, I., Calculation of fluid-fluid and solid-fluid phase equilibria in binary mixtures at high pressures, *Ber. Bunsenges. Phys. Chem.*, 88, 791–796, 1984.
6. Vidal, J., Phase equilibria and density calculations for mixture in the critical range with simple equation of states, *Ber. Bunsenges. Phys. Chem.*, 88, 784–791, 1984.
7. Brennecke, J. and Eckert, C., Phase equilibria for supercritical fluid process design, *AIChE J.*, 35, 1409–1427, 1989.
8. Brunner, G., Selectivity of supercritical compounds and entrainers with respect to model substances, *Fluid Phase Equilibria*, 10, 289–298, 1983.
9. de la Fuente, J.C.B., Mabe, G.D., Brignole, E.A. and Bottini, S.B., Phase equilibria in binary mixtures of ethane and propane with sunflower oil, *Fluid Phase Equilibria*, 101, 247–257, 1994.
10. Soave, G., Equilibrium constants from a modified Redlich-Kwong equation of state, *Chem. Eng. Sci.*, 27, 1197–1203, 1972.
11. Heidemann, R.A. and Kokal, S.L., Combined excess free energy models and equations of state, *Fluid Phase Equilibria*, 56, 17–37, 1990.
12. Espinosa, S., Fornari, T., Bottini, S. and Brignole, E., Phase equilibria in mixtures of fatty oils and derivatives with near critical fluids using the GC-EOS model, *J. Supercrit. Fluids*, 23, 91–102, 2002.
13. Ferreira, O., *Modelling of association effects by group contribution: Application to natural products,* Ph.D. Thesis, Univ. de Porto, Portugal, 2003.
14. Michelsen, M.L., A modified Huron-Vidal mixing rule for cubic equations of state, *Fluid Phase Equilibria*, 60, 213–219, 1990.
15. Dahl, S. and Michelsen, M.L., High-pressure vapor-liquid equilibrium with a UNIFAC-based equation of state, *AIChE J.*, 36, 1829–1836, 1990.
16. Skjold-Jørgensen, S., Gas solubility calculations II. Application of a new group-contribution equation of state, *Fluid Phase Equilibria*, 16, 317–351, 1984.
17. Skjold-Jørgensen, S., Group contribution equation of state (GC-EOS): A predictive method for phase equilibrium computations over wide ranges of temperature and pressures up to 30 MPa, *Ind. Eng. Chem. Res.*, 27, 110–118, 1988.
18. Bottini, S.B., Fornari, T. and Brignole, E., Phase equilibrium modeling of triglycerides with near critical solvents, *Fluid Phase Equilibria*, 158–160, 211–218, 1999.
19. Gros, H.P., Bottini, S.B. and Brignole, E., A group contribution equation of state for associating mixtures, *Fluid Phase Equilibria*, 116, 537–544, 1996.
20. Ferreira, O., Brignole, E.A. and Macedo, E.A., Modeling of phase equilibria for associating mixtures using an equation of state, *J. Chem. Thermodynamics*, 36, 1105–1117, 2004.
21. Chapman, W.G., Gubbins, K.E., Jackson, G. and Radosz, M., New reference equation of state for associating liquids, *Ind. Eng. Chem. Res.*, 29, 1709–1721, 1990.
22. Holderbaum, T. and Gmehling, J., PSRK: A Group Contribution Equation of State Based on UNIFAC, *Fluid Phase Equilibria*, 70, 251–270, 1991.
23. Espinosa, S., Foco, G., Bermudez, A. and Fornari, T., Revision and extension of the group contribution equation of state to new solvent groups and higher molecular weight alkanes, *Fluid Phase Equilibria*, 172, 129–143, 2000.

24. Michelsen, M.L., Calculation of phase envelopes and critical points for multicomponent mixtures, *Fluid Phase Equilibria*, 4, 1–10, 1980.
25. Kravanja, Z. and Grossmann, I.E., Multilevel-hierarchical MINLP synthesis of process flowsheets, *Comput. & Chem. Eng.*, 21, S421–S426, 1997.
26. Gros, H.P., Díaz, S. and Brignole, E.A., Near-critical separation of aqueous azeotropic mixtures: Process synthesis and optimization, *J. Supercrit. Fluids*, 12, 69–84, 1998.
27. Diaz, S., Espinosa, S. and Brignole, E.A., Citrus peel oil deterpenation with supercritical fluids: Optimal process and solvent cycle design, *J. Supercrit. Fluids*, 35, 49–61, 2005.
28. van Konynenburg, P.H. and Scott, R.L., Critical lines and phase equilibria in binary van der Waals mixtures, *Phil. Trans.*, 298, 495–540, 1980.
29. Lucks, K.D., The occurrence and measurement of multiphase equilibria behavior, *Fluid Phase Equilibria*, 29, 209–224, 1986.
30. Coorens, H.G.A., Peters, C.J. and De Swaan Arons, J., Phase equilibria in binary mixtures of propane and tripalmitin, *Fluid Phase Equilibria*, 40, 135–151, 1988.
31. Peters, C.J., *Supercritical fluids: Fundamentals for application. Multiphase equilibria in near-critical solvents*, Kluwer Academic Publisher. Editors: Kiran, E., and Levelt Sengers, M.H., 1994.
32. Peters, C.J. and Gauter, K., Occurrence of holes in ternary fluid multiphase systems of near-critical carbon dioxide and certain solutes, *Chem. Rev.*, 99, 419–431, 1999.
33. Koenen, H-E. and Gaube, J., Temperature dependence of excess thermodynamic properties of binary mixtures of organic compounds, *Ber. Bunsenges. Phys. Chem.*, 86, 31–36, 1982.
34. Horizoe, H., Tanimoto, T., Yamamoto, I. and Kano, Y., Phase equilibrium study for the separation of ethanol-water solution using subcritical and supercritical hydrocarbon solvent extraction, *Fluid Phase Equilibria*, 84, 297–320, 1993.
35. Brignole, E.A., Andersen, P.M. and Fredenslund, A., Supercritical fluid extraction of alcohols from water, *Ind. Eng. Chem. Res.*, 26, 254–261, 1987.
36. Wiebe, R., The binary system carbon dioxide-water under pressure, *Chem. Rev.*, 29, 475–481, 1941.
37. Coan, C.R. and King, A.D., Jr., Solubility of water in compressed carbon dioxide, nitrous oxide, and ethane., *J. Am. Chem. Soc.*, 93, 1857–1862, 1971.
38. Kobayashi, R. and Katz, D., Vapor-liquid equilibria for binary hydrocarbon-water systems, *Ind. and Eng. Chem.*, 45, 440–446, 1953.
39. Zabaloy, M., Mabe, G., Bottini, S.B. and Brignole, E.A., The application of high water-volatilities over some liquefied near-critical solvents as a means of dehydrating oxychemicals, *Fluid Phase Equilibria*, 5, 186–191, 1992.
40. Reverchon, E. and Adami, R., Nanomaterials and supercritical fluids, *J. Supercrit. Fluids*, 37, 1–22, 2005.
41. Martin, A., *Precipitation processes with supercritical carbon dioxide: mathematical modeling and experimental validation*, Ph.D. Thesis, Universidad de Valladolid, Spain, 2005.
42. Martin, A. and Cocero, M.J., Numerical modeling of jet hydrodynamics, mass transfer, and crystallization kinetics in the SAS process, *J. Supercrit. Fluids,* 32, 203–219, 2004.
43. Amaro-González, D., Mabe, G., Zabaloy, M. and Brignole, E.A., Gas antisolvent crystallization of organic salts from aqueous solutions, *J. Supercrit. Fluids*, 17, 249–258, 2000.
44. Simoes, P.C. and Brunner, G., Multicomponent phase equilibria of an extra-virgin olive oil in supercritical carbon dioxide, *J. Supercrit. Fluids,* 9, 75–81, 1996.

45. Peter, S., *Supercritical Fluid Technology in Oil and Lipid Chemistry. Chapter VI: Supercritical fractionation of lipids*, Editors: King, J.W. and List, G.R., AOCS Press, Illinois, 65–100, 1996.
46. Espinosa, S., Díaz, S. and Brignole, E.A., Thermodynamic modeling and process optimization of supercritical fluid fractionation of fish oil fatty acid ethyl esters. *Ind. Eng. Chem. Res.*, 41, 1516–1527, 2002.
47. King, J.W. and List, G.R., *Supercritical fluid technology in oil and lipid chemistry*, Editors: King, J.W. and List, G.R., AOCS Press, Illinois, 1996.
48. Hegel, P.E., Mabe, G.D.B., Pereda, S., Zabaloy, M.S. and Brignole, E.A., Phase equilibria of near critical CO_2 + propane mixtures with fixed oils in the LV, LL, and LLV region, *J. Supercrit. Fluids*, 37, 316–322, 2006.
49. Härröd, M., van den Hark, S., Holmqvist, A. and Moller, P., Hydrogenation at supercritical single-phase conditions, *ISSAF - 4th International Symposium On High Pressure Process Technology And Chemical Engineering*, Venice, Italy, 2002.
50. Baiker, A., Supercritical fluids in heterogeneous catalysis, *Chem. Rev.*, 99, 453–473, 1999.
51. Chouchi, D., Gourgouillon, D., Courel, M., Vital, J. and Nunes da Ponte, M., The influence of phase behavior on reactions at supercritical conditions: The hydrogenation of alfa-pinene, *Ind. Eng. Chem. Res.*, 40, 2551–2554, 2001.

2 Supercritical Extraction Plants

Equipment, Process, and Costs

Jose L. Martínez and Samuel W. Vance

CONTENTS

2.1 Introduction 25
2.2 Supercritical Fluid Extraction: Process Description 26
2.2.1 Supercritical Fluid Extraction of Compounds from a Solid Matrix ... 28
2.2.1.1 Processing Parameters in the Supercritical Extraction of Solids 30
2.2.2 Supercritical Fluid Extraction of Compounds from a Liquid Feed ... 31
2.3 Supercritical Fluid Processing Plants: Equipment Design 34
2.3.1 Overview 34
2.3.2 Vessels 35
2.3.3 Pumps and Compressors 37
2.3.4 Heat Exchangers 38
2.3.5 Piping and Valves 39
2.3.6 Control Systems 41
2.4 Industrial Process Implementation 42
2.5 Conclusions 48
References 48

2.1 INTRODUCTION

In the last decade, supercritical fluid technology has reemerged, mainly due to a dramatic rise in the research and development activities focused on innovative approaches as well as new trends in the pharmaceutical, food, and chemical sectors. In the food industry, these new trends include an increased preference for natural products over synthetic ones and regulations related to nutritional and toxicity levels of the active ingredients. On the other hand, consumers are taking a more proactive role in maintaining their health, which has driven a new generation of products on the market addressing disease prevention. These trends have made supercritical fluid technology a primary alternative to traditional solvent extraction for the extraction and fractionation of active ingredients.

The objective of this chapter is to provide a review of supercritical fluid extraction, describing the process and discussing the influence of the process parameters. Moreover, this chapter is intended to give an overview of the main components of a supercritical extraction plant as well as the steps involved in process commercialization.

2.2 SUPERCRITICAL FLUID EXTRACTION: PROCESS DESCRIPTION

A supercritical fluid extraction process consists of two steps: extraction of the components soluble in a supercritical solvent and separation of the extracted solutes from the solvent. The extraction can be applied to a solid, liquid, or viscous matrix. Based on the objective of the extraction, two different scenarios can be considered:

1) Carrier material separation. In this case, the feed material constitutes the final product after undesirable compounds are removed—for example, dealcoholization of alcohol beverages, removal of off-flavors, or decaffeination of coffee.
2) Extract material separation. The compounds extracted from the feed material constitute the final product—for example, essential oil or antioxidant extraction.

The separation of soluble compounds from the supercritical fluid can be carried out by modifying the thermodynamic properties of the supercritical solvent or by an external agent (Figure 2.1). In the first case, the solvent power is modified by manipulating the operating pressure or temperature. In the second case, the separation can be carried out by adsorption or absorption. The more common method decreases the operating pressure by an isoenthalpic expansion, which provides a reduction of the fluid density and therefore a reduction of the solvent power. If separation takes place by manipulating the temperature, two situations may occur, depending on the solubility of the dissolved compounds. If solubility increases with temperature at constant pressure, a decrease in temperature will decrease the solubility and separate the compounds dissolved in the supercritical solvent. If solubility decreases with an increase in temperature at constant pressure, an increase in temperature will separate the compounds from the supercritical fluid solvent. If the separation is carried out by an auxiliary agent, such as an adsorbent, no significant pressure change occurs, so the differential pressure across the pump is much lower. This type of process implies lower operating costs; however, the recovery of the extract from the adsorbent is often very difficult. To overcome this disadvantage of high losses of the extract, the adsorption step may be replaced by an absorption step. The extract dissolved in the supercritical solvent is absorbed by a wash fluid in a countercurrent flow using a packed column or spray tower under pressure. Separation of solutes by adsorption and absorption has been applied in the decaffeination of coffee [1, 2].

One of the main advantages of supercritical fluids is the ability to modify their selectivity by varying the pressure and temperature (i.e., modifying fluid density). Therefore, supercritical fluids are often used to extract selectively or separate specific compounds from a mixture. One procedure is by a fractional extraction process.

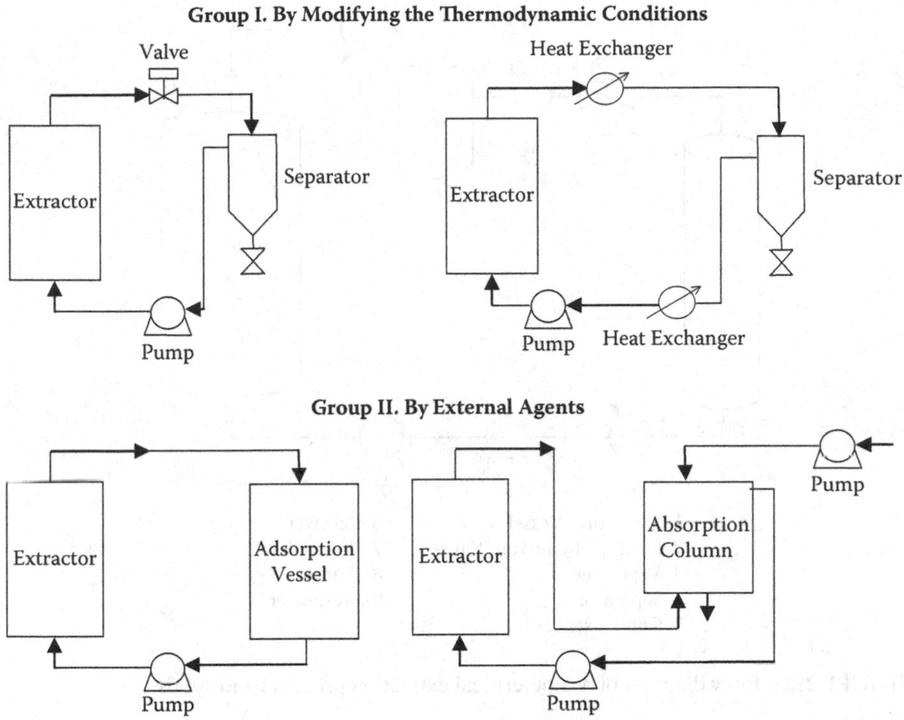

FIGURE 2.1 Basic scheme of supercritical extraction process.

In this case, the extraction is carried out in two stages. During the first stage, a relatively low fluid density is selected, which allows extraction of the compounds that are soluble at low pressure. Then, the residue is further extracted at high fluid density to recover heavier compounds (e.g., dealcoholization of cider [3]). Another example of fractional extraction consists of removal of nonpolar fractions in the first stage by using a supercritical solvent and the removal of a more polar fraction from the residue in the second stage by adding a cosolvent (e.g., extraction of active ingredients from grape seed [4]).

Another procedure to selectively extract or separate specific compounds from a mixture is sequential depressurization [5]. In this case, both fractions (light and heavy) are simultaneously extracted by using high-density fluid. Then the supercritical solvent and the extract pass through multiple depressurization steps, allowing fractional separation. In the first depressurization stage, the heavier fraction is collected; the volatile or light fraction is collected in the last stage. Two depressurization steps are generally used, although in some specific cases, three separation steps have been used. This method is commonly used in the extraction of spices, where the solubility of oleoresin and essential oil fractions in a supercritical solvent vary significantly with pressure and temperature. Generally, the extraction takes place at high pressures (40 to 60 MPa), so both fractions are soluble in the supercritical solvent. The separation or collection of the oleoresin fraction takes place in the first separator by

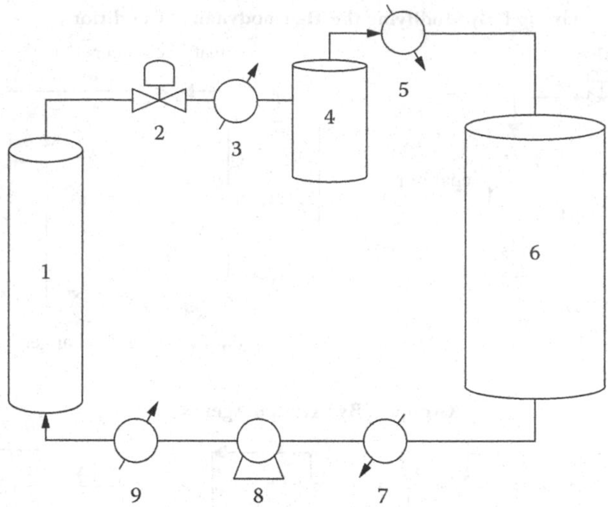

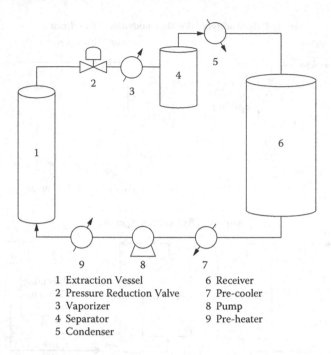

1 Extraction Vessel
2 Pressure Reduction Valve
3 Vaporizer
4 Separator
5 Condenser
6 Receiver
7 Pre-cooler
8 Pump
9 Pre-heater

FIGURE 2.2 Flow diagram of a supercritical extraction process from solids.

reducing the extraction pressure to intermediate pressure (15 to 20 MPa). Under such operating conditions, the aromatic fraction remains in the supercritical phase. After leaving the first separation, the pressure is further reduced and the essential oils are collected in the second separator. This type of process has been successfully applied in multiple products. In some cases, both fractions are desirable (e.g., oleoresin and essential oils, color and pungent fraction), whereas in others, only one of the fractions has commercial interest.

2.2.1 Supercritical Fluid Extraction of Compounds from a Solid Matrix

Most of the development and industrial implementation in supercritical fluid extraction has been performed on solid feed materials. Figure 2.2 illustrates a general flow diagram of a supercritical extraction process from solids. The solvent is subcooled prior to the pump, assuring a liquid phase to avoid cavitation issues. The pressurized solvent is heated above its critical temperature to the extraction temperature prior to the extraction vessel. The extraction vessel, which is filled with the feed material, is electrically or water heated to the extraction temperature. The supercritical solvent flows through the fixed bed and the soluble compounds are extracted from the carrier material. The supercritical fluid plus the extract leaves the extraction vessel from the top, through a pressure reduction valve. The solvent power decreases with pressure reduction, so the compounds precipitate. To assure total precipitation, the supercritical solvent is heated above the saturation temperature to reach the gas phase. Under those conditions, the solvent power is negligible. Then the material is collected in a separator while the solvent in gas phase leaves the separator vessel from the top and

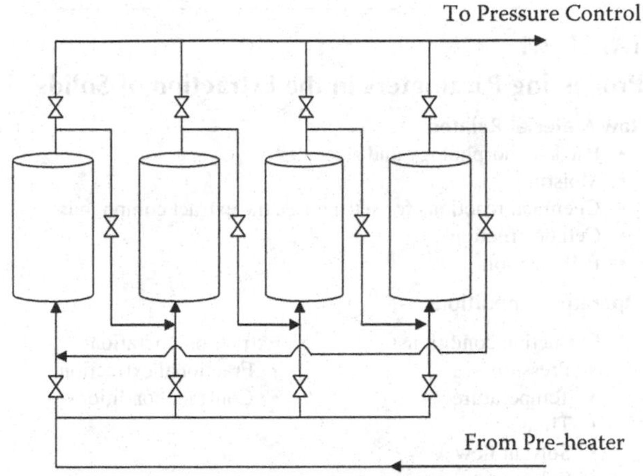

FIGURE 2.3 Scheme of cascade operation of multiple extraction vessels for extraction of solids.

is recirculated back to the extraction vessel. Once the raw material is fully extracted, the following steps are required in the extraction vessels:

- Depressurization
- Opening of the extraction vessel
- Unloading the spent material
- Loading with fresh material
- Closing the extraction vessel
- Pressurizing to operating conditions

One of the most difficult aspects is attaining continuous feed of the solids and continuous discharge at a high pressure extraction vessel. Generally, the solid feed material is handled by using preloaded baskets. From an industrial or commercial point of view, the use of only one extraction vessel, even with a quick-opening closure that allows for rapid opening and closing, is not economically viable. Therefore, multiple extractions vessels operating in a countercurrent flow are required. Figure 2.3 shows a general scheme of a cascade extraction with four extraction vessels. In this case, once the raw material in the first extraction vessel is fully extracted, the vessel is taken out from the process by valving. Once the vessel is depressurized, emptied, and refilled, it enters the process line as the last extraction vessel. The second extractor is the next one to be isolated of the process line. Operating this way, the fresh supercritical solvent extracts first the raw material that is partially exhausted and in the final extraction step, the supercritical solvent extracts fresh raw material. This configuration provides higher solvent loading (amount of material extract/amount of solvent). The objective is to maximize the solvent loading (i.e., to maintain the supercritical solvent saturated or close to the saturation point).

Since the extractors are operated batch wise, a critical factor is to shorten the charge and discharge cycle times. Therefore, a cap automation mechanism with

TABLE 2.1
Processing Parameters in the Extraction of Solids

Raw Material Related	
• Particle morphology and size	
• Moisture	
• Chemical reactions for setting free the extract compounds	
• Cell destruction	
• Pelletization	
Operating Conditions	
• Extraction conditions:	• Extraction operation:
• Pressure	• Fractional extraction
• Temperature	• Constant conditions
• Time	
• Solvent flow	
• Solvent-feed ratio	
• Separation conditions:	• Separation operation:
• Pressure	• Single stage
• Temperature	• Fractional separation

quick opening closure, as well as fast depressurization, and unloading/loading sequence are critical in the design of a supercritical extraction plant.

2.2.1.1 Processing Parameters in the Supercritical Extraction of Solids

Parameters affecting the supercritical fluid extraction of solids are listed in Table 2.1. The influence of the process parameters can be summarized as follows:

- Solubility of compounds increases by increasing the extraction pressure at constant temperature.
- At pressure close to the critical pressure, solubility of the compounds increases by decreasing the temperature. However, at high pressures, solubility of compounds increases by increasing the temperature. This crossover effect is due to the competing effects of the reduction in solvent density and the increase of the vapor pressure. The latter has marked influence at higher pressures. The pressure at which the crossover effect occurs depends on the type of compounds to extract. The crossover range for most of the compounds takes place between 20 and 35 MPa.
- The separation conditions depend on the solubility of the compounds at different pressures and temperatures as well as whether a fractionation of extract is carried out by sequential depressurization steps. Generally the separation pressure is carried out at 5–6 MPa. For essential oils or volatile fractions, the separation takes place at 3 to 5 MPa and low temperatures to maximize the recovery of the top notes components. For oils, the separation can take place at 15 to 20 MPa due to their low solubility in supercritical carbon dioxide (CO_2) under those conditions.
- The solvent-feed ratio depends on many factors, such as concentration of the solute in the feed material, solubility in the supercritical solvent, type

of feed material, and distribution of the compound in the feed material. Low solvent-feed ratios imply lower operating costs and higher production capacity. Generally, the industrial processes target solvent-feed ratios lower than 30. However, higher solvent-feed ratios are justified for high added-value products. In specific cases, a solvent-feed ratio higher than 100:1 has been reached for commercial applications.

- High solvent flow rates imply high operating and capital costs. However, they could increase production capacity. The solvent flow rate or the residence time of the solvent in the extraction vessel must be optimized. A high residence time implies a long batch time. Conversely, a short residence time may result in shorter contact time between the solvent and solute, resulting in a loading of the solvent much lower than the saturation concentration at the selected operating conditions. Linear velocities ranging from 1 to 5 mm/s are commonly used in the supercritical fluid extraction process.
- The size and morphology of the solid material have a direct effect on the mass transfer rate. In general, increasing the surface area increases the extraction rate. Therefore, smaller particle size or geometry (such as flakes) generally favors higher mass transfer, decreasing the batch time as well as diffusion controlled process. If the soluble substances are located in rigid structures inside of the solid matrix, the size reduction breaks this structure so it will be easily accessible for the solvent. However, very small particles favor a channeling effect, which decreases the extraction rate. Particle size needs to be evaluated case by case based on the type of material to be processed. In the case of processing of spices and seeds, particle size is generally between 30 and 60 Mesh.
- Similarly to particle size, moisture content must be evaluated case by case. High content of moisture is usually not desirable because moisture acts as a mass transfer barrier. On the other hand, moisture expands the cell structure, facilitating the mass transfer of the solvent and the solute through the solid matrix (e.g., in seeds and beans). For instance, the influence of moisture between 3% to 10% generally has no significant impact on the mass transfer of edible oil from seeds.

2.2.2 Supercritical Fluid Extraction of Compounds from a Liquid Feed

When feed material is in a liquid state, extraction is typically carried out in a countercurrent column. The dense material (liquid) is introduced from the middle or the top of the column, and the material with lower density (solvent) is introduced from the bottom of the column. This continuous process leads to lower operating costs than those incurred with extraction from a solid matrix. A general process flow diagram is shown in Figure 2.4. The separation steps and regeneration of the solvent is similar to the extraction from solids.

Similar to the conventional countercurrent column processes, the contact between phases is favored by random or structured packing material. Additionally, reflux of extract improves selectivity in the extraction process. The extract and solvent leave the column from the top, while the heavier material, or raffinate, is collected from

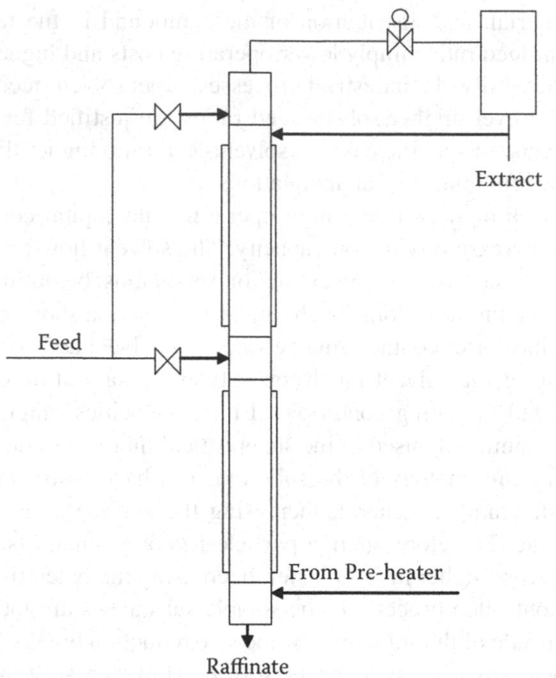

FIGURE 2.4 Flow diagram of a supercritical extraction process from liquids.

the bottom. The countercurrent column is heated electrically or with a hot water jacket and the extraction process can take place at constant temperature or with a controlled temperature gradient. The latter process provides an internal reflux that increases selectivity.

Process design is based on phase equilibrium data, which determine the number of theoretical stages necessary to perform a specific separation; height of the column, which is related to mass transfer or height equivalent to a theoretical plate, and diameter of the column, which determines the capacity. The latter parameter is related to hydrodynamic behavior of the mixture in contact with the packing.

In cases where the viscosity of the liquid is very high, the extraction process requires intensive and uniform contact between the feed and the solvent. This contact can be carried out by mechanical mixing or by nebulizing the viscous material through a nozzle. In the case of mechanical mixing, the mixer can be magnetically or mechanically coupled. In the latter case, the driving shaft is actuated outside of the vessel directly by a motor, whereas in the first case, it is actuated by magnetic fields. The torque generated by magnetic coupling is lower; however, there is no rotary seal and there is no need for lubrication. Direct drive couplings rotate the outer housing and the magnetic field then rotates the driven magnets secured to the mixer shaft. When the mixer is mechanically coupled, a shaft design providing high torque at high pressures, with extended lifetime of bearings and seals is required. Additionally the mixers must be properly designed based on the specific application. Figure 2.5 shows a mechanical coupling developed by Thar Technologies that operates at 69 MPa.

FIGURE 2.5 Mechanical mixing using a mechanical coupling. Design pressure: 69 MPa (Courtesy of Thar Technologies, Pittsburgh).

In a mechanical mixing process, the viscous liquid or melted product is pumped into the vessel. By adding the supercritical solvent, the viscosity of the product decreases, which facilitates mixing and reduces the torque required. The supercritical solvent flows through the viscous material extracting the soluble compounds. The supercritical fluid and the extract leave the extraction vessel from the top. Once the extraction is complete, the material left can be discharged from the bottom of the extraction vessel. If the viscosity of the remaining material is still high, the supercritical fluid assists in removing the material through the opening. This process can be applied to material with very high viscosity at atmospheric pressure.

Another alternative for processing viscous liquid material involves intensive contact between both phases (i.e., mixing and nebulizing the mixture). In this case, the viscous material and the supercritical solvent are mixed and sprayed through a nozzle. The supercritical solvent reduces the viscosity of the feed and therefore decreases the interfacial tension. By spraying through a nozzle, an atomization takes place, creating very fine droplets with a very large surface area and a high contact between both phases. The supercritical solvent extracts the soluble material and the insolubles precipitate in the bottom of the extraction vessel. This process is favored when there is a significant difference in solubility between the compounds to be separated. The critical parameters in this process are the contact or mixing devices, spraying devices, vessel design, and solid removal from a pressurized vessel. A process using this concept has been successfully developed and industrially implemented in the deoiling of crude lecithin [6]. This is a continuous process in which

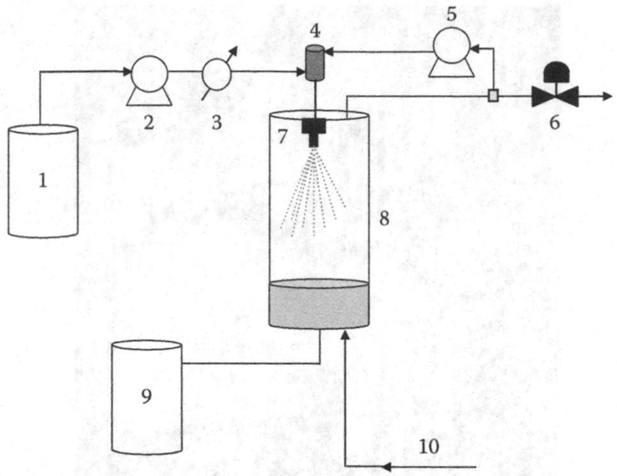

FIGURE 2.6 Scheme for a continuous process for de-oiling of crude lecithin. 1 Tank of crude lecithin, 2 Lecithin pump, 3 Preheater, 4 Mixer, 5 Recirculation pump, 6 Back pressure regulator, 7 Nozzle, 8 Extraction vessel, 9 Transfer vessel, 10 CO_2 inlet.

crude lecithin is pumped and mixed with supercritical CO_2 and then sprayed through a nozzle into a high pressure vessel. The neutral lipids are solubilized in the CO_2, while the polar lipids are precipitated in the bottom of the extraction vessel. The CO_2 and the neutral lipids leave the extraction vessel and the oil is recovered in the separator. The polar fraction in a powder form is continuously transferred to a storage tank (Figure 2.6). Some work was done in the early 1980s; however, it was never industrially scalable, mainly because the solvent-feed ratios used were extremely high and a batch process was used, so the production capacity was very low. Therefore, the plant size required to satisfy commercial demand had to be very large, which implied very high capital cost. The new process offers two significant advantages: (1) the process is continuous and (2) the solvent-feed ratio required is low. This is an example of industrial implementation of a supercritical process for a commodity product, so the operating costs must be comparable to that at conventional processing. The conventional process is a well-established process in the oil industry using acetone as a solvent. However, using acetone as a solvent forms acetone derivatives (mesityl oxide) with adverse effects for the deoiled lecithin due to its toxicity and off flavor. The oil industry has been searching for alternative methods but has not found an alternative process until now. A similar process can be applied to removal of residual solvents in the pharmaceutical industry [7].

2.3 SUPERCRITICAL FLUID PROCESSING PLANTS: EQUIPMENT DESIGN

2.3.1 Overview

Design and selection of equipment for a supercritical fluid processing (SFP) system requires consideration of some parameters and specifications that are unique to this type of plant. Many items that would be off-the-shelf for most plants are simply not available or not easily found for application to the operation or design conditions of the SFP environment and process requirements. For example, many SFP systems require sanitary design for food, nutraceutical, or pharmaceutical products and operating pressures substantially higher than normally found in food, nutraceutical, or pharmaceutical plants. Meeting these requirements entails investigation of vendors who can provide items that are suitable. Unique conditions may be encountered in regular operation or a major malfunction may occur that requires special materials of construction (e.g., metals can undergo brittle fracture in such environments). System capital cost must be closely controlled to be competitive with other systems.

2.3.2 Vessels

In general, SFP vessels are designed and manufactured in accordance with American Society of Mechanical Engineers (ASME) Section VIII Standards. In many cases, the process requires one, or sometimes two, full-diameter quick-opening closures for charging fresh feedstock or discharging spent feedstock. The closure mechanism is often automated to minimize downtime of a vessel when filling or emptying. There are a number of such closures proprietary to vessel designers or suppliers. Consideration must be given to methods of cleaning vessels between charges or emptying vessels when the solids plug or bridge. Vessels are jacketed or electrically traced for process temperature control. Vessel shape and aspect ratio must be carefully evaluated to minimize vessel costs without affecting performance. The most critical vessel design in the process is that of the extraction vessel. It requires the maximum design pressure and may be the most critical in selection of materials of construction. In many cases, special alloys of stainless steel or exotic metals can be used, but the actual selection of alloys and thicknesses may also depend on the ability to machine, forge, and weld vessel load-bearing components. Extraction vessels can be fabricated by forging, machining solid barstock, rolling and welding of plate, multiwall rolling and welding, composite multilayers, and casting.

Full diameter quick-opening vessel closures may utilize self-energizing seals, segmented rings, breech locking, flanges, and threaded caps. Most are proprietary designs. Examples of closures are shown in Figure 2.7.

Vessels other than the extraction vessels are usually designed for substantially lower operating pressures and temperatures, but the function of each vessel must be carefully considered in selecting the size and shape of separation vessels and process holding vessels. In some cases, the vessels may need special designs to keep them clean and minimize plugging and contamination.

At the beginning and end of the batch process cycle, the vessels may very well be at or near ambient pressures and temperatures, as materials are being transported to or from the SFP sections. But even for these areas, the vessels may still require adaptation for cleaning-in-place (CIP) or other cleaning methods.

In many cases, the feedstock is in particulate solid or pellet form. The extraction vessel would then be designed for (usually) batch charging and emptying of the extraction feedstock. In this case, an extraction vessel would be filled with feedstock, brought from ambient temperature and pressure to extraction pressure and

(a) (b)

FIGURE 2.7 Vessel closure types. a) Automated segment ring closure, b) Automated clamp closure (Courtesy of Thar Technologies, Pittsburgh).

FIGURE 2.8 Countercurrent column (10 m) of a supercritical process extraction plant (Courtesy of Tharex, Seoul).

temperature, extracted until the solute has been removed, and then depressurized, emptied, and recharged. The balance of the plant would be essentially continuous in operation with closed cycle solvent recirculation. Multiple extraction vessels would be used to approach a continuous operation.

In some cases, the feedstock is liquid and the extraction vessel may be a packed column operating continuously (Figure 2.8). In these situations, a truly continuous operation would be the norm, with reduction of pressure only done for product change or system shutdown and overhaul or maintenance. Some systems utilize supercritical

FIGURE 2.9 Dynamic axial column (30 cm ID) of a supercritical process chromatography plant (Courtesy of Thar Technologies, Pittsburgh).

fluid chromatography (SFC) in a packed column to achieve the separation of components targeting very high purities (95% to 99%). Figure 2.9 shows a process scale SFC using a dynamic axial chromatography column.

2.3.3 Pumps and Compressors

The second most critical equipment items are the pumps and compressors used for building system pressures and temperatures to the supercritical regions established in the process development studies. SFP flows are relatively low and pressures are relatively high, ranging from 5 to 120 liters per minute flow at 6 to 65 MPa. The process requires close control of temperatures, pressures, and flows. Pressures, in particular, require critical control because pressure fluctuations may make substantial difference in processing results and can result in overpressure devices shutting the process down and wasting both solutes and supercritical solvents. Such shutdowns result in poor production rates and unnecessary cost penalties for the system.

Most high-pressure pumps are multiplunger styles. Flow and pressure control commonly use some type of speed control, such as variable frequency speed. A further development includes diaphragm type pumps, which are actuated by plungers and more normal liquids that cause the displacement and flexing of the final pumping element (the diaphragm). The pumps have proprietary design features to provide suitable operation for the pressures and flows required. More standard pumps can be used for provision of makeup supercritical fluids and final product (extract) pumps.

Compressors may also be reciprocating pistons. In rare situations, more standard rotary compressors can be used, often in recovery of otherwise wasted supercritical fluid. Economics and process variables determine the extent of recovery of spent solvent fluids.

The fluid ends of plunger pumps or compressors require close tolerance machining of the plunger or piston and the cylinder to minimize leakage. Nonlubricated plungers are often chosen, with careful selection of materials to ensure low coefficients of friction and dimensional stability. Lubricants are avoided because they

FIGURE 2.10 High-pressure multiplunger pump. Design pressure: 96 MPa, Flow rate: 30 kg/min (Courtesy of Thar Technologies, Pittsburgh).

contaminate the process solvent and solute. O-rings, gaskets, and seals for reciprocating and rotating parts must be carefully designed. Materials must be compatible with the solvents and solutes. Absorption of solvent in the O-ring may present problems when the system is depressurized because the solvent may expand within the O-ring, causing disintegration, especially if depressurization is rapid. Unusual or exotic materials for the plunger and cylinder may be used as coatings or solid sections. Proprietary information is acquired by substantial equipment development and is carefully protected by designers and fabricators. Figure 2.10 shows a multiplunger pump with a design pressure of 96 MPa and a flow rate of 30 kg/min.

2.3.4 Heat Exchangers

Heat exchange equipment also presents unique problems for supercritical fluid systems due to the high pressures required in key parts of the process. Although heat exchangers in the process industries are a mature and very competitive technology, designs are not readily available at the pressures encountered. Also, special consideration must be given to cleaning of the process side heat exchange surface in the event of fouling with solutes and cleaning and disassembly of the exchanger for changeover to another product. Another special consideration in selection of heat exchanger types or styles is the risk analysis for heat exchanger tube failure. A leak or catastrophic failure may create a dry ice plug in the high pressure side (for CO_2 as the solvent), freezing of the heat exchange fluid, and overpressure of the low pressure heat transfer fluid piping system. Selection of overpressure safety devices must be carefully investigated by process risk analysis and Hazard and Operability (HAZOP) studies techniques for the system.

Removable heat exchanger heads are often desirable. The supercritical fluid solvent most often is on the tube side of shell-and-tube exchangers and design compromises must often be made between multitube tube diameter and number of tubes. The smaller the diameter of the tube, the more tubes are required to provide suitable fluid velocity through the exchanger but the less tubing and metal required for the heat transfer area. Smaller diameter tubes present additional problems for cleaning and for fastening to the tube sheets (usually by welding). Larger diameter tubes improve the fluid velocity and tube numbers but also may result in longer exchangers. As the tube length increases, differential linear expansion of the shell and tubes may require expansion joints or floating heads. These situations complicate the exchanger design and may add to the cost of design and fabrication.

In most process designs, ASME Section VIII Unfired Pressure Vessel Code Standards and Code Stamps are necessary. In some cases, a simpler design of the exchanger can be accomplished if the possibility of heat exchange surface fouling and plugging of the tubing interior can be minimized by carefully controlling processing conditions in the system.

Types of heat exchangers that can be used include shell-and-tube exchangers, double-pipe and multi-U-tube exchangers, double pipe coils, or simple coils in tanks. An example of shell-and-tube heat exchanger design is shown in Figure 2.11.

2.3.5 Piping and Valves

Selection of piping, fittings, and valves for SFP also requires special design specifications and criteria. The material to be used must be nonreactive with the supercritical fluid solvent and solutes in the process. The possibility of reactions between the solvent and the piping surfaces must be evaluated. The solvent may be substantially more aggressive in the supercritical fluid regime than would be true at lower pressures, so additional testing of materials may lead to more expensive alloys to minimize such reactions. Where possible, high strength alloys (with higher allowable stress than the typical 300 Series stainless steels) are the choice for overall cost and process suitability. Since flow rates for most supercritical fluid systems are much lower than in more conventional systems, pipe diameters and suitable high-pressure fittings are smaller while maintaining appropriate flow velocities in the interconnecting piping or tubing. Piping cost is thus minimized. However, the conventional threaded joints or "standard" flanges are not cost effective. In most cases, special high-pressure fittings, couplings, and the like will be the selection of choice, both for convenience and economic reasons. Special high-pressure couplings are shown in Figure 2.12.

Valving is another unique area for the process. Two types of valving must be considered: (1) isolation valving and (2) flow control valving. Isolation valving most often includes plug valves or butterfly valves for leak-proof on-off service. These valves commonly have metal-on-metal sealing surfaces where low friction is desired. Special coatings (sprayed, vapor deposition, or composition) may be used to avoid seizing or galling. Selection of dissimilar metals or metal oxides or carbides with high hardness values and good machining properties improves performance. However, pairing of materials and selection of designs for the operating environment is still more art than science. So specialty high-pressure valve companies are the vendors of choice. Flow control valves are also a specialty item at supercritical fluid operating pressures. In some cases, pressure drop through the control valve may be at critical flow or with a phase change when passing through the valve.

FIGURE 2.11 Shell-and-tube heat exchanger (Courtesy of Thar Technologies, Pittsburgh).

These conditions require specialized knowledge of the effects of the high velocity through the valve orifice and the possible presence of two-phase flow through the valve. Again, in selecting materials for packing glands and valve stem seals, care must be taken to select an appropriate elastomer or composite that will not absorb high pressure solvent during operation and then fracture or fail when the pressure is released. The supercritical fluid solvent may vaporize and expand in the packing or O-ring, with subsequent destruction of the seal. At the range of pressures under consideration, dimensional stability and elimination of creep flow are also necessary.

As SFP system throughputs become larger, automated control and isolation valves become more attractive. Pneumatic or hydraulic operators and, occasionally, electrically operated modulating operators may be required to minimize the down time for the plant. Fail-safe operation must be the order of the day.

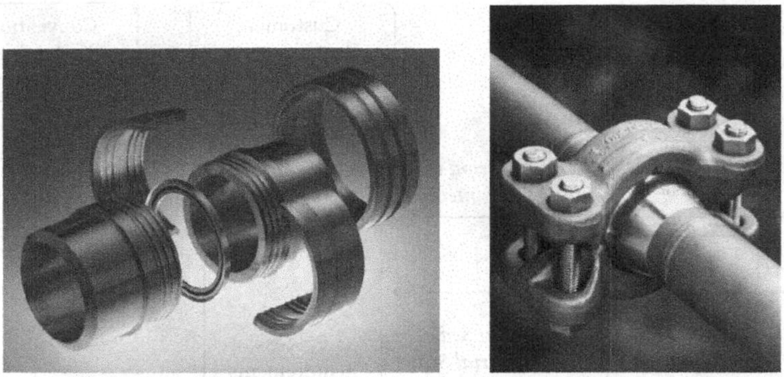

(a) (b)

FIGURE 2.12 High-pressure couplings: a) DUR O LOK (Courtesy of BETE Fog Nozzle, Inc., Greenfield, MA), b) Grayloc (Courtesy of Grayloc Products, Houston, TX).

2.3.6 Control Systems

As should be obvious by this point, many of the areas of control for SFP systems require special consideration because of the high operating pressures (with the resulting high potential energy in the process) and the possible hazards for both operating personnel and plant integrity. Control system failures in more conventional processing plants would not present the possible hazards to operators and damage to the equipment and hazards beyond the plant area. Relatively small variations in process conditions can be reflected in substantial variations in system pressures and phase transitions. So control response to these variations must be rapid and effective in damping the results within the system.

Selection of primary sensors must be carefully made with provisions for sensor failure, leakage, or error. Redundancy must be considered and carefully thought through. Even pressure gauges and temperature elements must be examined. Pressure gauges or transducers may require liquid seals or thermowells to permit isolation and replacement while the system is operating. Temperature sensors can be thermocouples or resistance temperature detectors (RTD) sensors, but response time must be weighed against thermowell isolation. Gauges should have blowout discs. Level sensors must be accurate and reliable. Pump and compressor flow rates are commonly measured by mass flowmeters (Coriolis meters) and flow controlled by frequency modulation of the connected motor. Overall system shutdown is controlled by distributed control with computer capability. System conditions at startup and shutdown of the process must be thoroughly thought through with a HAZOP review.

Summarizing, the foregoing description of the factors that must be considered in the specification and selection of the hardware that goes into a SFP plant shows that the unique mechanical system requirements must be carefully made and reviewed. Each system has some characteristics that must be evaluated based on the particular environment and process chemistry for that product or commodity. In some cases, the plant will be a multipurpose or multiproduct plant. Each purpose or product must be considered to establish the single product that would control the mechanical specifications of each item of equipment to be selected.

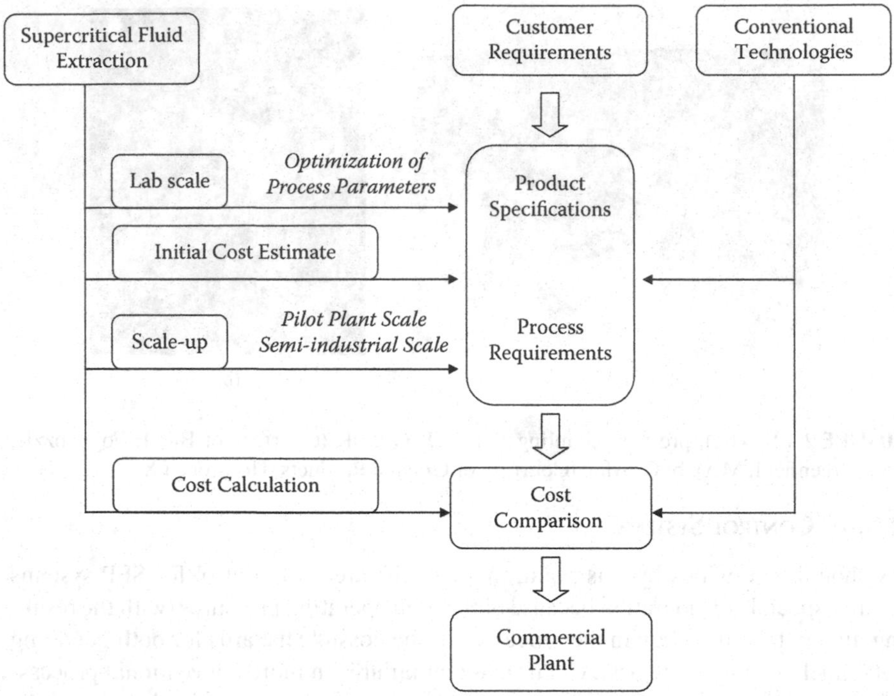

FIGURE 2.13 Work flow for industrial process implementation.

2.4 INDUSTRIAL PROCESS IMPLEMENTATION

Process development responds to a specific requirement of a company or to a market demand for a specific product. In some cases, the current process shows significant weakness even for a well-positioned product in the market. So, the processing companies that are aware of the limitations or constraints search for new processes to strengthen the product, resulting in a leading market position and projecting higher revenues. In other cases, traditional technologies do not offer a satisfactory solution to a specific problem. Additionally, there is a continuous search to reduce production costs.

A general workflow for an industrial process implementation is illustrated in Figure 2.13. The first step is to prove that the technology is capable of meeting the product specifications and process requirements defined by the customer or the market. In terms of product specifications, requirements generally are a minimum concentration or purity of specific compounds and minimum extraction or recovery efficiency. Regarding process requirements, the main constraints are maximum operating temperatures, type of pretreatment or material conditioning, and acceptable cosolvents to be used.

In the case of supercritical fluid extraction, the first point to be addressed is if a compound to be extracted is soluble in the supercritical solvent or if the solvent will be selective to fractionate or separate a mixture of compounds. The thermodynamic data required are the solubility of the specific compound in the

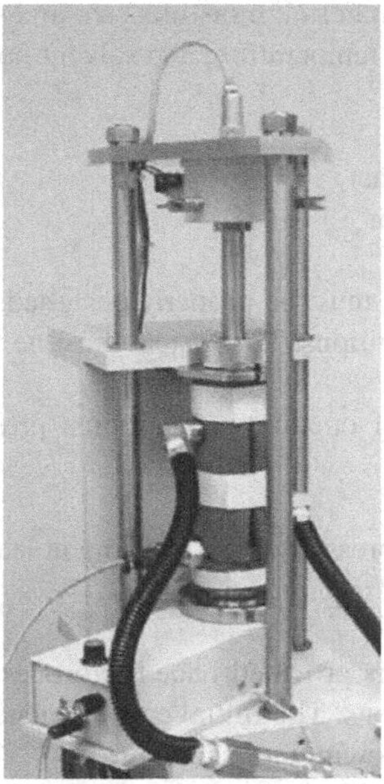

FIGURE 2.14 Phase equilibrium analyzer (Courtesy of Thar Technologies, Pittsburgh).

supercritical fluid as a function of pressure, temperature and solute concentration, partition coefficients, and selectivity or separation factors. A very extensive database of phase equilibria and solubility data for binary systems has been generated over the last two decades and can be used as a reference. However, in some cases, the published data are questionable.

A rapid way to determine solubility data or phase transitions for a binary mixture (specific compound and supercritical solvent) is by a phase equilibrium analyzer (Figure 2.14). This is a static method whereby the solute and solvent are loaded into a high-pressure vessel. This vessel consists of a variable volume high-pressure view cell with an integral stirrer, water jacket, and video system. Once the mixture is compressed to a single phase for a selected temperature, slow movement of the cell piston at a controlled rate slowly decreases the pressure until a second phase appears. By observing the video output of the system, it is possible to determine the cloud point for the sample at the current pressure and temperature. Additional experimental data are obtained by modifying the temperature and repeating the experimental procedure. The main advantages of this method are rapid generation of data and visible confirmation of dissolution; in addition, no sampling is required, no extraction efficiency is involved, and a minimum amount of solute is used.

Once the thermodynamic data are obtained, the next step is to evaluate the extraction of that specific compound from the original sample, which is generally a multicomponent mixture in a supercritical fluid extraction system at bench scale to optimize process parameters. The process parameters to optimize are listed here:

- Conditioning the raw material: moisture, size and shape, etc.
- Kinetic data; pressure, temperature, and solvent flow rate effect:
    - Extraction yield
    - Extraction time
    - Quality of the extract
- Fractionation conditions

The lab or bench system must be properly designed to be versatile and cover a wide range of operating conditions. The system must be able to:

- Cover a wide range of operating conditions: pressure, temperature, and flow rates
- Use cosolvents
- Perform sequential depressurization by using at least two separation stages
- Continuously log process parameter data

An initial cost estimate is provided once the process development satisfies the customer product specifications. The initial cost estimated is calculated using scale up methods based on the following information:

- Customer production requirements (i.e., amount of material to process per year, working days per year, and working hours per day)
- Raw material (i.e., particle size and shape, concentration of the product)
- Optimized process parameters (i.e., extraction pressure and temperature, solvent flow rate, residence time, kinetics of the process, bulk density of the feed material, and separation pressure and temperature)

This cost estimation provides to the customer the following information: operating costs ($/kg of feed, or $/kg of final product), plant size, and plant configuration. At this point, the customer determines if the supercritical process will meet their budget and if the investment and operating costs are comparable with or better than conventional technologies. If so, the next step is to scale up the process to pilot plant or semi-industrial scale. The objectives to accomplish in this stage, meeting all products specifications, are:

- Verification of the process parameters selected in the lab scale and their optimization if required
- Optimization of utility requirements
- Optimization of recirculation parameters of the solvent and addressing any issues related to material handling

Reducing operating costs requires minimizing energy requirements, which also implies a reduction in the associated capital cost of the auxiliary equipment. These costs are directly related to the recirculating costs of the solvent. Recycling of the solvent depends on the separating conditions of the substance from the extraction fluid. Typically, recycling is performed at low separation pressures and the solvent

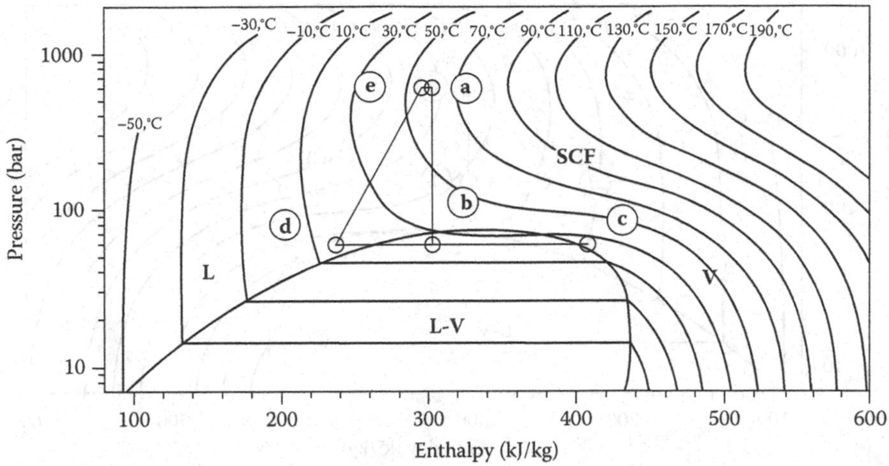

FIGURE 2.15 Diagram P-H for a low-pressure recirculation solvent using pump.

is recycled in the liquid state. Figure 2.15 shows the solvent cycle in a pressure-enthalpy diagram. In this case, the separation of the soluble compounds from the supercritical solvent is achieved by isoenthalpic throttling (a-b) followed by heating (b-c). At c, the solvent is in gas state, so the solvent power is negligible. Then the solvent is subcooled (c-d) and pumped to the extraction pressure (d-e). The solvent is heated to the extraction temperature (e-a). A chiller unit is used in order to cool down and condense the extraction fluid before it enters the pump. However, if extraction takes place at pressures below 30 MPa, the recycling of the solvent as gas, replacing the pump by a compressor, generally results in energy savings. For instance, in extraction of essential oils, where the extraction conditions are generally carried out in the pressure range of 8 to 20 MPa and in the temperature range of 35°C to 50°C, the solvent recycled in a gas state is more energy efficient.

As previously mentioned, at pressures higher than 30 MPa, solvent recycling in liquid stage is more efficient. Under these operating conditions, an alternative to provide a higher energy efficiency solvent cycle is addition of a compressor into the system after separation of the extracted compounds [8]. After expanding the mixture to form a two-phase region and heating the mixture, so that the solvent becomes a single gas phase, the solvent is compressed to a pressure higher than critical pressure by a compressor. Then the solvent is subcooled before entering the pump. Two advantages are obtained using this method. The first is that, instead of using a chiller, the solvent can be cooled down with water from a cooling tower. Second, the mechanical energy required of the pump is lower. The energy savings of recycling the solvent using both a pump and a compressor, compared with the more traditional process, depends on extraction conditions but could be up to 65%.

In cases where the solubility of the specific compound in the extraction fluid is very low, typically lower than 0.5%, at a pressure higher than the critical pressure of the solvent, the solvent can be recycled in a supercritical state providing additional energy savings. For instance, in the extraction of seed oils using supercritical CO_2, the separation of the oil in the separator should be carried out under supercritical

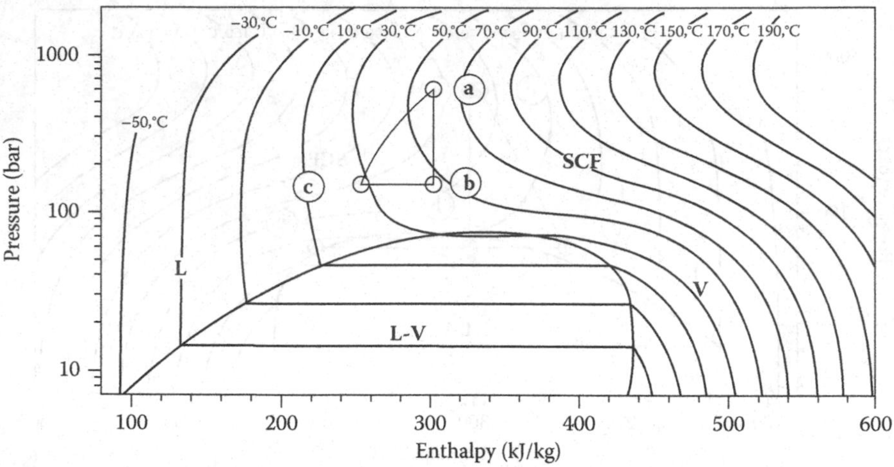

FIGURE 2.16 Diagram P-H for a supercritical recirculation solvent.

TABLE 2.2
Solvent Cycle at Different Operating Conditions

Solubility of Solute at P > Pc	Solvent Cycle	Extraction Pressure	Solvent Recycling State	Main Components
> 0.5%	Low pressure recycling	Pext > 30 MPa	Liquid	Pump
			Gas	Compressor + pump
		Pext < 30 MPa	Liquid	Pump
			Gas	Compressor
< 0.5%	High pressure recycling	Pext > 30 MPa	Supercritical	Expansion valve + pump
			Supercritical	Turbine + pump (energetically coupling)

conditions at a pressure less than 20 MPa and the solvent recycled in supercritical state. The solubility of the oil at pressures below 20 MPa is generally less than 0.3%. Additionally, the supercritical CO_2 could be recompressed directly without the need of subcooling the CO_2. There is not a change phase that could create cavitation in the pump. Table 2.2 summarizes the solvent cycle at different extraction conditions based on the solubility of the extract. However, because the compressibility of supercritical CO_2 is higher than that of liquid CO_2, it reduces pump capacity. Figure 2.16 shows the solvent cycle of CO_2 regenerated under supercritical conditions.

Chordia and Martinez [8] describe an alternative method of providing higher energy savings for recycling the solvent in the supercritical state. In this case, high pressure recycling is realized by replacing the expansion valve with a turbine and energetically coupling the turbine to the pump. Based on the operating conditions, energy savings of up to 60% can be reached. Table 2.2 summarizes the solvent

TABLE 2.3
Case Study: Cost and Revenue Estimates

	Flaxseed	Astaxanthin	Ginger
Amount to be processed (MT/year)	3,000	50	3,000
Concentration of the product (%)	13	2.5	5
Number of days/year	300	300	300
Number of hours/day	24	24	24
Estimated equipment cost ($)	2,500,000	1,500,000	4,500,000
Operating costs[1] ($/kg of feed)	0.23	5.57	0.43
ROI	3.2	1.0	1.6

[1] Includes power consumption, CO_2 losses, maintenance, and labor

cycle recommended for energy savings at different extraction conditions based on solubility of the extract.

An additional factor that must be considered to reduce operating costs is to minimize solvent losses. The losses take place from the extraction vessel during the depressurization for unloading the spent material, as well as in the separator when withdrawing the extract. Most of the CO_2 losses occur in the depressurization process. Generally, depressurization involves two steps: a) The extraction vessel is depressurized down to the receiver pressure (5 to 6 MPa) and b) The rest of the solvent is vented to the atmosphere. An alternative to reduce CO_2 losses is to use a compressor and condense additional CO_2. In this case, once the pressure in the extraction vessel equals the pressure in the receiver (5 to 6 MPa), the compressor compresses the CO_2 until the residual pressure in the extraction vessel reaches 0.2 to 0.5 MPa. Then the residual CO_2 is released into the atmosphere. In places where CO_2 cost is high, investment in capital cost of additional equipment to reduce CO_2 losses makes economic sense.

An additional operating cost is labor. The personnel required to operate an industrial SFE plant depends on the size, configuration of the plant, batch time, and automation of the plant. In general, a supervisor and two operators are required for a fully automated large plant. The labor cost, as well as the cost of CO_2, is highly dependent on the geographical location. After scaling-up the process and definition of the operating parameters and configuration, the final cost estimate is made. While the industrial plant is built, tolling is a preferable step in many cases. This step is used for multiple purposes: formulation of different products, market evaluation, and launching the product into the market while the industrial plant is under construction.

The capital cost of a SFE plant depends on many factors, such as the number of vessels, design pressure, size of the vessels, flow rate, automation, and Good Manufacturing Price (GMP) compliance. Generally, the capital cost of the SFE plant is higher than the traditional or conventional extraction plant, while the operating costs are lower. However, to compare properly the capital costs of SFE plant versus traditional extraction process, is necessary to take into account all the associate equipment used in the conventional extraction process, such as distillation or evaporation systems for solvent recovery; as well as associate costs in building requirements, instrumentation and electrical connections to meet explosion proof. On the other hand, the selling price of the products obtained by SFE is also higher (both extract and raffinate).

Table 2.3 illustrates the estimated costs and revenues obtained for a supercritical extraction plant used in three different industrial sectors: edible oil, spices, and algae. The operating cost includes power consumption, CO_2 losses, maintenance, and labor. In all cases, an extraction efficiency greater than 95% was achieved and the return of investment was less than 4 years. Based on the typical production requirements selected, the plant size varies significantly between the three examples. In the first case, the example considered is the recovery of flax seed oil from the cake after mechanical pressing. Flax seed oil is considered a specialty oil. Specialty oils are typically processed by mechanical pressing of seeds. However, this processing technique normally leaves a high percentage of residual oil (RO) in the cake (5 to 15wt,%). In most cases, the cake is used as animal feed. Supercritical fluid extraction provides a solvent-free method for recovering the RO from the cake [9]. The oil extracted by supercritical fluid extraction has a higher content of phytosterol and vitamin E than the oil obtained by mechanical pressing [10]. Additionally, the meal can be further processed to obtain concentrated or isolate proteins. Both the oil and meal processed with supercritical fluid extraction meet the demands of the nutraceutical and organic markets. Similar processes can be applied to recover other specialty oils, such as borage and evening primrose. Those specialty oils have higher sale values, which imply shorter return on investment (ROI).

In the second case, the case study is the extraction of astaxanthin from a microalgae (*Hematococcus pluvialis*). Astaxanthin content ranges from 1% to 3%. The concentration of astaxanthin in the extract can be much higher by SFE than by conventional solvent (acetone or hexane) by manipulating the selectivity of the supercritical solvent. Extracts with astaxanthin content ranging from 5% to 10% can be obtained. Even though the operating costs are high, the ROI is shorter because the sale price of the extract is very high.

In the third case, the spice selected was ginger. The extraction of ginger oleoresin and essential oil is carried out by sequential depressurization. In this case, the combined sales of both fractions provide a ROI in less than 2 years.

2.5 CONCLUSIONS

Supercritical fluid technology is considered by the nutraceutical and pharmaceutical sectors as a viable technology to satisfy customer demands by replacing conventional technologies as well as providing solutions that traditional technologies cannot provide. To successfully implement this technology on the industrial scale, it is necessary to understand the technology, focusing on the proper design of the plant components and optimization of the process parameters that provide minimum operating costs. Some examples have been presented, showing that this technology has been successfully applied to commodity products. There is a future trend to implement this technology as part of a process line combined with traditional processes—for example, SFE + conventional extraction + SFC, SFE + SFC, conventional process + supercritical drying.

REFERENCES

1. Zosel, K., U.S. Patent, 3,806,619, 1974.
2. Zosel, K., U.S. Patent, 4,247,570, 1981.
3. Medina, I. and Martinez, J.L., Dealcoholation of cider by supercritical extraction with carbon dioxide, *J. Chem. Tech. Biotech.*, 68(1), 14, 1997.
4. Martinez, J.L., Ashraf-Khorassani, M. and Chordia, L., Supercritical extraction process of grape seed oil and phenolic compounds, AICHE annual meeting, San Francisco, 2003.
5. Stahl, E., Quirein, K-W. and Gerar, D., *Dense Gases for Extraction and Refining,* Springer-Verlag, Berlin, 1988.
6. Chordia, L., Martinez, J.L. and Desai, B., U.S. Patent application 20050170063.
7. Martinez, J.L., Removal of residual solvents by supercritical fluids, AAPS 2004, Baltimore, 2004.
8. Chordia, L. and Martinez, J.L., U.S. Patent application 20050194313.
9. Chordia, L. and Martinez, J.L., U.S. Patent 7,091,366, 2006.
10. Martinez, J.L., *Recovery of residual specialty oils after mechanical press using supercritical fluid extraction,* 8th International Symposium on Supercritical Fluids, Kyoto, 2006.

3 Supercritical Fluid Extraction of Specialty Oils

Feral Temelli, Marleny D. A. Saldaña, Paul H. L. Moquin, and Mei Sun

CONTENTS

3.1 Introduction 52
3.2 Bioactives in Specialty Oils 52
3.2.1 Carotenoids 56
3.2.2 Polyunsaturated Fatty Acids (PUFAs) 57
3.2.3 Squalene 58
3.2.4 Sterols 58
3.2.5 Tocols 59
3.3 Extraction of Different Types of Specialty Oils 61
3.3.1 Nut Oils 62
3.3.1.1 Factors Affecting Extraction Yield 62
3.3.1.2 Characterization of Products Extracted by SC-CO_2 69
3.3.1.3 Comparison with Conventional Methods 72
3.3.2 Seed Oils 72
3.3.2.1 Factors Affecting Extraction Yield 76
3.3.2.2 Characterization of Products Extracted by SC-CO_2 78
3.3.2.3 Comparison with Conventional Methods 80
3.3.3 Cereal Oils 80
3.3.3.1 Factors Affecting Extraction Yield 80
3.3.3.2 Characterization of Products Extracted by SC-CO_2 83
3.3.3.3 Comparison with Conventional Methods 84
3.3.4 Fruit and Vegetable Oils 84
3.3.4.1 Factors Affecting Extraction Yield 86
3.3.4.2 Characterization of Products Extracted by SC-CO_2 89
3.3.4.3 Comparison with Conventional Methods 90
3.4 Future Trends 90
3.5 Conclusions 91
References 91

3.1 INTRODUCTION

Some plant-based oils are classified as specialty oils due to their high concentrations of bioactive components with demonstrated health benefits. In general, they are comprised of triacylglycerols with a fatty acid composition rich in unsaturates and minor components such as tocols (tocopherols and tocotrienols), carotenoids, sterols, and squalene. Such oils include nut oils (almond, hazelnut, peanut, pecan, pistachio, and walnut), seed oils (borage, flax, evening primrose, grape, pumpkin, and rosehip), cereal oils (amaranth, rice bran, and oat and wheat germ), and fruit and vegetable oils (buriti fruit, carrot, olive, and tomato). Even though the demand for specialty oils is growing at a rapid pace, they are still considered a niche market compared to the large-volume commodity oils. In general, specialty oils are sold in the form of capsules, targeting the dietary supplement market, as well as gourmet oils. Similar to commodity oils, specialty oils are also produced using conventional methods of mechanical pressing and/or solvent extraction. Even though cold pressing at temperatures below 60°C is used extensively in the specialty oil market, cold pressing is limited in terms of oil recovery and the high levels of residual oil left in the meal. On the other hand, conventional solvent extraction depends on the use of organic solvents such as hexane, which needs to be removed via subsequent evaporation. The heat applied for solvent removal may be detrimental to heat-labile bioactive components. In addition, government regulations on the use of organic solvents are getting stricter and the safety of residual organic solvents in the final product is being questioned.

Supercritical fluid extraction technology is growing at a rapid pace because it can overcome many of the disadvantages associated with conventional technologies and meet the consumer demand for "natural" products. The supercritical solvent of choice for food applications has been supercritical carbon dioxide (SC-CO_2). Advantages of processing with SC-CO_2 include low processing temperatures; minimal thermal degradation of the minor components of interest; ease of separation of extraction solvent, resulting in no solvent residue left in the product; and the fact that processing in the CO_2 environment minimizes undesirable oxidation reactions, which is especially beneficial for the sensitive bioactive components of specialty oils such as sterols, tocols, carotenoids, and polyunsaturated fatty acids (PUFAs).

The objectives of this chapter are to review some of the recent findings related to the health benefits of bioactive components present in specialty oils and the use of SC-CO_2 extraction technology for the recovery of specialty oils from different plant sources, such as nuts, seeds, cereals, fruits, and vegetables, with an emphasis on the effects of various sample preparation and extraction parameters on the yield and characteristics of the oils obtained.

3.2 BIOACTIVES IN SPECIALTY OILS

Of the large variety of bioactive compounds present in natural sources, this chapter focuses only on the carotenoids, PUFAs, squalene, sterols, and tocols (tocopherols and tocotrienols) found mainly in specialty oils. The main chemical and physical properties of these bioactive components are summarized in Table 3.1 [1, 2].

TABLE 3.1
Physical Properties of Bioactive Compounds [1, 2]

Bioactive Compound	Formula	Molecular Weight	Melting Point (°C)	Boiling Point (°C)	Solubility[a]	Structure
Carotenoids						
β-Carotene	$C_{40}H_{56}$	536.87	183	—	sl EtOH, chl; s eth., ace, bz	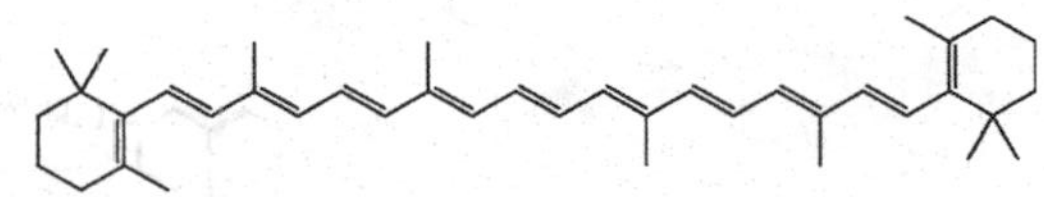
Lycopene	$C_{40}H_{56}$	536.87	175	—	sl EtOH, peth; s eth; vs bz, chl, CS_2	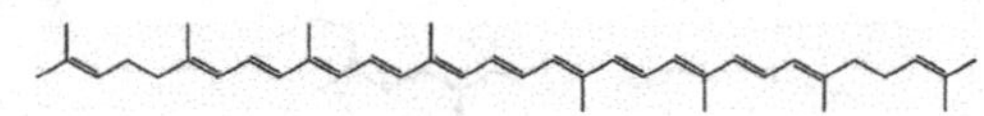
Lutein	$C_{40}H_{56}O_2$	568.87	196	—	vs bz, eth, EtOH, peth	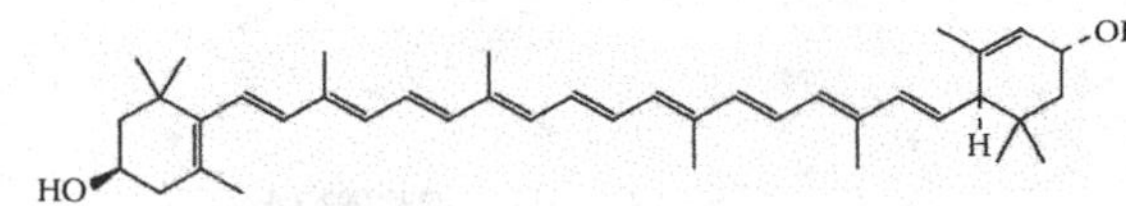
Tocols						
α-Tocopherol	$C_{29}H_{50}O_2$	430.71	3.0	210[b]	s EtOH, eth, ace, chl	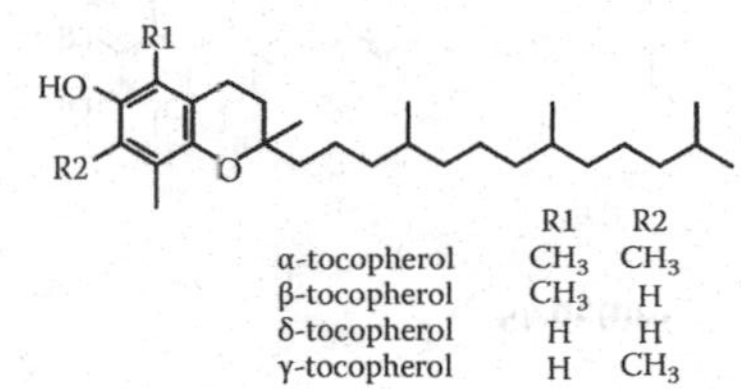
β-Tocopherol	$C_{28}H_{48}O_2$	416.68	—	205[b]		
δ-Tocopherol	$C_{27}H_{46}O_2$	402.65	—	150[c]		
γ-Tocopherol	$C_{28}H_{48}O_2$	416.68	–1.5	205[b]		

continued

TABLE 3.1 (continued)
Physical Properties of Bioactive Compounds [1, 2]

Bioactive Compound	Formula	Molecular Weight	Melting Point (°C)	Boiling Point (°C)	Solubility[a]	Structure
Tocols (continued)						
α-Tocotrienol	$C_{29}H_{44}O_2$	424.67	—	—	s EtOH, eth., chl, ace, oil	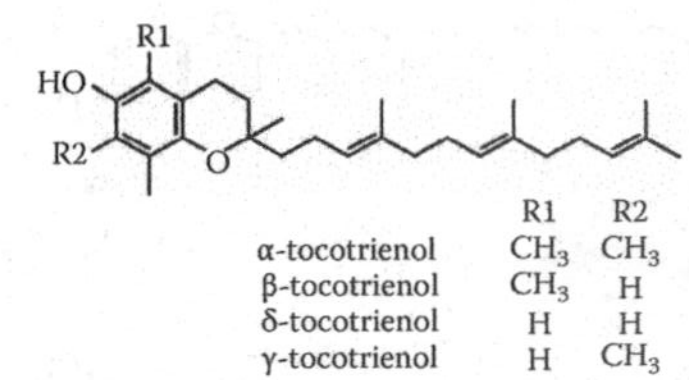
β-Tocotrienol	$C_{28}H_{42}O_2$	410.64				
δ-Tocotrienol	$C_{28}H_{42}O_2$	410.64				
γ-Tocotrienol	$C_{27}H_{40}O_2$	396.01				
Sterols						
Campesterol	$C_{28}H_{48}O$	400.68	157.5	—	—	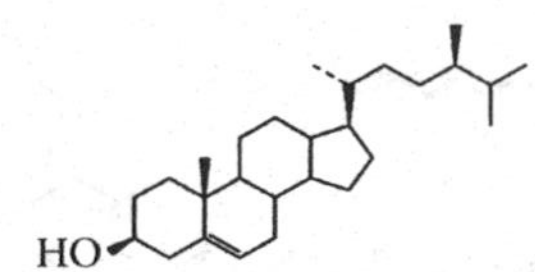
Stigmasterol	$C_{29}H_{48}O$	412.69	170	—	vs bz, eth, EtOH	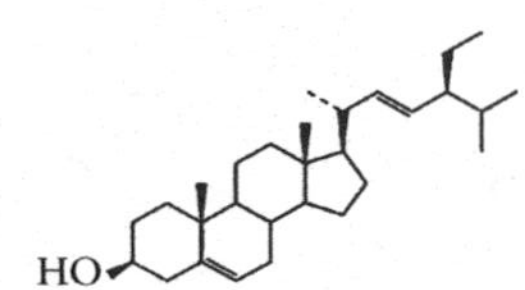

β-Sitosterol	$C_{28}H_{50}O$	414.71	137	—	s EtOH, eth, HOAc	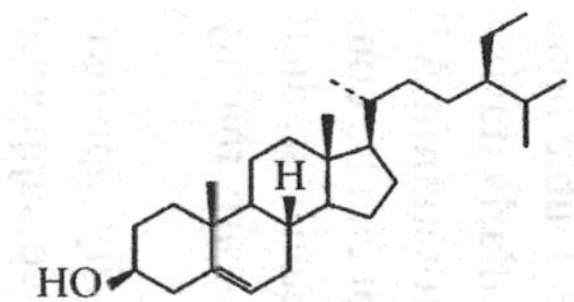
Hydrocarbon						
Squalene	$C_{30}H_{50}$	410.72	–4.8	280[d]	sl EtOH; s eth, ace, ctc	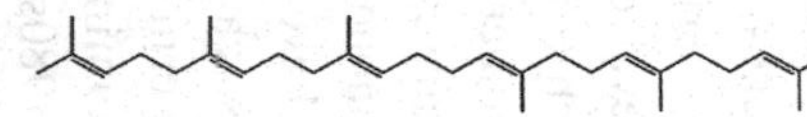
Fatty Acids						
Linoleic acid	$C_{18}H_{32}O_2$	280.45	–7	229[e]	vs ace, bz, eth, EtOH	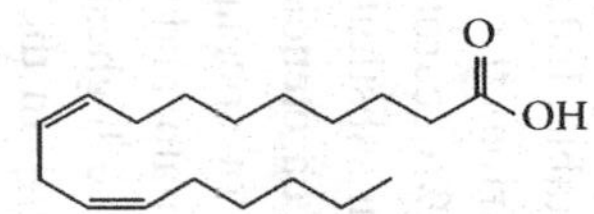
α-Linolenic acid	$C_{18}H_{30}O_2$	278.43	–11	230–232[f]	s EtOH, eth; sl bz	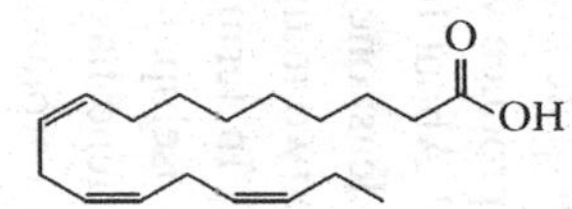
γ-Linolenic acid	$C_{18}H_{30}O_2$	278.43	—	—	—	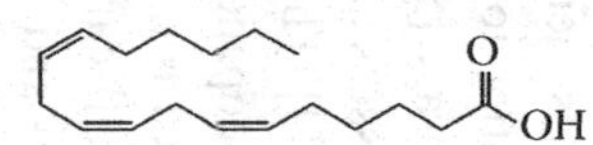

[a] sl: slightly soluble, s: soluble, vs: very soluble; ace: acetone, bz: benzene, chl: chloroform, ctc: carbon tetrachloride, EtOH: ethanol, eth: diethyl ether, HOAc: acetic acid, peth: petroleum ether, [b] Boiling point at 0.0133 kPa, [c] Boiling point at 0.000133 kPa, [d] Boing point at 2.266 kPa, [e] Boiling point at 2.133 kPa, [f] Boiling point at 0.133 kPa.

3.2.1 Carotenoids

Carotenoids represent a group of over 600 fat-soluble pigments [3]. These pigments are responsible for the bright yellow, orange, and red colors of fruits, roots, flowers, fish, invertebrates, birds, algae, bacteria, molds, and yeast. Some carotenoids are also present in green vegetables, where their color is masked by chlorophyll [4]. Carotenoids are typically divided into two classes: carotenes, which are C40 polyunsaturated hydrocarbons, and xanthophylls, oxygenated derivatives of carotenes. Carotenoid compounds are colored due to their high level of conjugated double bonds, which also makes them quite unstable. Indeed, each conjugated double bond can undergo isomerization to produce various *trans/cis* isomers, particularly during food processing and storage [5]. About 10% of carotenoids are called "provitamin A," indicating that they possess at least one unsubstituted β-ionone ring that can be converted into vitamin A [4]. The two main carotenoids that have been heavily studied are β-carotene and lycopene. In terms of specialty oils, carotenoids are mainly present in buriti fruit, carrot, rosehip, tomato, and wheat germ oils.

Of all carotenoids, β-carotene has the highest provitamin A activity, approximately twice that of α- and γ-carotene [4]. In the early 1980s, evidence supported β-carotene as a chemopreventive agent [6]. Thus, β-carotene was the subject of a number of studies, such as the Alpha-Tocopherol Beta-Carotene Cancer Prevention (ATBC), β-Carotene and Retinol Efficacy Trial (CARET), and the Physician's Health Study. The ATBC trial concluded that β-carotene supplementation did not help smokers who previously had a heart attack; in fact, their risk of fatal coronary heart disease actually increased [7, 8]. The CARET study [9] showed that supplementation with β-carotene and vitamin A had no benefit for current and recent ex-smokers and male asbestos-exposed workers and that it may increase the incidence and risk of death due to lung cancer, cardiovascular disease, and any other cause. Finally, the Physician's Health Study concluded that β-carotene intake rendered neither benefit nor harm in terms of cancer, cardiovascular disease, stroke, or overall mortality [10].

Lycopene, although lacking provitamin A activity, is known to be one of the most potent antioxidants among the digestible carotenoids. Its highly conjugated molecular structure is responsible for the bright red color of ripe tomatoes as well as the pigmentation of watermelons, pink grapefruits, apricots, and papayas [3, 11]. A number of studies have shown that lycopene could play a protective role in the development of atherosclerosis [12, 13]. As well, *in vivo* and *in vitro* studies have shown that it has a hypocholesterolemic effect, thereby suggesting that lycopene could attenuate atherogenesis and reduce the risk of cardiovascular disease [14]. Some studies have found that lycopene intake could lower the risk of prostate cancer, while others reported no protective effect [3]. However, a case study showed that high consumption of tomatoes and tomato-based food products reduced stomach cancers [15]. According to Omoni and Aluko [14], the complex interaction among the potentially beneficial compounds found in tomatoes might contribute to their anticancer properties. According to Rao and Shen [16], the recommended daily intake of lycopene is 5 to 10 mg/day. It is interesting to note that approximately 80% of dietary lycopene comes from tomatoes and that processed tomatoes have a higher level of lycopene than raw tomatoes because heat treatment and homogenization of tomatoes enhance the availability of lycopene [17–19]. Cohn et al. [20] compared the consumption of synthetic lycopene and lycopene in processed tomatoes and found that the availability was the same. Concerning health claims, in November 2005, the U.S. Food and Drug Administration (FDA) allowed the following qualified health claim: "Very limited and preliminary scientific research suggests that eating one-half to one cup of tomatoes and/or tomato sauce a week may reduce the risk of prostate cancer. FDA concludes that there is little scientific evidence supporting this claim" [21].

3.2.2 Polyunsaturated Fatty Acids (PUFAs)

PUFAs are fatty acids that contain two or more double bonds in the carbon chain. Most PUFAs are essential fatty acids and have to be provided to the body through the diet. They are usually classified as ω-3 and ω-6, depending on the position of the first double bond from the methyl end of the carbon chain. α-linolenic (ALA), eicosapentaenoic (EPA), docosapentaenoic acid (DPA), and docosahexaenoic (DHA) acids are examples of ω-3 PUFAs, whereas linoleic acid (LA) and γ-linolenic acid (GLA) are examples of ω-6 PUFAs. The main source of EPA, DPA, and DHA are fish oils (see Chapter 5). Because the focus of this chapter is plant-derived specialty oils, only PUFAs such as LA, ALA, and GLA will be discussed. With regard to specialty oils, PUFAs are found mainly in almond, apricot, hazelnut, peanut, walnut, borage, evening primrose, pumpkin, and rice bran oils.

LA, ALA, and GLA are essential fatty acids that human enzymes can transform into the PUFAs required by the body [22]. For instance, LA is converted by the human body into arachidonic acid. The latter is transformed into eicosanoids and prostaglandins, which are important mediators in cardiovascular disease [22]. A lack of LA can lead to fatty liver, skin lesions, and reproductive failure [23]. ALA, on the other hand, is converted to EPA and DHA [23]. DHA is a major component of the phospholipid membranes of the brain and retinas; therefore, a lack of DHA causes abnormal function [24]. When the body experiences a lack of ω-3 in the diet along with an increase in ω-6, the lack of ω-3 tends to be accentuated, which may lead to inhibition of the synthesis of DHA from ALA [25]. Thus, it is important to keep the ratio of ω-6 to ω-3 balanced in the diet. Some studies report a reduction in cardiovascular disease risk associated with higher ALA intake [26–28]. However, some investigators are still trying to prove otherwise. Such debate is clearly illustrated in the multiple letters published in the *American Journal of Clinical Nutrition* [29, 30].

It has been well established that intake of LA and GLA increases the tissue biosynthesis of 1-series prostaglandins, which in turn suppresses inflammation [31]. Clinical studies have also shown that administration of GLA can reduce pain and swelling in rheumatoid arthritis [32, 33]. GLA is also said to be a "conditionally essential fatty acid for the skin" [34]. Furthermore, a diet rich in EPA and GLA was deemed beneficial for patients with acute lung injury [35]. Based on the Lyon Diet Heart Study, dietary intake of ALA should be about 1.8 to 2 g/day [22]. The best sources of ALA are canola oil and algae. Nuts are a good source of not only ALA but also LA. Due to the close association between ALA and LA, care should be taken not to consume large amounts of LA-rich oils, such as soybean, sunflower, and walnut oils [22]. The dose of GLA fed during a study on rheumatoid arthritis was 1.4 g/day, which was reported to be well tolerated by the patients [32]. The average recommended intake of PUFAs is 7% of total energy intake [36]. However, consuming excessive levels of PUFAs without proper intake of antioxidants is not recommended since PUFAs are prone to oxidation, which may play a role in carcinogenesis [37] and other diseases.

3.2.3 Squalene

Squalene is a lipid that was originally obtained from shark liver oil. It is also found in olive, palm, and wheat germ oils [38]. A number of animal studies showed that dietary squalene has distinct anticarcinogenic effects. It was shown that squalene presents inhibitory action in carcinogenesis models of skin [39, 40], colon [41], and lung [42] cancer. However, it does not present chemopreventive activity [43]. One reported side effect of squalene is that of a 51-year-old man with esophageal cancer who developed severe exogenous lipoid pneumonia after eating large doses of squalene [44].

Besides its anticarcinogenic activity, squalene prevents lipid peroxidation in human skin surface [45] and is useful in treating conditions resulting from inadequate immune response [46]. It is also useful as a cytoprotectant (medication that combats ulcers by increasing mucosal protection) in cyclophosphamide-induced toxicities [47] and low-dose squalene (860 mg) coadministration with low-dose pravastatin (10 mg) further enhances the efficacy of pravastatin as a cholesterol-lowering drug [48].

In the human body, squalene is the precursor to important sterols such as cholesterol [49, 50]. Thus, it was originally thought that increased squalene consumption would actually increase blood cholesterol levels. This does not seem to be the case when 0.5 g of squalene is consumed per day. Indeed, Miettinen and Vanhanen [51] observed an increase in total blood cholesterol concentrations in male subjects after a dietary supplementation of 1 g/day of squalene for 9 weeks, but when the dose was reduced to 0.5 g/day, the blood sterol levels went back to normal. It is true that 60% to 80% of dietary squalene is absorbed through the intestine [52] and that a substantial amount of this squalene is converted to cholesterol in the human body. However, a reasonable increase in squalene consumption appears to significantly increase the fecal excretion of cholesterol [53].

3.2.4 Sterols

The main sterols in plant materials are sitosterol, campesterol, and stigmasterol [54]. They are mainly found in the specialty oils of acorn, hazelnut, walnut, cherry, grape, pumpkin, and rice bran. Over the years, it has been well established that a high dietary intake of phytosterols lowers blood cholesterol levels by competing with dietary and biliary cholesterol during intestinal absorption [55–57]. However, recently, Plat and Mensink [58] speculated that because phytosterols are more readily oxidized by free radicals than cholesterol, they could increase the level of oxidized low-density lipoproteins (LDLs), which form atherosclerotic plaques in arteries. At this time, little information is available to prove such a claim. On the other hand, phytosterols are not recommended for individuals who are suffering from sitosterolemia, an inheritable disorder that increases the absorption of cholesterol and phytosterols [55].

Fortunately, this disorder is quite rare; Björkhem et al. [59] knew of 45 patients with this disease in 1998.

Phytosterols have also been the subject of much investigation for properties other than their ability to lower cholesterol. In the 1980s, consumption of sitosterol was shown to reduce colon cancer in rats [60]; however, there is still no strong and consistent evidence that the same would be true in humans [61]. Some studies report that sterols may have an effect on immune function and that they could prevent the subtle immunosuppression experienced by marathon runners [62, 63]. Finally, animal studies reported that phytosterols could inhibit the inflammatory response [64] and that they could cause insulin-releasing properties for diabetics [65]. According to the Scientific Committee on Food, the average amount of phytosterols in the Western diet is 150 to 400 mg [66]. However, the recommended dose of phytosterols to reduce plasma LDL-cholesterol levels by 5% to 15% is 1.3 to 2 g/day [67, 68]. In order to achieve such levels and comply with the FDA-approved health claim on the role of plant sterol or plant stanol esters in reducing the risk of coronary heart disease [69], food manufacturers have introduced various functional food products with added phytosterols. In those countries where such products are currently marketed, the success is so great that authorities are now worried about consumers eating too much phytosterols. The concern is legitimate because a study showed that daily consumption of 3.8 to 4.0 g of plant sterol esters can significantly lower serum concentrations of various carotenoids and tocopherols [70].

3.2.5 Tocols

Tocopherols and tocotrienols make up the tocols family of vitamin E compounds, which must be obtained from the diet because humans cannot synthesize them [71]. Tocols are found in almond, hazelnut, pecan, walnut, flax, buriti fruit, tomato, rice bran, and wheat germ oils. The difference between tocopherols and tocotrienols lies in the phytyl chain attached to a chromanol ring: the phytyl chain is saturated in tocopherols, whereas the phytyl chain in tocotrienols has three double bonds [72]. These compounds represent a group of four isomers with varying numbers and position of methyl groups on the chromanol ring: α-, β-, γ-, and δ-tocopherol and α-, β-, γ-, and δ-tocotrienol.

Although all of these tocol isomers are absorbed through the intestine in the human body, it is believed that only α-tocopherol contributes toward meeting the human vitamin E requirement [73]. The reasoning behind this claim is that intracellular vitamin E content and distribution are regulated by different proteins binding specifically to α-tocopherol [74]. Another factor contributing to this school of thought is that α-tocopherol is the most potent naturally occurring scavenger of reactive oxygen and nitrogen species [72]. However, evidence is building on the importance of consuming a mixture of the whole family of vitamin E compounds since they may have additive and synergistic activities that support broader beneficial biological functions [75]. They may also act synergistically with other naturally occurring compounds commonly found in fruits and vegetables [72]. An overview of the current scientific literature reveals the importance of tocopherols and tocotrienols as the major fat-soluble antioxidants [73]. Indeed, these molecules can scavenge free radicals in the body, thereby preventing them from damaging cell membranes and genetic material and changing the character of fats and proteins [76]. One example is the protection that vitamin E grants to PUFAs, which are especially vulnerable to destructive oxidation [73]. Even though vitamin E is viewed as a potent antioxidant, in vitro studies have shown that vitamin E may have pro-oxidant effects at high dosages [77, 78]. Interestingly, it appears that α-tocopherol alone can have a pro-oxidant effect [79]; however, in the presence of γ- and δ-tocopherol, the pro-oxidant effect of α-tocopherol seems to diminish [75].

The antioxidant properties of mixed isomers of tocols could also be beneficial in preventing the onset of atherosclerosis [80, 81]. This medical condition is in part due to free radicals oxidizing LDLs, which in turn form atherosclerotic plaques on the surfaces of artery walls. By scavenging free radicals, tocopherols block this process [80]. Tocotrienols are also believed to be useful in the prevention and treatment of atherosclerosis in people with type 2 diabetes [82]. Besides its antioxidant properties, α-tocopherol acts as a regulator of gene expression that lowers the build-up of oxidized LDLs in arteries [83, 84]. Unlike the literature on atherosclerosis, the scientific literature on the effects of vitamin E on cardiovascular disease is divided. Some plainly state that there is no concrete evidence that vitamin E reduces cardiovascular-related mortality, particularly in high-risk individuals [85], whereas others show a reduction in mortality [86, 87] or no effect on cardiovascular events and death [75, 88–92].

The application of vitamin E in cancer treatment has also been studied. Although animal studies were successful in showing that vitamin E inhibited carcinogenesis and ultraviolet-induced deoxyribonucleic acid damage [93–96] and preclinical data revealed that it might stimulate an antitumor immune response [97], clinical trials have given mixed results [72]. Some studies on patients with stage I and II head and neck cancer who were fed supplements of α-tocopherol showed a higher incidence of second primary cancers and a lower degree of cancer-free survival [98]. Also, a few studies showed that α-tocopherol was of little to no benefit in preventing lung cancer [99]. However, α-tocopheryl succinate was found to be tumor specific in prostate and breast tissues [100]. There is also evidence that γ-tocopherol may help prevent and treat colon [101] and prostate [102] cancer.

A number of studies report that vitamin E can benefit individuals with osteoarthritis [103–106]. Supplementation of 400 mg/day of vitamin E had a beneficial analgesic and anti-inflammatory effect, with a low incidence of side effects [107]. As well, supplementation with mixed isomers of vitamin E was advantageous [108]. Some research has shown that α-tocopherol supplementation delays or prevents Alzheimer's disease (AD) diagnosis of elderly individuals with signs of mild cognitive impairment [109]; however, others report no clear benefits of vitamin E in the treatment of people with AD [110–112]. Unrelated to AD, a study on nursing home residents showed that supplemental vitamin E reduced the incidence and duration of respiratory infections [113].

The FDA has approved the following health claim for dietary supplements containing vitamin E and/or vitamin C [114]: "Some scientific evidence suggests that consumption of antioxidant vitamins may reduce the risk of certain forms of cancer. However, FDA has determined that this evidence is limited and not conclusive." Some work has also been performed to achieve qualified health claims for vitamin E supplements against cardiovascular disease. However, this health claim was refused based on insufficient evidence [115, 116]. The optimum consumption level of vitamin E for the general population is an interesting consideration that is still open to debate. According to Health and Welfare Canada, the Recommended Daily Intake (RDI) for vitamin E should be 5 to 10 mg; the American Dietetic Association sets the Recommended Dietary Allowance (RDA) at 15 mg/day of α-tocopherol for adults and 19 mg/day for women who are breastfeeding [73]. Because not all sources of vitamin E have the same biological activity, one has to keep in mind that 1 international unit (IU) of natural vitamin E corresponds to 0.67 mg of α-tocopherol, whereas 1 IU of synthetic vitamin E corresponds to 0.45 mg of α-tocopherol [73]. In addition, the growing evidence showing the benefits of mixed isomers of tocopherols and tocotrienols should probably be reflected in the RDI and RDA. Although vitamin E was considered to be nontoxic for many years, it is now known that an overdose of vitamin E can interfere with blood clotting, especially when taken along with anticoagulant medication or with acetylsalicylic acid [117]. Furthermore, a recent meta-analysis including 19 studies with more than 135,000 patients showed that more than 400 IU (270 mg) per day of vitamin E supplementation to patients (ages 47 to 84 years) who mostly had chronic diseases might have increased all-cause mortality [118].

3.3 EXTRACTION OF DIFFERENT TYPES OF SPECIALTY OILS

The specialty oils rich in bioactives can be extracted from many plant sources. These extracted oils are mainly mixtures of triglycerides, free fatty acids (including PUFAs), monoglycerides, diglycerides, and other minor components, such as tocols, carotenoids, sterols, and squalene. For specialty oils to be extracted from different plant materials using SC-CO_2, they have to be soluble in SC-CO_2. The solubility behavior of major and minor lipid components in SC-CO_2 has been previously reviewed and correlated [119, 120]. Solubility is a strong function of SC-CO_2 density and the properties of the solute, such as molecular weight, polarity, and vapor pressure. All the lipid components of interest in specialty oils are soluble in SC-CO_2 to different extents, depending on temperature and pressure conditions. Generally, solubility of lipids in SC-CO_2 decreases with an increase in polarity and molecular weight, thus following the order: fatty acid esters, fatty acids, and triglycerides [120].

SC-CO_2 extraction of specialty oils from various sources has been studied quite extensively. The extraction efficiency and the characteristics of the products are affected by several parameters, such as particle size and moisture content of the feed material, extraction temperature and pressure, solvent flow rate, extraction time, and the use of a cosolvent. Therefore, the following discussion emphasizes the impact of these processing parameters on the yield and composition of specialty oils obtained from nuts, seeds, cereals, fruits, and vegetables. Even though some nuts are seeds within the fruit of the plant, they are classified in this chapter as nuts based on consumption; for example, almond is classified as a nut because it is consumed as a snack upon roasting, whereas apricot kernel is classified as a seed. In addition, some fruits and vegetables like tomatoes and grapes are also included in the discussion of seeds because their seeds were used in extraction studies.

3.3.1 Nut Oils

Nuts provide protein, fiber, essential fatty acids, and vitamins. Their pleasant flavor and aroma and unique texture lead to their popularity as snacks. However, their high fat content limits their consumption due to consumer concerns about their high caloric content. Besides, nuts are prone to oxidation due to the presence of high levels of PUFAs. Even though they have low oxidation stability, nut oils have been commercialized in some countries as highly nutritious specialty oils. Defatted nuts also have market value as low-fat products; therefore, the proper treatment of nuts before and after extraction needs to be considered. Numerous studies on SC-CO_2 extraction of nuts, including acorn [121], almond [122–125], hazelnut [126–128], peanut [129–133], pecan [134–136], pistachio [137], and walnut [138–140], have been reported and are summarized in Table 3.2.

3.3.1.1 Factors Affecting Extraction Yield

3.3.1.1.1 Sample Preparation

Most of the nuts were extracted fresh, but some were roasted [122, 141]. Femenia et al. [122] reported that the roasted nuts had higher oil content due to water loss and partial degradation of protein and probably pectin as well during roasting. However, it was difficult to extract the oil from roasted almonds, possibly due to the formation of new links among cell wall polymers, thus reducing porosity and strengthening the wall structure [122].

a) **Particle size:** The nuts are generally ground to small particles to increase surface area and shorten the path lengths over which the solutes must travel to reach the bulk fluid phase and therefore facilitate the extraction of the nut oil. Thus, particle size impacts extraction kinetics of the oil, which is present as released oil on the surface of the particles as well as unreleased oil inside the particles. The extraction rate is dictated initially by the solubility of the free oil in SC-CO_2 (fast extraction period) and later by the diffusion of oil inside the particles (slow extraction period). In general, when fresh solvent comes in contact with the feed material, the free oil on the surface is quickly solubilized and extracted. Extraction rate is fast and limited by equilibrium solubility, as represented by the initial linear portion of the extraction curve. When the oil on the surface of the particles is depleted, SC-CO_2 has to diffuse into the particles and solubilize the oil and SC-CO_2+oil has to diffuse out, which is a slow process driven by the oil concentration gradient; thus the extraction curve approaches a constant value asymptotically. Özkal et al. [126] reported a yield of 0.51 g oil/g hazelnut at the end of the fast extraction period and only 0.01 g oil/g hazelnut during the slow extraction period. Therefore, the extraction could be stopped after the fast extraction period. The oil yield obtained at the end of the fast extraction period

TABLE 3.2
Extraction of Bioactive Compounds from Nuts Using SC-CO_2

Raw Material	Feed (g)	Sample Preparation: Particle Size (mm)	Sample Preparation: H_2O (%)	Bioactive Compound	Extraction Conditions: T (°C)	Extraction Conditions: P (MPa)	Extraction Conditions: Flow rate	Extraction Conditions: Time (min)	Extraction Conditions: Cosolvent	Recovery[a] (%)	Ref.
Acorn	n.i.	0.27	n.i.	Oleic acid, linoleic acid, β-sitosterol, stigmasterol, campesterol, tocopherols	40	18	1.5×10^{-2} m/min[b]	n.i.	None	n.i.	121
Almond	1500–2000	n.i.	n.i.	Unsaturated oil	50	33	333.3–666.7 g/min	n.i.	None	n.i.	122
	4000	n.i.	n.i.	Unsaturated oil	60	48.2	n.i.	n.i.	None	n.i.	123
	3000–4000	Milled, broken, whole	n.i.	Tocopherols, oleic acid, linoleic acid	35, 40, 50	35, 45, 55	166.7, 333.3, 500 g/min	n.i.	None	n.i.	124
	n.i.	0.3, 0.7, 1.9	n.i.	Unsaturated oil	40	35	12, 23.8 g/min	n.i.	None	n.i.	125
Hazelnut	4	< 0.85	3	Oleic acid, linoleic acid	40, 50, 60	30, 37.5, 45	1, 3, 5 g/min	10	None	34	126
	5	1–2	3	Oleic acid, linoleic acid	40, 50, 60	30, 45, 60	2×10^{-3} L/min	250–300	None	59	127
	50	0.7	n.i.	Unsaturated oil, β-sitosterol, α-tocopherol	35–48	18–23.4	$2.7–4.3 \times 10^{-2}$ m/min	240	None	>95	128

continued

TABLE 3.2 (continued)
Extraction of Bioactive Compounds from Nuts Using SC-CO_2

Raw Material	Feed (g)	Sample Preparation: Particle Size (mm)	Sample Preparation: H_2O (%)	Bioactive Compound	Extraction Conditions: T (°C)	P (MPa)	Flow rate	Time (min)	Cosolvent	Recovery[a] (%)	Ref.
Peanut	1000 (h.e.), 500 (v.e.)	h.e.: 0.86–1.68 v.e.: 0.86–1.19, 3.35–4.75	n.i.	Unsaturated oil	h.e.: 25–100; v.e.: 25–120	h.e.: 27.5–69; v.e.: 14–55	h.e.: 20 L/min; v.e.: 40 L/min	v.e.: 180	None	h.e.: 50; v.e.: 99	129
	5,579	0.864–1.18, 1.18–1.7, 1.7–2.36, 2.36–3.35, 3.35–4.75	5, 9,15	Unsaturated oil	25, 55, 75, 95	27.5, 41.5, 55	40 and 60 L/min	180	None	99	130
	50,860	Ground: 0.864–1.18, 1.18–1.7, 1.7–2.36, 2.36–3.35, 3.35–4.75 Flakes: 1.27	n.i.	Unsaturated oil	25	55	40 L/min	180	None	n.i.	131
	1	n.i.	n.i.	Unsaturated oil	40–80	13.8–55.1	n.i.	n.i.	None	n.i.	132
	n.i.	n.i.	4.2–5.1	Unsaturated oil	50–65	35–50	n.i.	n.i.	None	n.i.	133
Pecan	5–6	Halves	4	Unsaturated oil, tocopherols	40, 80	41.3, 55.1, 68.9	n.i.	160	None	77	134
	20	Halves	4.9, 6.4, 7.4, 11	Unsaturated oil	75	62	3 L/min	60	None	n.i.	135
	90	Halves, pieces	Halves: 4.8; Pieces: 4.1	Unsaturated oil	45, 62, 75	41.3, 55.1, 62, 66.8	1, 1.5, 2, 2.5, 3, 4, 7.5 L/min	60	None	n.i.	136
Pistachio	10	1–1.68	n.i.	Unsaturated oil	50, 60, 70	20.7, 27.6, 34.5	2.6 g/min	n.i.	10% EtOH+	66.1[c]	137

Walnut	n.i.	0.01, 0.05, 0.1, 0.5	n.i.	Linoleic acid, oleic acid, linolenic acid, β-sitosterol, campesterol, α-tocopherol	35, 40, 45, 48	18, 20, 22, 23.4	n.i.	n.i.	None	95	138
	n.i.	Pieces	3	Linolenic acid, unsaturated oil	80	68.9	150 g/min	n.i.	None	n.i.	139
	n.i.	Pieces	n.i.	Unsaturated oil	80	68.9	150 g/min	n.i.	None	n.i.	140

T: temperature, P: pressure; n.i.: not indicated; h.e.: horizontal extraction, v.e.: vertical extraction,

[a] Recovery (g extract/g oil in feed material × 100), [b]superficial velocity, [c]yield (g/100 g feed material).

+ Cosolvent added continuously into SC-CO_2 at the level (%, w/w) indicated.

was dependent on particle size [126]. For example, the oil yield for hazelnut with a particle size of 1.5 mm was about half (0.22 g oil/g nut) of that from particles less than 0.85 mm (0.51 g/g). In another study, using whole, broken (4 to 8 mm), and milled (0.5 to 3 mm) almond, Leo et al. [124] observed that the extraction yield increased with reducing particle size (50, 350, and 800 g/kg, respectively) under similar operating conditions. In addition, oil recovery increased from 36% to 82% when the particle size of peanut was reduced from 3.35 and 4.75 mm to 0.86 and 1.19 mm [130]. There is a practical limit to grinding to smaller particles due to the oily nut particles sticking on the sieves [130].

b) **Moisture and equilibration time:** Moisture has a great impact on oil extraction because the kernel expands with moisture absorption, resulting in a more permeable cell membrane, allowing both oil and CO_2 to pass more readily. However, excess water can also impede the diffusion of oil and have a negative effect on oil accessibility. The amount of pecan oil extracted after 48 hours of moisture equilibration was approximately 30% higher than that obtained after 1 hour [135]. Water was coextracted with the oil and increased linearly with the initial moisture content of pecans [135]. The moisture content of the extracted oil was 0.7% and 11.7% at initial moisture levels of 3.5% and 12%, respectively, in the pecan. The oil obtained was cloudy with a yellowish color. A high level of water in the extracted oil is not desirable since it negatively impacts its stability. Moisture affects not only extraction efficiency and yield but also the physical structure of the nuts [135]. Breakage of the kernels during the extraction can be avoided by adjusting the moisture content to a certain level, for example, 8% to 11% (w/w) for peanuts [133]. Passey et al. [133] tested soaking, steaming, and humidification as pretreatment methods for peanuts and found that soaking and steaming were as effective as humidification in preventing the breakage of the kernels, but they caused browning, loss of water solubles, and low rate of extraction. Moisture content also affects pecan breakage [135]. With moisture absorption, the kernel becomes soft and pliable, in part because the moisture affects the plasticization of proteins and carbohydrates and alters the physical properties of the tissues. Extraction of pecan after 48 hours equilibration produced less breakage than extraction at 1 hour at moisture contents of 6.1% and 7.7% [135]. This was due to the water in the kernel being more evenly distributed after a longer equilibration time. However, the effect of the equilibration time decreased at 8.5% moisture and was negligible at 11.6% [135]. This can be explained by the osmotic pressure caused by the water around the kernel.

3.3.1.1.2 Extraction Parameters

a) **Temperature and pressure:** Most of the nut oil extractions were performed at a temperature range of 35°C to 100°C and a pressure range of 9 to 70 MPa. The solubility of oil in SC-CO_2 is mainly determined by the SC-CO_2 density and the volatility of the oil components. In general,

$SC\text{-}CO_2$ density increases with pressure at constant temperature and decreases with temperature at constant pressure, where the density decrease becomes smaller at higher pressures. On the other hand, the volatility of oil components increases with temperature. These two opposing effects of temperature on density and volatility lead to the well-established crossover behavior of solubility isotherms. A temperature increase may also cause breakdown of cell structure and increase the diffusion rate of the oil in the particles, therefore accelerating the extraction process [130]. It is also important to consider that oil composition varies widely among the different types of nuts and thus differences in functional groups and fatty acid composition are responsible for the differences in volatility, solubility in $SC\text{-}CO_2$, and crossover pressure. Solubility increases with temperature and pressure above the crossover pressure. These general trends of the effects of temperature and pressure on oil solubility and yield are reflected in some studies. For example, the solubility of peanut oil in $SC\text{-}CO_2$ decreased with temperature at pressures below 35 MPa and increased at higher pressures [130]. Increasing the temperature from 29°C to 91°C increased the initial extraction rate from 15 to 129 mg peanut oil/L CO_2 at 55 MPa; however, at 27.5 MPa, increasing the temperature from 27°C to 100°C decreased the initial extraction rate from 7.6 to 0.5 mg oil/L CO_2 [130]. Increasing the pressure from 41.3 to 55.1 MPa increased the pecan oil yield from 14.3% to 21.3% at 45°C and from 17.5% to 31.5% at 75°C. However, a further increase in pressure to 66.8 MPa only slightly increased the oil yield to 21.5% and 32.4% at 45°C and 75°C, respectively [136]. The positive effect of pressure on oil yield was also observed at high temperatures. Similarly, raising the pressure from 17.7 to 68.9 MPa yielded 100% and 200% more oil from pecans at 40°C and 80°C, respectively [134].

The rate of depressurization following an extraction affects the breakage of the nut kernel, as demonstrated for the pecan kernel [135]. When the vessel was depressurized from 62 MPa in 20 min there was no breakage, whereas a significant amount of breakage occurred during the 10-min test. During the 20 min depressurization, the pressure in the extraction vessel was dropped from 62 MPa to about 7 MPa within the first minute and was around 2 MPa after 10 min and 0 MPa after 20 min. This suggested that the final stages of depressurization were crucial in causing pecan breakage. When the extraction vessel was opened immediately after depressurizing the reactor, the particles jumped around and popping sounds were heard, suggesting that most of the break-up occurred as the CO_2-saturated particles were depressurized [130]. Gradual pressurization and depressurization were also necessary to minimize damage to the walnut pieces [139]. The rupture or breakage of the cells occurs due to the phase change of CO_2. A rapid depressurization to atmospheric pressure forms liquid CO_2 as well as dry ice. Therefore, the CO_2 trapped inside the solid matrix expands, causing breakage in the cells. Slow depressurization with appropriate level of heating to overcome the Joule-Thomson effect and to maintain the temperature above 31°C assures that the CO_2 will be in gas phase and avoid any breakage.

b) **Flow rate and flow direction:** Increasing solvent flow rate results in an increased solvent-to-feed ratio and decreased mass transfer resistance. The extracted pecan oil yield increased from 8.8% to 21.5% with an increase in flow rate from 1 to 4 L/min (measured at standard temperature and pressure, STP); however, with a further increase to 7.5 L/min, the oil yield changed only from 21.5% to 21.7% [136]. The effect of flow rate was also studied for the extraction of hazelnut oil [127]. At a low pressure (15 MPa), an increase in flow rate from 0.5 to 2 mL/min (measured at extractor pressure and 10°C) did not cause a significant difference in the extraction yield of hazelnut oil; however, at a high pressure (30 MPa), the oil yield increased more than three-fold [127]. This might be due to the low solubility of oil in SC-CO_2 at low pressures. Another important factor that affects extraction yield is the direction of the CO_2 flow. The vertical extractor produced a higher total oil recovery and a more uniformly extracted peanut meal sample than the horizontal extractor [130]. This might be caused by the meal settling in the horizontal extractor and leaving a lower resistance flow path along the top of the meal, leading to channeling and insufficient contact between the fresh solvent and peanut meal. Similarly, the flow direction greatly affected the solubility of the peanut oil in SC-CO_2 at both 55 MPa/75°C and 55 MPa/95°C, with the downward flow resulting in a higher solubility [130]. This is probably because of the large temperature gradient between the top and bottom of the extractor in the upward system, which might be due to the incoming fresh CO_2 cooling the inside of the extractor. With downward flow, it was possible to maintain the temperature throughout the experiment, which was attributed to the balancing of density-induced convection effects with downward flow [130]. Thus, it was possible to maintain the constant extraction rate for a longer period.

c) **Extraction time:** Due to the physical structure of the nut, the penetration of the solvent and the diffusion of the unreleased oil in the particles are very slow. Therefore, extraction time is usually limited to the fast extraction period since the amount of oil recovered in the slow extraction period is negligible. The duration of the fast extraction period is also inversely related to particle size. However, the extraction rate and the duration of the fast extraction period are also affected by temperature, pressure, flow rate, and cosolvent addition. For example, the durations of the fast extraction period for the hazelnut oil extractions conducted at 50°C and 45 MPa were 50 and 60 min for particle sizes of less than 0.85 mm and 1.5 mm, respectively [126]. When the extraction conditions were changed to 40°C, 37.5 MPa, and 5 g/min flow rate, the fast extraction period was 50 min for the nut particles of less than 0.85 mm in size [126]. Fast extraction period decreased from 183 to 64 and 32 min with a pressure increase from 30 to 45 and 60 MPa at 40°C. On the other hand, it decreased from 64 to 33 min with a temperature increase from 40°C to 50°C at 45 MPa [127].

d) **Use of cosolvent:** A higher yield can be obtained by adding a small amount of cosolvent, such as ethanol. For example, when 10% ethanol was added to CO_2 (w/w), the extraction yield of pistachio oil at 60°C increased by 230% at 34.5 MPa and by 750% at 20.7 MPa compared to that obtained with CO_2 alone [137]. In addition, the extraction yield of pistachio oil using 10% ethanol as a cosolvent at 60°C was higher at 34.5 MPa than that at 20.7 MPa.

3.3.1.2 Characterization of Products Extracted by SC-CO2

3.3.1.2.1 Chemical Composition

The fatty acid composition of the nut oils is presented in Table 3.3. The main fatty acids in the extracted nut oils are linolenic (Ln), linoleic (L), oleic (O), and palmitic (P) acids forming triglycerides like LLL, OLL, LLLn, OOO, and POO, the amounts of which are dependent on the type of nut. The fatty acid composition of SC-CO_2-extracted oils exhibited minor differences in comparison to oils obtained from the feed material. However, a small increase in the percentage of oleic and stearic acids was detected when about 65% of the almond oil was extracted [122], indicating that SC-CO_2 extraction may result in minor modifications of the fatty acid profile of the extracted oils. On the other hand, no fractionation was detected during extractions of hazelnut oil, as the fatty acid composition of the three hazelnut oil factions obtained during the first 30 min, between 70 and 120 min, and after 120 min was similar to that of the oil extracted with hexane [127]. There was also no significant difference in the fatty acid composition of the pecan oils obtained at extraction times between 15 and 480 min [136]. The tocopherol content of the fat-reduced (25% and 40%) walnuts was significantly lower than that in the full-fat nuts [139]. The nuts after extraction had increased protein, mineral, and carbohydrate content due to the reduction in their oil content. The extracted almond flakes with 86.5% oil removal had approximately twice as much protein, carbohydrates, and minerals as raw almonds [123]. The protein content of 25% and 40% fat-reduced walnuts increased from 14% in the full-fat nut to 21% and 27%, respectively [139]. The cell structure of the almond was gradually modified as the extraction progressed [122]. When 15% of the oil was extracted, the pectin was affected with no modification of cellulose and hemicelluloses. When 35% of the oil was extracted, marked changes could be observed in both pectin and hemicelluloses; while at 57% and 64% oil extraction, the sample exhibited major cell wall disruption [122]. Similar observations were reported for walnuts, where lipid extraction beyond 40% resulted in collapsed cell walls and fracture and powdering of walnut pieces. Minerals such as calcium and magnesium were also affected by the extraction, especially after 65% of the oil was extracted. When these divalent cations were removed from the calcium-pectin complex, the cross-links between the galacturonic acid units of adjacent pectin chains or between the pectins and other polymers were destroyed. A mass balance loss of 2% to 20% was reported by Goodrum and Kilgo [130], who attributed it to the loss of volatile organics and water vapor in the exhaust CO_2 stream. In ground peanuts, the higher the moisture content, temperature, and pressure, the greater was the mass loss [130]. As expected, a greater mass of volatile components was lost in the exhaust

TABLE 3.3
Nut Fat Content and Fatty Acid Composition[a] of Nut Oils

Raw Material	Fat Content (%, w/w)	Fatty Acid Content[a–d] C16:0	C16:1	C18:0	C18:1	C18:2	C18:3	Ref.
Acorn	12.1	13.42–13.44[a]	0.07–0.08	2.5	64.81–65.42	16.43–17.07	0.52–0.57	121
Almond	57.0	7.87–8.48[b]	0.56–0.63	1.58–1.68	69.25–70.31	19.65–20.16	—	122
	54.5	6.60–7.10[b]	0.50–0.60	1.7–2.2	68–73	17.7–22	—	124
Hazelnut	66.2	5.27–6.01[c]	0.17–0.20	2.19–2.45	82.65–85.18	6.27–8.42	0.08–0.09	128
	56.0	5.86–5.99[c]	—	2.14–2.17	79.34–79.62	11.37–11.45	—	127
Pecan	58.5	6.20–10.10[c]	—	2.9–3.2	60.4–65.7	23.1–25.5	1–3.3	136
Walnut	69.0[e]	6.90–8.10[d]	—	1.5–2.0	16.5–16.7	60.9–61.2	12.5–13.7	140
	71.0	6.08–6.49[a]	0.07	2.1–2.13	20.98–21.22	56.46–56.88	13.16–13.41	138

Palmitic acid (C16:0), Palmitoleic acid (C16:1), Stearic acid (C18:0), Oleic acid (C18:1), Linoleic acid (C18:2), Linolenic acid (C18:3).
[a]mol %, [b]wt %, [c]not indicated, [d]GC Area %, [e]Fat content reported in [139].

CO_2 stream at higher temperatures. As well, losses due to inefficiencies in the collection of solutes prior to CO_2 exhaust should not be overlooked.

3.3.1.2.2 Other Quality Attributes

The color of the residual meal and the extracted oil are affected by the degree of extraction. With higher oil removal, the color of the pecan and peanut kernel residues was lighter, as most of the pigments are fat soluble [134, 141]. The L-value (a measure of lightness) of the fat-reduced walnuts was higher than that of the full-fat nuts, indicating that the fat-reduced walnuts had a whiter appearance [139]. In another example, a gradual color change on the residual meal was observed, with meal located at the solvent outlet being darker and having more oil [130]. The color of meal residue is also affected by temperature and cosolvent addition. The peanut meal became darker with temperature increase at constant pressure. At 75°C, the meal located at the reactor inlet changed from chalk white to light brown and finally to dark brown when temperature reached 95°C [130]. Pecan kernels appeared more red and less yellow after extraction, and this trend increased with temperature [134]. When pure SC-CO_2 was used, only a small change in original pistachio color was observed. However, when ethanol was added as a cosolvent, an almost white pistachio residue was produced due to the extraction of chlorophyll [137]. Color of the residue is also affected by pretreatment of the sample. Compared with soaked peanuts, humidified peanuts had the closest color of oil and peanuts [133].

The color of the extracted oil is affected by temperature, pressure, and cosolvent addition. Palazoglu and Balaban [137] showed increasing color intensities for the extracted pistachio oils with an increase in pressure. The color of the oil extracted at 60°C and 34.5 MPa was dark yellow, whereas that obtained at 50°C and 27.5 MPa was lighter. Oil extracted at 70°C and 20.7 MPa with 5% ethanol addition was yellowish green and that at 70°C and 34.5 MPa was green. The crude peanut oil color ranged from yellow to dark brown as extraction temperature increased from 25°C to 120°C [130]. The extracted pecan and walnut oils were all amber in color [134, 139], whereas the almond oil was yellow [124].

The texture (hardness) of the pistachio nut changed significantly after SC-CO_2 extraction. A sensory crunchiness test indicated that the harder the pistachio, the crunchier it is [137]. The shear-compression force of peanuts increased with extraction time, which indicates that the defatted peanut has a crispy texture [141]. The hardness of the walnut decreased with the fat content of the nut (full fat, 25% and 40% fat-reduced) [139]. Also, flavor intensity was reduced after SC-CO_2 extraction, which might be attributed to the removal of the flavor compounds. The aroma and flavor intensity, fracturability, and moistness of peanuts all decreased with increasing extraction time [141].

The peanut meal volume decreased after extraction and the decrease in volume increased with temperature and pressure. This bulk volume reduction was attributed to the compaction of the meal by high pressure and the elimination of oil from the meal. At 25°C, the meal crumbled when touched. At 100°C and 55 MPa, the meal was a coherent mass, which was difficult to fracture by hand [130]. The bulk density of the extracted almonds was considerably less than that of the raw materials, about 41% less for the diced almonds and 54% to 59% less for the flaked almonds [123].

Few studies evaluated the oxidative stability of nuts after SC-CO_2 extraction. The fat-reduced walnuts (25% and 40%) had a lower peroxide value (PV) than that of the full-fat walnuts when stored at 25°C and 40°C for 8 weeks. After 5 weeks at 40°C, the 25% fat-reduced walnuts had a higher PV than the 40% fat-reduced nuts, but there was no significant difference throughout the 25°C storage condition [139].

3.3.1.3 Comparison with Conventional Methods

Nut oils are traditionally recovered mechanical pressing of organic solvents. The mechanical pressing process caused considerable splitting (12% to 43%) and breakage (3.6% to 46%) of peanuts [133]. Furthermore, about 5% of water-soluble sugars and proteins were lost in the soaking step following the pressing, which can expand the peanut back to its original size and shape. Similarly, a considerable amount of fat-soluble vitamins and other valuable constituents, such as phospholipids, are removed during hexane extraction [134]. It has been demonstrated that polar phospholipids play an important role in lipid stability. Organic solvent extraction also leaves undesirable solvent residue in the final products [134, 139]. On the other hand, the walnut oil extracted by SC-CO_2 had a higher amount of tocopherols (405.7 μg/g oil) compared with the oil extracted with hexane (303.2 μg/g oil) [138]. The oil extracted by SC-CO_2 was also clearer than that obtained by hexane, indicating the need for less refining [121, 128, 138]. However, the SC-CO_2-extracted oil showed greater susceptibility to oxidation.

SC-CO_2 extraction is also used to obtain shea nut oil. The traditional process involves pouring hot molten shea oil into 10% Fuller's earth containing hot acetone, cooling and precipitating the polyisoprenoid gum onto the earth, and then filtering to remove the earth and gum [142]. However, with SC-CO_2 extraction, a clean and high-quality oil can be obtained with the high-molecular-weight polyisoprenoids left in the extraction vessel as a rubber-like mass [142]. In addition, the extracted oil has low levels of free fatty acids, monoglycerides, diglycerides, iron, triterpene acetate, and triterpene cinnamate, which is a major advantage for its use in the confectionary industry as a cocoa butter substitute.

3.3.2 Seed Oils

Numerous studies on the extraction of specialty oils from seeds, such as apricot [143–145], borage [146–148], cherry [149], echium [148], evening primrose [150–152], flax [153], grape [154–158], hiprose [159], *Hybrid hibiscus* [160], milk thistle [161], munch [162], pumpkin [163], rosehip [164–168], sea buckthorn [169–171], sesame [172], and tomato [173], using SC-CO_2 have been reported and are summarized in Table 3.4. In addition to SC-CO_2, some studies also used propane as the high-pressure solvent. For example, *Silybum marianum* seed oil extraction rate using propane was found to be higher than that obtained with CO_2, while using 10 times less propane than CO_2 [161]. However, the tocopherol content of the oil obtained by CO_2 (0.085%) was higher than that obtained by propane (0.02%) [161]. Thus, propane might not be an appropriate solvent for tocopherol extraction. To attain complete oil extraction from rosehip seeds, a solvent-to-feed ratio of 3 was used for propane/CO_2 mixture at

TABLE 3.4
Extraction of Bioactive Compounds from Seeds Using SC–CO_2

Raw Material	Feed (g)	Sample Preparation: Particle Size (mm)	Sample Preparation: H_2O (%)	Bioactive Compound	Extraction Conditions: T (°C)	P (MPa)	Flow rate	Time (min)	Cosolvent	Recovery[a] (%)	Ref.
Apricot	5	< 0.85	3.9	Oleic acid, linoleic acid	40, 50, 60	15, 30, 45, 52.5, 60	< 0.5 g/min	n.i.	None	n.i.	143
	5	n.i.	3.9	Oleic acid, linoleic acid	40, 50, 60	30, 37.5, 45	2, 3, 4 g/min	15	0, 1.5, 3% EtOH+	85	144
	5	< 0.425, < 0.85, 0.92, 1.5	3.9	Linoleic acid	40, 50, 60, 70	30, 37.5, 45, 52.5, 60	1, 2, 3, 4, 5 g/min	n.i.	0, 0.5, 1, 1.5, 3% EtOH+	n.i.	145
Borage	10	n.i.	n.i.	Linolenic acid, stearidonic acid	40	10–35	0.5 L/min	n.i.	0, 0.5–2% caprylic acid methyl ester+	51.5[c]	146
	40	0.5, 0.75, 1, 1.5	0, 1.8, 7.4	Linolenic acid	10, 40, 60	5–35	0.5–2 L/min	180	None	29[c]	147
	150	0.65	n.i.	Linolenic acid	10, 25, 40, 55	6, 10, 20, 30	0.03–0.2 g/min	n.i.	None	n.i.	148
Cherry	n.i.	n.i.	10.38	Linoleic acid, sterols	40, 60	18, 20, 22	1.2, 3.6, 4.8×10^{-2} m/min[b]	n.i.	None	n.i.	149
Echium	150	0.65	n.i.	Stearidonic acid, linolenic acid;	10, 25, 40, 55	6, 10, 20, 30	0.03–0.2 g/min	n.i.	None	n.i.	148
Evening primrose	0.8	<0.5	n.i.	γ-Linolenic acid	35–60	8–71	1×10^{-3} L/min	100	None	95	150
	50	<0.355	n.i.	γ-Linolenic acid	40, 50, 60	20, 30, 50, 70	18 g/min	n.i.	None	> 95	151

continued

TABLE 3.4 (continued)
Extraction of Bioactive Compounds from Seeds Using SC-CO_2

		Sample Preparation			Extraction Conditions						
Raw Material	Feed (g)	Particle Size (mm)	H_2O (%)	Bioactive Compound	T (°C)	P (MPa)	Flow rate	Time (min)	Cosolvent	Recovery[a] (%)	Ref.
	50	<0.355	n.i.	γ-Linolenic acid	40, 50, 60	20, 30, 50, 70	9, 18, 27 g/min		None	n.i.	152
Flax	5	n.i.	n.i.	Linolenic acid, tocopherols	50, 70	21, 35, 55	1, 3 L/min	180	None	74	153
Grape	176	<1.25	n.i.	Linoleic acid, tocopherols, phytosterols, squalene	65	37	60 g/min	360	None	13.6[c]	154
	20	n.i.	6.8	n.i.	40	28	0.5–1 L/min	n.i.	None	n.i.	155
	40	0.35, 0.75, 1.5, 2.83	0.3, 1.1, 2.4, 6.3	Unsaturated oil	10, 40, 60	5, 10, 20, 30	0.5, 1, 1.5, 2 L/min	300	None	92	156
	3	n.i.	n.i.	Phenolics	40	20, 30	n.i.	n.i.	None	n.i.	157
	4, 500	0.25–0.42, 0.42–0.841, 0.841–2	n.i.	Unsaturated oil (linoleic acid)	35, 40, 45	20, 25, 30, 40	0.4 mL/min; 4 L/min	120, 210, 300	None	100	158
Hiprose	13	0.42, 0.79, 1.03	n.i.	Linoleic acid	40, 50, 70	10.3, 20.6, 41.3, 68.9	1, 2, 4, 6 g/min	n.i.	None	7.4[c]	159
Hybrid hibiscus	5	0.1	n.i.	Phytosterols	80	53.7	2×10^{-3} mL/min	50	None	20[c]	160
Milk thistle	30	n.i.	n.i.	Tocopherols	25, 40, 60, 80	10, 20, 30	n.i.	n.i.	None	20.5[c]	161
Munch	150	0.05–0.25	n.i.	β-Dimorphecolic acid (DA)	45	30	n.i.	n.i.	None	> 95	162

Pumpkin	20	0.36	n.i.	PUFAs, sterols, tocopherols	35–45	18–20	0.03–0.1 m/min[b]	120	None	n.i.	163
Rosehip	10	n.i.	n.i.	Carotenoids, unsaturated oil	28, 35	10, 25	1–1.5 L/min	n.i.	Propane+	n.i.	164
	26	0.85–2.36, 0.425–0.85, 0.15–0.425	n.i.	Unsaturated oil	40	30	11.4 g/min	n.i.	None	39.2[c]	165
	100	n.i.	n.i.	Unsaturated oil, flavonoids, carotenoids	40, 50, 60	30, 40, 50	21 g/min	n.i.	None	7.1[c]	166
	210	n.i.	n.i	Linoleic acid, α-linolenic acid	40, 60, 80	30, 50, 70	n.i.	n.i.	None	n.i.	167
	26	n.i.	n.i.	Unsaturated oil	40, 50	30, 40	4, 8, 12, 18, 24 g/min	60–90	None	n.i.	168
Sea buckthorn	3–40	n.i.	n.i.	Unsaturated oil	25, 40, 60	9.6, 17.4, 27	1 L/min	n.i.	None	n.i.	169
	120	<0.491, 0.491–0.643, 0.643–1.033, >1.033	n.i.	Linoleic acid	30, 35, 40, 45	15, 20, 25, 30	0.83–3.33 L/min	270	None	n.i.	170
	n.i.	0.5–1	13.1	Linoleic acid	30, 35, 40, 50	15, 20, 25, 30	1.7, 3.3, 5, 6.7 L/min	300	None	n.i.	171
Sesame	10	n.i.	n.i.	Oleic acid, linoleic acid	50, 60, 70	20.7, 27.6, 34.5	n.i.	n.i.	0, 5, 10% EtOH+	89.4	172
Tomato	4.5	0.27	n.i.	Unsaturated oil	40, 55, 70	10.8, 17.6, 24.5	2.8 g/min	480	None	n.i.	173

[a] Recovery (g extract/g oil in feed material × 100), [b]superficial velocity, [c]yield (g/100 g feed material), +cosolvent added continuously into SC-CO_2 a the level (%, w/w) indicated.

28°C and 10.0 MPa, while a ratio of only 1 was sufficient for propane alone at 25°C and 5.0 MPa [174].

3.3.2.1 Factors Affecting Extraction Yield

3.3.2.1.1 Sample Preparation

Unlike extracting oil from nuts, which have spherical and elongated hexagonal cell structures containing oil, it is nearly impossible to extract oil from uncracked seeds due to their unique physical structures. The oil in nuts is easily accessible by diffusion into the cellulosic structure; however, the seed coat is almost impermeable [155]. The oil in the rosehip or hiprose seeds, for example, is contained in the oil-bearing structures, which are enclosed in a thick and highly lignified testa [159, 165]. The hiprose may be contained in long microscopic channels, which are protected by a lignin structure that is probably too compact to allow effective diffusion of the supercritical fluid in a reasonable time [159]. Therefore, grinding the seeds is an important step of sample preparation, which can break the intact channels and expose the oil in the channels to the extraction solvent.

a) **Particle size:** As expected, the oil yield increased with decreased particle size. For example, when hiprose seed particles ranging from 1.03 to 0.79 and 0.42 mm were used, the oil yield increased from 4.9% to 5.2% and 7.4%, respectively [159]. Similar particle size effects were observed when grape [156] and borage [147] oils were extracted from their seeds. Particles with diameters less than 0.35 mm were suggested for grape seed extraction [156]. With a decrease in the particle size of sea buckthorn seeds from between 0.50–2.36 to 0.43–1.00 mm, the duration of the extraction process was reduced from 6 to 3 hours [175].
b) **Moisture and equilibration time:** The moisture content of the grape seed (6.3%, 2.4%, 1.1%, 0.3%) was modified by drying the ground sample for different lengths of time (0, 2, 4, 6.3 hours) [156]. The extraction yield was not significantly affected by the moisture content of the grape seeds. However, the sample exposed to the longer drying time (6.3 hours) had a slightly lower oil yield, which might be due to the evaporation of volatile constituents [156]. Similarly, the extraction yield of the borage seed oil was also not significantly affected by the drying process (partially and almost fully dehydrated); however, the moisture content of the borage seed (0%, 1.8%, 7.4%) had a negative impact on the extraction yield, with the higher moisture content samples resulting in lower extraction yields [147].

3.3.2.1.2 Extraction Parameters

a) **Temperature and pressure:** In general, extraction yield increased with pressure at constant temperature. For example, increasing the pressure from 10 to 15, 20, and 25 MPa at 40°C resulted in an increase in borage oil yield from 0.1% to 5.6%, 15.2% and 21.9%, respectively. However, only a slight increase in the yield (21.6% to 24.3%) was observed with a pressure increase from 30 to 35 MPa [146]. A similar trend (34.4% to 91.3%, respectively) was observed during extraction of evening primrose seeds at 60°C and pressures of 20 and 30 MPa [151]. At constant temperature, increasing the pressure to 50 MPa improved the oil yield slightly to 97.2%, while a further increase in pressure to 70 MPa had almost no effect on the yield (97.7%) [151]. Not only the extraction yield but also the extraction rate increased with pressure. In addition, with an increase in pressure from 20.6 to 41.2 and 68.9 MPa, the extraction time for hiprose decreased from about 30 to 15 and 5 min, respectively [159], since triglycerides are solubilized to a greater extent at higher pressures in SC-CO_2. More than 94% of the available oil from evening primrose seeds was extracted in only 14 min [151]. As discussed in Section 3.3.1.1.2, temperature has a negative effect on solubility and extraction yield at low pressures but a positive effect at higher pressures due to the crossover of the solubility isotherms. Moreover, the negative effect of temperature at low pressures seems to be greater than the positive effect at high pressures. For example, with an increase in temperature from 40°C to 50°C at 20 MPa, the evening primrose oil yield dropped from 66.1% to 59.6%, with a further drop to 34.4% at 60°C. However, at a higher pressure of 50 MPa, the oil yield increased from 96.8% to 97.5% with a temperature increase from 40°C to 50°C and then slightly dropped to 97.2% at 60°C [151]. A similar trend was reported for the extraction of *Silybum marianum* oil using SC-CO_2. In this case, the seed oil yield decreased from 19.9% to 5.2% with a temperature increase from 25°C to 80°C at 20 MPa but increased from 15.3% to 20.5% at 30 MPa [161]. In fact, the pressure increase seems to have a more substantial effect on the oil yield than the temperature increase. Sesame oil yield increased only from 5.4% to 11.8% when temperature was increased from 50°C to 70°C at 27.6 MPa; however, the yield increased from 3.6% to 31.3% when pressure was increased from 20.7 to 34.5 MPa [172].

b) **Flow rate and flow direction:** The extraction yield of borage oil increased with flow rate [147]. Similarly, 66% and 74% of flax oil was obtained at a flow rate of 1 and 3 L/min (measured at STP), respectively [153]. Similar to the nut oils, flow direction was reported to affect the extraction, with downward flow being more favorable. Sovova et al. [155] reported that grape seed oil extraction was retarded when the supercritical solvent flow was upward through the bed (due to natural convection). It is therefore advantageous to operate laboratory extraction units in the downward flow mode.

c) **Extraction time:** The duration of the fast extraction period depends on the plant type and variety. For example, two varieties of grape seeds resulted in low oil yields (5.9% and 6.1%) throughout the fast extraction period of 60 min with a very little amount of additional oil extracted at extended extraction times [154]. The other four grape varieties containing intermediate levels of oil (9.4% to 10.7%) were nearly completely extracted within 60 min, whereas the high oil content variety (13.6%) required 120 min.

d) **Use of cosolvent:** Addition of 10% ethanol as a cosolvent greatly enhanced the extraction yield of sesame oil. At 27.6 MPa and 50°C, the recovery increased from 5.4% to 74.1% upon addition of ethanol into SC-CO_2, whereas at 70°C, the yield increased from 11.8% to 89.4%, respectively [172]. At 40°C, when caprylic acid methyl ester was used as a cosolvent (0%, 0.5%, 1%, and 2%), the borage oil yield increased from 0.1% to 2.9%, 3.1% and 6.7% at 10 MPa; from 15.2% to 20.7%, 30.0% and 36.6% at 20 MPa; and from 21.6% to 30.7%, 38.2% and 51.5% at 30 MPa [146]. At a low pressure of 10 MPa, the amount of coextracted solvent was high, ranging from 51.1% to 73.0% and 79.7% of the total fatty acid methyl ester at 0.5, 1%, and 2% addition of caprylic acid methyl ester [146]. When pressure was increased to 20 MPa, the amount of coextracted solvent was reduced to 11.9%, 20.8%, and 39.8% of the total fatty acid methyl ester at 0.5%, 1%, and 2%, respectively, due to the higher recovery of triglycerides. It was shown that this cosolvent can be easily removed from the final extract at low pressures.

3.3.2.2 Characterization of Products Extracted by SC-CO_2

3.3.2.2.1 Chemical Composition

The fatty acid composition of the seed oils is summarized in Table 3.5; where the unsaturated fatty acids (mainly C18:1, C18:2, and C18:3) accounted for more than 90% of the total fatty acids. Fatty acid composition of the oil obtained by Soxhlet extraction was similar to that of the SC-CO_2 extract for grape seed [154]. In addition, fatty acid profiles of the grape seed oils obtained at different temperatures (35°C and 40°C) and pressures (30 and 40 MPa) were also similar [158]. However, Szentmihalyi et al. [164] found that the lower temperature SC-CO_2 extraction resulted in higher levels of oleic and linoleic acids in the rosehip oil compared to those in the Soxhlet extract. Dauksas et al. [146] found that the linolenic acid content of the borage seed oil increased from 16.2% to 20.1% with a pressure increase from 10 to 20 MPa; however, it slightly decreased to 18.5% with a further increase to 35 MPa. The fatty acid composition of the borage oil obtained using various concentrations of caprylic acid methyl ester as a cosolvent at different pressures was different; however, no clear trends were established [146]. The fatty acid compositions as well as the ratios between the saturated and unsaturated fatty acids (11:89) of the *Hippophae rhamnoides* L. seed oils extracted using SC-CO_2 and hexane were similar [177]. Although there was no difference between the fatty acid compositions of SC-CO_2- and hexane-extracted grape seed oils, SC-CO_2-extracted fractions obtained at 30, 60, 120 and 180 min were different [156]. The α-linolenic acid content of the SC-CO_2-extracted flax oil was higher than that of solvent-extracted oil; however, the saturated and monounsaturated fatty acids were higher in the solvent extract [153].

3.3.2.2.2 Other Quality Attributes

The SC-CO_2-extracted evening primrose oil had a yellow color and its intensity increased with pressure, approaching the deep yellow color of the hexane-extracted oil [151]. The yellow intensity of the borage oil extracts also increased with pressure [146]. Similarly, the SC-CO_2 extracted sea buckthorn seed oil was a clear, yellow-brown liquid at room temperature, whereas the pulp flake oil was red and semisolid [175]. Odabasi and Balaban [172] found that the sesame oil from SC-CO_2 extraction appeared clear. However, when 5% ethanol was used as a cosolvent, the oil extract

TABLE 3.5
Seed Fat Content and Fatty Acid Composition of Seed Oils

Raw Material	Fat Content (%, w/w)	Fatty Acidy Content[a–d] C16:0	C16:1	C18:0	C18:1	C18:2	C18:3	Ref.
Apricot	48.1	5.22–5.71[c]	0.6–0.78	1–1.3	67.37–68.07	24.84–25.11	—	144
Borage	29.0	13.32[c]	0.19	4.58	19.78	39.57	22.56	147
Cherry	8.5	5.26[a]	0.27	2.15	32.64	40.84	1.1	149
Echium[e]	30.0	6–8[b]	—	3–5	15–19	14–18	37–45	148
Evening primrose	27.5	4.75–6.7[c]	—	1.46–1.87	4.85–5.4	75.27–77.07	(γ-) 9.22–10.36; (α-) 0.13–0.16	150
Flax	38.0	5.7–5.8[d]	—	4.0–4.2	14.1–14.4	12.8–13.4	60.5–61	153
Grape	10–15	6.28–8.26[b]	0.06–0.15	3.6–5.22	12.71–18.47	67.56–73.223	0.44–1.12	154
Hybrid hibiscus	8.5–20	14.8–27.0[b]	0–0.7	0–4.1	17.8–47.3	42.6–64	0–0.7	160
Lunaria[e]	37.0	1–3[b]	—	1–2	16–20	8–10	2–4	148
Munch	22.8	1.9–2.4[b]	—	1.5–1.9	17.5–22.1	12.3–14.9	0.7–1	162
Neem[e]	45.0	13–15[c]	—	14–19	50–60	8–16	—	176
Pumpkin	43.5	14.9[c]	0.25	3.19	13.5	59.9	1.64	163
Rosehip	8.0	3.6–7.87[d]	—	2.45–3.27	16.25–22.11	35.94–54.75	20.29–26.48	164
	18.5	8.85[b]	—	2.42	21.75	40.09	—	177
Sesame	—	9.23–10.53[c]	—	4.17–5.3	37.6–38.48	44.17–45.85	—	172
Tomato	29	10.9–15.79[b]	1.15–3.81	4.97–11.98	19.9–24.13	39.69–53.24	4.58–5.5	173

Palmitic acid (C16:0), Palmitoleic acid (C16:1), Stearic acid (C18:0), Oleic acid (C18:1), Linoleic acid (C18:2), Linolenic acid (C18:3).
[a]mol %, [b]wt %, [c]not indicated, [d]GC % Area, and [e]Fatty acid composition reported for the new material.

was clear at low temperature but cloudy at high temperature. With 10% ethanol addition, all samples were cloudy regardless of the temperature. This could be attributed to the coextraction of phospholipids, waxes, and pigments that occurs with ethanol addition. Therefore, the selection and concentration of cosolvent addition must be optimized based on desirable product quality attributes.

3.3.2.3 Comparison with Conventional Methods

Hexane extraction tends to give a higher oil yield than SC-CO_2 extraction due to the extraction of undesirable compounds that must be removed during refining [147]. Gomez and de la Ossa [156] reported that the grape seed oil yield from SC-CO_2 extraction was 6.9%, while that with hexane was 7.5%. This is due to the hexane extraction being nonselective for triglycerides since hexane can also extract free fatty acids, phospholipids, pigments, and unsaponifiables. Bozan and Temelli [153] also found that the oil yield obtained from flaxseed with SC-CO_2 after a 3 h extraction was 21–25%, whereas the petroleum ether extraction gave a 38% yield. In addition, the SC-CO_2-extracted oils from pumpkin and *Hippophae rhamnoides* L. seeds were clearer than those extracted by hexane [163]. Bernardo-Gil et al. [149] also reported that cherry seed oil extracted with SC-CO_2 was clearer than the hexane extract, minimizing the need for further refining. Nevertheless, the pumpkin seed oil extracted with hexane was better protected against oxidation (induction time = 8.3 h) compared with the SC-CO_2–extracted oil (induction time = 4.2 h) [163].

3.3.3 Cereal Oils

To date, the cereal oils that have been studied are amaranth [178–180], oat [181], rice bran [182–187], wheat germ [188–191], and wheat plumule [192] oil, which are summarized in Table 3.6.

3.3.3.1 Factors Affecting Extraction Yield

3.3.3.1.1 Sample preparation

Cereal grains, which in general have a moisture content of 3% to 12%, may not require a drying step prior to extraction.

a) **Particle size:** Extraction yields obtained using original (0.75 mm) and the milled (0.3 mm) wheat germ were similar [188]. However, according to Panfili et al. [190], wheat germ oil recovery increased from 57% to 92% when the particle size was reduced from 0.5 to 0.35 mm. A similar trend was reported for the SC-CO_2 extraction of oil from *Amaranthus* grain, where very little oil was extracted from whole grains [179].
b) **Moisture:** In the case of *Amaranthus* grain, moisture contents of 0%, 5%, and 10% had no significant effect on the oil and squalene extraction yields [179]. As the maximum moisture content of the harvested *Amaranthus* grain is 10%, drying is not a necessary step. However, in wheat germ extraction, the tocopherol yield increased with a decrease in moisture content from 11.5% to 8.2% and 5.1% [189], but decreased with a further

TABLE 3.6
Extraction of Bioactive Compounds from Cereals Using SC-CO_2

	Sample Preparation				Extraction Conditions						
Raw Material	Feed (g)	Particle Size (mm)	H_2O (%)	Bioactive Compound	T (°C)	P (MPa)	Flow rate	Time (min)	Cosolvent	Recovery[a] (%)	Ref.
Amaranth (*Amaranthus cruentus*)	40	n.i.	n.i.	Linoleic acid, oleic acid	40, 45, 50	10, 20, 25, 30	3.3, 6.8, 8.5 g/min	n.i.	None	n.i.	178
	60	n.i.	n.i.	Squalene	40, 50, 60, 70	15, 20, 25, 30	1, 2, 3, 5 L/min	120	None	4.8[c]	179
(*Amaranthus caudatus*)	5	< 0.2	n.i.	Tocopherols, squalene	40	20.2, 40.5	2 L/min	15	None	n.i.	180
Oat	1500, 185	n.i.	n.i.	Oleic acid, linoleic acid	40	25, 35	n.i.	300	20% EtOH[++]	n.i.	181
Rice bran	20	n.i.	3.1	Tocopherols	40	14.7–34.3	n.i.	n.i.	None	22[c]	182
	300	n.i.	10.1	Unsaturated oil, tocopherols, sterols, oryzanol	0, 20, 40, 60	17, 24, 31	41.7 g/min	360	None	96.8	183
	300	n.i.	8.48	Unsaturated oil, tocopherols, sterols, oryzanol	40, 45, 50	8.6, 9.9, 11.2	58.3 g/min	240	None	4.1[c]	184
	30	0.5	n.i.	PUFAs, tocopherols, squalene	40, 50, 70	20.7, 27.6, 34.5, 41.3	n.i.	480	None	80	185
	20	n.i.	n.i.	α-Tocopherol, γ-oryzanol	40, 60, 80	34.5, 51.7, 68.9	1.1 g/min	n.i.	None	24.7[c]	186

continued

TABLE 3.6 (continued)
Extraction of Bioactive Compounds from Cereals Using SC-CO_2

		Sample Preparation			Extraction Conditions						
Raw Material	Feed (g)	Particle Size (mm)	H_2O (%)	Bioactive Compound	T (°C)	P (MPa)	Flow rate	Time (min)	Cosolvent	Recovery[a] (%)	Ref.
Rice bran	800	0.25, 0.5, 0.75	n.i.	Linoleic and palmitic acids, sitosterol and campesterol	50, 60	10, 20, 30, 40	300 g/min	180	None	20[c]	187
Wheat germ	25	0.3, 0.75	n.i.	Tocopherols, linoleic acid	10–60	5–30	0.5–2 L/min	180	None	>95[c]	188
	5	n.i.	n.i.	Tocopherols	35, 40, 45, 50	13.7, 20.7, 27.6, 34.5, 41.3	1–3 × 10^{-3} L/min	90	None	2.2[c]	189
	600–700	0.35–0.5	n.i.	Linolenic, linoleic acids, β-carotene, lutein, and zeaxanthin, tocopherols, tocotrienols	55	25, 38	0.5 L/min	180	None	92	190
	20	Flakes: 2.1	n.i.	Tocopherols	10, 25, 40, 50	20.2	300–400 L/min	n.i.	None	95	191
Wheat plumule	n.i.	n.i.	n.i.	PUFAs	35, 45, 55, 65	10, 15, 20, 30	1.2, 1.8, 2.4, 3 L/min	60–480	None	n.i.	192

T: temperature, P: pressure, n.i.: not indicated.

[a] Recovery (g extract/g oil in feed material × 100), [b]superficial velocity, [c]yield (g/100 g feed material), ++Cosolvent added to sample before extraction at the level (%, w/w) indicated.

reduction in moisture to 4.3%. This might be attributed to the shrinking of the germ particles.

3.3.3.1.2 Extraction Parameters

a) **Temperature and pressure:** During wheat germ extraction, tocopherol yield increased slightly with a pressure increase from 13.8 to 27.6 MPa at 40°C, while a further increase in pressure to 34.5 and 41.4 MPa did not improve the yield significantly [189]. On the other hand, tocopherol yield decreased with temperature at pressures below 26.2 MPa and increased with temperature at pressures above 26.2 MPa. Similarly, amaranth oil yield increased with pressure in the range of 15 to 25 MPa. Also, temperature had an adverse effect on the amaranth oil yield at the pressure range of 15 to 25 MPa but had a positive effect at 30 MPa [179]. Similarly, another study on amaranth seed oil showed that extraction yield and rate decreased with temperature at 10 MPa and increased with temperature at 20 and 30 MPa [178]. Although the amaranth oil yield varied significantly with pressure, the yields of squalene in the amaranth oil at 40°C and different pressures were very close, ranging from 0.24 to 0.27 g/100 g grain [179]. On the other hand, the amount of squalene extracted from rice bran decreased with CO_2 density and the maximum amount was 2.9% at 70°C and 20.7 MPa [185]. The SC-CO_2 density has a different effect on fatty acids. As the reduced density was increased at constant temperature, the amount of linoleic acid extracted increased but that of oleic acid decreased [185].

b) **Flow rate:** The amaranth oil yield and initial extraction rate both increased with flow rate from 1 to 2 L/min, but there were no differences at flow rates of 2, 3 and 5 L/min (measured at STP) under the same extraction conditions [179]. At flow rates above 2 L/min, the extraction rate was reduced markedly after 1 h of extraction and only a small amount of oil was extracted in the following 2 h. Similarly, in the wheat germ oil extractions, the oil yield and extraction rate both increased with flow rate from 0.5 to 1 and 1.5 L/min (measured at STP), while a further increase of flow rate to 2 L/min did not change either the oil yield or the extraction rate [188].

c) **Extraction time:** Not many studies involving cereal oils evaluated the extraction yield as function of extraction time. As expected, the extraction yield increased with extraction time for wheat plumule oil [192].

d) **Use of cosolvent:** The addition of ethanol as a cosolvent slightly decreased the free fatty acid content from 5.6% to 4% in the case of oat oil, but increased the phosphorus content from less than 1 ppm to 80 ppm probably due to the recovery of polar phospholipids [181].

3.3.3.2 Characterization of Products Extracted by SC-CO_2

3.3.3.2.1 Chemical Composition

Fatty acid composition of the SC-CO_2 extracted cereal oils is presented in Table 3.7. The main fatty acids in the rice bran and wheat germ oils are oleic, linoleic, and palmitic acids [185, 187, 188]. The fatty acid compositions of SC-CO_2- and hexane-extracted oils were similar and there were no differences among the extracts obtained at different extraction conditions [188]. Crude rice bran oil contains 70% triglycerides, 7% free fatty acids, 3.6% fatty acid esters, and some minor components, such as oryzanol [193], α-tocopherol, and β-sitosterol. The α-tocopherol content of rice bran oil was reported to be 284 mg/kg [182] and 1050.8 to 1279.3 mg/kg [186]. In addition, Shen et al. [184] found that the SC-CO_2-extracted rice bran oil contained 0.23 mg/g oil of α-tocopherol and 18.5 mg/g oil of β-sitosterol. Tocopherol and carotenoids are two important minor components in wheat germ. α-Tocopherol was reported to be the most abundant tocopherol isomer at 1329 mg/100 g wheat germ [189] and at 2123 mg/kg wheat germ oil [190]. Lutein and zeaxanthin were the most abundant carotenoids in wheat germ, at 47.7 and 37.3 mg/kg oil, respectively [190]. Squalene, an important minor compound in amaranth oil, was reported to be present at 5.27% (40°C, 25 MPa) to 15.3% (50°C, 20 MPa) in the extracts obtained [179].

3.3.3.2.2 Other Quality Attributes

The SC-CO_2-extracted rice bran and wheat germ oils both had light color compared with those extracted by hexane [182, 191]. However, the SC-CO_2-extracted oil was very unstable compared with hexane-extracted oil [182].

3.3.3.3 Comparison with Conventional Methods

Because solvent extraction is less selective than SC-CO_2 extraction, it generally results in a high total extract yield, leading to reduced concentrations of desirable bioactives. For example, wheat germ oil yield from SC-CO_2 extraction was slightly lower (7.3% to 8.0%) than that obtained with hexane (8.6%), whereas the tocopherol content of the SC-CO_2-extracted oil was higher [188]. The yield of total tocopherols from wheat germ obtained by SC-CO_2 was higher than those obtained by solvent extraction [189]. Extraction of vitamin E by hexane and chloroform/methanol took about 960 and 140 min, respectively, whereas SC-CO_2 extraction required only 90 min. The operating temperature for SC-CO_2 extraction was lower (40°C) than those of hexane and chloroform/methanol (70°C and 65°C) extraction [189]. By choosing suitable extraction conditions, some compounds can be selectively extracted by SC-CO_2, but not by solvent extraction. Squalene can be extracted from rice bran by SC-CO_2 but not by using chloroform and methanol [185]. About 80% of PUFAs was extracted by SC-CO_2, whereas only 60% recovery was obtained with solvent extraction [185]. A high squalene yield (0.31 g/100 g *Amaranthus* grain) and concentration (15.3% in extract) was obtained at 50°C and 20 MPa using SC-CO_2, while the squalene concentration in the solvent extract was only 6% [179].

3.3.4 Fruit and Vegetable Oils

Buriti fruit [194], carrot [195-198], cloudberry [199], hiprose fruit [174], olive husks [200], tomato [201-207] are the main fruits and vegetables studied (Table 3.8). Lycopene is the dominant carotenoid (85% to 90% of total carotenoids) in the tomato extract [207], whereas β-carotene (60% to 80%) is the major one in carrot extract. Squalene, tocopherol, and sterols are the main bioactive components found in olives.

TABLE 3.7
Fatty Acid Composition of Cereal Oils

Raw Material	Fatty Acid Content[a]									Ref.
	C14:0	C16:0	C16:1	C18:0	C18:1	C18:2	C18:3	C20:0	C20:1	
Amaranth	—	12.32–17.94	—	2.71–4.66	23.85–32.88	43.66–47.48	—	0.38–1.54	—	180
Oat	0.1–0.2	13.6–15.7	—	1.4–1.6	38.3–43.7	39.1–42.2	1.0–1.5	—	0.8–0.9	181
Rice bran	—	16.5–17.9	—	1.1–1.4	38.8–41.4	37.8–40.4	1.5–1.7	0.3–0.6	0.4–0.6	183
Wheat germ	—	18.09–18.95	0.22–0.24	0.49–0.73	13.69–16.52	57.1–58.99	6.46–8.51	—	—	188

Palmitic acid (C16:0), Palmitoleic acid (C16:1), Stearic acid (C18:0), Oleic acid (C18:1), Linoleic acid (C18:2), Linolenic acid (C18:3), Arachidic acid (C20:0), Eicosenoic acid (C20:1).

[a] Units not indicated.

3.3.4.1 Factors Affecting Extraction Yield

3.3.4.1.1 Sample Preparation

Drying of samples prior to extraction is necessary, as carrots and tomatoes contain 80% to 95% moisture. Grinding is also needed to achieve small particle size. Olive pomace and husk are by-products of olive oil production. To further recover the valuable bioactive compounds by SC-CO_2 extraction, pretreatment of such by-products the conventional fruit and vegetable processing industry is necessary.

a) **Particle size:** During extraction of freeze-dried carrots, a higher extraction yield was obtained with smaller carrot particles [196, 198]. The total carotenoid yield increased from 1109.9 to 1369.6 and 1503.8 μg/g dry carrot when the particle size was decreased from 1–2 mm to 0.5–1 mm and 0.25–0.5 mm, respectively [198].
b) **Moisture:** Moisture had different effects on the carotenoids yield. The α- and β-carotene yields increased with decreasing level of moisture in the feed material, while the lutein yield decreased [198]. The lutein yield decreased from 55.3 to 29.9, 19.3 and 13.0 μg/g dry carrot with a decrease in moisture from 84.6 to 48.3, 17.5 and 0.8%, while the α- and β-carotene yields increased from 184.1 to 323.0, 442.3 and 599.0 μg/g, and from 354.2 to 547.8, 668.3 and 891.7 μg/g dry carrot, respectively [198]. On the other hand, only trace amounts of lycopene were extracted when the tomato feed material contained 50% to 60% moisture [207]. This can be explained by the fact that water can act as a cosolvent for the extraction of relatively polar compounds, like lutein, whereas the presence of water is not favorable for the relatively nonpolar lycopene and carotenes.

3.3.4.1.2 Extraction Parameters

a) **Temperature and pressure:** Lycopene extraction yield increased with pressure from 33.5 MPa to 45 MPa at a constant temperature of 66°C and increased with temperature from 45°C to 66°C at a constant pressure of 45 MPa [207], because SC-CO_2 density increases with pressure at constant temperature and solubility increases with temperature above the crossover pressure. Temperature greatly affects the extraction rate at pressures above the cross-over pressure. Using tomato skin [202], the extraction rate and yield were greatly increased at 110°C, resulting in 96% lycopene recovery in 40 min and 100% recovery in 50 min. However, only about 20% and 30% recovery were achieved in 80 min at 60°C and 85°C, respectively. Pressure also affected the composition of the extracts as the recovery of *trans*-lycopene increased and that of *cis*-lycopene decreased with CO_2 density [205]. Therefore, the fractionation of *trans*-lycopene is possible when optimum CO_2 density is chosen as the lycopene isomers have different solubilities in SC-CO_2.
b) **Flow rate:** Total carotenoids yield increased with flow rate [198], ranging from 934.8 to 1332.3 μg/g and 1973.6 μg/g dry carrot at CO_2 flow rates of 0.5, 1, and 2 L/min (measured at STP), respectively. However, the lycopene

TABLE 3.8
Extraction of Compounds from Fruits and Vegetables Using SC-CO_2

Raw Material	Sample Preparation			Bioactive Compound	Extraction Conditions					Recovery[a] (%)	Ref.
	Feed (g)	Particle Size (mm)	H_2O (%)		T (°C)	P (MPa)	Flow rate	Time (min)	Cosolvent		
Buriti fruit	n.i.	n.i.	11	Carotenoids, tocopherols	40, 55	20, 30	18.6, 25.8 g/min	n.i.	None	7.8[c]	194
Carrot	2	0.5–1	0.8	α-,β-Carotene, lutein	40, 50	12–33	1.2 L/min	480	None	n.i.	195
	n.i.	0.26, 0.47, 1.12	n.i.	Carotenoids	40, 50, 60	7.8–29.4	n.i.	n.i.	1, 3, 5% EtOH[+]	n.i.	196
	2000	n.i.	n.i.	carotenes, phenolics, phytosterols, linolenic acid	45–50	35–38	n.i.	120–180	None	n.i.	197
	2	0.25–0.5, 0.5–1, 1–2	0.8, 17.5, 48.7, 84.6	α-,β-carotene, lutein	40, 55, 70	27.6, 41.3, 55.1	0.5, 1, 2 L/min	240	0, 2.5, 5% canola oil[+]	0.2[c]	198
Cloudberry	42	n.i.	n.i.	Unsaturated oil, β-carotene, tocopherols	40, 60	9, 10, 12, 15, 30	n.i.	n.i	None	n.i.	199
Hiprose fruit	n.i.	0.36	n.i.	Tocopherols, carotenoids	35	25	1–1.5 L/min	n.i.	None	100	174
Olive husks	n.i.	0.4	n.i.	Unsaturated oil (oleic acid)	35–57	10.4–18	n.i.	n.i.	None	n.i.	200

continued

TABLE 3.8 (continued)
Extraction of Bioactive Compounds from Fruits and Vegetables Using SC-CO_2

Raw Material	Feed (g)	Sample Preparation		Bioactive Compound	Extraction Conditions					Recovery[a] (%)	Ref.
		Particle Size (mm)	H_2O (%)		T (°C)	P (MPa)	Flow rate	Time (min)	Cosolvent		
Tomato	0.5	0.05–0.25	n.i.	Phytoene, phytofluene, ξ-carotene, β-carotene, lycopene	40, 50, 60	8–26	4×10^{-3} L/min	30	None	n.i.	201
	0.3	n.i.	n.i.	Lycopene	60, 85, 110	40.5	1.5×10^{-3} L/min	50	Acetone, MeOH, EtOH, hexane, dichloromethane, water[++]	100	202
	n.i.	n.i.	n.i.	Lycopene	45–80	35–38	n.i.	120–180	None	55	203
	3	n.i.	n.i.	Lycopene, tocopherols	32–86	13.8–48.3	2.5×10^{-3} L/min	n.i.	None	61	204
	0.5	n.i.	n.i.	Lycopene	40	8–28	4×10^{-3} L/min	n.i.	None	n.i.	205
	20	n.i.	n.i.	Lycopene	40	32	n.i.	n.i.	None	n.i.	206
	3000	1	n.i.	Lycopene	45–70	33.5–45	133.3–333.3 g/min	120–480	1–20% hazelnut oil[++]	60	207

T: temperature, P: pressure, n.i.: not indicated.

[a] Recovery (g extract/g oil in feed material × 100), [b]superficial velocity, [c]yield (g/100 g feed material), [+]cosolvent added to sample before extraction at the level (%, w/w) indicated, [++]cosolvent added to sample before extraction at the level (%, w/w) indicated.

yield decreased as flow rate was increased from 2.5 to 15 mL/min (measured at extraction temperature and pressure) [204]. Compared with the lycopene recovery of 38.8% (or yield of 4.59 μg/g raw material) obtained at a flow rate of 2.5 mL/min, only 8% recovery (~1 μg/g or less yield) was obtained at a flow rate greater than 10 mL/min (measured at extraction temperature and pressure) [204]. With flow rates from 0.875 to 1.25 L/min (measured at STP), the olive husk oil yield increased, whereas the yield decreased at higher flow rates [208]. The decrease may be attributed to the short residence time of CO_2 in the extractor and therefore the CO_2 leaving the extractor not being saturated with oil. A low flow rate (1.8 g/min) produced a smaller amount of squalene but at a higher concentration, whereas a high flow rate (5.4 g/min) produced a higher amount of squalene at a lower concentration in the extract [209].

c) **Use of cosolvent:** Acetone, ethanol, methanol, hexane, dichloromethane, and water have been compared as cosolvents in SC-CO_2 by mixing the cosolvent with the sample prior to extraction [202] and it was shown that all cosolvents tested except water increased lycopene recovery. In fact, water showed a negative effect, decreasing lycopene recovery to 2%. Ethanol increased recovery but decreased extraction rate. All the other cosolvents studied not only increased the lycopene yield but also improved the extraction rate to varying degrees [202]. The use of vegetable oils as a cosolvent for the recovery of carotenoids from vegetables was recently developed [198, 203, 207]. For example, hazelnut oil was chosen by Vasapollo et al. [207] because of its low acidity, which can prevent the degradation of lycopene during extraction. Lycopene yield increased with hazelnut oil addition as a cosolvent, but the extract was more diluted at higher amounts of oil [207]. For the extraction without cosolvent addition, the lycopene recovery was practically maintained below 10% from 2 to 5 hours extraction time, while in the presence of hazelnut oil, the lycopene recovery increased to about 20% in 5 hours and 30% in 8 hours. Sun and Temelli [198] added canola oil in a continuous manner into SC-CO_2 for the recovery of carotenoids from carrot. The extraction yield with SC-CO_2 without canola oil addition for α-carotene was 137 to 330.4 μg/g and β-carotene was 171.7 to 386.6 μg/g feed material at different temperatures and pressures, while the yields more than double to 288.0–846.7 μg/g and 333.8–900.0 μg/g feed for α- and β-carotene, respectively, upon addition of canola oil. The major advantage of using vegetable oils as cosolvents is the elimination of organic solvent addition, which needs to be removed later, and the fact that the oil enriched in bioactives can be used as is in a variety of product applications.

3.3.4.2 Characterization of Products Extracted by SC-CO_2

3.3.4.2.1 Chemical Composition

Fatty acid composition of oils extracted from various fruits and vegetables is shown in Table 3.9. Trilinolein (LLL) is the main triglyceride present in carrot oil followed

TABLE 3.9
Fatty Acid Composition of Vegetable Oils

Raw Material	Fatty Acid Conent								Ref.
	C14:0	C16:0	C16:1	C18:0	C18:1	C18:2	C18:3	C20:1	
Carrot	0.4	16.6	1.8	1.8	11.6	60.1	4.9	0.4	197
Tomato	1.0[b]	1.5	0.6	69.0	5.8	1.3	—	11.4	206
	1.2[c]	3.8	3.5	18.6	4.4	3.4	—	18.9	206

Myristic acid (C14:0), Palmitic acid (C16:0), Palmitoleic acid (C16:1), Stearic acid (C18:0), Oleic acid (C18:1), Linoleic acid (C18:2), Linolenic acid (C18:3), Eicosenoic acid (C20:1).

[a] wt %, [b]Separation vessel 1, [c]Separation vessel 2.

by LLP, LLO, POL, and OOP [197]. Linoleic acid is the main fatty acid, followed by palmitic acid in both carrot and tomato oils [197, 204]. The fatty acid composition of tomato extract obtained by SC-CO_2 was similar to that of chloroform extract. But the extracts obtained by SC-CO_2 at different temperature and pressure conditions had different fatty acid compositions, which were due to the differences in the solubilities of linoleic and palmitic acids at different conditions [204].

3.3.4.2.2 Other Quality Attributes

The yellow-orange color of carrot oil was mainly contributed by the carotenes, which are fat-soluble pigments [197].

3.3.4.3 Comparison with Conventional Methods

Carrot oil extracted by SC-CO_2 had higher carotenes (1,850 mg/kg) than that of commercial carrot oil (170 mg/kg) [197]. It also had a high sterol content (30.2 mg/kg), which was 17-fold higher than that in commercial carrot oil (1.7 mg/kg). The squalene concentration of olive oil in the SC-CO_2 extract was 10 times higher than that obtained with solvent extraction. However, this enrichment was accompanied by a drop in the overall extracted squalene quantities [209]. SC-CO_2 extraction produced superior olive husk oil in terms of oil acidity, PV, and phosphorus content [208]; therefore, a simpler refining process would be required.

3.4 FUTURE TRENDS

The literature reviewed in this chapter demonstrates the feasibility of using SC-CO_2 for the recovery of specialty oils from a variety of plant materials. As shown, it is essential to study each plant material individually because the pretreatment of feed material and optimum extraction conditions are dependent on the structure and composition of the specific plant material. The majority of these studies have been carried out at laboratory scale, and pilot-scale SC-CO_2 extraction studies are lacking. Even though some applications have already reached commercial scale, additional pilot-scale studies would provide important data necessary for scale-up and economic feasibility assessment. For SC-CO_2 technology to be adopted more widely, its economic viability and advantages over conventional techniques must be proven for each application. Pilot-scale studies may show that despite initial high capital costs, operating costs would be lower and the overall feasibility can be proven at certain scales of operation. In addition, supercritical technology allows the possibility of coupling an extraction operation with column fractionation under supercritical conditions to further concentrate the bioactive components of interest. As well, the residual meal following extraction of specialty oils can be evaluated for other high-value end uses since degradation of meal is minimized when SC-CO_2 is used as the solvent. On the other hand, more research is needed to investigate the quality attributes of SC-CO_2-extracted specialty oils, such as oxidative stability, chemical composition, stability of bioactive components throughout extraction and storage, and the flavor profile and consumer acceptability of such oils. The advantages of SC-CO_2 extraction over conventional solvent extraction need to be better communicated to consumers.

3.5 CONCLUSIONS

Specialty oils are traditionally recovered by mechanical pressing or extraction using organic solvents. The disadvantages of these conventional techniques are the high level of residual oil in the pressed meal, undesirable solvent residue left in the product, and degradation of fat-soluble bioactive components. SC-CO_2 extraction is a promising technology that overcomes these disadvantages for the recovery of specialty oils rich in bioactive components such as carotenoids, PUFAs, squalene, sterols, and tocols from different plant sources. Extensive research carried out with a large variety of plant materials—such as nuts, seeds, cereals, fruits, and vegetables—has shown that SC-CO_2 is effective in recovering specialty oils rich in bioactive compounds. The extraction efficiency in terms of yield and recovery as well as the composition of specialty oils are affected by different factors, such as sample preparation (particle size and moisture content) and extraction parameters (temperature, pressure, solvent flow rate, extraction time, and use of a cosolvent). These parameters also have an impact on various quality attributes, such as color, flavor, and oxidative stability of the extracted oil and texture of the residual meal. In general, the color of the residual meal became lighter as more oil was removed because most of the pigments are fat soluble. SC-CO_2 extraction produced superior oil with respect to oil acidity and peroxide value. However, more research on quality attributes like oxidative stability would be beneficial to better elucidate the effect of the use of SC-CO_2 on the extraction of specialty oils. Ethanol has been used as a cosolvent in numerous studies to enhance the efficiency of extraction; however, the fact that additional heat treatment is needed to remove ethanol from the final product should not be overlooked because heat treatment can be detrimental to the sensitive bioactive components of interest.

REFERENCES

1. Physical constants of organic compounds, in *CRC Handbook of Chemistry and Physics, Internet Version 2007, (87th Edition)*, Lide, D.R., Ed., Taylor and Francis, Boca Raton, FL, 2006.
2. Kamal-Eldin, A. and Appelqvist, L., The chemistry and antioxidant properties of tocopherols and tocotrienols, *Lipids,* 31, 671, 1996.
3. Giovannucci, E., A review of epidemiologic studies of tomatoes, lycopene, and prostate cancer, *Exp. Biol. Med.,* 227, 852, 2002.

4. Groff, J.L., Gropper, S. S. and Hunt, S.M., *Advanced Nutrition and Human Metabolism.* West Pub. Co., Minneapolis, MN, p. 575, 1995.
5. Rao, A.V. and Agarwal, S., Role of antioxidant lycopene in cancer and heart disease, *J. Am. Coll. Nutr.,* 19, 563, 2000.
6. Goodman, G.E. et al., The beta-carotene and retinol efficacy trial: incidence of lung cancer and cardiovascular disease mortality during 6-year follow-up after stopping β-carotene and retinol supplements, *J. Natl. Cancer Inst.,* 96, 1743, 2004.
7. Heinonen, O.P. and Albanes, D., The effect of vitamin E and beta carotene on the incidence of lung cancer and other cancers in male smokers, *New Engl. J. Med.,* 330, 1029, 1994.
8. Rapola, J.M. et al., Randomised trial of α-tocopherol and β-carotene supplements on incidence of major coronary events in men with previous myocardial infarction, *Lancet,* 349, 1715, 1997.
9. Omenn, G.S. et al., Effects of a combination of beta carotene and vitamin A on lung cancer and cardiovascular disease, *New Engl. J. Med.,* 334, 1150, 1996.
10. Hennekens, C.H. et al., Lack of effect of long-term supplementation with beta-carotene on the incidence of malignant neoplasms and cardiovascular disease, *New Engl. J. Med.,* 334, 1145, 1996.
11. Shi, J. and Le Maguer, M., Lycopene in tomatoes: Chemical and physical properties affected by food processing, *Crit. Rev. Food Sci.,* 40, 1, 2000.
12. Klipstein-Grobusch, K. et al., Serum carotenoids and atherosclerosis: The Rotterdam Study, *Atherosclerosis,* 148, 49, 2000.
13. Rissanen, T.H. et al., Serum lycopene concentrations and carotid atherosclerosis: The Kuopio Ischaemic Heart Disease Risk Factor Study, *Am. J. Clin. Nutr.,* 77, 133, 2003.
14. Omoni, A.O. and Aluko, R.E., The anti-carcinogenic and anti-atherogenic effects of lycopene: A review, *Trends Food Sci. Technol.,* 16, 344, 2005.
15. Franceschi, S. et al., Tomatoes and risk of digestive-tract cancers, *Int. J. Cancer,* 59, 181, 1994.
16. Rao, A.V. and Shen, H., Effect of low dose lycopene intake on lycopene bioavailability and oxidative stress, *Nutr. Res.,* 22, 1125, 2002.
17. Stahl, W. and Sies, H., Uptake of lycopene and its geometrical isomers is greater from heat-processed than from unprocessed tomato juice in humans, *J. Nutr.,* 122, 2161, 1992.
18. Van Het Hof, K.H. et al., Carotenoid bioavailability in humans from tomatoes processed in different ways determined from the carotenoid response in the triglyceride-rich lipoprotein fraction of plasma after a single consumption and in plasma after four days of consumption, *J. Nutr.,* 130, 1189, 2000.
19. Porrini, M., Riso, P. and Testolin, G., Absorption of lycopene from single or daily portions of raw and processed tomato, *Brit. J. Nutr.,* 80, 353, 1998.
20. Cohn, W. et al., Comparative multiple dose plasma kinetics of lycopene administered in tomato juice, tomato soup, or lycopene tablets, *Eur. J. Nutr.,* 43, 304, 2004.
21. Center for Food Safety and Applied Nutrition, *Qualified health claims: Letter regarding tomatoes and prostate cancer (lycopene health claim coalition),* http://www.cfsan.fda.gov/~dms/qhclyco2.html, U.S. Food and Drug Administration, Rockville, MD, 2005.
22. De Lorgeril, M. et al., Alpha-linolenic acid in the prevention and treatment of coronary heart disease, *Eur. Heart J.,* 3, 2001.
23. Connor, W.E., Neuringer, M. and Reisbick, S., Essential fatty acids: The importance of n-3 fatty acids in the retina and brain, *Nutr. Rev.,* 50, 21, 1992.
24. Neuringer, M., Connor, W.E. and Lin, D.S., Biochemical and functional effects of prenatal and postnatal ω-3 fatty acid deficiency on retina and brain in rhesus monkeys, *Proc. National Academy of Sci. U.S.A.,* 83, 4021, 1986.

25. Connor, W.E., α-Linolenic acid in health and disease, *Am. J. Clin. Nutr.,* 69, 827, 1999.
26. Julius, U., Fat modification in the diabetes diet, *Exp. Clin. Endocrinol. Diabetes,* 111, 60, 2003.
27. De Lorgeril, M. et al., Mediterranean alpha-linolenic acid-rich diet in secondary prevention of coronary heart disease, *Lancet,* 343, 1454, 1994.
28. Ascherio, A. et al., Dietary fat and risk of coronary heart disease in men: Cohort follow up study in the United States, *Brit. Med. J.,* 313, 84, 1996.
29. Lanzmann-Petithory, D. et al., Primary prevention of cardiovascular diseases by α-linolenic acid (multiple letters), *Am. J. Clin. Nutr.,* 76, 1456, 2002.
30. Renaud, S.C. et al., The beneficial effect of α-linolenic acid in coronary artery disease is not questionable (multiple letters), *Am. J. Clin. Nutr.,* 76, 903, 2002.
31. Iliev, E., Tsankov, N. and Broshtilova, V., Short communication: Omega-3, -6 fatty acids in the improvement of psoriatic symptoms, *Semin. Integr. Med.,* 1, 211, 2003.
32. Leventhal, L.J., Boyce, E.G. and Zurier, R.B., Treatment of rheumatoid arthritis with gamma-linolenic acid, *Ann. Intern. Med.,* 119, 867, 1993.
33. Zurier, R.B. et al., Gamma-linolenic acid treatment of rheumatoid arthritis: A randomized, placebo-controlled trial, *Arth. Rheum.,* 39, 1808, 1996.
34. Muggli, R., Systemic evening primrose oil improves the biophysical skin parameters of healthy adults, *Int. J. Cosmetic Sci.,* 27, 243, 2005.
35. Singer, P. et al., Benefit of an enteral diet enriched with eicosapentaenoic acid and gamma-linolenic acid in ventilated patients with acute lung injury, *Crit. Care Med.,* 34, 1033, 2006.
36. Zock, P.L. and Katan, M.B., Linoleic acid intake and cancer risk: A review and meta-analysis, *Am. J. Clin. Nutr.,* 68, 142, 1998.
37. Broitman, S.A. et al., Polyunsaturated fat, cholesterol, and large bowel tumorigenesis, *Cancer,* 40, 2455, 1977.
38. Merck Research Laboratories, *The Merck Index,* 12th ed., Whitehouse Station, NJ, 1741, 1996.
39. Murakoshi, M. et al., Inhibition by squalene of the tumor-promoting activity of 12-O-tetradecanoylphorbol-13-acetate in mouse-skin carcinogenesis, *Int. J. Cancer,* 52, 950, 1992.
40. Desai, K.N., Wei, H. and Lamartiniere, C.A., The preventive and therapeutic potential of the squalene-containing compound, Roidex, on tumor promotion and regression, *Cancer Lett.,* 101, 93, 1996.
41. Rao, C.V., Newmark, H.L. and Reddy, B.S., Chemopreventive effect of squalene on colon cancer, *Carcinogenesis,* 19, 287, 1998.
42. Smith, T.J. et al., Inhibition of 4-(methylnitrosamino)-1-(3-pyridyl)-1-butanone-induced lung tumorigenesis by dietary olive oil and squalene, *Carcinogenesis,* 19, 703, 1998.
43. Scolastici, C., Ong, T.P. and Moreno, F.S., Squalene does not exhibit a chemopreventive activity and increases plasma cholesterol in a Wistar rat hepatocarcinogenesis model, *Nutr. Cancer,* 50, 101, 2004.
44. Choi, H.S. et al., Severe exogenous lipoid pneumonia following ingestion of large dose squalene: Successful treatment with steroid, *Tuber. Respir. Dis.,* 60, 235, 2006.
45. Kohno, Y. et al., Kinetic study of quenching reaction of singlet oxygen and scavenging reaction of free radical by squalene in n-butanol, *Biochim. Biophys. Acta-Lipids Lipid Metab.,* 1256, 52, 1995.
46. Lewkowicz, N. et al., Biological action and clinical application of shark liver oil, *Polski Merkuriusz Lekarski,* 20, 598, 2006.
47. Senthilkumar, S. et al., Effect of squalene on cyclophosphamide-induced toxicity, *Clin. Chim. Acta,* 364, 335, 2006.

48. Chan, P. et al., Effectiveness and safety of low-dose pravastatin and squalene, alone and in combination, in elderly patients with hypercholesterolemia, *J. Clin. Pharmacol.*, 36, 422, 1996.
49. Goodwin, T.W., Biosynthesis of sterols, in *Lipids: Structure and Function,* Vol. 4, Stumpf, P.K. and Conn, E. E., Eds., Academic Press, London, pp. 485–507, 1980.
50. Lehninger, A.L., Nelson, D.L. and Cox, M.M., Lipids, in *Principles of Biochemistry,* 2nd ed., Worth Publishers, New York, pp. 240–267, 1993.
51. Miettinen, T.A. and Vanhanen, H., Serum concentration and metabolism of cholesterol during rapeseed oil and squalene feeding, *Am. J. Clin. Nutr.,* 59, 356, 1994.
52. Smith, T.J., Squalene: Potential chemopreventive agent, *Expert Opin. Inv. Drug.,* 9, 1841, 2000.
53. Strandberg, T.E., Tilvis, R.S. and Miettinen, T.A., Metabolic variables of cholesterol during squalene feeding in humans: Comparison with cholestyramine treatment, *J. Lipid Res.,* 31, 1637, 1990.
54. Fernandez, P. and Cabral, J.M.S., Phytosterols: Applications and recovery methods—Review, *Bioresour. Technol.*, 98, 2335, 2007.
55. Piironen, V. et al., Plant sterols: Biosynthesis, biological function and their importance to human nutrition, *J. Sci. Food Agric.*, 80, 939, 2000.
56. Norméń, L. et al., Soy sterol esters and β-sitostanol ester as inhibitors of cholesterol absorption in human small bowel, *Am. J. Clin. Nutr.,* 71, 908, 2000.
57. Jones, P.J.H. et al., Cholesterol-lowering efficacy of a sitostanol-containing phytosterol mixture with a prudent diet in hyperlipidemic men, *Am. J. Clin. Nutr.,* 69, 1144, 1999.
58. Plat, J. and Mensink, R.P., Plant stanol and sterol esters in the control of blood cholesterol levels: Mechanism and safety aspects, *Am. J. Cardiol.,* 96, 2005.
59. Björkhem, I., Boberg, K.M. and Leitersdorf, E., Inborn errors in bile acid biosynthesis and storage of sterols other than cholesterol, in *The Online Metabolic and Molecular Bases of Inherited Disease,* Scriver, C.R. et al., Eds., McGraw-Hill, New York, 2002.
60. Raicht, R.F., Cohen, B.I. and Fazzini, E.P., Protective effect of plant sterols against chemically induced colon tumors in rats, *Cancer Res.,* 40, 403, 1980.
61. De Jong, A., Plat, J. and Mensink, R.P., Metabolic effects of plant sterols and stanols (Review), *J. Nutr. Biochem.,* 14, 362, 2003.
62. Bouic, P.J.D. et al., The effects of β-sitosterol (BSS) and β-sitosterol glucoside (BSSG) mixture on selected immune parameters of marathon runners: Inhibition of post marathon immune suppression and inflammation, *Int. J. Sports Med.,* 20, 258, 1999.
63. Bouic, P.J.D. and Lamprecht, J.H., Plant sterols and sterolins: A review of their immune-modulating properties, *Altern. Med. Rev.,* 4, 170, 1999.
64. Gupta, M.B., Gupta, G.P. and Bhargava, K.P., Anti-inflammatory and anti-pyretic activities of β-sitosterol, *Planta Med.,* 39, 157, 1980.
65. Ivorra, M.D. et al., Antihyperglycemic and insulin-releasing effects of β-sitosterol 3-β-D-glucoside and its aglycone, β-sitosterol, *Arch. Int. Pharmacodyn. Ther.,* 296, 224, 1988.
66. Scientific Committee on Food, General view of the scientific committee on food on the long-term effects of the intake of elevated levels of phytosterols from multiple dietary sources, with particular attention to the effects on β-carotene, vol. SCF/CS/NF/DOS/20 ADD 1 Final, European Commission, Brussels, Belgium, 2002.
67. Hallikainen, M.A., Sarkkinen, E.S. and Uusitupa, M.I.J., Plant stanol esters affect serum cholesterol concentrations of hypercholesterolemic men and women in a dose-dependent manner, *J. Nutr.,* 130, 767, 2000.
68. Law, M., Plant sterol and stanol margarines and health, *Brit. Med. J.,* 320, 861, 2000.

69. U.S. Department of Health and Human Services, *FDA authorizes new coronary heart disease health claim for plant sterol and plant stanol esters,* http://www.fda.gov/bbs/topics/ANSWERS/ANS01033.html, Food and Drug Administration, Fishers Lane, Rockville, MD, 2000.
70. Plat, J. and Mensink, R.P., Effects of diets enriched with two different plant stanol ester mixtures on plasma ubiquinol-10 and fat-soluble antioxidant concentrations, *Metabolism,* 50, 520, 2001.
71. Rupérez, F.J. et al., Chromatographic analysis of α-tocopherol and related compounds in various matrices, *J. Chromatogr. A,* 935, 45, 2001.
72. Tucker, J.M. and Townsend, D.M., Alpha-tocopherol: Roles in prevention and therapy of human disease, *Biomed. Pharmacother.,* 59, 380, 2005.
73. Insel, P., Turner, E.R. and Ross, D., Vitamins: Vital keys to health, in *Discovery Nutrition,* 2nd ed., Jones and Bartlett Publishers, Sudbury, MA, pp. 317–365, 2006.
74. Hacquebard, M. and Carpentier, Y.A., Vitamin E: Absorption, plasma transport and cell uptake, *Curr. Opin. Clin. Nutr.,* 8, 133, 2005.
75. Saldeen, K. and Saldeen, T., Importance of tocopherols beyond α-tocopherol: Evidence from animal and human studies, *Nutr. Res.,* 25, 877, 2005.
76. Jacob, R.A. and Burri, B.J., Oxidative damage and defense, *Am. J. Clin. Nutr.,* 63, 1996.
77. Abudu, N. et al., Vitamins in human arteriosclerosis with emphasis on vitamin C and vitamin E, *Clin. Chim. Acta,* 339, 11, 2004.
78. Bowry, V.W. and Stocker, R., Tocopherol-mediated peroxidation. The prooxidant effect of vitamin E on the radical-initiated oxidation of human low-density lipoprotein, *J. Am. Chem. Soc.,* 115, 6029, 1993.
79. Witting, P.K., Bowry, V.W. and Stocker, R., Inverse deuterium kinetic isotope effect for peroxidation in human low-density lipoprotein (LDL): A simple test for tocopherol-mediated peroxidation of LDL lipids, *FEBS Lett.,* 375, 45, 1995.
80. Institute of Medicine, Food, and Nutrition Board, *Dietary reference intakes for vitamin C, vitamin E, selenium, and carotenoids,* National Academy Press, Washington, D.C., 2000.
81. Naito, Y., Shimozawa, M. and Yoshikawa, T., Tocotrienols and atherosclerosis, *J. Clin. Biochem. Nutr.,* 34, 121, 2004.
82. Baliarsingh, S., Beg, Z.H. and Ahmad, J., The therapeutic impacts of tocotrienols in type 2 diabetic patients with hyperlipidemia, *Atherosclerosis,* 182, 367, 2005.
83. Teupser, D., Thiery, J. and Seidel, D., α-Tocopherol down-regulates scavenger receptor activity in macrophages, *Atherosclerosis,* 144, 109, 1999.
84. Devaraj, S., Hugou, I. and Jialal, I., α-Tocopherol decreases CD36 expression in human monocyte-derived macrophages, *J. Lipid Res.,* 42, 521, 2001.
85. Vivekananthan, D.P. et al., Use of antioxidant vitamins for the prevention of cardiovascular disease: Meta-analysis of randomised trials, *Lancet,* 361, 2017, 2003.
86. Knekt, P. et al., Antioxidant vitamin intake and coronary mortality in a longitudinal population study, *Am. J. Epidemiol.,* 139, 1180, 1994.
87. Kushi, L.H. et al., Dietary antioxidant vitamins and death from coronary heart disease in postmenopausal women, *New Engl. J. Med.,* 334, 1156, 1996.
88. Stephens, N.G. et al., Randomised controlled trial of vitamin E in patients with coronary disease: Cambridge Heart Antioxidant Study (CHAOS), *Lancet,* 347, 781, 1996.
89. Virtamo, J. et al., Effect of vitamin E and beta carotene on the incidence of primary nonfatal myocardial infarction and fatal coronary heart disease, *Arch. Intern. Med.,* 158, 668, 1998.
90. Tognoni, G. et al., Low-dose aspirin and vitamin E in people at cardiovascular risk: A randomised trial in general practice, *Lancet,* 357, 89, 2001.

91. Valagussa, F. et al., Dietary supplementation with n-3 polyunsaturated fatty acids and vitamin E after myocardial infarction: Results of the GISSI-Prevenzione trial, *Lancet,* 354, 447, 1999.
92. Yusuf, S., Vitamin E supplementation and cardiovascular events in high-risk patients, *New Engl. J. Med.,* 342, 154, 2000.
93. Shklar, G. et al., Regression by vitamin E of experimental oral cancer, *J. National Cancer Inst.,* 78, 987, 1987.
94. Krol, E.S., Kramer-Stickland, K.A. and Liebler, D.C., Photoprotective action of topically applied vitamin E, *Drug Metab. Rev.,* 32, 413, 2000.
95. Neuzil, J. et al., Induction of cancer cell apoptosis by α-tocopheryl succinate: Molecular pathways and structural requirements, *FASEB J.,* 15, 403, 2001.
96. Woutersen, R.A., Appel, M.J. and Van Garderen-Hoetmer, A., Modulation of pancreatic carcinogenesis by antioxidants, *Food Chem. Toxicol.,* 37, 981, 1999.
97. Shklar, G. and Schwartz, J., Tumor necrosis factor in experimental cancer regression with alpha-tocopherol, beta-carotene, canthaxanthin and algae extract, *Eur. J. Cancer Clin. Oncol.,* 24, 839, 1988.
98. Taylor, P.R. and Greenwald, P., Nutritional interventions in cancer prevention. *J. Clin. Oncol.,* 23, 333, 2005.
99. Caraballoso, M. et al., Drugs for preventing lung cancer in healthy people. *Cochrane Database Syst. Rev.,* 2003.
100. Israel, K. et al., Vitamin E succinate induces apoptosis in human prostate cancer cells: Role for Fas in vitamin E succinate–triggered apoptosis, *Nutr. Cancer,* 36, 90, 2000.
101. Campbell, S.E. et al., Comparative effects of RRR-alpha- and RRR-gamma-tocopherol on proliferation and apoptosis in human colon cancer cell lines, *BMC Cancer,* 6, 13, 2006.
102. Moyad, M.A., Brumfield, S.K. and Pienta, K.J., Vitamin E, alpha- and gamma-tocopherol, and prostate cancer, *Semin. Urol. Oncol.,* 17, 85, 1999.
103. Hirohata, K. et al., Treatment of osteoarthritis of the knee joint at the state of hydroarthrosis, *Kobe Med. Sci.,* 11(supplement), 65, 1965.
104. Doumerg, C., Etude clinique experimentale de l'alpha-tocopheryle-quinone en rheumatologie et en reeducation, *Therapeutique,* 45, 676, 1969.
105. Machtey, I. and Ouaknine, L., Tocopherol in osteoarthritis: A controlled pilot study, *J. Am. Geriatr. Soc.,* 26, 328, 1978.
106. Blankenhorn, G., Clinical efficacy of Spondyvit® (vitamin E) in activated arthroses. A multicenter, placebo-controlled, double-blind study, *Zeitschrift fur Orthopadie und Ihre Grenzgebiete,* 124, 340, 1986.
107. Scherak, O. et al., Therapy with high doses of vitamin E in patients with osteoarthritis, *Zeitschrift fur Rheumatologie,* 49, 369, 1990.
108. Jordan, J.M. et al., A case-control study of serum tocopherol levels and the alpha- to gamma-tocopherol ratio in radiographic knee osteoarthritis: The Johnston County Osteoarthritis Project, *Am. J. Epidemiol.,* 159, 968, 2004.
109. Grundman, M., Vitamin E and Alzheimer disease: The basis for additional clinical trials, *Am. J. Clin. Nutr.,* 71, 2000.
110. Tabet, N., Birks, J. and Grimley, E.J., Vitamin E for Alzheimer's disease. *Cochrane Database Syst. Rev.,* 2000.
111. Luchsinger, J.A. et al., Antioxidant vitamin intake and risk of Alzheimer disease, *Arch. Neurol-Chicago,* 60, 203, 2003.
112. Kalmijn, S. et al., Polyunsaturated fatty acids, antioxidants, and cognitive function in very old men, *Am. J. Epidemiol.,* 145, 33, 1997.
113. Meydani, S.N., Han, S.N. and Hamer, D.H., Vitamin E and respiratory infection in the elderly, *Ann. N.Y. Acad. Sci.,* 1031, 214, 2004.

114. Center for Food Safety and Applied Nutrition, *Qualified health claims subject to enforcement discretion,* http://www.cfsan.fda.gov/~dms/qhc-sum.html, U.S. Food and Drug Administration, Rockville, MD, 2005.
115. Center for Food Safety and Applied Nutrition, *Letter regarding dietary supplement health claim for vitamin E and heart disease,* U.S. Food and Drug Administration, Rockville, MD, 2001. http://www.cfsan.fda.gov/~dms/ds-ltr16.html
116. Trumbo, P.R., The level of evidence for permitting a qualified health claim: FDA's review of the evidence for selenium and cancer and vitamin E and heart disease, *J. Nutr.,* 135, 354, 2005.
117. Horwitt, M.K., Critique of the requirement for vitamin E, *Am. J. Clin. Nutr.,* 73, 1003, 2001.
118. Miller III, E.R. et al., Meta-analysis: High-dosage vitamin E supplementation may increase all-cause mortality, *Ann. Intern. Med.,* 142, 2005.
119. Güclü-Üstündağ, O. and Temelli, F., Correlating the solubility behavior of minor lipid components in supercritical carbon dioxide, *J. Supercrit. Fluids,* 31, 235, 2004.
120. Güclü-Üstündağ, O. and Temelli, F., Correlating the solubility behavior of fatty acids, mono-, di-, and triglycerides, and fatty acid esters in supercritical carbon dioxide, *Ind. Eng. Chem. Res.,* 39, 4756, 2000.
121. Lopes, I.M.G. and Bernardo-Gil, M.G., Characterisation of acorn oils extracted by hexane and by supercritical carbon dioxide, *Eur. J. Lipid Sci. Technol.,* 107, 12, 2005.
122. Femenia, A. et al., Effects of supercritical carbon dioxide (SC-CO_2) oil extraction on the cell wall composition of almond fruits, *J. Agric. Food Chem.,* 49, 5828, 2001.
123. Passey, C.A. and Gros-Lois, M., Production of calorie-reduced almonds by supercritical extraction, *J. Supercrit. Fluids,* 6, 255, 1993.
124. Leo, L. et al., Supercritical carbon dioxide extraction of oil and α-tocopherol from almond seeds, *J. Sci. Food Agric.,* 85, 2167, 2005.
125. Marrone, C. et al., Almond oil extraction by supercritical CO_2: Experiments and modelling, *Chem. Eng. Sci.,* 53, 3711, 1998.
126. Özkal, S.G. et al., Response surfaces of hazelnut oil yield in supercritical carbon dioxide, *Eur. Food Res. and Technol.,* 220, 74, 2005.
127. Özkal, S.G., Salgin, U. and Yener, M.E., Supercritical carbon dioxide extraction of hazelnut oil, *J. Food Eng.,* 69, 217, 2005.
128. Bernardo-Gil, M.G. et al., Supercritical fluid extraction and characterisation of oil from hazelnut, *Eur. J. Lipid Sci. Technol.,* 104, 402, 2002.
129. Goodrum, J.W. and Kilgo, M.B., Peanut oil extraction using compressed CO_2, *Energy Agric.,* 6, 265, 1987.
130. Goodrum, J.W. and Kilgo, M.B., Peanut oil extraction with SC CO_2: Solubility and kinetic functions, *Trans. ASAE.,* 30, 1865, 1987.
131. Goodrum, J.W. and Kilgo, M.B., Modeling of liquid CO_2 extraction of peanut oil in a fixed bed, *Trans. ASAE,* 31, 926, 1988.
132. Chiou, R.R.Y. et al., Partial defatting of roasted peanut meals and kernels by supercritical CO_2 using semicontinuous and intermittently depressurized processes, *J. Agric. Food Chem.,* 44, 574, 1996.
133. Passey, C.A. and Patil, N.D., *Process for preparing low-calorie nuts,* U.S. Patent 5290578, 1994.
134. Zhang, C. et al., Feasibility of using supercritical carbon dioxide for extracting oil from whole pecans, *Trans. ASAE,* 38, 1763, 1995.
135. Li, M., Bellmer, D.D. and Brusewitz, G.H., Pecan kernel brekage and oil extracted by supercritical CO_2 as affected by moisture content, *J. Food Sci.,* 64, 1084, 1999.
136. Alexander, W.S., Brusewitz, G.H. and Maness, N.O., Pecan oil recovery and composition as affected by temperature, pressure, and supercritical CO_2 flow rate, *J. Food Sci.,* 62, 762, 1997.

137. Palazoglu, T.K. and Balaban, M.O., Supercritical CO_2 extraction of lipids from roasted pistachio nuts, *Trans. ASAE,* 41, 679, 1998.
138. Oliveira, R., Rodrigues, M.F. and Bernardo-Gil, M.G., Characterization and supercritical carbon dioxide extraction of walnut oil, *J. Am. Oil Chem. Soc.,* 79, 225, 2002.
139. Crowe, T.D. and White, P.J., Oxidation, flavor, and texture of walnuts reduced in fat content by supercritical carbon dioxide, *J. Am. Oil Chem. Soc.,* 80, 569, 2003.
140. Crowe, T.D. and White, P.J., Oxidative stability of walnut oils extracted with supercritical carbon dioxide, *J. Am. Oil Chem. Soc.,* 80, 575, 2003.
141. Santerre, C.R., Goodrum, J.W. and Kee, J.M., Roasted peanuts and peanut butter quality are affected by supercritical fluid extraction, *J. Food Sci.,* 59, 382, 1994.
142. Turpin, P.E., Coxon, D.T. and Padley, F.B., The fractionation of shea nut oil using supercritical carbon dioxide: A feasibility study for the extraction of oil from gum, *Fett Wiss. Technol.,* 92, 179, 1990.
143. Özkal, S.G., Yener, M.E. and Bayindirli, L., The solubility of apricot kernel oil in supercritical carbon dioxide, *Int. J. Food Sci. Technol.,* 41, 399, 2006.
144. Özkal, S.G., Yener, M.E. and Bayindirli, L., Response surfaces of apricot kernel oil yield in supercritical carbon dioxide, *Lebensm.–Wiss. Technol.,* 38, 611, 2005.
145. Özkal, S.G., Yener, M.E. and Bayindirli, L., Mass transfer modeling of apricot kernel oil extraction with supercritical carbon dioxide, *J. Supercrit. Fluids,* 35, 119, 2005.
146. Dauksas, E., Venskutinis, P.R. and Sivik, B., Supercritical fluid extraction of borage (*Borago officinalis* L.) seeds with pure CO_2 and its mixture with caprylic acid methyl ester, *J. Supercrit. Fluids,* 22, 211, 2002.
147. Gomez, A.M. and de la Ossa, E.M., Quality of borage seed oil extracted by liquid and supercritical carbon dioxide, *Chem. Eng. J.,* 88, 103, 2002.
148. Gaspar, F. et al., Solubility of echium, borage, and lunaria seed oils in compressed CO_2, *J. Chem. Eng. Data,* 48, 107, 2003.
149. Bernardo-Gil, G. et al., Extraction of lipids from cherry seed oil using supercritical carbon dioxide, *Eur. Food Res. Technol.,* 212, 170, 2001.
150. Gawdzik, J. et al., Supercritical fluid extraction of oil from evening primrose (*Oenothera paradoxa* H.) seeds, *Chemia Analityczna,* 43, 695, 1998.
151. Favati, F., King, J.W. and Mazzanti, M., Supercritical carbon dioxide extraction of evening primrose oil, *J. Am. Oil Chem. Soc.,* 68, 422, 1991.
152. King, J.W., Cygnarowicz, M.L. and Favati, F., Supercritical fluid extraction of evening primrose oil kinetic and mass transfer effects, *Ital. J. Food Sci.,* 9, 193, 1997.
153. Bozan, B. and Temelli, F., Supercritical CO_2 extraction of flaxseed, *J. Am. Oil Chem. Soc.,* 79, 231, 2002.
154. Beveridge, T.H.J. et al., Yield and composition of grape seed oils extracted by supercritical carbon dioxide and petroleum ether: Varietal effects, *J. Agric. Food Chem.,* 53, 1799, 2005.
155. Sovova, H., Kucera, J. and Jez, J., Rate of the vegetable oil extraction with supercritical CO_2 -II. Extraction of grape oil, *Chem. Eng. Sci.,* 49, 415, 1994.
156. Gomez, A.M., Lopez, C.P. and de la Ossa, E. M., Recovery of grape seed oil by liquid and supercritical carbon dioxide extraction : a comparison with conventional solvent extraction, *Chem. Eng. J.,* 61, 227, 1996.
157. Murga, R. et al., Extraction of natural complex phenols and tannins from grape seeds by using supercritical mixtures of carbon dioxide and alcohol, *J. Agr. Food Chem.,* 48, 3408, 2000.
158. Cao, X.L. and Ito, Y.C., Supercritical fluid extraction of grape seed oil and subsequent separation of free fatty acids by high-speed counter-current chromatography, *J. Chromatogr. A,* 1021, 117, 2003.
159. Reverchon, E., Kaziunas, A. and Marrone, C., Supercritical CO_2 extraction of hiprose seed oil: Experiments and mathematical modelling, *Chem. Eng. Sci.,* 55, 2195, 2000.

160. Holser, R.A. and Bost, G., *Hybrid hibiscus* seed oil compositions, *J. Am. Oil Chem. Soc.*, 81, 795, 2004.
161. Hadolin, M. et al., High pressure extraction of vitamin E–rich oil from *Silybum marianum*, *Food Chem.*, 74, 355, 2001.
162. Muuse, B.G., Cuperus, F.P. and Derksen, J.T.P., Extraction and characterization of *Dimorphotheca pluvialis* seed oil, *J. Am. Oil Chem. Soc.*, 71, 313, 1994.
163. Bernardo-Gil, M.G. and Lopes, L.M.C., Supercritical fluid extraction of *Cucurbita ficifolia* seed oil, *Eur. Food Res. Technol.*, 219, 593, 2004.
164. Szentmihalyi, K. et al., Rosehip (*Rosa canina* L.) oil obtained from waste hip seeds by different extraction methods, *Bioresour. Technol.*, 82, 195, 2002.
165. del Valle, J. M. and Uquiche, E.L., Particle size effects on supercritical CO_2 extraction of oil-containing seeds, *J. Am. Oil Chem. Soc.*, 79, 1261, 2002.
166. del Valle, J. M. et al., Comparison of conventional and supercritical CO_2-extracted rosehip oil, *Brazil. J. Chem. Eng.*, 17, 335, 2000.
167. Eggers, R., Ambrogi, A. and von Schnitzler, J., Special features of SCF solid extraction of natural products: Deoiling of wheat gluten and extraction of rosehip oil, *Brazil. J. Chem. Eng.*, 17, 329, 2000.
168. del Valle, J.M. et al., Supercritical CO_2 processing of pretreated rosehip seeds: effect of process scale on oil extraction kinetics, *J. Supercrit. Fluids,* 31, 159, 2004.
169. Stastova, J. et al., Rate of the vegetable oil extraction with supercritical CO_2 - III. Extraction from sea buckthorn, *Chem. Eng. Sci.*, 51, 4347, 1996.
170. Yin, J. et al., Modeling of supercritical fluid extraction from *Hippophae rhamnoides* L. seeds, *Sep. Sci. Technol.*, 38, 4041, 2003.
171. Yin, J.Z. et al., Experiments and numerical simulations of supercritical fluid extraction for *Hippophae rhamnoides* L. seed oil based on artificial neural networks, *Ind. Eng. Chem. Res.*, 44, 7420, 2005.
172. Odabasi, A.Z. and Balaban, M.O., Supercritical CO_2 extraction of sesame oil from raw seeds, *J. Food Sci. Technol.*, 39, 496, 2002.
173. Roy, B.C., Goto, M. and Hirose, T., Temperature and pressure effects on supercritical CO_2 extraction of tomato seed oil, *Int. J. Food Sci. Technol.*, 31, 137, 1996.
174. Illés, V. et al., Extraction of hiprose fruit by supercritical CO_2 and propane, *J. Supercrit. Fluids,* 10, 209, 1997.
175. Yakimishen, R., Cenkowski, S. and Muir, W.E., Oil recoveries from sea buckthorn seeds and pulp, *Appl. Eng. Agric.*, 21, 1047, 2005.
176. Mongkholkhajornsilp, D. et al., Supercritical CO_2 extraction of nimbin from neem seeds - A modelling study, *J. Food Eng.*, 71, 331, 2005.
177. Yin, J.Z. et al., Analysis of the operation conditions for supercritical fluid extraction of seed oil, *Sep. Purif. Technol.*, 43, 163, 2005.
178. Westerman, D. et al., Extraction of Amaranth seed oil by supercritical carbon dioxide, *J. Supercrit. Fluids,* 37, 38, 2006.
179. He, H.P., Corke, H. and Cai, J.G., Supercritical carbon dioxide extraction of oil and squalene from *Amaranthus* grain, *J. Agric. Food Chem.*, 51, 7921, 2003.
180. Bruni, R. et al., Wild *Amaranthus caudatus* seed oil, a nutraceutical resource from Ecuadorian flora, *J. Agric. Food Chem.*, 49, 5455, 2001.
181. Fors, S.M. and Eriksson, C.E., Characterization of oils extracted from oats by supercritical carbon dioxide, *Lebens.–Wiss. Technol.*, 23, 390, 1990.
182. Zhao, W. et al., Fractional extraction of rice bran oil with supercritical carbon dioxide, *Agric. Biol. Chem.*, 51, 1773, 1987.
183. Shen, Z. et al., Pilot scale extraction of rice bran oil with dense carbon dioxide, *J. Agr. Food Chem.*, 44, 3033, 1996.
184. Shen, Z. et al., Pilot scale extraction and fractionation of rice bran oil using supercritical carbon dioxide, *J. Agric. Food Chem.*, 45, 4540, 1997.

185. Kim, H.J. et al., Characterization of extraction and separation of rice bran oil rich in EFA using SFE process, *Sep. Purif. Technol.*, 15, 1, 1999.
186. Perretti, G. et al., Improving the value of rice by-products by SFE, *J. Supercrit. Fluids,* 26, 63, 2003.
187. Danielski, L. et al., A process line for the production of raffinated rice oil from rice bran, *J. Supercrit. Fluids,* 34, 133, 2005.
188. Gomez, A.M. and de la Ossa, E.M., Quality of wheat germ oil extracted by liquid and supercritical carbon dioxide, *J. Am. Oil Chem. Soc.,* 77, 969, 2000.
189. Ge, Y. et al., Extraction of natural vitamin E from wheat germ by supercritical carbon dioxide, *J. Agric. Food Chem.,* 50, 685, 2002.
190. Panfili, G. et al., Extraction of wheat germ oil by supercritical CO_2: Oil and defatted cake characterization, *J. Am. Oil Chem. Soc.,* 80, 157, 2003.
191. Taniguchi, M. et al., Extraction of oils from wheat germ with supercritical carbon dioxide, *Agric. Biol. Chem.,* 49, 2367, 1985.
192. Zhang, X.W. et al., Supercritical carbon dioxide extraction of wheat plumule oil, *J. Food Eng.,* 37, 103, 1998.
193. Dunford, N.T. and King, J.W., Phytosterol enrichment of rice bran oil by a supercritical carbon dioxide fractionation technique, *J. Food Sci.,* 65, 1395, 2000.
194. de Franca, L. F. et al., Supercritical extraction of carotenoids and lipids from buriti (*Mauriti flexuosa*), a fruit from the Amazon region, *J. Supercrit. Fluids,* 14, 247, 1999.
195. Saldaña, M.D.A. et al., Comparison of the solubility of β-carotene in supercritical CO_2 based on a binary and a multicomponent complex system, *J. Supercrit. Fluids,* 37, 342, 2006.
196. Goto, M., Sato, M. and Hirose, T., Supercritical carbon dioxide extraction of carotenoids from carrots, in Yano, T., Matsuno, R. and Nakamura, K., Eds., *Developments in Food Engineering, Proceedings of the Sixth International Congress on Engineering and Food,* Blackie Academic & Professional, London, 1994, 835–837.
197. Ranalli, A. et al., Characterization of carrot root oil arising from supercritical fluid carbon dioxide extraction, *J. Agric. Food Chem.,* 52, 4795, 2004.
198. Sun, M. and Temelli, F., Supercritical carbon dioxide extraction of carotenoids from carrot using canola oil as a continuous co-solvent, *J. Supercrit. Fluids,* 37, 397, 2006.
199. Manninen, P., Pakarinen, J. and Kallio, H., Large-scale supercritical carbon dioxide extraction and supercritical carbon dioxide countercurrent extraction of cloudberry seed oil, *J. Agr. Food Chem.,* 45, 2533, 1997.
200. Esquivel, M.M., Bernardo-Gil, M.G. and King, M.B., Mathematical models for supercritical extraction of olive husk oil, *J. Supercrit. Fluids,* 16, 43, 1999.
201. Gomez-Prieto, M.S., Caja, M.M. and Santa-Maria, G., Solubility in supercritical carbon dioxide of the predominant carotenes in tomato skin, *J. Am. Oil Chem. Soc.,* 79, 897, 2002.
202. Ollanketo, M. et al., Supercritical carbon dioxide extraction of lycopene in tomato skins, *Eur. Food Res. Technol.,* 212, 561, 2001.
203. Shi, J., *Separation of carotenoids from fruits and vegetables,* European Patent Appl. WO0179355, 2001.
204. Rozzi, N.L. et al., Supercritical fluid extraction of lycopene from tomato processing byproducts, *J. Agric. Food Chem.,* 50, 2638, 2002.
205. Gomez-Prieto, M.S. et al., Supercritical fluid extraction of all *trans*-lycopene from tomato, *J. Agric. Food Chem.,* 51, 3, 2003.
206. Ruiz del Castillo, M.L. et al., Lipid composition in tomato skin supercritical fluid extracts with high lycopene content, *J. Am. Oil Chem. Soc.,* 80, 271, 2003.
207. Vasapollo, G. et al., Innovative supercritical CO_2 extraction of lycopene from tomato in the presence of vegetable oil as co-solvent, *J. Supercrit. Fluids,* 29, 87, 2004.

208. de Lucas, A., Rincon, J. and Gracia, I., Influence of operating variables on yield and quality parameters of olive husk oil extracted with supercritical carbon dioxide, *J. Am. Oil Chem. Soc.*, 79, 237, 2002.
209. Stavroulias, S. and Panayiotou, C., Determination of optimum conditions for the extraction of squalene from olive pomace with supercritical CO_2, *Chem. Biochem. Eng. Q.*, 19, 373, 2005.

4 Extraction and Purification of Natural Tocopherols by Supercritical CO_2

Tao Fang, Motonobu Goto, Mitsuru Sasaki, and Dalang Yang

CONTENTS

4.1 Background ... 104
4.1.1 Problems with Concentrating Tocopherols Using Molecular Distillation ... 104
4.1.2 Pretreatment before Concentrating Tocopherol ... 105
4.1.3 Fundamental Research on Concentrating Tocopherols ... 106
4.2 Main Experimental Materials ... 107
4.3 Analytical Methods ... 107
4.4 Correlation for Experimental Data ... 107
4.5 Binary Phase Equilibria ... 107
4.5.1 Apparatus and Procedure ... 107
4.5.2 High Pressure View Cell ... 110
4.5.3 Phase Equilibrium Properties ... 110
4.5.4 Solubility ... 113
4.5.5 Distribution Coefficient ... 113
4.6 Ternary Phase Equilibria ... 118
4.6.1 Apparatus and Procedure ... 118
4.6.2 Influences of Pressure and Temperature on Phase Equilibrium ... 120
4.6.3 Separation Factor between Tocopherol and Methyl Oleate ... 122
4.6.4 Equilibrium Lines ... 123
4.6.5 Phase Behavior of ME-DOD ... 124
4.7 Separation with Supercritical CO_2 Fractionation ... 126
4.7.1 Fractionation Apparatus and Procedure ... 127
4.7.2 Pretreatment Result and Composition of ME-DOD ... 129
4.7.3 Effect of the Initial Pressure ... 131
4.7.4 Effect of the Final Pressure ... 132
4.7.5 Composition of Tocopherol Concentrate ... 134

4.7.6 Viscosity Comparison 134
4.7.7 Application in Commercial Production 136
4.8 Conclusions 136
References 138

4.1 BACKGROUND

Tocopherols, commonly known as vitamin E, are known for antioxidative activity which has been widely applied in the fields of food, medicine, and cosmetics. Figure 4.1 illustrates the molecular structures of tocopherols. The main source of natural tocopherols is deodorizer distillate (DOD), a byproduct of the edible oil refining process that is rich in tocopherols and sterols [1].

4.1.1 Problems with Concentrating Tocopherols Using Molecular Distillation

Molecular distillation, also called short-path distillation, has been applied to the commercial production of tocopherols from DOD [2–5]. It is characterized by high vacuum in the distillation space, short exposure of the distilled liquid to the operating temperatures, and short distance between the evaporator and the condenser (20 to 70 mm) [3].

However, on the basis of project investigation, we found some problems in the commercial production of tocopherols using molecular distillation. First, the process of molecular distillation is generally performed with multistage distillators (3 to 5 units) at high vacuum (0.1 to 10 Pa) and high temperature (433 to 503 K). Noticeably, a high quality vacuum pump is absolutely indispensable for ensuring enough vacuum condition for each distillator. Also, it is obvious that the high quality pump used for commercial production, the distillator, with fine structure and strict operation conditions, leads to relatively high equipment investment and operation cost.

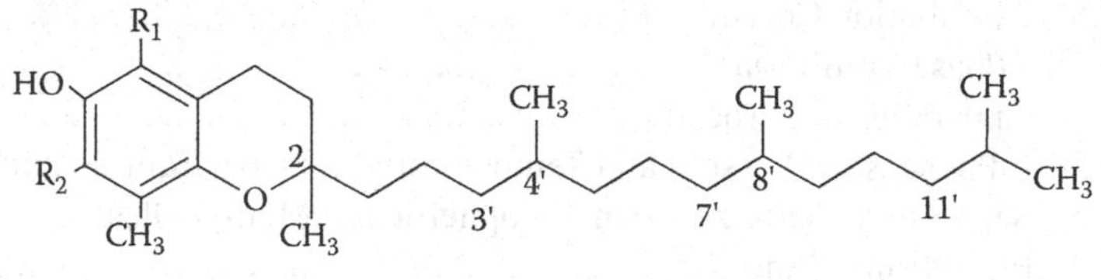

Substituents		Notation
R_1	R_2	
CH_3	CH_3	α-tocopherol
CH_3	H	β-tocopherol
H	CH_3	γ-tocopherol
H	H	δ-tocopherol

FIGURE 4.1 Molecular structure of tocopherols. (From Brunner, G., *Gas Extraction: An Introduction to Fundamentals of Supercritical Fluids and the Application to Separation Processes,* Darmstadt: Sternkopff, Springer, New York, 1994. With permission.)

Second, according to our communication with some companies employing molecular distillation, it is generally difficult to keep all pressures in the distillator constant at low levels (0.1 to 10 Pa) during the long operation. Consequently, the operation temperature has to be increased to compensate for the decrease in vacuum degree with a view to stabilizing the concentration of the fractions. As a result, such fluctuations in temperature and pressure lead to an unstable quality of the tocopherol product. In addition, the definition of molecular distillation is not very accurate because there is no effect of rectification caused by fraction condensation and refluence. As we know, the effect of rectification is the main difference between distillation and vaporization. Furthermore, the phenomenon of entrainment probably occurs without rectification and influences on the separation selectivity.

Third, the general opinion is that the short contact or residence time of the product at high temperatures during molecular distillation does not cause any degradation of the product and does not affect the quality of the product. However, Mau [5] researched on concentrating tocopherols by molecular distillation and reported the existence of tocopherol dimmers and other degradation products at 433 to 493 K, even though pressure was lower than 0.133 Pa, and the total recovery of tocopherols in all fractions and residue was only about 75.16% of the initial amount of tocopherols in feed.

Finally, in the commercial operation of molecular distillation with three to five distillation units, the residence time of tocopherols, the main compound in the second or third fraction, is no less than 1 hour, which is not a short time for a separation operation.

Because thermal degradation of tocopherols is commonly caused by high processing temperature [6], development of new alternative isolation techniques, including supercritical fluid extraction (SFE), has been desired.

4.1.2 Pretreatment before Concentrating Tocopherol

To modify the composition of raw material, a process of pretreatment is generally necessary. Pretreatment involves two steps (esterification and methanolysis) that convert free fatty acids (FFAs) and glycerides (Gly.), respectively, into fatty acid methyl esters (FAMEs). The main objective is to modify the composition of DOD and to increase the solubility of soybean DOD in supercritical carbon dioxide (CO_2) extraction. In the published literature on chemical modification of DOD, esterification was carried out with the catalysts sulfuric acid (H_2SO_4) [5, 7–9, 12], hydrochloric acid (HCL) [10], or Na [11]. Additionally, some researchers added a second reaction of methanolysis with the catalysts sodium methoxide ($NaOCH_3$) [8, 12] and sodium hydroxide (NaOH) [9]. In our lab work, H_2SO_4 and $NaOCH_3$ were selected as the catalysts of methyl esterification and methanolysis, respectively. Figure 4.2 shows the pretreatment process of DOD. After each reaction, the mixture was washed with hot water until it became neutral. Finally, the mixture was held at low temperature and, as a result, most of the sterols crystallized and were removed by filtration. After pretreatment, the oil, methyl esterified DOD (ME-DOD), was obtained, which contains mainly FAMEs (70% to 80%), tocopherols (10% to 15%), and impurities (such as residual sterols, glycerides, squalene, pigments, and long chain paraffins, comprising in total about 10% to 15%).

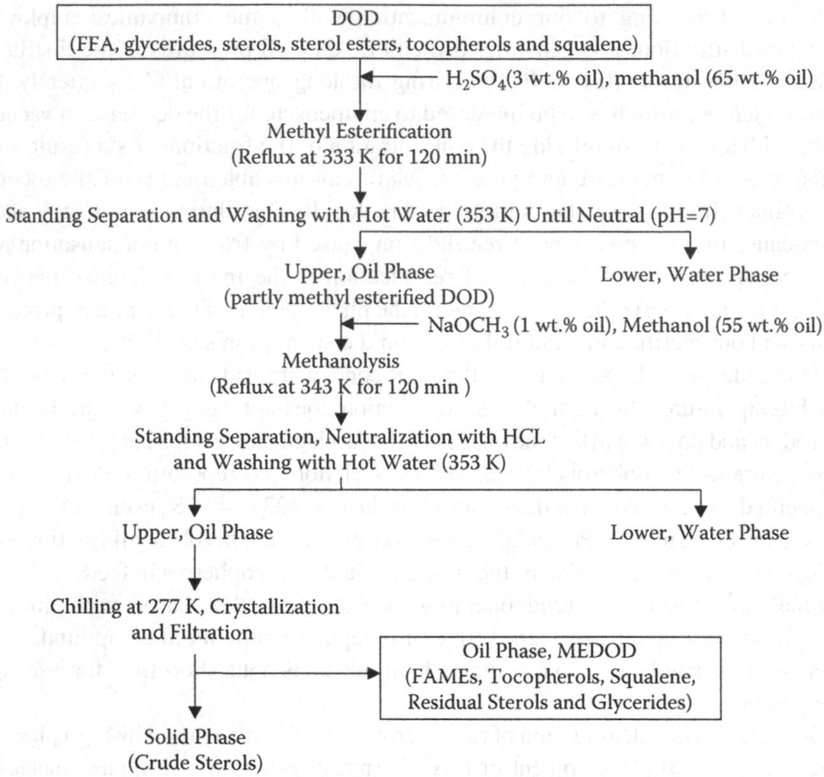

FIGURE 4.2 Pretreatment process for preparing ME-DOD from DOD. (From Fang, T., Goto, M., Wang, X., Ding, X., Geng, J., Sasaki, M. and Hirose, T., *J. Supercritical Fluids,* 40, 50, 2007. With permission.)

4.1.3 Fundamental Research on Concentrating Tocopherols

Supercritical CO_2 is relatively suitable as a separating solvent for extracting some fat-soluble components. Although some researchers tried to concentrate tocopherols from DOD by supercritical CO_2 [6–16], the operation parameters, especially pressure, vary. For example, Zhao et al. [7] concentrated 75% tocopherols at 12 MPa, whereas Lee et al. [10] reported that 40 MPa could be used to obtain 40% extract. King et al. [14] combined SFE with supercritical fluid chromatography (SFC) for concentrating tocopherols, and the optimized conditions were 25 MPa/353 K for SFE and 25 MPa/313 K for SFC. Generally, if pressure is low, satisfactory selectivity can be obtained but productivity is low. Contrarily, higher pressure leads to lower selectivity and higher productivity. To find more reasonable operating conditions, phase equilibria of ME-DOD + CO_2 must be clarified.

When concentrating natural tocopherols from ME-DOD, the important step is to remove FAMEs, which contribute more than 70% of ME-DOD. To explore the reasonable operation conditions for this step, the complex system of ME-DOD + CO_2 was initially regarded as a pseudo-ternary (methyl oleate + tocopherol + CO_2) system. The reason for choosing methyl oleate is that the ME-DOD applied in our research is from the soybean oil industry and the main FAMEs are methyl oleate and methyl linoleate (about 75% to 80% of all FAMEs); the latter is similar to methyl oleate in physicochemical properties [17].

As far as phase equilibrium is concerned, two binary systems of α-tocopherol + CO_2 and methyl oleate + CO_2 were measured and correlated. Then the ternary system (methyl oleate + tocopherol + CO_2) and the realistic system (ME-DOD + CO_2) were investigated. Finally, the phase equilibrium data were analyzed and accordingly the separation of natural tocopherols from ME-DOD was carried out with supercritical CO_2 fractionation.

4.2 MAIN EXPERIMENTAL MATERIALS

CO_2 was supplied by the Uchimura Sanso Co., Ltd. (Osaka, Japan), with a purity of 99.97%. Methyl oleate and Dl-α-tocopherol were obtained from Wako Pure Chemical Industries, Inc. (Tokyo, Japan) with purities of ≥ 98%. DOD (9.23% tocopherols) was supplied by Kaidi Fine Chemical Industrial Co., Ltd. (Wuhan, Hubei Province, P. R. China), and its composition is illustrated in Table 4.1. ME-DOD (10.19% tocopherols) was prepared by DOD according to the procedure shown in Figure 4.2.

4.3 ANALYTICAL METHODS

The approximate contents of FFAs and glycerides, including monoglycerides, diglycerides, and triglycerides) were calculated from acid and saponification values (A.V. and S.V., respectively) and expressed as the contents of oleic acid and triolein [18].

Analyses of tocopherols, sterols, and FAMEs were performed with high performance liquid chromatography (HPLC) and gas chromatograph with flame ionization detector (GC-FID), respectively [19–21]. The composition of ME-DOD and tocopherol concentrate was determined with gas chromatograph-mass spectrometry (GC-MS). Additionally, the viscosity of the samples obtained in the separation experiment were measured with an AR1000 rheology meter (TA Instruments Co., Ltd, England) [21].

4.4 CORRELATION FOR EXPERIMENTAL DATA

The Soave-Redlich-Kwong (SRK) EOS [22] with the Adachi-Sugie (AS) mixing rule [23] was used to correlate the experimental data. The SRK EOS is the modification of the simple Redlich-Kwong EOS, with which the vapor pressure curve can be reproduced well. This procedure of correlation was completed by PE 2000, which was developed by Pfohl, Petkov, and Brunner and contains some common EOS and proved capability to obtain a convergence similar to that from ASPEN software [24].

4.5 BINARY PHASE EQUILIBRIA

4.5.1 Apparatus and Procedure

The experimental apparatus used in this work is called "gas-liquid alternating circulation system," which is used to simultaneously measure the compositions in both

TABLE 4.1
Characteristics of DOD

A.V.* (mg KOH/g)	S.V.* (mg KOH/g)	FFA* (%)	Gly.* (%)	Tocopherols (%)	Isomer Percentage (%)			Sterols (%)	Isomer Percentage (%)		
					α-	β-+γ-	δ-		Campesterol	Stigmasterol	β-Sitosterol
95.4	147.1	48.1	27.2	9.23	11.73	60.03	28.24	9.45	33.5	23.5	43.0

* According to AOCS methods [18], the approximate contents of free fatty acid (FFA) and glycerides (Gly. including monoglycerides, diglycerides, and triglycerides) were calculated from acid and saponification values (A.V. and S.V., respectively) and expressed as the contents of oleic acid and triolein.

Source: Fang, T., Goto, M., Wang, X., Ding, X., Geng, J., Sasaki, M. and Hirose, T., *J. Supercritical Fluids,* 40, 50, 2007. With permission.

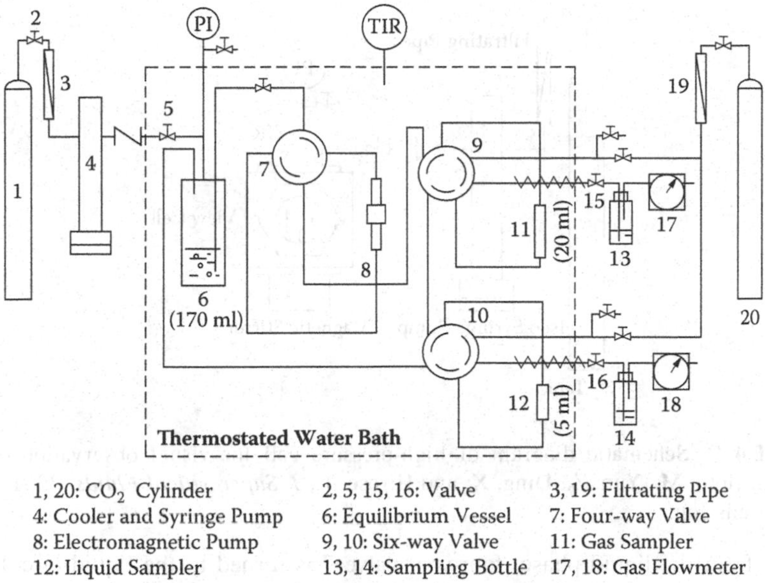

FIGURE 4.3 Schematic diagram of experimental apparatus. (From Fang, T., Goto, M., Yun, Z., Ding, X. and Hirose, T., *J. Supercritical Fluids*, 30, 1, 2004. With permission.)

liquid and gas phases. As shown in Figure 4.3, the equilibrium system is immersed in a water bath, with two stirrers for keeping the system's temperature uniform with the precision temperature control of +0.5 K.

About 70 mL of pure component is initially charged into equilibrium vessel 6 (170 mL, max. pressure 30 MPa), and then CO_2 flows into the apparatus from cylinder 1, passing through valve 2 and filtrating pipe 3, the cooler and syringe pump 4 (Isco 260 D, max. pressure 57.71 MPa, Teledyne Isco Inc., Lincoln, NE, USA), valve 5, and then into equilibrium vessel 6. After the pressure and temperature reach the required values, valve 5 is closed and then magnetic pump 8 is powered on, which can keep the fluid flowing upward at about 4 mL per minute. By rotating four-way valve 7, gas (light phase) or liquid (heavy phase) is chosen as the circulating fluid in the system. For example, the circulating fluid in Figure 4.3 is the gas phase. After the gas phase is kept circulating for 1.5 hours, six-way valve 9 is turned and the fluid in sampler 11 (20 mL) is taken as the gas sample. After valve 15 is opened and adjusted, the CO_2 in the gas slowly passes through sampling bottle 13 and flowmeter 17 (precision 0.01 L), which records the amount of CO_2. The pipes between valve 9 and 15 and bottle 13 are all heated by an electronic heater. The compounds precipitated in sampler 11 are washed into bottle 13 with *n*-hexane and rinsed with about 10 mL *n*-hexane to avoid the sample loss, since the solutes may form aerosol particles and then pass through the collecting bottle with CO_2 fluid. After every run, the residual *n*-hexane in the sample is removed from cylinder 20 with CO_2 and then merged into the sample battle. The *n*-hexane solution is quantified by electronic balance (precision 0.1 mg) and analyzed by GC (for methyl oleate) or HPLC (for α-tocopherol). From the amounts of CO_2 and chromatographic data, the composition in the gas phase is calculated. As for the

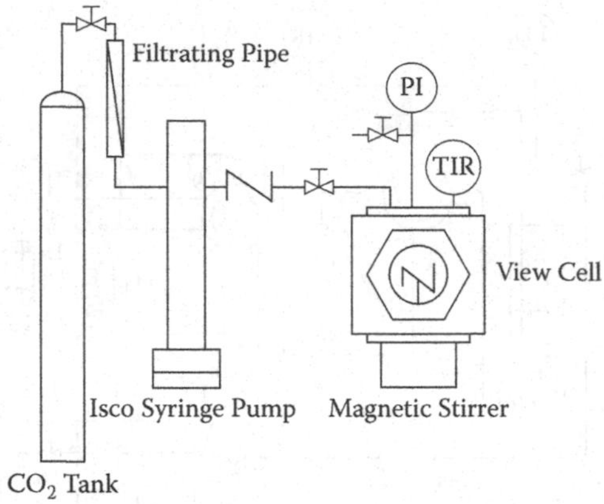

FIGURE 4.4 Schematic diagram of high-pressure cell for visual observation. (From Fang, T., Goto, M., Yun, Z., Ding, X. and Hirose, T., *J. Supercritical Fluids,* 30, 1, 2004. With permission.)

sample from the liquid phase, four-way valve 7 is turned to the liquid circulation. After circulating for 1.5 hours, six-way valve 10 is turned for sampling. By a similar method to the gas phases, the liquid composition is also obtained.

This structure of our apparatus prevents pressure fluctuation during sampling. To obtain more accurate results, all data represent mean values of three samplings at a uniform condition, and the relative standard deviations are within 0.005 mol% for gas composition (Y_l) and 0.5 mol% for liquid composition (X_l).

4.5.2 High Pressure View Cell

For observing the equilibrium systems, a view cell (30 mL, max. pressure 30 MPa, Akico Co., Tokyo, Japan) was employed in our experiment, as shown in Figure 4.4. A magnetic stirrer is coupled with the cell and the temperature is controlled by electric heaters embedded inside the cell's wall. The Isco pump was used for pressurizing the system.

4.5.3 Phase Equilibrium Properties

The binary phase equilibrium data for methyl oleate + CO_2 and α-tocopherol + CO_2 were measured and correlated at 313.15 to 353.15 K in the pressure ranges of 5 to 23 MPa for methyl oleate and 5 to 30 MPa for α-tocopherol [19]. The isotherms of 313.15, 333.15, and 353.15 K for methyl oleate + CO_2 and α-tocopherol + CO_2 were measured over the pressure ranges of 5 to 23 MPa and 5 to 30 MPa, respectively. The experimental results and their correlated data are shown in Figure 4.5 and Figure 4.6, in which the data measured by other researchers [25–37] are also illustrated.

As for the phase equilibrium of methyl oleate + CO_2, at 313.15 K and 333.15 K, our experimental data in Figure 4.5 agree well with the data reported by other researchers, except the data reported by Cheng et al. [25]. Cheng et al. found an unusual waist-shape curve at the vicinity of the critical pressure of CO_2, an interesting

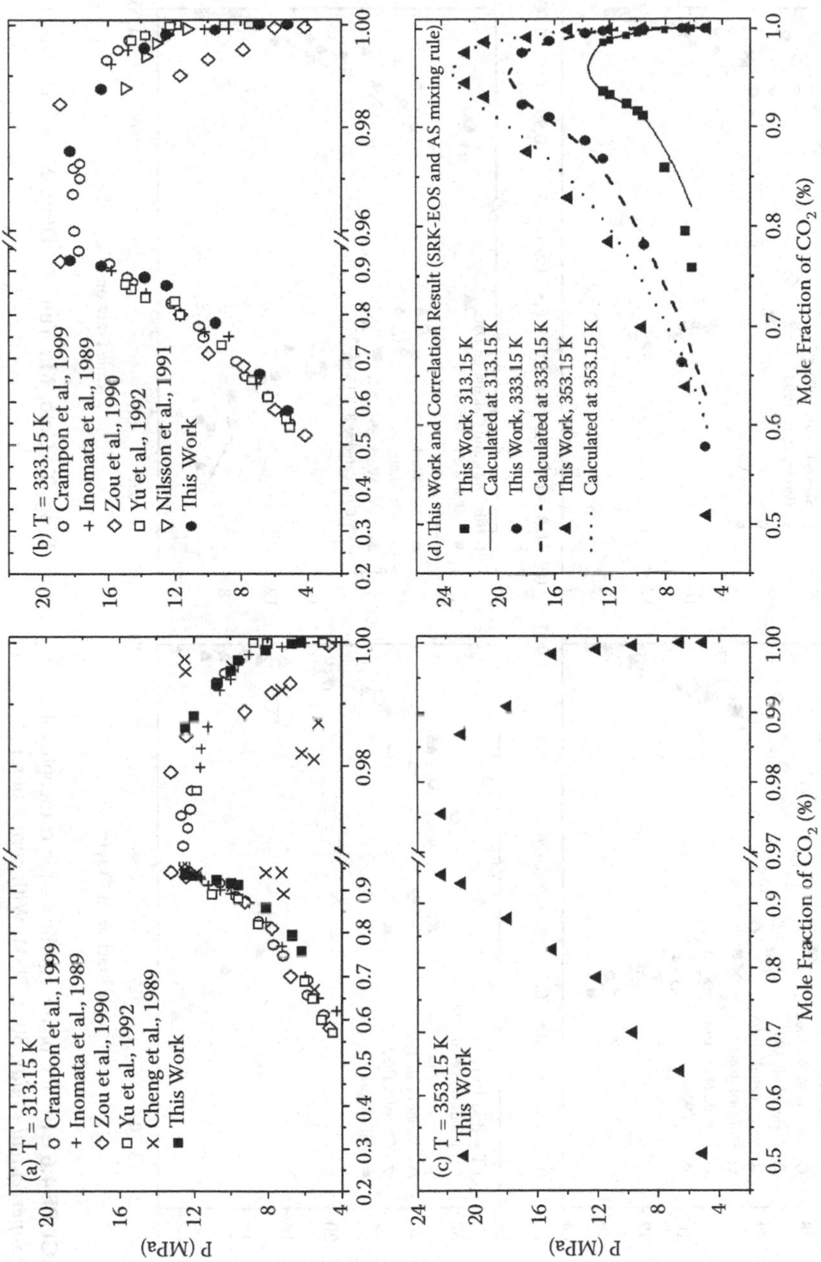

FIGURE 4.5 Binary phase equilibria for methyl oleate + CO_2. (From Fang, T., Goto, M., Yun, Z., Ding, X. and Hirose, T., *J. Supercritical Fluids*, 30, 1, 2004. With permission.)

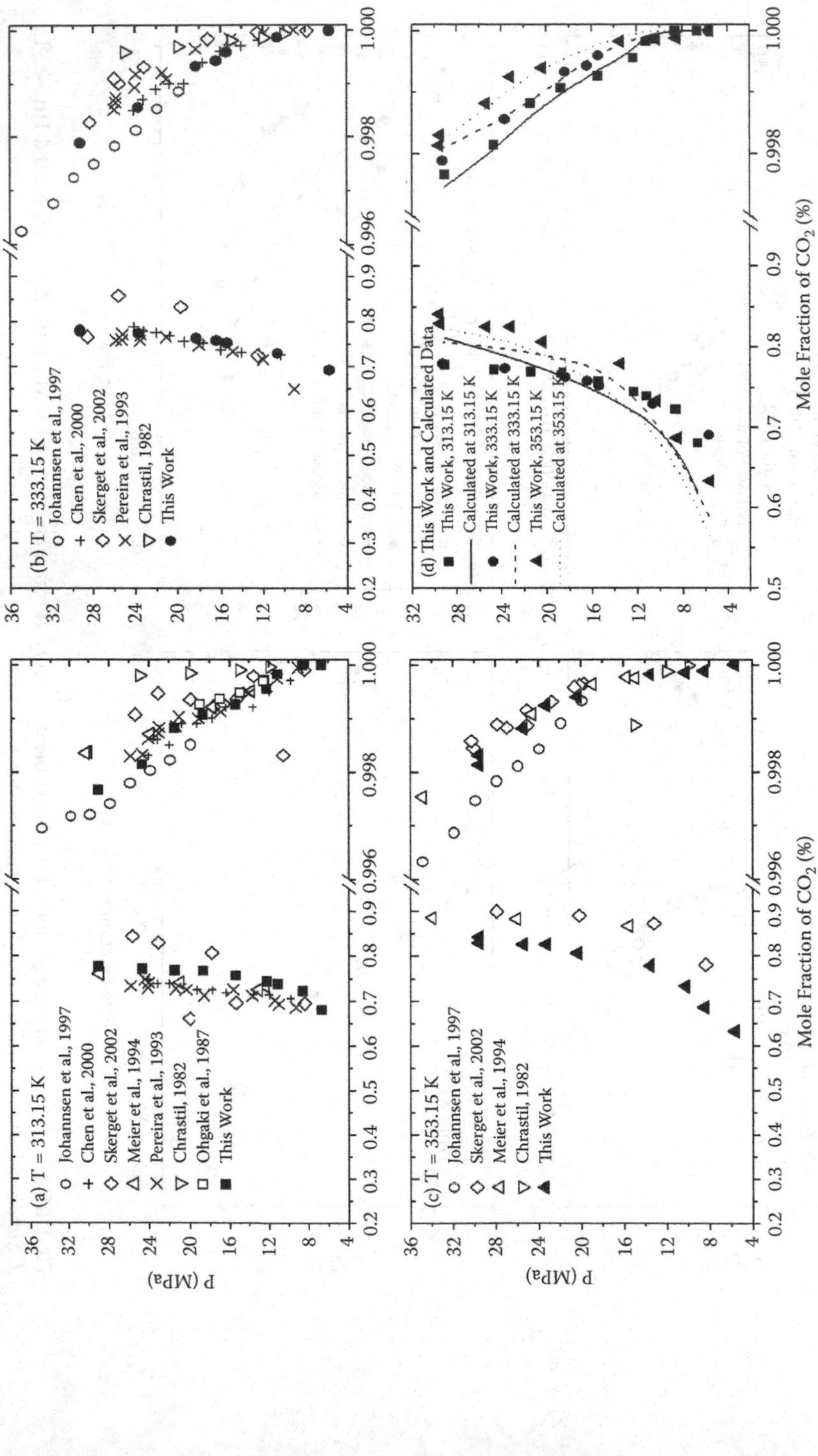

FIGURE 4.6 Binary phase equilibria for α-tocopherol + CO_2. (From Fang, T., Goto, M., Yun, Z., Ding, X. and Hirose, T., *J. Supercritical Fluids*, 30, 1, 2004. With permission.)

phenomenon that was not verified by other researchers' and our data. Additionally, in the case of 353.15 K, no data are reported in the literature.

For the system of α-tocopherol + CO_2, Figure 4.6a–c illustrates that our measured data match with the literature data at three temperatures. The results, especially the composition data in gas phase, are remarkably different from author to author. The possible reasons for the errors were thought to be high viscosity of α-tocopherol, pressure change, and aerosols caused by sudden depressurization [34]. Additionally, the data for liquid phase are not as abundant as those for gas phase.

Our data illustrate that both methyl oleate and α-tocopherol mole fractions in gas phase increase as pressure increases at constant temperature. Meanwhile, the CO_2 fraction in liquid phase rises with increasing pressure. The equilibrium concentration of methyl oleate in CO_2 is always much higher than the equilibrium concentration of α-tocopherol in CO_2. Moreover, the influence of temperature on gas composition is contrary to that of pressure. In addition, for methyl oleate + CO_2, with the increase of temperature, the CO_2 fraction in liquid obviously decreases, whereas for α-tocopherol + CO_2 at 313.15 K and 333.15 K, the liquid composition changes slightly with temperature. On the other hand, at 353.15 K, the CO_2 mole fraction in liquid is slightly larger than those at other temperatures with pressures higher than 12 MPa.

In addition, an ideal correlation for our experimental data, as shown in Figure 4.5d and 4.6d, can be obtained by the SRK EOS with the AS mixing rule, where the average deviation obtained is lower than 0.12% for gas and 7% for liquid.

4.5.4 Solubility

The solubilities of methyl oleate and α-tocopherol in supercritical CO_2 were calculated from the gas phase composition. The results are shown in Figure 4.7. The solubilities of the two compounds are presented as a function of the CO_2 density.

Two effects are observed. At constant temperature, solubility increases with increasing density. This is probably due to the increasing solvent power of CO_2 at higher density. At constant density, a rise of temperature results in an increase of solubility due to the increase in vapor pressure of the solutes. Similar phenomena were reported by Johannsen and Brunner [35].

4.5.5 Distribution Coefficient

Equilibrium data of two phases can be used to calculate the distribution coefficient, which was defined by:

$$K_i = \frac{y_i}{x_i} \tag{4.1}$$

where y_i and x_i are mole fractions of component i in gas and liquid phase, respectively.

Calculation results are shown in Figure 4.8. As density increases at constant temperature, the distribution coefficient of methyl oleate increases significantly. In the case of methyl oleate + CO_2, when pressure was more than 12.5 MPa at 313.15 K,

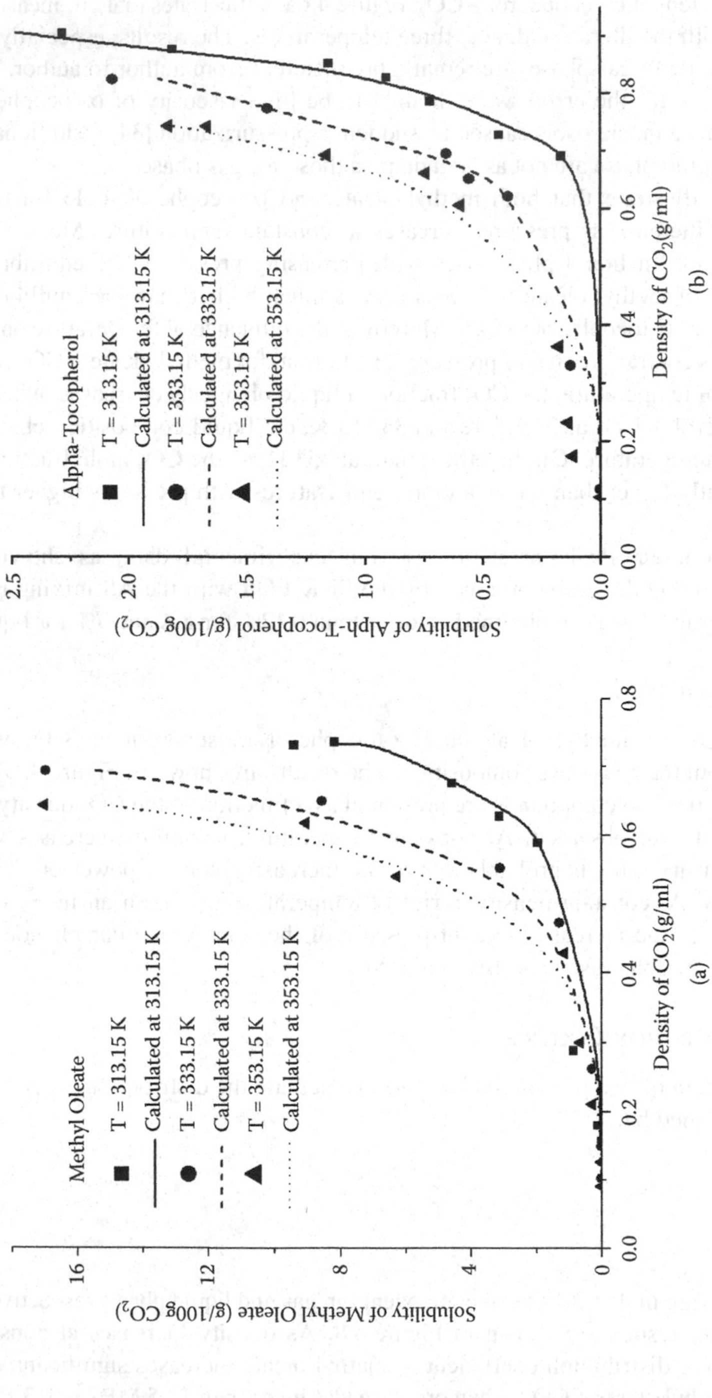

FIGURE 4.7 The solubilities of methyl oleate and α-tocopherol in CO_2. (From Fang, T., Goto, M., Yun, Z., Ding, X. and Hirose, T., *J. Supercritical Fluids*, 30, 1, 2004. With permission.)

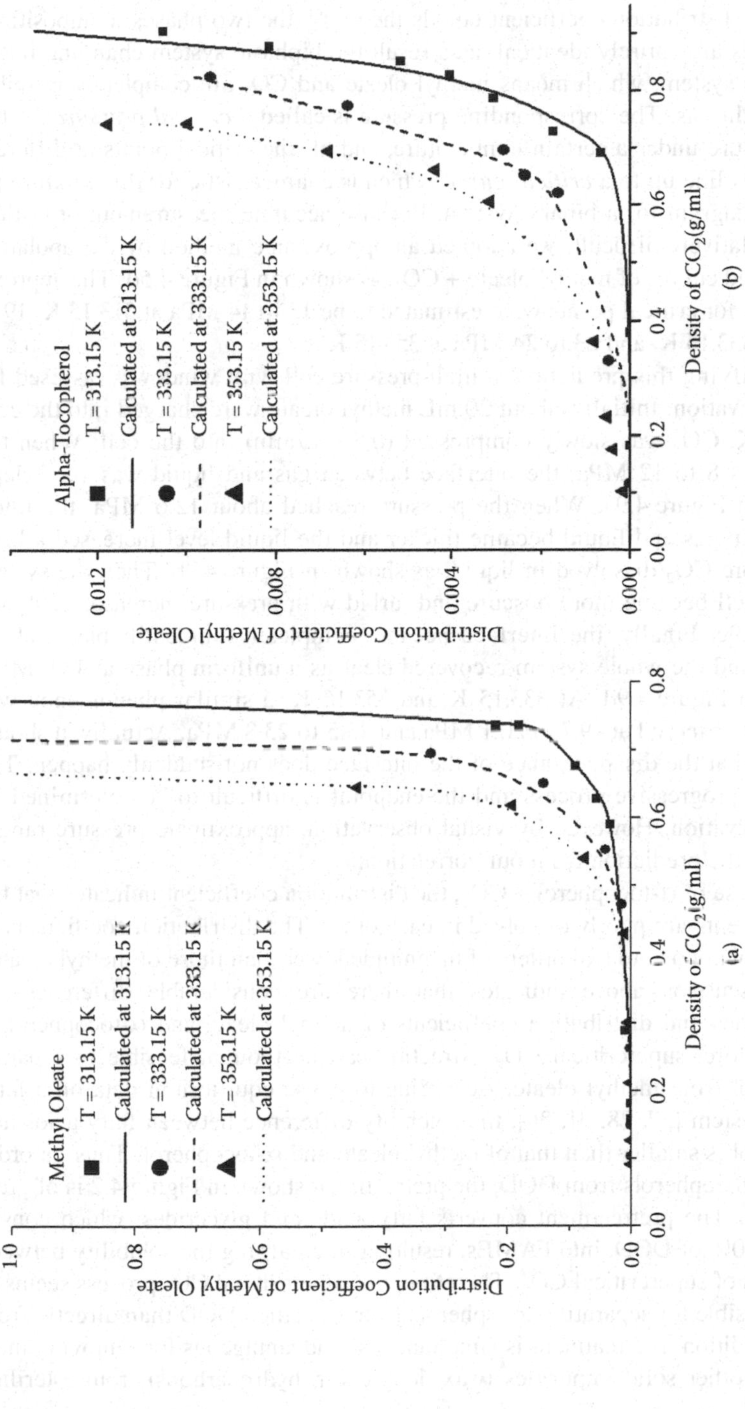

FIGURE 4.8 Distribution coefficients of methyl oleate and α-tocopherol in CO_2. (From Fang, T., Goto, M., Yun, Z., Ding, X. and Hirose, T., *J. Supercritical Fluids*, 30, 1, 2004. With permission.)

18.5 MPa at 333.15 K, and 22.5 MPa at 353.15 K, there is no remarkable difference in equilibrium compositions between the two phases. This means that in the experimental range investigated, critical points for the binary mixture probably exist. If the distribution coefficient equals the unity, the two phases' compositions and densities are entirely identical, as a result the biphasic system changing into a monophasic system, which means methyl oleate and CO_2 are completely miscible at such conditions. The corresponding pressure is called a *critical pressure* for the binary mixture under a certain temperature, and all the critical points at different temperatures line up to a *critical curve,* which is characteristic for this mixture [1] in a P-T-x diagram for a binary system. Because accurate measurement of critical points is relatively difficult, we adopted an approximate method by extrapolating the correlated curves of methyl oleate + CO_2, as shown in Figure 4.5d. The approximate ranges for critical points were estimated to be 13 to 14 MPa at 313.15 K, 19 to 20 MPa at 333.15 K, and 23 to 24 MPa at 353.15 K.

For verifying this prediction, a high-pressure cell with windows was used for visual observation. Initially, about 20 mL methyl oleate were charged into the cell. At 313.15 K, CO_2 was slowly compressed (0.5 mL/min) into the cell. When the pressure was 8 to 12 MPa, the interface between gas and liquid was very clear, as shown in Figure 4.9a. When the pressure reached about 12.6 MPa, the interface between gas and liquid became thicker and the liquid level increased a little because more CO_2 dissolved in liquid, as shown in Figure 4.9b. Then the system inside the cell became more obscure and turbid with pressure increase, as shown in Figure 4.9c. Finally, the interface became completely unidentifiable at about 13.7 MPa, and the whole system recovered clear as a uniform phase at 13.9 MPa, as shown in Figure 4.9d. At 333.15 K and 353.15 K, a similar phenomenon was respectively observed at 19.7 to 20.1 MPa and 23.5 to 23.8 MPa. Actually, it should be noticed that the disappearance of the interface does not suddenly happen. The change is a progressive process and the endpoint is difficult to be determined by visual observation. However, by visual observation, approximate pressure ranges are close to the prediction from our correlation.

In the case of α-tocopherol + CO_2, the distribution coefficient indicates that the two components are poorly dissolved in each other. The distribution coefficients of α-tocopherol are one to two orders of magnitude lower than those of methyl oleate.

The discussion above indicates that there are considerable differences in the solubilities and distribution coefficients of methyl oleate and α-tocopherol in CO_2. Therefore, supercritical CO_2 extraction can be thought feasible to separate α-tocopherol from methyl oleate. According to phase equilibrium data on a fatty acid-CO_2 system [27, 28, 31, 36], the solubility difference between fatty acids and α-tocopherol is smaller than that of methyl oleate and α-tocopherol. Thus, in order to separate tocopherols from DOD, the pretreatment shown in Figure 4.2 is of great significance. The pretreatment converts fatty acids and glycerides, which consist of 70% to 80% of DOD, into FAMEs, resulting in enlarging the solubility between components of supercritical CO_2. Therefore, the supercritical CO_2 process seems to be more feasible for separating tocopherols from esterified DOD than directly from DOD. In addition, pretreatment is simultaneously advantageous for removing most sterols and other solid impurities (wax, long-chain hydrocarbons) from esterified

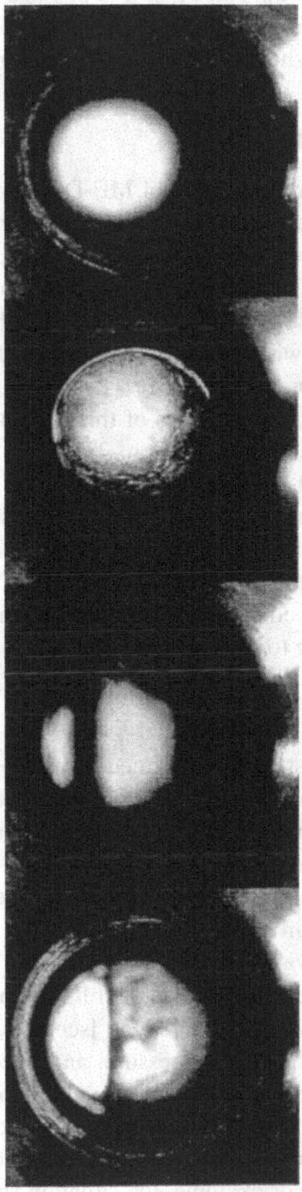

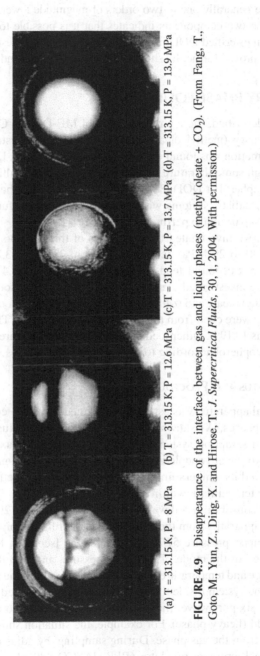

(a) T = 313.15 K, P = 8 MPa (b) T = 313.15 K, P = 12.6 MPa (c) T = 313.15 K, P = 13.7 MPa (d) T = 313.15 K, P = 13.9 MPa

FIGURE 4.9 Disappearance of the interface between gas and liquid phases (methyl oleate + CO_2). (From Fang, T., Goto, M., Yun, Z., Ding, X. and Hirose, T. *J. Supercritical Fluids*, 30, 1, 2004. With permission.)

DOD [5, 7–10] because the solubility of sterols is far lower in fatty acid methyl esters than that in fatty acids.

Compared with methyl oleate, the solubility and distribution coefficient of α-tocopherol are generally one to two orders of magnitude lower. Such a large difference between the two components indicates that it is possible to concentrate natural tocopherols from esterified DOD in the experimental range investigated. This conclusion needs to be proved further by study of the ternary system and the realistic system.

4.6 TERNARY PHASE EQUILIBRIA

As previously described, the complex system of ME-DOD + CO_2 can be regarded as a pseudo-ternary (methyl oleate + tocopherol + CO_2) system. Regretfully, no published information was found on phase equilibrium of ME-DOD in supercritical CO_2, even though some literature reported on ternary and multicomponent systems involving α-tocopherol or DOD [38–41]. Thus, we measured the phase behaviors for the ternary and realistic systems with the view of providing fundamental information for further separation experiments and process design.

In this part, we investigated the influences of three factors on phase behavior: pressure (from 10 to 29 MPa), temperature (from 313.15 to 353.15 K), and initial feed composition. Six initial feed compositions (0, 10.19, 32.44, 50.46, 71.93, and 100 mass%) were investigated. Among these, 0% and 100% stood for the pure compositions of methyl oleate and α-tocopherol, respectively. Their corresponding phase equilibrium data were cited from the binary data in part 4.5.3. The feed composition of ME-DOD was 10.19%. Other feed compositions were prepared by mixing methyl oleate and α-tocopherol according to different proportions.

4.6.1 Apparatus and Procedure

An experimental apparatus was established for measuring the compositions of both liquid and gas phases. As shown in Figure 4.10, the apparatus consisted of feed, equilibrium, and sampling systems. A view cell (30 mL, max. pressure 30 MPa, Akico Co., Tokyo, Japan) coupled with a magnetic stirrer was employed as the equilibrium vessel, and its temperature was controlled with an electric heater capable of maintaining the temperature within ±0.1 K.

Initially the equilibrium cell was charged with about 15 to 20 mL feed. CO_2 then flowed into the apparatus from the CO_2 tank via the filtering pipe and syringe pump (ISCO 260 D, max. pressure 57.71 MPa, Teledyne Isco Inc., Lincoln, NE, USA), which is operated in the mode of constant pressure, and into the equilibrium cell. After the pressure and temperature reach the required values, the magnetic stirrer was turned on and the system inside the equilibrium cell was stirred for at least 2 hours. By rotating the six-port valve, the samples were converted to and from gas (light phase) and liquid (heavy phase). For example, the situation shown in Figure 4.10 is a sample taken from the gas phase. During sampling, by adjusting the microswitch of the digital back pressure regulator (BPR, JASCO 880-81, JASCO International Co. Ltd., Tokyo, Japan), gaseous CO_2 slowly passed through the sampling bottle and flowmeter, which recorded the amount of CO_2 (defined as compound 3) consumed.

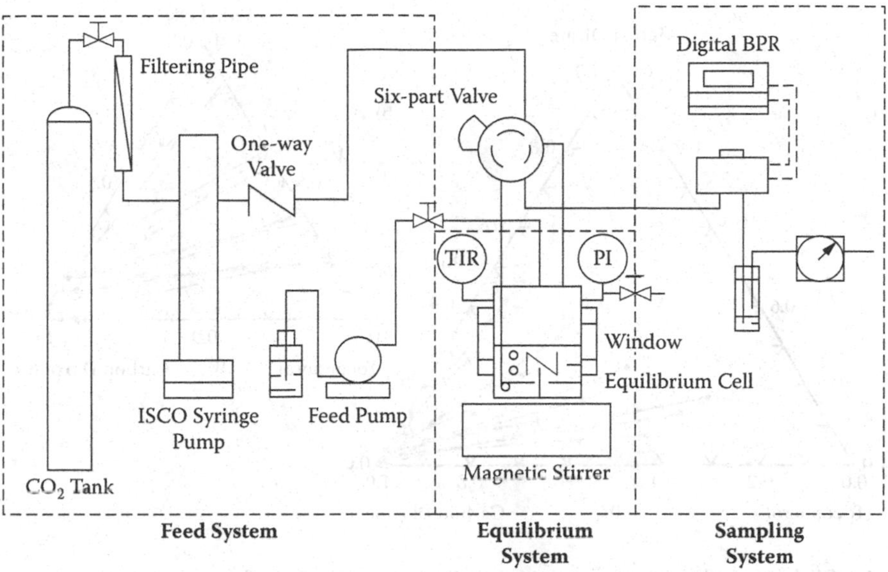

FIGURE 4.10 Schematic diagram of the phase equilibrium apparatus. (From Fang, T., Goto, M., Sasaki, M. and Hirose, T., *J. Chem. Eng. Data*, 50, 390, 2004. With permission.)

The pipes connected to the six-port valve, cell, and digital BPR were heated by electronic heaters. Their temperatures were controlled in a manner similar to that of the equilibrium cell. Additionally, because of the effect of adiabatic expansion, the fluid temperature decreased greatly and suddenly during sampling. Consequently, some solutes or CO_2 may have condensed at the outlet of the BPR, contributing more or less to measurement inaccuracy. To avoid such phenomena, the BPR outlet was heated and maintained at a temperature of 371 K. In addition, about 10 mL *n*-hexane was initially loaded in the sample bottle because the solutes may have formed aerosol particles and pass through the collecting bottle with the CO_2 fluid [34]. The *n*-hexane solution was quantified by electronic balance (precision 10^{-4} g) and analyzed by GC (for methyl oleate, compound 1) and HPLC (for tocopherol, compound 2). According to the amount of CO_2 consumed and chromatographic data, the gas composition (y_1, y_2, y_3) was calculated. By a similar method, the liquid composition (x_1, x_2, x_3) was obtained. Additionally, because similar system lines were used for sampling from two phases, when the six-port valve was switched for sampling from another phase, the fluid from the cell was kept flowing without sampling for about 1 to 2 minutes, in order to avoid carryover of the samples. Another key point was to ensure that the sample was taken from an equilibrium system by sampling only when the liquid-gas interface was clearly visible.

The structure of our apparatus avoided pressure fluctuation during the equilibrium and sampling steps because the equilibrium cell pressure was maintained at the set values (the required values) by setting the ISCO pump in the mode of constant pressure. More importantly, during sampling, the CO_2 flow rate should be kept relatively low by adjusting the BPR microswitch. In our experiment, the CO_2 flow rate

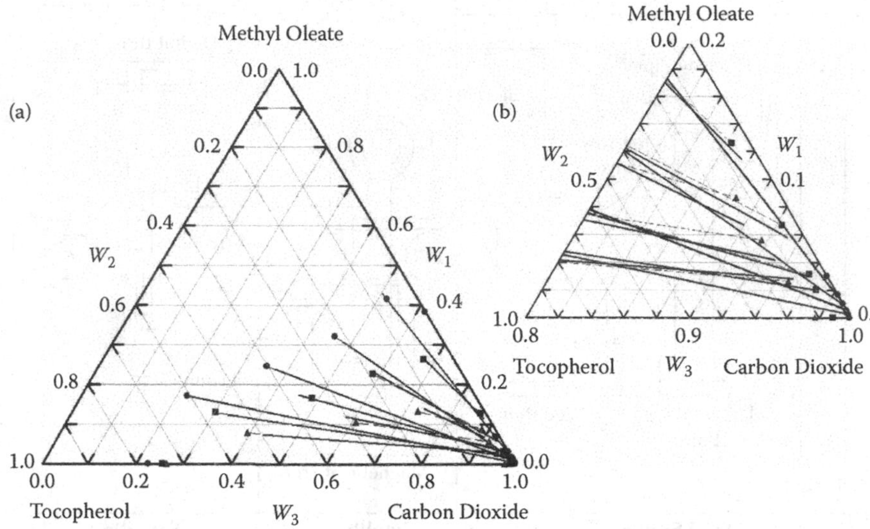

FIGURE 4.11 Influence of pressure on the phase equilibria of (w_1 methyl oleate + w_2 tocopherol + w_3 CO_2) at T = 313.15 K: (a) Liquid-gas equilibria; (b) The equilibrium compositions in gas; ●, experimental data at P = 10 MPa; ■, experimental data at P = 20 MPa; ▲, the experimental data at P = 29 MPa; –, the correlated tie lines at P = 10 MPa; –, the correlated tie lines at P = 20 MPa; –, the correlated tie lines at P = 29 MPa. (From Fang, T., Goto, M., Sasaki, M. and Hirose, T., *J. Chem. Eng. Data*, 50, 390, 2004. With permission.)

was maintained at lower than 20 mL·min^{-1} and 5 mL·min^{-1} (at 0.1 MPa and room temperature) during sampling from gas and liquid, respectively.

To obtain more accurate results, all data represented mean values of three samplings at uniform conditions with an uncertainty of ±0.001 mass fraction for gas composition and ±0.002 for liquid composition, respectively.

For gas phase sampling, the CO_2 amount was adjusted to about 2 L (at 0.1 MPa and room temperature) at experimental pressures lower than 20 MPa since low solubility at low pressures was a main reason for experimental error; however, for higher pressures (≥ 20 MPa), the amount of CO_2 was about 1 L (at 0.1 MPa and room temperature). For liquid phase, the CO_2 amount was adjusted to about 0.01 to 0.02 L (at 0.1 MPa and room temperature). The contents of methyl oleate and tocopherol in the liquid sample were generally high and required dilution before chromatographic analysis.

The samples dissolved in *n*-hexane were analyzed by GC to determine the concentration of methyl oleate. Analysis of tocopherol was performed by HPLC [19, 20].

4.6.2 Influences of Pressure and Temperature on Phase Equilibrium

The isotherms at 313.15, 333.15, and 353.15 K for the ternary system of methyl oleate (1) + tocopherol (2) + CO_2 (3) were measured over the pressure range from 10 to 29 MPa. At 313.15 K, the composition data at 10, 20, and 29 MPa were drawn in a triangular diagram, as shown in Figure 4.11. Obviously, the two-phase region, which is surrounded by the equilibrium data, shrinks with increasing pressure. In other words,

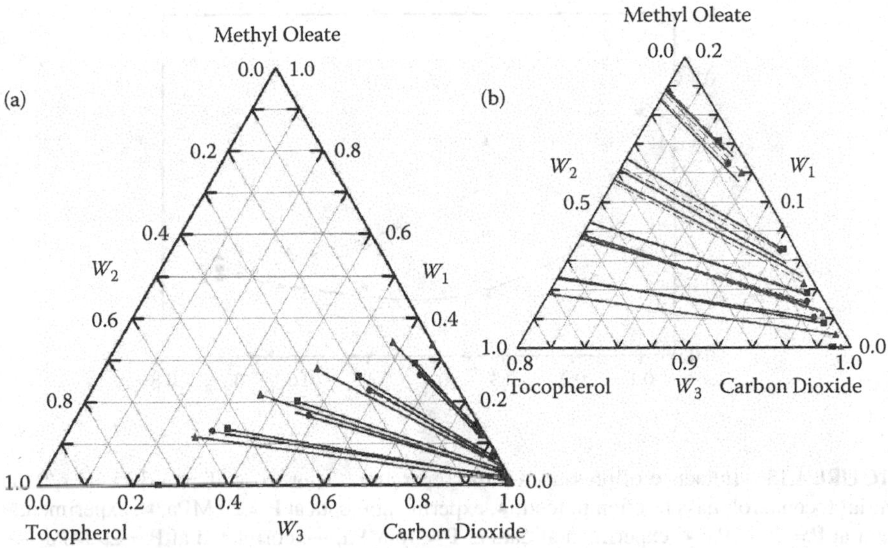

FIGURE 4.12 Influence of temperature on the phase equilibria of (w_1 methyl oleate + w_2 tocopherol + w_3 CO_2) at P = 20 MPa: (a) Liquid-gas equilibria; (b) The equilibrium compositions in gas; ●, the experimental data at T = 313.15 K; ■, the experimental data at T = 333.15 K; ▲, the experimental data at T = 353.15 K; –, the correlated tie lines at T = 313.15 K; –, the correlated tie lines at T = 333.15 K; –, the correlated tie lines at T = 353.15 K. (From Fang, T., Goto, M., Sasaki, M. and Hirose, T., *J. Chem. Eng. Data*, 50, 390, 2004. With permission.)

the mutual solubility of the components increased. Figure 4.11a shows the CO_2 mass fraction in liquid rises with pressure increase, while the CO_2 mass fraction in gas is reduced, as shown in Figure 4.11b, which means the solubilities of other components in gas increase. At lower pressures (10 MPa), because both methyl oleate and tocopherol have limited miscibility in CO_2, the ternary phase behavior reveals a phase equilibrium of ternary type II [1]. Thus, there is no critical point. At higher pressures (20, 29 MPa), because methyl oleate and CO_2 are completely miscible, the two-phase area is ternary type I, which is characterized by a critical point where the two phases become identical. In the type I system, higher solubility in the gas can be reached than in ternary type II system.

In addition to the measured data, Figure 4.11 shows the tie lines connecting with the equilibrium data in the liquid and gas phases. The tie lines were correlated with the SRK EOS and the AS mixing rule. Characteristically, the gradient of the equilibrium tie lines gradually changes from one side line of the triangle to the other. This means that with the increase of methyl oleate mass fraction in feed, phase behavior tends to be close to that of the binary system of methyl oleate + CO_2.

Figure 4.12 shows the influence of temperature on phase equilibrium at 20 MPa. Obviously, the influence of temperature is contrary to that of pressure. With increasing temperature, the two-phase area expands, as shown in Figure 4.12a. In addition, the phase equilibria are of ternary type I at 313.15 and 333.15 K, and then at 353.15 K the phase equilibrium develops into ternary type II, where the binary critical pressure for CO_2 + methyl oleate is greater than 20 MPa. Noticeably, the

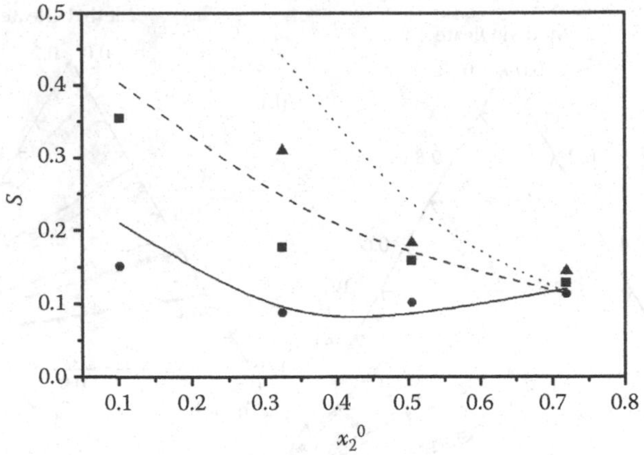

FIGURE 4.13 Influence of pressure on the separation factor (S) at T = 313.15 K: x_2^0, the initial tocopherol mass fraction in feed; ▲, experimental data at P = 29 MPa; ■, experimental data at P = 20 MPa; ●, experimental data at P = 10 MPa; ····, correlated at P = 29 MPa; ---, correlated at P = 20 MPa; —, correlated at P = 10 MPa. (From Fang, T., Goto, M., Sasaki, M. and Hirose, T., *J. Chem. Eng. Data*, 50, 390, 2004. With permission.)

influence of temperature on the gas composition seems to be less significant than that of the liquid composition at 20 MPa, as shown in Figure 4.12b.

4.6.3 Separation Factor between Tocopherol and Methyl Oleate

According to the measured gas and liquid composition data, the separation factor (S) between tocopherol (compound 2) and methyl oleate (compound 1) was calculated by:

$$S = (y_2/x_2)/(y_1/x_1) \tag{4.2}$$

where y_i and x_i are mass fractions of component i in gas and liquid, respectively.

The separation factor represents the process selectivity for separating methyl oleate from tocopherol. In detail, a lower value indicates higher selectivity, whereas a higher value indicates that it is more difficult to separate the two compounds under certain conditions. Furthermore, when the separation factor equals unity, the composition in gas is similar to that in liquid and the supercritical CO_2 process cannot separate methyl oleate from tocopherol.

Figure 4.13 and Figure 4.14 show the influences of pressure and temperature on the separation factor, respectively. In Figure 4.13, both the experimental data and correlated curve illustrate that, at a constant temperature, the separation factor increases as pressure increases, except for one part at 10 MPa. However, as shown in Figure 4.14, the influence of temperature is contrary to that of pressure. As mentioned above, a lower separation factor indicates a higher selectivity; accordingly, the tendencies shown in Figure 4.13 and Figure 4.14 indicate that low pressure and high temperature lead to high selectivity, which is advantageous for separating methyl oleate from tocopherol with supercritical CO_2. In addition, at constant pressure

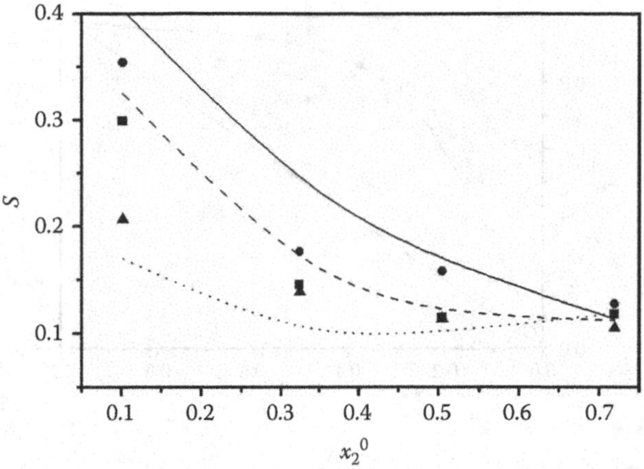

FIGURE 4.14 Influence of temperature on the separation factor (S) at P = 20 MPa: x_2^0, the initial tocopherol mass fraction in feed; ●, experimental data at T = 313.15 K; ■, experimental data at T = 333.15 K; ▲, experimental data at T = 353.15 K; —, correlated at T = 313.15 K; ---, correlated at T = 333.15 K; ····, correlated at T = 353.15 K. (From Fang, T., Goto, M., Sasaki, M. and Hirose, T., *J. Chem. Eng. Data*, 50, 390, 2004. With permission.)

and temperature, the separation factor increases as the initial tocopherol content (x_2^0) decreases. Moreover, this tendency is more obvious as pressure increases. A lower content of tocopherol means a higher content of methyl oleate since the feed mainly consisted of methyl oleate and tocopherol, among which the former is more soluble in supercritical CO_2. Consequently, more methyl oleate in the feed generates more tocopherol for distribution in supercritical CO_2, resulting in an increase in the separation factor. In other words, methyl oleate acts as the cosolvent for tocopherol. A similar phenomenon was also reported by Bamberger et al. [42], who measured the solubilities of fatty acids, pure triglycerides, and triglyceride mixtures in supercritical CO_2. They found that the solubilities of the less soluble triglycerides in mixtures like tripalmitin were enhanced by the presence of more soluble triglycerides, like trilaurin. In this situation, the more soluble compounds were said to be acting as the cosolvents.

4.6.4 Equilibrium Lines

Using the obtained equilibrium data, equilibrium lines were drawn in the pressure and temperature ranges investigated. Figure 4.15 and Figure 4.16 illustrate the equilibrium lines at 313 K and 20 MPa, respectively. The data were represented as the mass fraction of methyl oleate (CO_2 free basis).

As shown in Figure 4.15 and Figure 4.16, either a pressure increase or a temperature decrease moves the equilibrium line closer to the diagonal line. In fractionation designing, this tendency means that an increase in theoretical stages is accompanied by a decrease in process selectivity. Such a tendency agrees well with that of illustrated by the separation factor.

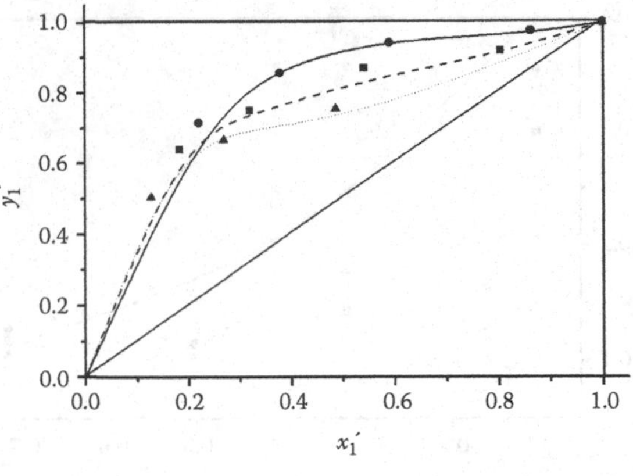

FIGURE 4.15 Equilibrium lines at T = 313.15 K: x_1', mass fraction of methyl oleate in liquid (CO_2 free basis); y_1', mass fraction of methyl oleate in gas (CO_2 free basis); ●, experimental data at P = 10 MPa; ■, experimental data at P = 20 MPa; ▲, experimental data P = 29 MPa; —, correlated at P = 10 MPa; ---, correlated at P = 20 MPa; ····, correlated at P = 29 MPa. (From Fang, T., Goto, M., Sasaki, M. and Hirose, T., *J. Chem. Eng. Data*, 50, 390, 2004. With permission.)

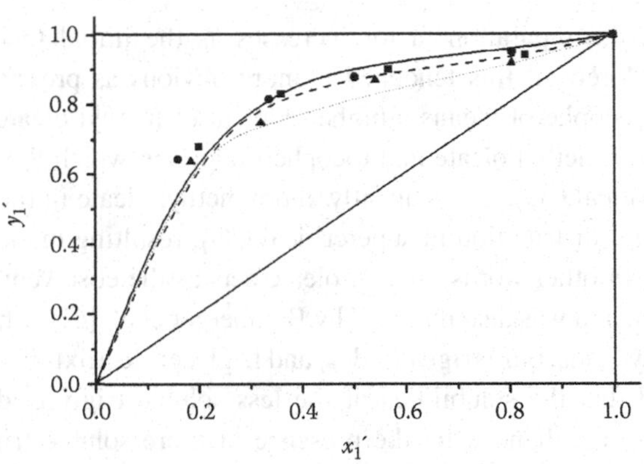

FIGURE 4.16 Equilibrium lines at P = 20 MPa: x_1', mass fraction of methyl oleate in liquid (CO_2 free basis); y_1', mass fraction of methyl oleate in gas (CO_2 free basis); ●, experimental data at T = 353.15 K; ■, experimental data at T = 333.15 K; ▲, experimental data T = 313.15 K; —, correlated at T = 353.15 K; ---, correlated at T = 333.15 K; ····, correlated at T = 313.15 K. (From Fang, T., Goto, M., Sasaki, M. and Hirose, T., *J. Chem. Eng. Data*, 50, 390, 2004. With permission.)

4.6.5 Phase Behavior of ME-DOD

During the measurement of phase equilibrium, the realistic system of ME-DOD was taken as the 10.19% (tocopherol mass fraction) feed composition. Also, the ME-DOD system was the raw materials for concentrating natural tocopherols; thus, its phase behavior is discussed separately. Figure 4.17 shows the influence of pressure and temperature on the separation factor.

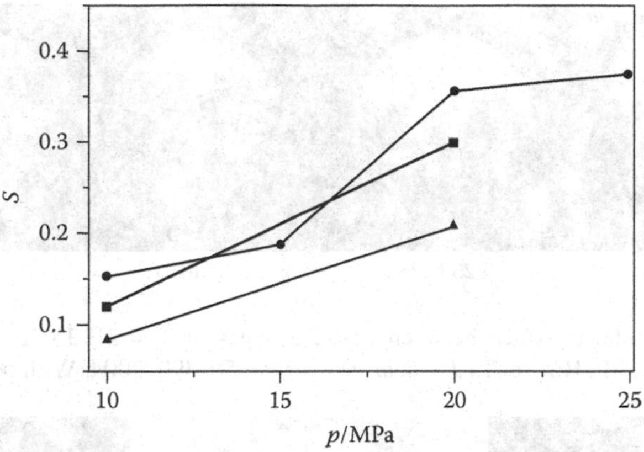

FIGURE 4.17 Separation factor (S) of ME-DOD in supercritical CO_2: ●, experimental data at T = 313.15 K; ■, experimental data at T = 333.15 K; ▲, experimental data T = 353.15 K. (From Fang, T., Goto, M., Sasaki, M. and Hirose, T., *J. Chem. Eng. Data*, 50, 390, 2004. With permission.)

The trends illustrated in Figure 4.17 are similar to those in Figure 4.13 and Figure 4.14. Noticeably, the separation factors at pressures lower than 20 MPa are relatively small. For instance, at 313.15 K, the separation factor remained lower than 0.2 for all pressures lower than 15 MPa. As pressure increases, the separation factor greatly increases, reaching 0.35 at 20 MPa. The increase of temperature offsets the effect of pressure to some extents. On the basis of the property, a separation strategy seems to be reasonable and feasible. A fractionation column is necessary for the ME-DOD liquid system. First of all, low pressure (15 to 20 MPa) was used in combination with a temperature distribution in the column to separate FAMEs like methyl oleate. Then the pressure was increased to separate tocopherol from other impurities. This procedure needs to be verified by a fractionation operation in which operation parameters can be determined and optimized.

During our experiments, some phenomena of the realistic system of ME-DOD + CO_2 were observed through the visual equilibrium cell. Figure 4.18 shows the liquid-gas interface changes that occur as pressure increases. The interface increasingly changes from clear to obscure, with the interface finally disappearing at high pressure about 29 MPa. At high pressure, a critical point probably exists that causes the whole system to become entirely miscible. In this situation, the liquid and gas compositions are identical and the separation factor equals unity. It should be noticed that the disappearance of the interface does not happen suddenly. The change is a progressive process and an accurate endpoint is difficult to determine by visual observation. The critical pressure at 313.15 K is approximately estimated in the pressure range from 27.8 to 29.0 MPa by visual observation.

Figure 4.19 shows the changes in the feed situation when ME-DOD was charged by pump into the equilibrium cell at different pressures and at a constant 5 mL/min. At low pressure (5 MPa), ME-DOD could smoothly flow into the equilibrium cell, but at high pressures, the charged ME-DOD resembles drops or fog. Moreover the rate for flowing downward at a high pressure was slower than that at low pressure.

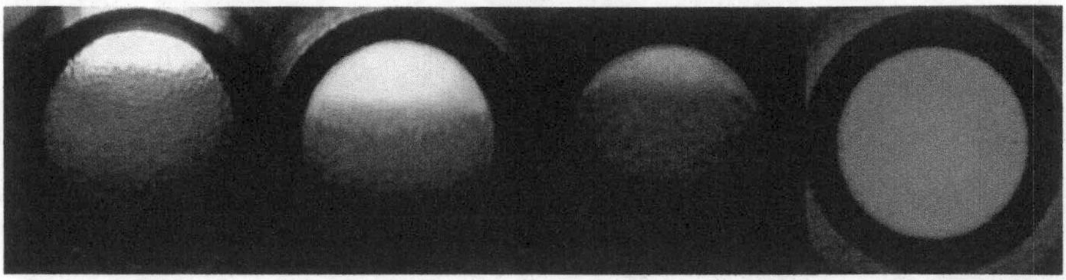

FIGURE 4.18 The interface between liquid and gas at T = 313.15 K. (From Fang, T., Goto M., Sasaki, M., Hirose, T., *J. Chem. Eng. Data*, 50, 390, 2004. With permission.)

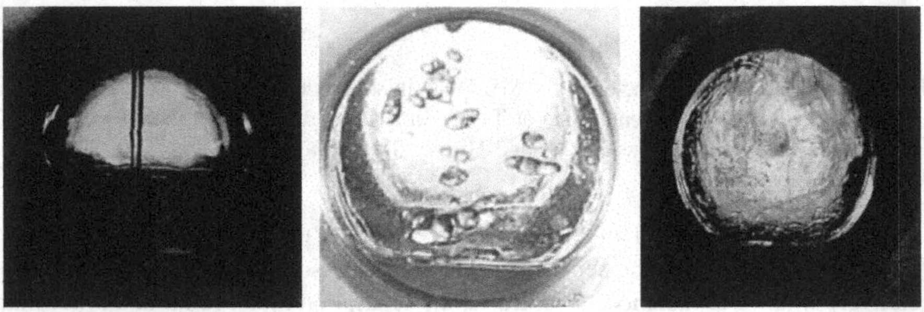

FIGURE 4.19 Feed situation of ME-DOD at different pressures (feed rate = 5 mL/min and T = 313.15 K). (From Fang, T., Goto, M., Sasaki, M. and Hirose, T., *J. Chem. Eng. Data*, 50, 390, 2004. With permission.)

We hypothesized that the main reason for this was the smaller difference in density between ME-DOD and supercritical CO_2 at high pressure; for example, at 20 MPa and 313.15 K, the density of CO_2 is 0.840 g/mL and that of ME-DOD is 0.865 g/mL. Such a small difference in density is likely to cause the drop or fog phenomenon, even though the clear interface between liquid and gas remained. This phenomenon should be considered when designing a continuously countercurrent operation.

4.7 SEPARATION WITH SUPERCRITICAL CO_2 FRACTIONATION

As described in section 4.1, the important step in concentrating natural tocopherols from ME-DOD is to remove the FAMEs, which contribute more than 70% of ME-DOD. FAMEs are important chemical materials in biofuel, metal-cutting oil, and cleaning agent production, as well as in the synthesis of other fatty acid products [17]. In section 4.6, the fundamental research on ternary and realistic phase equilibria has established a preliminary separation strategy, which must be tested through a supercritical CO_2 fractionation operation.

A fractionation column is necessary for the ME-DOD liquid system. First, low pressure (the initial pressure) is used in combination with a temperature gradient along the column to separate the FAMEs. Then, the pressure is increased to separate the tocopherols from other impurities.

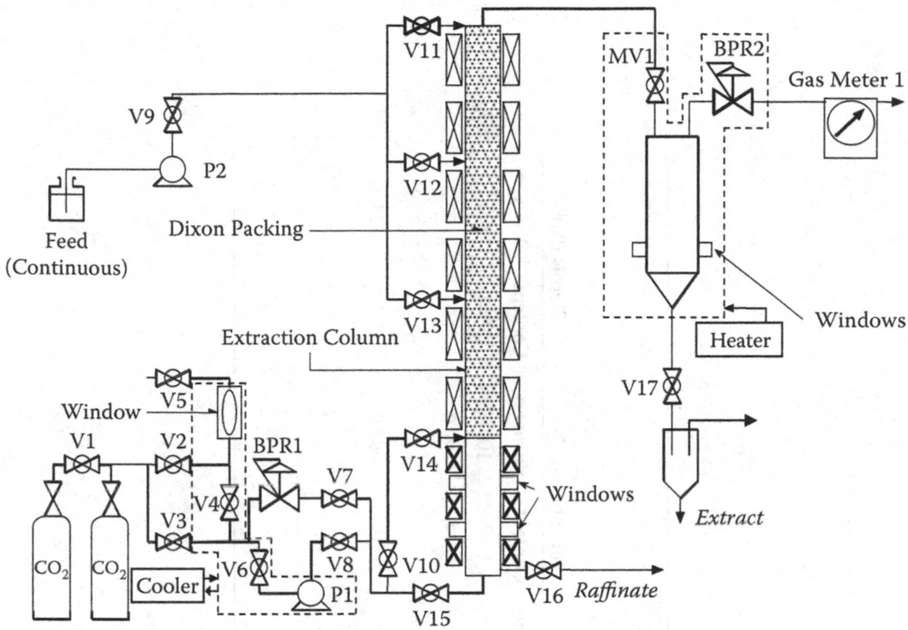

FIGURE 4.20 Schematic diagram of fractionation apparatus. (From Fang, T., Goto, M., Wang, X., Ding, X., Geng, J., Sasaki, M. and Hirose, T., *J. Supercritical Fluids,* 40, 50, 2007. With permission.)

4.7.1 Fractionation Apparatus and Procedure

A fractionation system was rebuilt from a supercritical CO_2 apparatus [43]. The experimental setup consisted of a countercurrent contact column (2.4 m × 20 mm i.d., 750 mL) and a separator (600 mL) for the top product, as shown in Figure 4.20. The column was packed with stainless steel 3 mm Dixon Packing (Naniwa Special Wire Netting Co., Ltd., Tokyo) over a length of 1.8 m. The separation experiment was carried out in semicontinuous countercurrent operation with a maximum feed of 200 g.

Before each run, about 120 g ME-DOD was charged into the column at 40±1 g/h so that an abundance of raw material had accumulated at the column bottom. Consequently, the fresh CO_2 fluid could come into contact with sufficient ME-DOD at the start of the fractionation operation. This ensures that each experiment began at a steady-state condition, which means both the feed and top fraction flow in a continuous situation with relatively stable flow rates. Fresh CO_2 was charged into the column through valve 14 (V14), the pressures of the column and separator were adjusted by BPR 1 and BPR 2, respectively. The temperature gradient of the column was concurrently adjusted by eight proportional integral differential controllers. The separator conditions were maintained at 3.8 to 4.0 MPa and 333 K. When pressure and temperature reached the required values, CO_2 was introduced from the column bottom by opening V15 and simultaneously closing V14, indicating the starting point of the fractionation operation. During continuous operation, the

TABLE 4.2
Result of the Scale-Up Pretreatment Process

Material and Products	Mass (Kg)	FFA (%)[a]	FAMEs (%)[b]	Glycerides (%)[a]		Tocopherols (%)[c]		Sterols (%)[b]	
				Content	Recovery	Content	Recovery	Content	Recovery
DOD	353.6	48.1	0	27.2	100	9.23	100	9.45	100
PME-DOD*	351.5	1.2	53.58	22.3	81.5	9.19	99.0	9.51	100.2
Crude sterols	47.6	/	/	/	/	/	/	54.7	77.9
ME-DOD	310.8	0.7	71.28	6.1	19.7	10.19	97.0	2.7	25.2

*: "Partly methyl esterified DOD," which is obtained by methyl esterification
[a]: Determined and calculated by the AOCS methods [18]
[b]: Determined and calculated by GC analysis [21]
[c]: Determined and calculated by HPLC analysis [21]

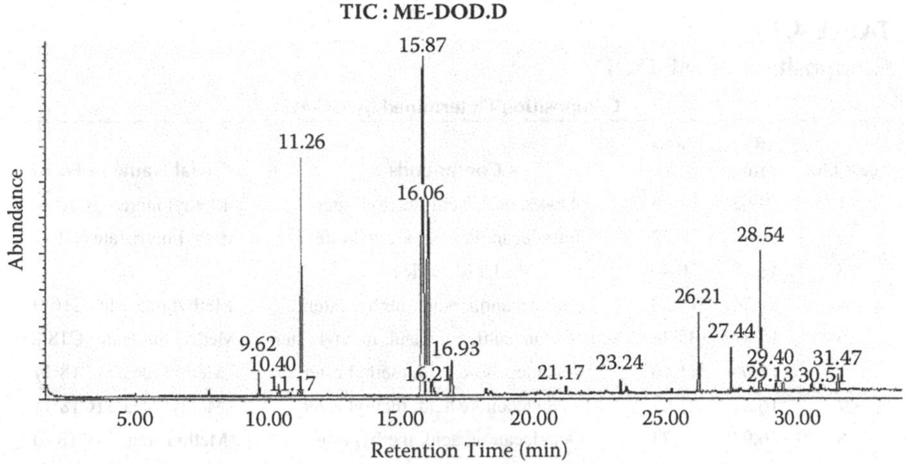

FIGURE 4.21 GC-MS TIC chromatography of ME-DOD. (From Fang, T., Goto, M., Wang, X., Ding, X., Geng, J., Sasaki, M. and Hirose, T., *J. Supercritical Fluids,* 40, 50, 2007. With permission.)

ME-DOD feed location was changed by switching on/off V11, V12, and V13. Additionally, to achieve different solvent-to-feed (S/F) ratios, the flow rate of ME-DOD was maintained at a constant 40±1 g/h while the flow rate of CO_2 was varied by adjusting micrometering valve 1 (MV 1). The total maximum feed used for each run was about 200 g, including the initial feed of 120 g before fractionation operation. In the interval between consecutive runs, the residual materials were removed out of the column by opening the bottom valve (V16) and releasing the column pressure.

4.7.2 Pretreatment Result and Composition of ME-DOD

Before the fractionation experiment, ME-DOD was prepared from DOD according to the procedure shown in Figure 4.2. Table 4.2 lists the result of the scale-up experiment with 350 Kg DOD. First of all, our pretreatment method does not cause obvious damage to tocopherols as the tocopherols' recovery in ME-DOD was 97% and simultaneously about 74.8% of sterols were removed. Noticeably, the total sterols' recovery in crude sterols and ME-DOD was larger than 100% because some sterols were released from sterol esters during pretreatment. Additionally, after methyl esterification, most FFA was converted into FAMEs, while the conversion rate of glycerides was only 18.5%, indicating that methyl esterification is not enough to simplify the complicated system of DOD and methanolysis is necessary for converting glycerides into FAMEs. Finally, as for the two products, ME-DOD and crude sterols, their total amounts were a little larger than that of initial feed because a little water was mixed into the ME-DOD after the step of washing.

The composition of ME-DOD analyzed with GC-MS is shown in Figure 4.21 and Table 4.3. A total of 18 compounds were identified. The composition was 83.16% FAMEs, 3.55% squalene, 11.18% tocopherols, and 2.11% sterols. The main FAMEs were methyl palmitate (14.57%), methyl linoleate (39.23%), and methyl oleate

TABLE 4.3
Composition of ME-DOD

Composition Determined by GC-MS

Peak No.	RT (min)	Area (%)	Compounds	Trivial Name of FAME
1	9.62	1.08	Dodecanoic acid, methyl ester	Methyl laurate (C12:0)
2	10.40	0.72	Tetradecanoic acid, methyl ester	Methyl myristate (C14:0)
3	11.18	0.40	Unidentified	/
4	11.26	14.23	Hexadecanoic acid, methyl ester	Methyl palmate (C16:0)
5	15.87	39.23	9, 12-octadecadienoic acid, methyl ester	Methyl linoleate (C18:2)
6	16.06	21.16	9-octadecenoic acid, methyl ester	Methyl oleate (C18:1)
7	16.21	1.67	7-octadecenoic acid, methyl ester	Methyl oleate (C18:1)
8	16.93	3.73	Octadecanoic acid, methyl ester	Methyl stearic (C18:0)
9	21.17	0.37	13-docosenoic acid, methyl ester	Methyl brassidate (C22:1)
10	23.25	0.57	Docosanoic acid, methyl ester	Methyl behenate (C22:0)
11	26.21	3.55	Squalene	
12	27.44	2.28	δ-tocopherol	
13	28.54	6.98	γ-tocopherol	
14	29.13	0.74	β-tocopherol	
15	29.40	1.18	α-tocopherol	
16	30.51	0.69	Campesterol	
17	30.82	0.33	Stigmasterol	
18	31.47	1.09	β-sitosterol	

Tocopherols (%) Determined by HPLC and Isomers' Percentage

Tocopherols (%)	α-Tocopherol	β and γ-Tocopherols	δ-Tocopherol
10.19	12.05	61.57	26.38

Sterols (%) Determined by GC-FID and Isomers' Percentage

Sterols (%)	Campesterol	Stigmasterol	β-Sitosterol
2.71	32.11	23.18	44.71

FAMEs (%) Determined by GC-FID and Main FAMEs' Percentage

FAMEs	C16:0	C18:2	C18:1
71.28	17.66	45.52	28.28

Source: Fang, T., Goto, M., Wang, X., Ding, X., Geng, J., Sasaki, M. and Hirose, T., *J. Supercritical Fluids,* 40, 50, 2007. With permission.

(21.16%), and the three compounds made up 90.14% of all FAMEs. Because some compounds with high molecular weight, such as glycerides, sterol esters, pigments, and wax, could not be identified with the current analyses, the area percentages of compounds were not accurate and could not be used for quantification. Thus, HPLC and GC-FID were employed for determining the contents of tocopherols, sterols, and FAMEs, respectively. Table 4.3 also shows the analysis data obtained by HPLC and

GC-FID, in which the contents of tocopherols, FAMEs, and sterols were 10.19%, 71.28%, and 2.71%, respectively. Noticeably, most FFA and glycerides were converted into FAMEs and most sterols in DOD were removed through the pretreatment process. As a result, tocopherols were partly concentrated in ME-DOD, similar to the results reported by Lee et al. [10].

From the above analysis, ME-DOD contains FAMEs (about 70%), which must be removed in the first step at the initial pressure.

4.7.3 Effect of the Initial Pressure

For the greatest degree of tocopherol enrichment inside the column, the initial pressure was investigated so that most of the FAMEs were extracted with little tocopherol content. In the first 2 hours, about 80 g ME-DOD was charged into the column and the operation was in continuous countercurrent fractionation mode, where CO_2 was the continuous phase and the feed oil was the dispersed phase. After feeding, the operation was changed to batch fractionation mode. The extracted fractions were collected in the separator. Every 30 minutes or 1 hour, the fractions were removed from the separator and weighed until the total yield from the separator reached about 70 wt.% (140 g) of the total feed (200 g). At that point, the experiment was terminated.

The phase equilibrium data in section 4.6.4 indicated that the separation factor between FAMEs and tocopherols change markedly from 15 to 20 MPa. Consequently, pressures of 14, 16, and 18 MPa were investigated for these experiments, while other operation parameters were kept at similar values throughout. The column temperature gradient was set in a linear distribution from 313 K at the bottom to 348 K at the top, the S/F ratio was adjusted to 75 (the flow rates of CO_2 and ME-DOD were 3±0.05 Kg/h and 40±1 g/h, respectively), and the feed location was V13.

Figure 4.22 shows that the extraction yield of FAMEs was greatly influenced by the initial pressures, as higher pressure led to higher solubility and faster extraction. For instance, the extraction yield at 18 MPa reached more than 70% in 2.5 hours with about 7.5 Kg CO_2, while at 14 MPa, it took far more time (10 hours) and more CO_2 (30 Kg) for the extraction yield to reach the same level. On the other hand, higher pressure resulted in more tocopherols extracted together with FAMEs, with the highest tocopherol content being 3.2% at 18 MPa, which was about three times of that at 14 MPa. This trend agreed with the common rule that solubility generally increases with pressure. However, high pressure also resulted in decrease of the selectivity, which is a disadvantage for the separation [6, 10].

For detailed discussion, average tocopherols' content (ATC) of all fractions and the average oil loading (AOL) were used for evaluating the separation efficiency, with the former standing for the purity of the FAME product and the latter representing the process velocity. AOL is defined as the ratio between the total oil mass collected and the CO_2 consumed over a given time. Practically, AOL can be calculated from the slopes of the yield curves shown in Figure 4.22, and the results are listed in Table 4.4. As pressure increased, AOL increased from 0.48 g/100g CO_2 at 14 MPa to 1.97 g/100g CO_2 at 18 MPa. Similarly, ATC also increased, indicating that more tocopherols were solvated with the FAMEs in the supercritical CO_2. This

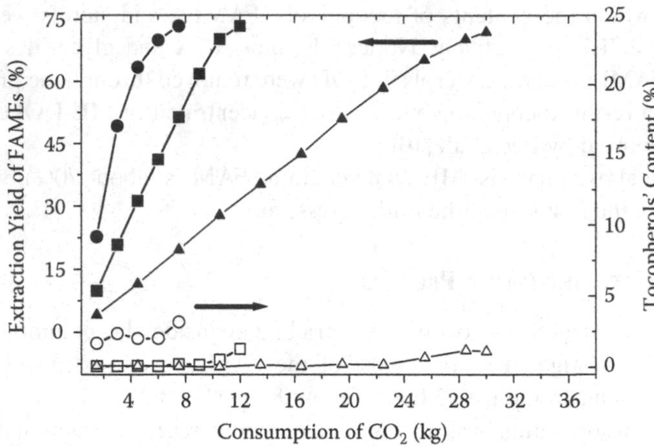

FIGURE 4.22 Effect of the initial pressure on the separation of FAMEs. Extration yield of FAMEs: ●, 18 MPa; ■, 16 MPa; ▲, 14 MPa; Tocopherols%: ○, 18 MPa; □, 16 MPa; △, 14 MPa. (Other parameters: the column temperature gradient of 313 to 348 K, S/F ratio = 75, the feed location at V13). (From Fang, T., Goto, M., Wang, X., Ding, X., Geng, J., Sasaki, M. and Hirose, T., *J. Supercritical Fluids,* 40, 50, 2007. With permission.)

TABLE 4.4
Effect of the Initial Pressure on the Separation of FAMEs

P (MPa)	Total Feed (g)	Total Yield (%)	AOL (g/100 g CO_2)	ATC (%)	Proportion of Tocopherol Isomers α	β+γ	δ
14	200	71.97	0.48	0.19	97.98	2.02	N.D.*
16	200	73.34	1.22	0.21	94.11	5.89	N.D.
18	200	73.82	1.97	2.11	16.61	71.56	11.83

*: N.D. means not determined.

Source: Fang, T., Goto, M., Wang, X., Ding, X., Geng, J., Sasaki, M. and Hirose, T., *J. Supercritical Fluids,* 40, 50, 2007. With permission.

influenced the quality of the FAME product and led to a loss of tocopherols enriched in the following process. Another phenomenon observed was that the proportion of tocopherol isomers in the fractions was greatly influenced by the initial pressure. The tocopherols at low pressures (14 and 16 MPa) were mainly α-tocopherol, whereas those present at 18 MPa were composed of the four isomers in proportions similar to that of the raw material. Hence, 16 MPa was selected as the initial pressure for separating FAMEs.

4.7.4 Effect of the Final Pressure

When the total yield reached about 70%, it meant that most FAMEs were separated from natural tocopherols, which enriched inside the column. Then the second step was commenced, wherein tocopherols were concentrated by increasing the column

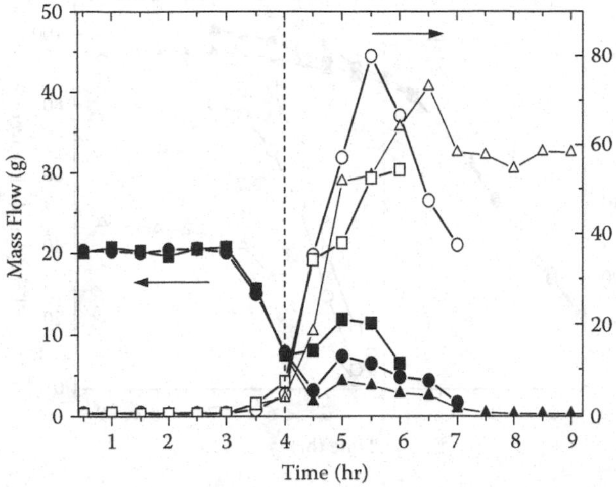

FIGURE 4.23 Effect of the final pressure on mass flow and tocopherols' content. Mass flow: ▲, 18 MPa; ●, 20 MPa; ■, 22 MPa; Tocopherols%: △, 18 MPa; ○, 20 MPa; □, 22 MPa. (Other parameters: the initial pressure = 16 MPa for 4 hours, S/F ratio = 75, the column temperature gradient of 313 to 348 K, the feed location at V12). (From Fang, T., Goto, M., Wang, X., Ding, X., Geng, J., Sasaki, M. and Hirose, T., *J. Supercritical Fluids*, 40, 50, 2007. With permission.)

pressure to a higher value (the final pressure). The experiment was terminated when the total yield reached about 85% to 90% feed, because it was found to be difficult to obtain more than 90% yield of feed in the experimental range and the extraction velocity was practically very slow (< 5 g/30 min). The total and tocopherol yields were calculated from the mass flow and tocopherol content in the fractions. We mainly investigated the influence of the final pressure on ATC and tocopherols' recovery (TR). Figure 4.23 shows the fraction change as a function of time at different final pressures, and Figure 4.24 illustrates the effect of pressure modes on total yield and TR. In the first 4 hours, there was no distinct change in the first step of pressure to 16 MPa. The mass flow was about 15 to 20 g/30 min (Figure 4.23) and, after 4 hours, the yield reached about 70 wt.% of the feed (Figure 4.24). Evidently, the mass flow decreased to less than 10 g/30 min. In addition, tocopherol yield at 16 MPa was only 3% to 5%, which is that favored concentration of tocopherols in the subsequent step at higher pressure. In the second step, at higher pressure, the mass flow and total yield increased with an increase in the final pressure, but the change in tocopherol content at 22 MPa was not as precipitous as that at 20 or 18 MPa (Figure 4.23). This indicated that, while a higher final pressure leads to an increase in the total solubility, it simultaneously results in a decrease in selectivity. The ATC of all fractions at 22 MPa was only 44.1%, although the corresponding TR was 82.8%. Both 18 MPa and 20 MPa resulted in ATCs greater than 50%, but the TR at 18 MPa was only 46.8% and lower than 81.3% at 20 MPa (Figure 4.24). Moreover, 18 MPa resulted in a longer fractionation process because of low solubility; for instance, the extracted fraction was always less than 5 g/30 min after 7 hours, and the experiment at 18 MPa was terminated at 9 hours because of too slow extraction. Therefore, 20 MPa was

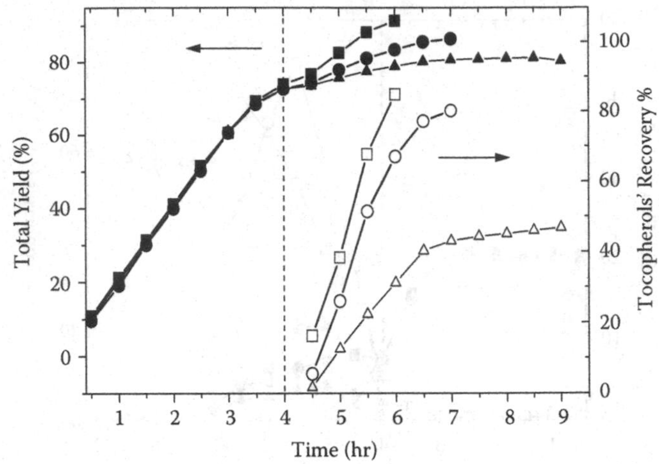

FIGURE 4.24 Effect of the final pressure on total yield and tocopherols' recovery. Total yield: ▲, 18 MPa; ●, 20 MPa; ■, 22 MPa; Tocopherols' recovery: △, 18 MPa; ○, 20 MPa; □, 22 MPa. (Other parameters: the initial pressure = 16 MPa for 4 hours, S/F ratio = 75, the column temperature gradient of 313 to 348 K, the feed location at V12). (From Fang, T., Goto, M., Wang, X., Ding, X., Geng, J., Sasaki, M. and Hirose, T., *J. Supercritical Fluids,* 40, 50, 2007. With permission.)

selected as the optimal final pressure especially, at 5.5 hours, a fraction with purity of 80.5% was obtained, which was the highest content obtained in our study.

4.7.5 Composition of Tocopherol Concentrate

According to the fractionation operation, it proved to be feasible for concentrating natural tocopherols from ME-DOD with supercritical CO_2 fractionation. With the optimized pressure parameters, about 29.1 g having 57.1% tocopherol concentrate (all fractions at 20 MPa) were obtained from 200 g ME-DOD. GC-MS was used to analyze the composition in the concentrate. The result is shown in Figure 4.25.

The area percentage of tocopherols was 68.8%, higher than the 57.1% determined by HPLC. The main reason was that there were some impurities with high molecular weight that could not be identified with the current GC-MS condition. Additionally, among the determinable compounds, FAMEs (C16:0, C18:2, C18:1, C18:0), squalene, and sterols were the main impurities, with the area percentages of 15.4%, 5.8%, and 10%, respectively.

4.7.6 Viscosity Comparison

In this work, DOD and ME-DOD were used as raw material and obtained different fractions by supercritical CO_2 fractionation. Besides the composition difference, these materials are different in physical properties. In particular, their viscosities attracted our interest since viscosity is a very important physical property commonly used in engineering design. For example, when designing the feeding system for ME-DOD, its viscosity data help determine whether preheating is necessary.

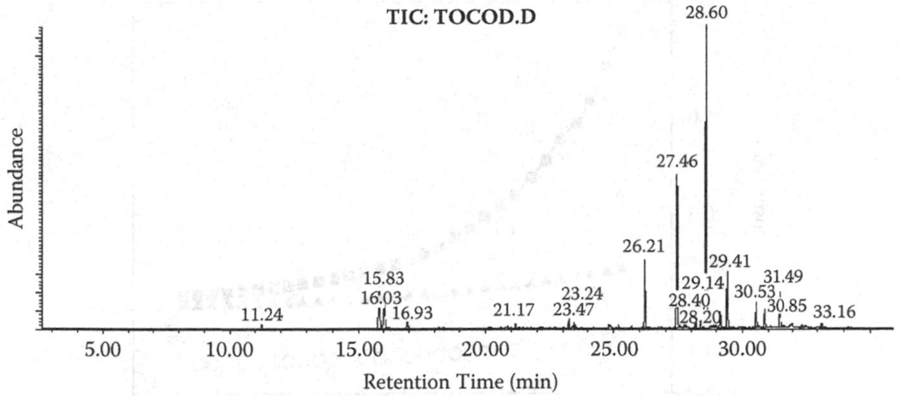

FIGURE 4.25 GC-MS TIC chromatography of 57.1% tocopherols. (From Fang, T., Goto, M., Wang, X., Ding, X., Geng, J., Sasaki, M. and Hirose, T., *J. Supercritical Fluids,* 40, 50, 2007. With permission.)

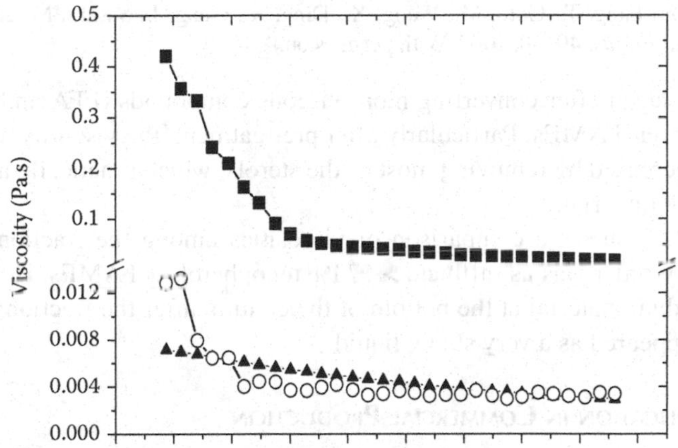

FIGURE 4.26 Comparison between the viscosities of DOD, ME-DOD, and H_2O. ■, DOD; ▲, ME-DOD; ○, H_2O. (From Fang, T., Goto, M., Wang, X., Ding, X., Geng, J., Sasaki, M. and Hirose, T., *J. Supercritical Fluids,* 40, 50, 2007. With permission.)

Figure 4.26 illustrates the viscosity changes of DOD, ME-DOD, and water as a function of temperature. When the temperature was varied from 293 to 303 K, the viscosities of the three samples decreased quickly and then the decrease tendency became relatively stable and slow at higher temperatures. Such characteristics indicated that the three samples were typically pseudoplastic fluids. In addition, the viscosity of ME-DOD is far smaller than that of DOD and similar to that of water. Thus, the pretreatment process shown in Figure 4.2 leads to two advantageous results for continuous fractionation process. One is that the converted ME-DOD has larger solubility in supercritical CO_2 than DOD; the other is that the viscosity

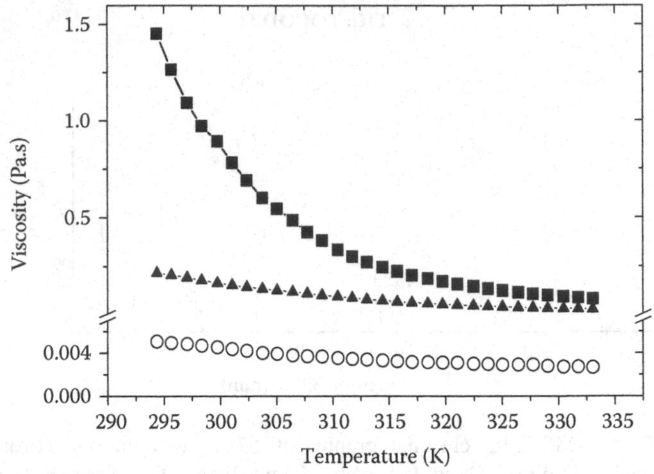

FIGURE 4.27 Comparison between the viscosities of the fractions obtained at different pressures. ■, Raffinate at 20 MPa; ▲, 57.1% tocopherols (fraction at 20 MPa); ○, FAMEs (fraction 16 MPa). (From Fang, T., Goto, M., Wang, X., Ding, X., Geng, J., Sasaki, M. and Hirose, T., *J. Supercritical Fluids,* 40, 50, 2007. With permission.)

is greatly reduced after converting more viscous compounds (FFA and glycerides) into less viscous FAMEs. Particularly after pretreatment, the viscosity of ME-DOD is further decreased by removing most of the sterols, which practically act as a kind of thickening material.

Figure 4.27 shows a comparison of viscosities among the fractions obtained. The viscosity order was as raffinate > 57.1% tocopherols > FAMEs. Here, raffinate was the residual material at the bottom of the column after the fractionation operation and it appeared as a very sticky liquid.

4.7.7 Application in Commercial Production

On the basis of the whole research, an industrial application was carried out in past years. Scale-up experiments with an 18 L column were done and the results were reported in literature [44]. In addition, a workshop (Figure 4.28) and a commercial-scale fractionation system of 350 L × 2 (Figure 4.29) were established in Kaidi Fine Chemical Industrial Co. Ltd. (Wuhan, Hubei Province, P. R. China) and the technology was industrialized with an annual process capacity of 750 t ME-DOD in the year of 2000. According to our work, three patents were applied and finally released for publication [45–47].

4.8 CONCLUSIONS

As the first step of the whole research, binary phase equilibrium data of methyl oleate + CO_2 and α-tocopherol + CO_2 were measured and correlated with the Soave-Redlich-Kwong EOS and Adachi-Sugie mixing rule. According to the obtained data, the solubility and distribution coefficient were calculated. Compared

FIGURE 4.28 Tocopherol concentration workshop established in Wuhan (Kaidi Fine Chemical Industrial Co., Ltd., Hubei, P. R. China).

FIGURE 4.29 Supercritical CO_2 fractionation system (350 L × 2) for concentrating natural tocopherols.

with methyl oleate, the solubility and distribution coefficient of α-tocopherol were generally one to two orders of magnitude lower. Such a large difference between the two components indicates that it is possible to concentrate natural tocopherols from ME-DOD in the experimental range investigated.

Second, ternary phase equilibrium data of methyl oleate + tocopherol + CO_2 were measured and correlated. On the basis of the experimental data and correlation results, the separation factor and equilibrium line were investigated. The discussion indicated that lower pressures and higher temperatures lead to a higher selectivity. Also, higher content of methyl oleate in the feed improves the distribution of tocopherols in gas, resulting in decreased selectivity. More noticeably, the experimental data on the realistic system of ME-DOD + CO_2 led to the formation of a separation strategy.

Finally, supercritical CO_2 fractionation was employed to concentrate tocopherols from ME-DOD. The initial pressure was investigated for separating FAMEs. For the following tocopherol concentration step, a final pressure of 20 MPa resulted in relatively high average tocopherol content (> 50%) and tocopherol recovery (about 80%). On the basis of the fundamental and separation research, an application in commercial production was also conducted in the past years. According to the obtained results, it can be concluded that supercritical CO_2 fractionation is technically feasible for concentrating natural tocopherols from methyl esterified DOD.

REFERENCES

1. Brunner, G., *Gas Extraction: An Introduction to Fundamentals of Supercritical Fluids and the Application to Separation Processes*, Darmstadt: Sternkopff, Springer, New York, 1994.
2. Ito, V.M., Martins, P.F., Batistella, C.B., Maciel Filho, R. and Wolf-Macie, M.R., Natural compounds obtained through centrifugal molecular distillation, *Appl. Biochem. & Biotech.*, 129–132, 716, 2006.
3. Martins, P.F., Batistella, C.B., Maciel Filho, R. and Wolf-Macie, M.R., Comparison of two different strategies for tocopherols enrichment using a molecular distillation process, *Ind. & Eng. Chem. Res.*, 45, 753, 2006.
4. Sumner, C.E., Jr., Barnicki, S.D. and Dolfi, M.D., Process for the production of sterol and tocopherol concentrates, U.S. Patent US 5424457, A 199506131995, 1995.
5. Mau, J. and Tsen, H., Investigation on the conditions for the preparation of high-purity vitamin E concentrate from soybean oil deodorizer distillate, *J. Chin. Agri. Chem. Soc. (Taipei)*, 33, 686, 1995.
6. Lucas, A., Martinez, E.O., Rincón J., Blanco, M.A. and Gracia, I., Supercritical fluid extraction of tocopherol concentrates from olive tree leaves, *J. Supercrit. Fluids*, 22, 221, 2002.
7. Zhao, Y., Sheng, G. and Wang, D., Pilot-scale isolation of tocopherols and phytosterols from soybean sludge in a packed column using supercritical carbon dioxide, in *Proc. 5th Int. Symp. on Supercritical Fluids*, Atlanta, Georgia, USA, April 8–12, 2000.
8. Zhou, Q., Sheng, G., Jiang, H. and Wu, M., Concentration of tocopherols by supercritical carbon dioxide with cosolvents, *Eur. Food Res. Tech.*, 219, 398, 2004.
9. Nagesha, G.K., Manohar, B. and Udaya Sankar, K., Enrichment of tocopherols in modified soy deodorizer distillate using supercritical carbon dioxide extraction, *Eur. Food Res. Tech.*, 217, 427, 2003.
10. Lee, H., Chung, B.H. and Park, Y.H., Concentration of tocopherols from soybean sludge by supercritical carbon dioxide, *J. Am. Oil Chem. Soc.*, 68, 571, 1991.

11. Shishikura, A., Fujimoto, K., Kaneda, T., Arai, K. and Saito, S., Concentration of tocopherols from soybean sludge by supercritical fluid extraction, *J. Jpn. Oil Chem. Soc.*, 37, 8, 1988.
12. Liu, Y., Fang, T. and Ding, X., Phase equilibrium for supercritical CO_2 and the methyl esterified product form soybean oil deodorizer distillate, *J. Food Lipids*, 13, 390, 2006.
13. Mendes, M.F., Pessoa, F.L.P. and Uller, A.M.C., An economic evaluation based on an experimental study of the vitamin E concentration present in deodorizer distillate of soybean oil using supercritical CO_2, *J. Supercrit. Fluids*, 23, 257, 2002.
14. King, J.W., Favati, F. and Taylor, S.L., Production of tocopherol concentrates by supercritical fluid extraction and chromatography, *Sep. Sci. Tech.*, 31, 1843, 1996.
15. Chang, C.J., Chang, Y.F., Lee, H., Lin, J. and Yang, P.W., Supercritical carbon dioxide extraction of high-value substances from soybean oil deodorizer distillate, in *Proc. 5th Int. Symp. on Supercritical Fluids*, Atlanta, Georgia, USA, April 8–12, 2000.
16. Brunner, G., Malchow, T., Stürken, K. and Gottschau, T., Separation of tocopherols from deodorizer condensates by countercurrent extraction with carbon dioxide, *J. Supercrit. Fluids*, 4, 72, 1991.
17. Swern, S., *Bailey's Industrial Oils and Fats,* John Wiley & Sons, New York, 1986.
18. American Oil Chemists' Society, Official Methods and Recommended Practices of the American Oil Chemists' Society, 17th ed., 2000, A.O.C.S., Washington, DC.
19. Fang, T., Goto, M., Yun, Z., Ding, X. and Hirose, T., Phase equilibria for binary systems of methyl oleate-supercritical CO_2 and α-tocopherol-supercritical CO_2, *J. Supercrit. Fluids,* 30, 1, 2004.
20. Fang, T., Goto, M., Sasaki, M. and Hirose, T., Phase equilibria for the ternary system methyl oleate + tocopherol + supercritical CO_2, *J. Chem. Eng. Data*, 50, 390, 2004.
21. Fang, T., Goto, M., Wang, X., Ding, X., Geng, J., Sasaki, M. and Hirose, T., Separation of natural tocopherols from soybean oil byproduct with supercritical carbon dioxide, *J. Supercrit. Fluids,* 40, 50, 2007.
22. Soave, G., Equilibrium constants from a modified Redlich-Kwong equation of state, *Chem. Eng. Sci.*, 27, 1197, 1972.
23. Adachi, Y. and Sugie, H., A new mixing rule-modified conventional mixing rule, *Fluid Phase Equilibrium*, 28, 103, 1986.
24. Weber, W., Petkov, S. and Brunner, G., Vapour-liquid equilibria and calculations using the Redlich–Kwong-Aspen equation of state for tristearin, tripalmitin, and triolein in CO_2 and propane, *Fluid Phase Equilibrium*, 158–160, 695, 1999.
25. Cheng, H., Zollweg, J.A. and Streett, W., Experimental measurement of supercritical fluid–liquid phase equilibrium, In *Supercritical Fluid Science and Technology, ACS Symposium Series 406*, Johnston, K.P. and Penninger, J.M.L., Eds., American Chemical Society, Washington, DC, 1989, 86.
26. Inomata, H., Kondo, T., Hirohama, S., Arai, K., Suzuki, Y. and Konno, M., Vapour-liquid equilibria for binary mixtures of carbon dioxide and fatty acid methyl esters, *Fluid Phase Equilibrium*, 46, 41, 1989.
27. Zou, M., Yu, Z.R., Kashulines, P., Rizvi, S.S.H. and Zollweg, L.A., Fluid-liquid phase equilibria of fatty acids and fatty acid methyl esters in supercritical carbon dioxide, *J. Supercrit. Fluids*, 3, 23, 1990.
28. Nilsson, W.B., Seaborn, G.T. and Hudson, J.K., Partition coefficients for fatty acid esters in supercritical fluid CO_2 with and without ethanol, *J. Am. Oil Chem. Soc.*, 69, 305, 1992.
29. Yu, Z-R., Rizvi, S.S.H. and Zollweg, J.A., Phase equilibria of oleic acid, methyl oleate, and anhydrous milk fat in supercritical carbon dioxide, *J. Supercrit. Fluids*, 5, 114, 1992.

30. Crampon, C., Charbit, G. and Neau, E., High-pressure apparatus for phase equilibria studies: Solubilities of fatty acid esters in supercritical CO_2, *J. Supercrit. Fluids*, 16, 11, 1999.
31. Chrastil, J., Solubility of solids and liquids in supercritical gases, *J. Phys. Chem.*, 86, 3016, 1982.
32. Oghaki, K., Tsukahara, I., Semba, K. and Katayama, T., A fundamental study of the extraction with a supercritical fluid. Solubilities of α-tocopherol, palmitic acid, and tripalmitin in compressed carbon dioxide at 25°C and 40°C, *Int. J. Chem. Eng.*, 29, 303, 1989.
33. Pereira, P.J., Goncalves, M., Coto, B., de Azevedo, E.G. and da Ponte, M.N., Phase equilibria of CO_2+DL-α-tocopherol at temperatures from 292 to 333 K and pressures up to 26 MPa, *Fluid Phase Equilibrium*, 91, 133, 1993.
34. Meier, U., Gross, F. and Trepp, C., High pressure phase equilibrium studies for the carbon dioxide/α-tocopherol (vitamin E) system, *Fluid Phase Equilibrium*, 92, 289, 1994.
35. Johannsen, M. and Brunner, G., Solubilities of fat-soluble vitamins A, D, E and K in supercritical carbon dioxide, *J. Chem. Eng. Data,* 42, 106, 1997.
36. Chen, C.C., Chang, C.M.J. and Yang, P.W., Vapor–liquid equilibria of carbon dioxide with linoleic acid, α-tocopherol, and triolein at elevated pressures, *Fluid Phase Equilibrium*, 175, 107, 2000.
37. Skerget, M., Kotnik, P. and Knez, Z., Phase equilibria in systems containing α-tocopherol and dense gas, *J. Supercrit. Fluids*, 26, 181, 2003.
38. Brunner, G., Malchow, T., Stuerken, K. and Gottschau, T., Separation of tocopherols from deodorizer condensates by countercurrent extraction with carbon dioxide, *J. Supercrit. Fluids*, 4, 72, 1991.
39. Stoldt, J., Saure, C. and Brunner, G., Phase equilibria of fat compounds with supercritical carbon dioxide, *Fluid Phase Equilibrium*, 116, 399, 1996.
40. Stoldt, J. and Brunner, G., Phase equilibrium measurements in complex systems of fats, fat compounds, and supercritical carbon dioxide, *Fluid Phase Equilibrium*, 146, 269, 1998.
41. Araujo, M. E., Machado, N. T. and Meireles, M.A., Modeling the phase equilibrium of soybean oil deodorizer distillates + supercritical carbon dioxide using the Peng-Robinson EOS, *Ind. Eng. Chem. Res.*, 40, 1239, 2001.
42. Bamberger, T., Erickson, J.C., Cooney, C.L. and Kumar, S.K., Measurement and model prediction of solubilities of pure fatty acids, pure triglycerides, and mixtures of triglycerides in supercritical carbon dioxide. *J. Chem. Eng. Data*, 33, 327, 1988.
43. Sato, M., Kondo, M., Goto, M., Kodama, A. and Hirose, T., Fractionation of citrus oil by supercritical countercurrent extractor with side-stream withdrawal, *J. Supercrit. Fluids*, 13, 311, 1998.
44. Fang, T., Goto, M., Liu, Q., Ding, X. and Hirose, T., Countercurrent extraction for the fractionation of natural tocopherols with supercritical CO_2, in *Proc. of the 6th Int. Symp. on Supercritical Fluids*, Tome 1, 425, Versailles, France, April 28–30, 2003.
45. Wang, X., Geng, J., Liu, C., Ding, X. and Fang, T., Process for fractionally extracting natural vitamin E by supercritical CO_2 fluid, China Patent CN1369487A, 2002.
46. Wang, X., Fang, T., He, F., Liu, C., Liu, S., Cao, Y. and Wu, X., Process for extracting vitamin E from plant-oil deodorizer distillates, China Patent CN1418877A, 2003.
47. Wang, X., Geng, J., Liu, C., Ding, X., and Fang, T., A novel column used for the continuously countercurrent operation of supercritical CO_2 fractionation, China Patent CN2561484Y, 2003.

5 Processing of Fish Oils by Supercritical Fluids

Wayne Eltringham and Owen Catchpole

CONTENTS

5.1 Introduction 141
5.2 Fish Oil Components: Sources, Properties, and Commercial Uses 142
5.2.1 Omega-3 Fatty Acids: Eicosapentaenoic Acid and Docosahexaenoic Acid 142
5.2.2 Squalene and Diacyl Glyceryl Ethers 144
5.2.3 Vitamin A (Retinol) 146
5.2.4 Wax Esters 146
5.3 Separation/Fractionation Technologies 147
5.3.1 Traditional Processing Methods 148
5.3.1.1 Distillation 149
5.3.1.2 Low-Temperature Crystallization 151
5.3.1.3 Urea Crystallization 152
5.3.1.4 Chromatographic Methods 156
5.3.1.5 Enzymatic Transformation 156
5.3.2 Supercritical Fluid Processing of Fish Oils 158
5.3.2.1 Phase Equilibria: Supercritical CO_2 and Fish Oil Components 158
5.3.2.2 Polyunsaturated Fatty Acid Processing 168
5.3.2.3 Squalene and DAGE Processing 176
5.3.2.4 Vitamin A Processing 178
5.3.2.5 Processing of Other Marine Oil Components 181
5.4 Summary 181
References 181

5.1 INTRODUCTION

Extraction and fractionation of fish oils has become a major area of research over the last 30 years because of the potential application of these extracts and fractions in the pharmaceutical, nutraceutical, and cosmetic industries. The major constituents of fish oils are triacylglycerides (TAGs). Minor components include free fatty acids, squalene, tocopherols, cholesterol, wax esters, sterol esters, phospholipids, diacylglycerides, diacylglycerol ethers, pigments, and vitamins A, D, and E. Oils of marine origin are significantly more complex than those from plants and terrestrial animals. The fatty acids that constitute TAGs in fish oils vary considerably according to degree of unsaturation, variety of chain length, and number of isomeric compounds. TAGs are typically made up of straight-chain fatty acids containing from 12 to 24 or more carbons, with the degree of unsaturation varying from zero to six double bonds. Fish oils often contain more than 60 different fatty acids, including isomers, which differ according to the position of unsaturation within the carbon chain. The complex composition of fish oils makes them difficult to process to concentrate specific fatty acids. High levels of unsaturation make the use of high-temperature processing methods problematic because of the susceptibility of these compounds to oxidative and thermal degradation. The variability in fatty acid composition of fish oil is highly dependent on fish species, season, feeding habits, part of the fish used (e.g., liver or flesh), and catch location. Some fatty acid profiles of selected fish oils are shown in Table 5.1 [1], which shows that liver oil composition is usually less saturated than that of the flesh.

This chapter discusses several important fish oil constituents, including their physical properties and applications in the nutritional, pharmaceutical, and cosmetic industries. Moreover, the chapter is intended to give an overview of the various laboratory-to-production scale techniques that can be used to extract and fractionate various fish oil components. A comparison of various extraction and fractionation techniques is discussed, including how some of the problems associated with "traditional" processing methods can be overcome using supercritical fluid (SCF) technologies.

5.2 FISH OIL COMPONENTS: SOURCES, PROPERTIES, AND COMMERCIAL USES

5.2.1 Omega-3 Fatty Acids: Eicosapentaenoic Acid and Docosahexaenoic Acid

Production of omega-3 (ω3) fatty acid concentrates continues to be a topic of interest for both the pharmaceutical and health food industries. Since the early studies on long-chain ω3-polyunsaturated fatty acids (PUFAs) by Burr and Burr [2], the health benefits of these compounds have been studied extensively. With the growing public awareness of the health benefits of ω3-PUFAs, the market for such products is expected to grow, increasing the demand for efficient production and isolation methods. The most widely studied ω3 fatty acids are eicosapentaenoic acid (EPA) and docosahexaenoic acid (DHA) (Figure 5.1). Extensive clinical findings on their effects on human physiology and their use in the prevention and treatment of diseases such as arteriosclerosis [3,4], thrombosis [5], arthritis [6], and several types of cancer [7,8] have been reported.

Omega-3 fatty acids of marine origin can reduce serum TAG levels [9]. This is of particular importance because raised levels of serum TAGs are considered to be a risk factor for coronary heart disease [10]. Urakaze et al. [5] showed that intravenous injections of an EPA-containing emulsion can increase the EPA content in plasma and platelet phospholipids. Their study showed that platelet aggregation is significantly depressed and that EPA-containing emulsions may be useful for patients requiring preventative care of thrombosis. Other medical studies report using ω3 fatty acids

TABLE 5.1
Fatty Acid Profiles of Selected Fish Oils

	Salt Water Oils										Freshwater Oils	
	Atlantic Cod	Atlantic Cod Liver	Spiny Dogfish	Spiny Dogfish Liver	Pacific Halibut	Pacific Herring	Mackerel	Menhaden	Striped Mullet	Pink Salmon	Lake Herring	Rainbow Trout
Fatty Acid	Weight (%) of Total Fatty Acids											
14:0	1.8	2.8	2.0	1.6	2.8	7.6	4.9	8.0	4.6	3.4	5.5	2.1
15:0	0.5	0.4	0.5	0.3	0.3	0.4	0.5	0.5	6.3	1.0	0.4	0.8
16:0	33.4	10.7	21.2	13.2	15.1	18.3	28.2	28.9	17.3	10.2	17.7	11.9
16:1	2.4	6.9	6.0	5.7	8.9	8.3	5.3	7.9	11.0	5.0	7.1	8.2
16:2	0.6	1.0	0.9	1.0	0.8	1.0	0.7	0.8	3.8	1.7	0.7	1.2
17:0	0.9	1.2	1.2	1.0	0.7	0.5	1.0	1.0	0.8	1.6	0.6	1.5
18:0	4.0	3.7	2.7	4.3	3.4	2.2	3.9	4.0	5.0	4.4	3.0	4.1
18:1	11.9	23.9	27.5	28.5	25.7	16.9	19.3	13.4	8.4	17.6	18.1	19.8
18:2 ω6	1.2	1.5	1.3	0.7	0.9	1.6	1.1	1.1	3.2	1.6	4.3	4.6
18:3	0.8	0.9	0.6	0.6	0.3	0.6	1.3	0.9	.4	1.1	3.4	5.2
18:4 ω3	1.2	2.6	0.7	0.8	3.6	2.8	3.4	1.9	3.0	2.9	1.8	1.5
19:0		0.6		0.7				0.9	1.5	1.6	0.8	0.9
20:1	1.6	8.8	5.8	1.5	8.0	9.4	3.1	0.9	0.7	4.0	1.2	3.0
20:4 ω6	3.2	1.0	2.5	0.8	2.5	0.4	3.9	1.2	2.6	0.7	3.4	2.2
20:5 ω3 (EPA)	12.4	8.0	7.9	3.7	10.1	8.6	7.1	10.2	7.5	13.5	5.9	5.0
22:1	0.7	5.3	4.1	10.3	5.1	11.6	2.8	1.7	0.7	3.5	2.8	1.3
22:6 ω3 (DHA)	21.9	14.3	10.4	6.5	7.9	7.6	10.8	12.8	13.4	18.9	13.3	19.0
Total saturated	40.6	19.4	27.6	21.1	22.3	29.0	38.5	43.3	35.5	22.2	28.0	21.3
Total monounsaturated	16.6	44.9	43.4	46.0	47.7	46.2	30.5	23.9	20.8	30.1	29.2	32.3
Total polyunsaturated	41.3	29.3	24.3	14.1	26.1	22.6	28.3	28.9	33.9	40.4	32.8	38.7
Total ω3	34.3	22.3	18.3	10.2	18.0	16.2	17.9	23.0	20.9	32.4	19.2	24.0

Source: Adapted with permission from *Journal of the American Oil Chemists' Society,* 41, 662. ©1964 American Oil Chemists' Society.

for the reduction of blood pressure [11,12]. A study by Kremer and coworkers [6] reported the effect of EPA intake in patients suffering from rheumatoid arthritis. After a 12-week diet high in polyunsaturated fat, low in saturated fat, and with a daily supplement of EPA (1.8g), patients noted a decrease in morning stiffness and joint pain. Other workers have also reported anti-inflammatory activity of ω3 fatty acids [13]. Researchers at the Paterson Institute (Manchester, UK) have identified a mechanism by which ω3 fatty acids may prevent the development of metastatic

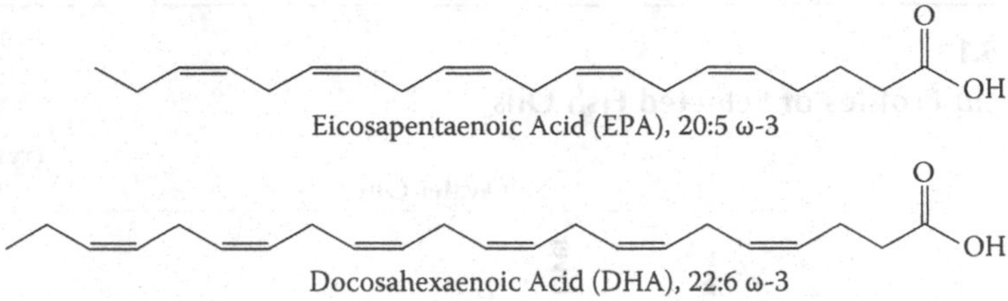

FIGURE 5.1 The chemical structures of EPA and DHA.

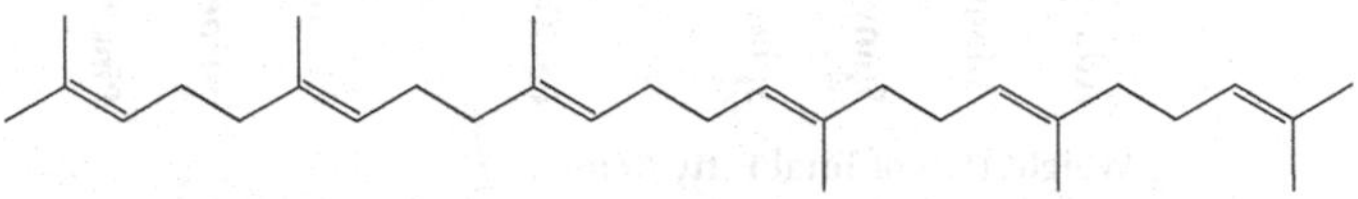

FIGURE 5.2 The chemical structure of squalene ($C_{30}H_{50}$).

prostate cancer [14]. They showed that the ω6 polyunsaturated fatty acid arachidonic acid (20:4 ω6) is a potent stimulator of malignant epithelial cellular invasion, which can increase the risk of prostate cancer development. They stated that the observed cellular invasion is inhibited by EPA and DHA in the ratios of ω3 1:2 ω6. Several other studies have also focused on the anticarcinogenic properties of EPA [15–17]. Omega-3 fatty acids, especially EPA, also show promise in the treatment of neuropsychiatric disorders, such as depression [18], schizophrenia [19], and anorexia nervosa [20].

Both EPA and DHA occur naturally in the body, where they have been shown to be important in membrane structure and function [21]. They are found in especially large amounts in brain cells, eyes, nerves, and adrenal glands. In particular, DHA is one of the most abundant constituents of brain structural lipids, where it has important effects on membrane order (fluidity), the activity of membrane-bound enzymes, and signal transduction. DHA is considered to be essential for the visual and neurological development of infants. The lipid fraction of human mother's milk contains DHA-to-EPA ratios of 4:1, with DHA content being 30 times more than the amount of DHA observed in cows' milk lipid. In the U.S., 80% of infant formulas contain DHA so that children receive this nutrient during the important phases of brain and nervous system growth and development.

5.2.2 Squalene and Diacyl Glyceryl Ethers

The natural occurrence of squalene was first reported in 1906 by Mitsumaru Tsujimoto, an industrial engineer who pioneered the chemistry of fats and oils in Japan. Thirty years later, Nobel laureate Paul Karrer described the chemical structure of squalene for the first time. Squalene is a 30-carbon isoprenoid (Figure 5.2) that is used commercially as an additive in pharmaceutical preparations, cosmetics, sunscreens, dyes, lubricants, and health foods [22, 23]. During the 1950s, squalene was found to occur naturally in the human body [24]. Human sebum, a natural product expressed by sebaceous glands, contains around 12% squalene. Sebum helps keep the skin supple and helps to maintain skin moisture levels. Squalene was later found to be

TABLE 5.2
Shark Liver Oil Compositions for Some Selected Species of Shark

Shark Species	Common Name	Squalene	DAGEs	TAGs	Other[d]	Ref.
Carcharhinus plumbeus[a]	Sandbar shark	0.0	0.0	83.0	17.0	36
Centrophorus scalpratus[b]	Endeavour shark	81.6	9.9	8.5	0.0	37
Centrophorus squamosus[c]	Leafscale gulper shark	70	11	18	1.0	36
Centroscymnus crepidater[b]	Long-nose velvet shark	73.0	20.0	5.0	2.0	37
Centroscymnus plunketi[b]	Plunket shark	0.9	76.6	22.5	0.0	37
Dalatias licha[c]	Kitefin shark	79	18	2	1.0	36
Deania calcea[b]	Platypus shark	69.6	1.6	10.8	18.0	37
Etmopterus granulosus[b]	Lantern shark	50.3	32.1	9.3	8.3	37
Hexanchus griseus[a]	Bluntnose sixgill shark	1.0	70.0	29.0	0.0	36
Somniosus pacificus[b]	Pacific sleeper shark	—	49.5	49.1	1.4	37
Squalus acanthias[c]	Piked dogfish	0.0	12.0	87.0	1.0	36

[a] Hawaiian waters, [b]Southern Australian waters, [c]Chatham Rise, New Zealand, [d]Including free fatty acids (FFA), phospholipids, sterols, pristane, wax esters, and sterol esters.

the precursor to cholesterol [25], which in turn is the precursor to a range of steroids and vitamin D.

Gloor and Karenfeld [26] found that the body's consumption of squalene increases when skin is exposed to ultraviolet (UV) radiation. Kohno and coworkers [27] showed that the first molecule targeted by UV radiation in the skin was squalene. These results led to more intense research on the interaction of squalene with UV radiation and, in 1995, Kohno et al. [28] demonstrated that squalene can prevent UV-induced oxidation of lipids in the skin. Radioprotective and detoxifying effects of squalene have also been reported [29]. Dietary squalene may have the potential to lower blood cholesterol levels [30, 31]. A number of research papers suggest that squalene shows potential as an anticarcinogen [32–34]. For a more thorough treatise on the properties of squalene, its action in metabolic pathways, and its possible preventative capabilities for human disease, the reader is directed to a work by Das [35].

The reports on the beneficial effects of squalene on human health have led to a commercial demand for this long-chain unsaturated hydrocarbon. Although squalene is naturally found in small quantities in olive oil and by-products from the refining of olive oil, wheat germ oil, rice bran oil, and yeast, the most abundant source by far is the livers of deep-sea sharks. Other major components found in the livers of deep-sea sharks include diacyl glyceryl ethers (DAGEs) and TAGs. Because sharks do not possess swim bladders, the presence of large quantities of low-density oils (squalene density 0.86 g cm^{-3}; DAGE density 0.89 g cm^{-3}) in their livers allows them to achieve and maintain buoyancy. Table 5.2 shows some typical compositions of shark liver oil for several species of shark [36, 37]. DAGEs are considered to be efficient in wound healing applications and in preventing the multiplication of bacteria. Some researchers have also suggested that DAGEs may aid in the reduction of certain types of cancer, promote formation of blood cells, and provide protection against radiation

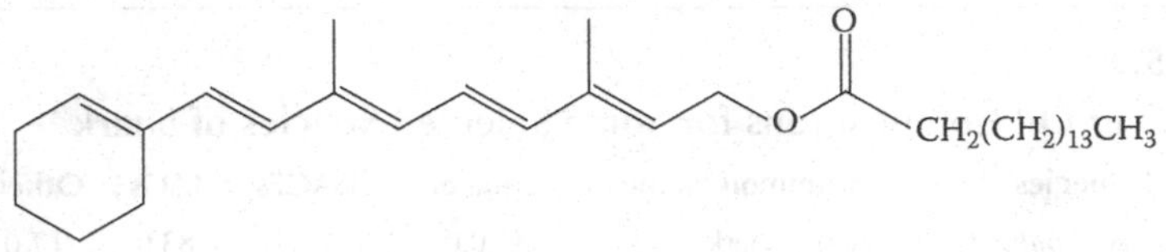

FIGURE 5.3 The chemical structure of vitamin A palmitate.

injury [38]. Shark liver oils with defined levels of DAGEs and squalene are currently sold as nutraceuticals.

5.2.3 Vitamin A (Retinol)

Vitamin A, a fat-soluble compound, is derived in the bodies of animals from β-carotene (also known as *provitamin A*). Vitamin A, carried in the blood by retinol-binding protein, has been identified as an essential nutrient necessary for growth, reproduction, and vision. In the body, it is oxidized to retinal, a key component of the visual system, and to retinoic acid, which effects gene expression via specific nuclear receptors. The discovery of an anticarcinogenic action of vitamin A by Saffiotto et al. [39] led to rapid growth in vitamin A research. In animal experiments, vitamin A and its metabolite, retinoic acid, were shown to have anticancer properties. Retinoic acid is now an ingredient used in anti-ageing cosmetics. Vitamin A–rich creams are used for severe cases of acne [40]. For reviews describing the discovery, structure elucidation, and roles of vitamin A in the human diet, the reader is directed to the publications of Wolf [41] and Underwood [42].

The importance of vitamin A in human health has lead to a requirement for dietary supplements in the form of capsules and food fortification. The livers of cartilaginous fish, such as sharks and rays, are a potentially valuable source of vitamin A. Vitamin A is found in fish liver oils almost entirely (96–100%) as esters [43], such as vitamin A palmitate (Figure 5.3).

5.2.4 Wax Esters

Oils of certain deep-sea species of fish are composed almost entirely of wax esters, which are esters of long-chain fatty acids and fatty alcohols. Fish species in the South Pacific whose oils are composed of wax esters include orange roughy, black oreo, and small-spined oreo (Table 5.3) [44]. The fatty acid portions of the esters have carbon chain lengths of C_{14}–C_{24} and are either saturated or monounsaturated. The fatty alcohols have carbon chain lengths of C_{16}–C_{24} and are mostly saturated (Table 5.4) [44]. Wax esters with unsaturation near the ester bond are more volatile than wax esters with unsaturation near the center of the aliphatic chain but have similar oxidative stability [45]. In teleost fish (those with a bony skeletal structure), the occurrence of wax esters in muscle tissue correlates better with sea depth and vertical migration patterns than with taxonomy [46].

In fish, wax esters appear to function as sources of energy, insulation, and aids to buoyancy and biosonar [47]. In industry, they have applications as special lubricants, cosmeceuticals, and chemical raw materials for the manufacture of soaps and

TABLE 5.3
Total Lipid Composition and Composition of the Wax Ester Fraction of Orange Roughy, Black Oreo, Small Spined Oreo, and Sperm Whale

Component	Orange Roughy	Black Oreo	Small Spined Oreo	Sperm Whale
	Weight (%) of Oil			
Wax esters	94.9	91.5	95.6	65.8
TAGs	3.1	4.8	2.5	30.1
Cholesterol/alcohols	1.0	2.7	1.5	4.0
Phospholipids	1.0	1.0	0.4	0.1
Total Carbon Number	Weight (%) of Wax Ester Component			
C_{26}				4.7
C_{28}				14.0
C_{30}	0.2	0.5	0.5	21.1
C_{32}	2.1	3.5	2.9	23.2
C_{34}	11.4	11.6	9.3	19.9
C_{36}	16.7	21.8	18.3	11.7
C_{38}	24.8	21.3	26.2	4.4
C_{40}	23.4	19.8	25.4	
C_{42}	14.8	10.8	12.8	
C_{44}	5.5	6.1	4.3	
C_{46}	1.1	4.6	0.3	

Source: Adapted with permission from *Journal of the American Oil Chemists' Society*, 59, 390. ©1982 American Oil Chemists' Society.

detergents [48]. Oil from the head of the sperm whale, which contains more than 65% wax esters (Table 5.3), is particularly suitable for these applications. Owing to the present status of the sperm whale as an endangered species, however, the oil is no longer an item of commerce and alternative sources have been investigated. A study of the chemical and physical properties of orange roughy oil showed that it could readily replace sperm whale oil [44]. The production of orange roughy oil has been commercialized in New Zealand, with several plants producing thousands of tonnes per year [49].

5.3 SEPARATION/FRACTIONATION TECHNOLOGIES

There are many methods for the fractionation of fish oils, but only a few are suitable for large-scale production. The suitable methods, which usually require conversion of TAGs to fatty acids or ethyl esters, include adsorption chromatography, fractional or molecular distillation, enzymatic splitting, low-temperature crystallization, urea complexation and SCF extraction/fractionation techniques. Standard oil refining technologies are not considered here.

TABLE 5.4
Percentages of Fatty Acids and Fatty Alcohols of the Whole Fish Wax Esters of Orange Roughy, Black Oreo, Small Spined Oreo, and Sperm Whale

	Orange Roughy % Fatty		Black Oreo % Fatty		Small Spined Oreo % Fatty		Sperm Whale % Fatty	
Component	**Acid**	**Alcohol**	**Acid**	**Alcohol**	**Acid**	**Alcohol**	**Acid**	**Alcohol**
				Saturated				
< 14:0							21.6	
14:0	1.2		4.1	1.9	6.8		9.4	8.0
15:0	< 0.1		0.8	< 0.1	0.7		0.9	1.4
16:0	1.0	7.3	15.5	20.8	8.1	9.4	5.1	39.5
17:0	0.7		1.1	0.8	3.8		0.4	1.1
18:0	0.3	8.1	3.2	2.3	3.7	0.9	1.5	7.7
19:0				0.6				0.2
20:0	< 0.1		0.2		< 0.1		< 0.1	
22:0	< 0.1		< 0.1		< 0.1		< 0.1	
24:0	< 0.1		< 0.1		< 0.1		< 0.1	
				Unsaturated				
14:1	0.5		0.3		0.4		19.6	
15:1	< 0.1		0.7		0.1		0.2	0.2
16:1	11.8		7.9		10.9		15.6	4.1
17:1	1.0		0.8	0.8	3.7		1.3	1.0
18:1	56.0	34.6	26.9	19.0	32.8	23.3	17.8	35.4
18:2	1.9		1.0		0.9		0.5	
20:1	17.8	30.6	15.8	29.6	16.5	33.7	3.9	1.4
22:1	7.8	13.8	11.6	20.3	9.5	31.6	1.4	
23:1	< 0.1		1.8		2.1		0.8	
24:1	< 0.1	5.4	8.3	3.9	< 0.1	1.1	< 0.1	

Source: Adapted with permission from *Journal of the American Oil Chemists' Society*, 59, 390. ©1982 American Oil Chemists' Society.

5.3.1 Traditional Processing Methods

Separation of fish oil fatty acids and esters using traditional methods is complicated by several factors. First, methods relying on differences in molecular weight, such as distillation, are hindered by the relatively small molecular weight differences in these compounds, especially when attempting to separate saturated and unsaturated fatty acids of the same chain length. Second, PUFAs are readily susceptible to degradation, oxidation, polymerization, and stereomutation, even at moderately elevated temperatures. Berdeaux et al. [50] reported the thermal degradation of PUFAs during the deodorization of fish oils. The high processing temperatures associated with vacuum distillation can be problematic because ω3 fatty acids are susceptible to oxidative deterioration due to their high degree of unsaturation. The primary oxidation products, lipid hydroperoxides, are especially unstable and can degrade to yield volatile secondary oxidation products. The secondary oxidation products can impart undesirable, unpleasant fishy odors and flavors to end products. The mixture of TAGs in fish oils is too complex for efficient isolation of individual components and often only modest enrichments can be expected from fractionation. Therefore, most efforts have been directed toward the fractionation of acids and their methyl/ethyl esters. Dealing with these single-chain compounds allows differences in chain length or degree of unsaturation to be effectively addressed. The acids and esters have greater volatility than TAGs, allowing the use of temperature-controlled separation methods, such as molecular distillation. After separation, specific TAGs or free acids can be reconstituted. Most separation methods, when used alone, can only separate fish oil acids into group fractions. Therefore, two or more procedures are often required to produce individual components in high purity.

5.3.1.1 Distillation

Distillation relies on differences in mixture component vapor pressures, which are strongly related to molecular weights for a homologous family of compounds. Enrichment is achieved by exploiting the differences in vapor pressure through countercurrent contacting of vapor and liquid phases in stages using plates or continuously using random or structured packing. If we assume a binary (or pseudobinary) mixture containing components A and B, then the degree of attainable enrichment is dependent on the ratio of the individual component vapor pressures, p_A and p_B. Assuming A is the more volatile component, the separation factor, $\alpha = p_A/p_B$, is greater than unity and some degree of separation can be achieved.

Figure 5.4 shows an idealized vapor-liquid composition diagram for a mixture of A and B that is being separated in a stage-wise distillation column. At the feed point of the distillation process, the concentration of A in the liquid phase is $\mathbf{X}_0$ and that in the vapor phase is $\mathbf{Y}_0$. The vapor-liquid equilibrium curve predicts that the initial concentration of A in the vapor phase is greater than that in the liquid phase by virtue of its greater volatility. If the vapor is condensed (condensation represented by the horizontal lines drawn between the vapor-liquid equilibrium curve and the auxiliary line in the stripping or enrichment section of the column), the resulting liquid will have a concentration of A equal to $\mathbf{Y}_0/\boldsymbol{K}_A = \mathbf{X}_1$, where $\boldsymbol{K}_A$ is the partition coefficient (K-value) of component A. The K-value of any component i is defined as:

$$K_i = \frac{y_i}{x_i} \tag{5.1}$$

where $\boldsymbol{y}_i$ is the concentration of component i in the vapor phase and $\boldsymbol{x}_i$ is the concentration of component i in the liquid phase.

Similarly, the vapor in equilibrium with the liquid phase of concentration $\mathbf{X}_1$ has a concentration of A equal to $\mathbf{Y}_1$. Again, the concentration of A in the vapor phase is enhanced due to its greater volatility. If this vapor is then condensed, the resulting

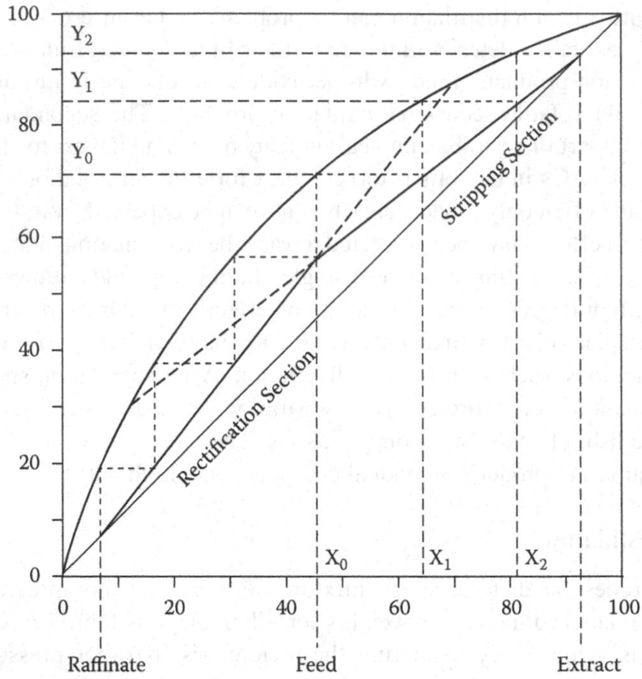

FIGURE 5.4 Separation of a hypothetical mixture of components A and B by distillation.

liquid has a mole percent concentration of A according to the same relationship—that is, $\mathbf{Y_1/}\boldsymbol{K_A}\mathbf{ = X_2}$. Similar relationships hold in the rectification section, where the "heavy" component B is concentrated. When these steps are carried out continuously in a multistage or continuously packed device, it is theoretically possible to obtain highly pure A as the extract and highly pure B as the raffinate. Alternatively, the evaporation and subsequent condensation steps can be carried out as a batch-wise fractional distillation process until the desired degree of enrichment with respect to A has been achieved.

Stout et al. [51] highlighted the practical difficulty of obtaining ω3-PUFAs in high concentrations in the natural TAG form. Molecular distillation of menhaden oil in its natural TAG form increased the EPA content in the residue from 16.0% to 19.5% and the DHA content from 8.4% to 17.3%. Carrying out the distillation using the menhaden oil ethyl esters, however, almost doubled the EPA concentration from 15.9% to 28.4%, whereas the concentration of DHA showed an almost fivefold increase, from 9% to 43.9%.

Fractional distillation of fish oils is preferentially carried out using fatty acid esters under reduced pressure (0.1 to 10 torr) since these, unlike free fatty acids, approximate ideality. Also, the greater volatility of the esters allows separation to be carried out at lower temperatures, which is important considering the thermal instability of the ω3 components. Boiling points of unsaturated esters are marginally lower than those of saturated esters of the same chain length. Therefore, for a given chain length, unsaturated esters are enriched during the early stages of fractionation and saturated ones are enriched in the end fractions. During the transition from one chain length to the next, the distillate contains a mixture of saturated lower chain length and unsaturated higher chain length components.

Although the greater volatility of the fatty acid esters allows the use of lower process temperatures (compared with temperatures required for free fatty acids), temperatures are still moderately high (typically 423 to 473 K), and exposure to distillation conditions over a prolonged period of time can be detrimental to the polyunsaturated constituents, causing hydrolysis, polymerization, isomerization, and thermal oxidation [52]. Privett and coworkers [53, 54] found considerable decomposition of arachidonate during slow distillation in a spinning band column. Fractionation of marine oil esters containing chain lengths of C_{20} or more is difficult because separation factors decrease with increasing molecular weight [55].

5.3.1.2 Low-Temperature Crystallization

Crystallization separations are based on the differences in composition of equilibrated liquid and solid phases. The process can be carried out using the crude liquid oil or in a solvent solution. An operation with the crude liquid oil requires slow cooling and slow agitation. This produces a slurry of solid and liquid components, the latter being enriched with PUFAs. When using a solution, the equilibrium is dependent on component solubilities. Common solvents of choice include methanol, acetone, petroleum ether, acetonitrile, nitromethane, and liquid propane. The formation of solid crystals inevitably results in entrapment of some liquid, and so separation factors are not high. The solubility of fats in organic solvents increases with increasing unsaturation and decreases with increasing molecular weight [56]. Singleton [57] and Stout et al. [51] have measured the solubility of several fatty acids in a variety of solvents. Their findings led to the following rules:

- Long-chain saturated fatty acids are less soluble than short-chain saturated fatty acids.
- Saturated acids are less soluble than monounsaturated and diunsaturated acids of the same chain length.
- *Trans* isomers are less soluble than *cis* isomers.
- Straight-chain acids are less soluble than branched-chain ones.
- Free fatty acids are always less soluble than their methyl ester counterparts.

The melting points of fatty acids vary considerably according to their degree of unsaturation. This wide variation in melting point can be exploited to enable the separation of saturated and unsaturated fatty acid compounds [58]. At low temperatures, long-chain saturated fatty acids (having higher melting points) crystallize out, leaving the PUFAs in solution. The free acids are prone to association in solution. Although this can be overcome by processing the methyl ester analogues, the practical advantage is limited because of the greater solubility of the esters. Also, the formation of eutectics prevents achievement of the degree of fractionation one would predict from solubility data alone. The choice of solvent and processing temperature must be considered carefully because these factors can have a pronounced effect on

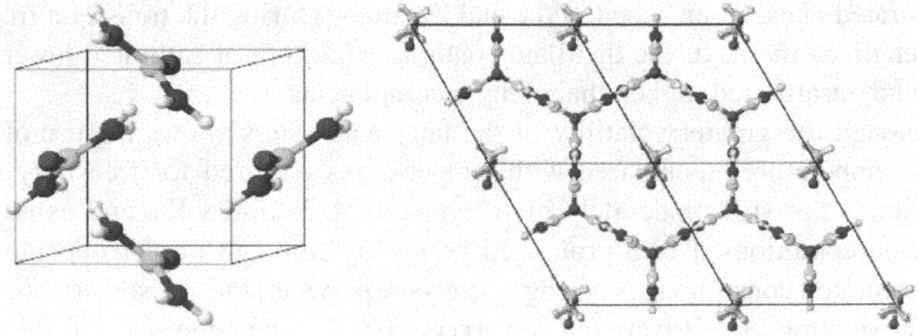

(a) Pure Urea (b) Palmitic Acid ($C_{16:0}$)/Urea Inclusion Complex

FIGURE 5.5 Crystal structures for pure urea and urea–fatty acid complexes.

the concentration of PUFAs obtained. There is also the possibility of separating fatty acids on the basis of degree of unsaturation [59]. DHA (or salts thereof) crystallizes at a lower temperature than EPA in solvents such as acetone. However, the very low temperatures required make the process somewhat impractical [60]. Until the introduction of SCF fractionation techniques, low-temperature crystallization was considered to be the method least detrimental to polyunsaturated fatty acids.

5.3.1.3 Urea Crystallization

Urea complexation with straight-chain compounds was first reported by Bengen [61]. While pure urea crystallizes in a tightly packed tetragonal structure (Figure 5.5a), in the presence of long straight-chain molecules, it crystallizes into a hexagonal structure to form an inclusion complex (Figure 5.5b) [62]. The urea complex is constructed of a spiral arrangement of urea molecules, and the straight-chain molecules are held in the hexagonal channels by van der Waals forces, London dispersion forces, or induced electrostatic interactions [63]. The hexagonal channel is wide enough to accept molecules with a diameter of around 5 Å, but molecules with greater diameters are not easily accommodated. The formation and stability of urea complexes is therefore governed by shape, size, and geometry.

The use of urea crystallization separates fatty acids mainly according to their degree of unsaturation. When urea crystallizes from a solution containing a mixture of fatty acids, the saturated and monounsaturated fatty acids are preferentially included in the complex while the polyunsaturated fatty acids remain in solution. This is because the presence of double bonds in fatty acids prevents the chain from orienting into the "ideal" geometry for complex formation. This causes an imbalance of the optimum intermolecular distances, which disrupts the net attractive forces and destabilizes the inclusion complex. The stability of urea inclusion complexes is lessened by shorter chain lengths and a higher number of double bonds [64]. *Trans* isomers form more stable complexes than the corresponding *cis* isomers, and compounds with conjugated double bonds form more stable adducts than those with methylene interrupted or isolated double bonds. The most influential processing variables affecting the extent of crystallization are the urea–fatty acid ratios and the process temperature. The urea–fatty acid ratio can be used to fractionate fatty acids according to differing degrees of unsaturation. When low concentrations of urea

TABLE 5.5
The Effect of Urea/Fatty Acid Ratio on the Noncomplexed Fatty Acid Composition Resulting from Urea Fractionation of Cod Liver Oil at 277 K

		Urea/Fatty Acid Ratio			
Fatty Acid	**Cod Liver Oil**	**1:1**	**2:1**	**3:1**	**4:1**
14:0	4.2	2.7	0.7	0.5	0.7
16:0	10.6	2.0	0.2	0.5	0.0
16:1 ω7	7.8	9.6	6.9	2.5	3.2
18:0	2.6	0.9	0.1	0.0	0.0
18:1 ω9	17.0	17.6	3.2	2.9	0.7
18:1 ω7	4.6	5.9	1.4	1.0	0.0
18:2 ω6	1.5	2.0	1.6	0.7	0.7
18:3 ω6	0.2	0.2	0.4	0.5	0.5
18:3 ω3	0.8	1.1	1.0	0.6	0.6
18:4 ω3 (SA)[a]	2.4	3.3	6.3	8.0	8.5
20:0	0.2	0.2	0.2	0.0	0.1
20:1 ω9	10.8	9.0	1.3	0.6	0.8
20:3 ω6	0.1	0.1	0.1	0.2	0.2
20:4 ω6	0.5	0.8	1.0	0.9	1.0
20:5 ω3 (EPA)	9.4	13.0	22.6	24.8	25.6
22:0	0.1	0.0	0.0	0.0	0.0
22:1 ω11	8.3	3.8	0.4	0.0	0.0
22:1 ω9	0.1	0.0	0.0	0.0	0.0
22:4 ω6	0.5	0.7	1.5	1.7	1.8
22:5 ω3	1.2	1.6	2.4	1.4	1.6
22:6 ω3 (DHA)	11.0	15.8	45.4	58.2	59.9
24:0	0.0	0.0	0.0	0.0	0.0

Source: Reprinted with permission from *Journal of the American Oil Chemists' Society*, 72, 575. ©1995 American Oil Chemists' Society.

[a] Stearidonic acid.

are used, the different fatty acids compete for complex formation according to the ability to form the most stable inclusion compound.

Robles Medina et al. [65] studied the effect of urea–fatty acid ratio on the fatty acid composition of urea filtrates (noncomplexed fatty acids) from cod liver oil. Table 5.5 shows their results from urea crystallization at 277 K and Figure 5.6 shows the effect of these ratios on various fatty acid groups of interest. With a urea–fatty acid ratio of 1:1, the saturated fatty acids were partially removed from solution, while the concentration of monounsaturates remained constant. Increasing the ratio to 2:1 facilitated maximum removal of the saturated acids along with partial removal of the monounsaturated acids. At a ratio of 3:1, the removal of saturated and

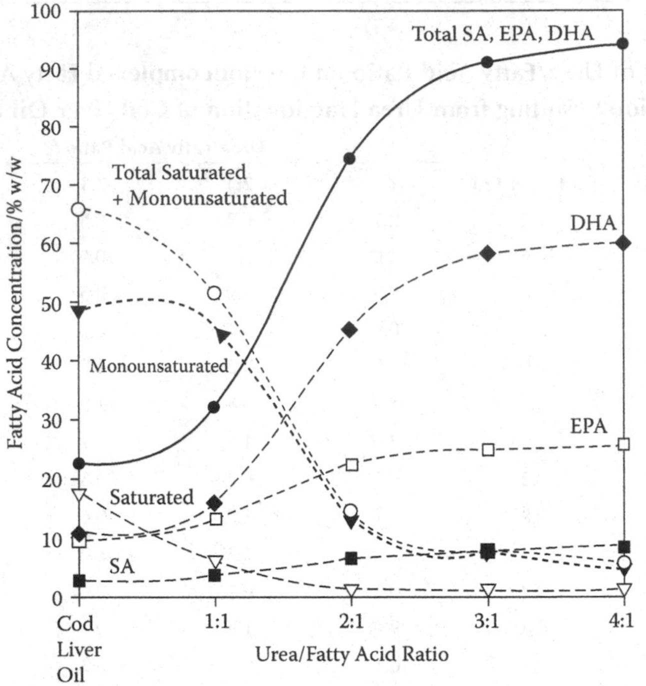

FIGURE 5.6 The effect of urea–fatty acid ratio on the noncomplexed fatty acid composition resulting from urea fractionation of cod liver oil at 277 K.

monounsaturated fatty acids were at a maximum. Increasing the ratio to 4:1 did not greatly increase the removal of the saturated and monounsaturated components from the solution. Increasing the urea–fatty acid ratios also increased the concentration of stearidonic acid (SA), EPA, and DHA in the filtrate. The maximum enrichment of PUFAs was also observed at an optimum urea–fatty acid ratio of 3:1, a phenomenon noted by other research groups [66–68]. In the same study, Robles Medina and coworkers [65] investigated the effect of temperature on the fatty acid composition of urea concentrates from cod liver oil using a urea–fatty acid ratio of 4:1; Figure 5.7 is compiled from their data. The filtrate concentrations of SA and DHA were found to be greatest at 261 K, the total ω3 fatty acid concentrations were found to be greatest at 277 K, and the concentration of EPA in the filtrate was maximized at around 293 K. Although different temperature values were reported, this trend on the influence of temperature change is similar to that observed by Willes and coworkers [66], who used urea fractionation and high-performance liquid chromatography (HPLC) to produce fractions rich in PUFAs from fish oils. During urea crystallization, they reported that the concentrations of SA and DHA in the filtrate were greatest at 268 K, the concentration of total ω3 in the filtrate was greatest at 283 K, and that of EPA was maximized at 288 K.

Wanasundara and Shahidi [69] optimized the production of ω3 fatty acid concentrates from seal blubber oil using urea complexation. They investigated the effects of the urea–fatty acid ratio, crystallization time, and crystallization temperature.

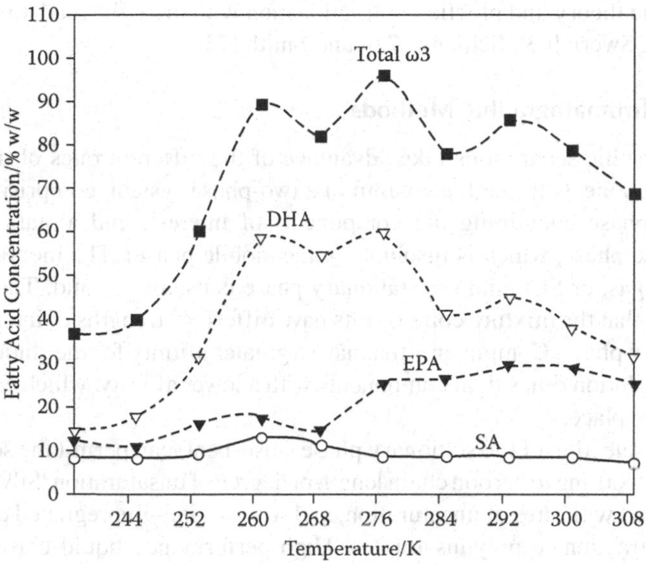

FIGURE 5.7 The effect of temperature on the noncomplexed fatty acid composition resulting from urea fractionation of cod liver oil using a urea–fatty acid ratio of 4:1.

The study was carried out using free fatty acids. Under optimum conditions, the maximum amount of total ω3 fatty acids (88.2%; 70.1% of which was DHA and 9.36% of which was EPA) from seal blubber oil was obtained at a urea–fatty acid ratio of 4.5:1, a crystallization time of 24 hours, and a crystallization temperature of 263 K.

There are several advantages to using urea complexation for the fractionation of fatty acids [70]:

- Large quantities of material can be handled using simple equipment.
- They can be carried out under mild conditions (e.g., room temperature).
- Biocompatible solvents such as ethanol can be used.
- Separations are often relatively efficient when compared with methods such as solvent extraction and fractional crystallization.
- It is a relatively inexpensive process.

Urea inclusion complexation is usually carried out in methanol or ethanol. One should be cautious when using methanol because it may cause methylation of some fatty acids during complex formation, producing a mixture of fatty acids and methyl esters [71]. Urea concentration should be near saturation since the concentration of recovered PUFAs from the filtrate decreases with decreasing urea concentration [70]. Using acetone, or higher hydrocarbons, as solvents for urea complexation should be avoided because it can compete with the fatty acids for inclusion [72]. Rather than being used as a stand-alone process, urea crystallization is often used to preconcentrate fish oil fatty acid mixtures prior to further processing. For a more thorough treatise of the theory and practice of fractionation with urea, the reader is directed to the works of Swern [63], Schlenk [72], and Smith [73].

5.3.1.4 Chromatographic Methods

Chromatographic separations take advantage of the different rates of migration of mixture components through a column in a two-phase system, comprising a mobile phase (the phase containing the components of interest) and a stationary phase (an immobile phase, which is insoluble in the mobile phase). The mobile phase can be a liquid, gas, or SCF and the stationary phase is usually a solid. The phases are chosen such that the mixture components have differing strengths of interaction with the stationary phase. Components that have a greater affinity for the stationary phase have longer elution times than components with a lower affinity, which enables separation to take place.

Careful selection of the stationary phase (adsorbent) can permit the separation of fatty acids according to carbon chain length or degree of unsaturation. Silver ions form a weak π-bond with sites of unsaturation, and so silver ion–impregnated columns can be used to fractionate polyunsaturates. High performance liquid chromatography [74] and silver resin chromatography [75] have been used for the preparation of ω3 concentrates. Teshima and coworkers [76] have used a silver nitrate–impregnated silica column to isolate EPA and DHA methyl esters from squid liver oil. Purities of 85% to 96% for EPA and 95% to 98% for DHA were reported with yields of 39% and 48%, respectively. Adlof and Emiken [75] enriched the ω3 content of commercial ω3-PUFA concentrates from 76.5% to 99.8% using isocratic elution from a silver resin column. Guil-Guerrero and Belarbi [77] purified EPA and DHA from cod liver oil using a silver nitrate–impregnated silica column. The oil was saponified and treated with urea, and the noncomplexed fatty acids were then converted to methyl esters before chromatographic processing. The column was eluted with a sequence of solvents. They managed to obtain a 64% yield of DHA with 100% purity. The recovery of EPA was 29.6%, with a final purity of 90.6%.

The purity of eluted fractions also depends on the choice of eluting solvent. Perrut [78] used methanol/water (90:10 v/v) to separate fish oil ethyl esters. Purities of 96% EPA and 85% DHA were achieved. Wille and colleagues [66] used the same methanol/ethanol (90:10 v/v) solvent system for the fractionation of fish oil methyl esters and produced EPA-rich and DHA-rich fractions with 86% and 83% purity, respectively. Despite the developments in chromatographic techniques to refine and concentrate fish oil components, the use of very large volumes of solvents, potential product solvent residues, loss of column resolution after repeated use, and the presence of potentially toxic silver residues are likely to hinder scale-up to production scale volumes.

5.3.1.5 Enzymatic Transformation

The hydrolysis or esterification of fatty acids can be catalyzed by lipases. These reactions can be carried out at low temperatures, which is beneficial considering the detrimental effects high temperatures can have on PUFAs. The direction and efficiency of the reaction depends on the conditions employed. Water content in the reaction medium is a crucial factor influencing the direction of the reaction. High water content shifts the chemical equilibrium toward hydrolysis, whereas lower water content shifts the equilibrium toward esterification. For esterification, the water content should be kept to a minimum to prevent partial hydrolysis of products, but at the same time, the water content should be sufficiently high in order to prevent enzyme deactivation [79].

The TAG form of PUFA is considered to be nutritionally more favorable than fatty acid esters because studies have shown that the intestinal absorption of ω3-PUFA esters is impaired [80, 81]. Also, TAGs are often promoted as being more "natural" than fatty acid esters, leading to a commercial demand of PUFAs in the TAG form. He and Shahidi [82] studied the glycerolysis of ω3-PUFAs obtained from seal blubber oil using *Chromobacterium viscosum* lipase. Up to 94% conversion was achieved with 13.8%, 43.1%, and 37.4% of monoglycerides, diglycerides, and triglycerides, respectively, in the product. Bottino and colleagues [83] reported the resistance of certain long-chain PUFAs of marine oils to lipase-catalyzed hydrolysis. The *cis* carbon-carbon double bonds present in some fatty acids result in bending of the carbon chain, causing the fatty acid terminal methyl group to lie close to the ester linkage. The methyl group in this instance produces a stearic hindrance to lipase-catalyzed hydrolysis. An increase in the number of double bonds further increases the stearic hindrance. However, there are some reports in which lipases from *Chromobacterium viscosum* and *Pseudomonas* sp. released EPA and DHA from triglycerides [84–86].

Gámez-Meza et al. [87] studied the concentration of EPA and DHA from sardine oil by enzymatic hydrolysis. They investigated five commercial lipases from *Pseudomonas* (three immobilized and two soluble). They found that the immobilized lipase preparation PS-CI (a lipase from *Pseudomonas* sp. immobilized on a chemically modified ceramic) provided the greatest degree of hydrolysis for EPA and DHA (81.5% and 72.3% from their initial content in the sardine oil after 24 hours) and attributed this observation to the higher protein content of this lipase. At the start of hydrolysis (3 hours), they noticed that the lipases displayed a significant preference for saturated fatty acids containing 14 to 16 carbon atoms. However, the resistance to release EPA and DHA decreased as the hydrolysis reaction progressed. Subsequent urea crystallization of the PS-CI hydrolyzed oil enriched EPA from 14.5% to 46.2% and DHA from 12.6% to 40.3%, with a 78.0% yield.

Schmitt-Rozieres et al. [88] studied the recovery of EPA and DHA from effluents of the sardine canning industry. The oily effluent component contained around 10% each of EPA and DHA. After the removal of solid particles, proteins, and peptides from the crude effluent, the resultant oil was hydrolyzed and EPA and DHA were enriched by selective enzymatic esterification. Using Lipozyme™, DHA was enriched up to 80% but no enrichment was observed for EPA. By immobilizing *Candida rugosa* lipase on an Amberlite IRC50 cation-exchange resin, a 30% enrichment of EPA was achieved.

Zuyi and Ward [89] studied the lipase-catalyzed alcoholysis of cod liver oil to concentrate ω3-PUFAs. They studied the effect of water content on reaction, indicating that the water content is influenced by the hydrophobicity of the reaction medium; the more hydrophobic the alcohol, the lower the water content that is required. Using isopropanol, they observed that the alcoholysis of triglycerides increased with increasing water content in the range 0% to 7.5% v/v. At higher water concentrations (> 10%) several negative effects, including enzyme destabilization, purification complications, and promotion of undesirable hydrolysis reactions, were observed. Carrying out the isopropanolysis with 5% water (v/v) at 383 K yielded a high concentration of monoglyceride containing 40% ω3-PUFAs.

Lipase-catalyzed transesterification for the concentration of PUFAs from fish oils has been shown to be a useful alternative to traditional esterification and distillation methods. With a conversion of 52%, Breivik and coworkers [90] were able produce a concentrate containing 46% EPA + DHA (the initial sardine oil contained 24.7% EPA + DHA) using *Pseudomonas* sp. lipase and a stoichiometric amount of ethanol without solvent, at room temperature. The resulting TAGs were isolated and converted to ethyl esters using either conventional chemical means or enzymatic conversion by immobilized *Candida antarctica* lipase. Urea fractionation of the resulting product increased the EPA + DHA content to around 85%. The use of organic solvents as reaction media for enzymatic conversions can be disadvantageous because this can lead to environmental and residual solvent issues associated with product purification. SCFs have been investigated as alternative solvents for this process.

5.3.2 Supercritical Fluid Processing of Fish Oils

Although various SCFs have been used to extract/fractionate fatty acids and their esters, carbon dioxide (CO_2) is by far the most commonly used solvent because of its availability, low cost, and nonreactivity. Using CO_2 limits the oxidation, decomposition, and polymerization of the PUFAs present in fish oils because separations occur under an inert atmosphere and processes can be carried out at moderately low temperatures. In addition, CO_2 is nontoxic, is nonflammable, and produces solvent-residue-free extracts, which is particularly important if the desired materials are for human consumption. The ability to modify solvent properties by manipulation of temperature and pressure or by the addition of a cosolvent gives supercritical CO_2 processes a unique advantage. Solvent properties can be tuned for specific separation problems offering greater versatility and flexibility over more conventional fractionation processes. Moreover, the increasing social awareness of the health benefits of certain fish oil components in the diet has increased commercial demand for these products in the food and nutraceutical industries. The increased demand has led to a reevaluation of processing methods and supercritical CO_2 methodologies provide solvent-free, "natural" products, which have wide consumer appeal. The remainder of this chapter focuses on the phase equilibria of CO_2 with various fish oil components as well as the various supercritical techniques employed to extract/fractionate and isolate these products. Phase equilibria data are of fundamental importance for optimal design of extraction and fractionation operations.

5.3.2.1 Phase Equilibria: Supercritical CO_2 and Fish Oil Components

This section provides a brief summary of solubility measurements and modelling of solubility and phase equilibria for fish oil components in supercritical CO_2. Fish oils are a complex mixture of lipid components belonging to several lipid classes, including acylglycerols, fatty acids, fatty acid esters, sterols, tocopherols, and hydrocarbons. Successful isolation using supercritical processes requires reliable information on the solubility behavior of the solutes of interest as affected by operating conditions and solute/solvent properties. Because complete predictive modelling of multicomponent phase behavior in supercritical systems is not yet realized, experimental data still play an essential role in process design and the development of both simple and rigorous thermodynamic models.

Simple empirical models based on the Chrastil model [91] have been widely used for prediction of solubility in CO_2 at a given temperature and density. The Chrastil correlation can only be used for the vapor-phase concentration of solutes and gives no information on the liquid-phase composition. Equation of state models, such as the Peng-Robinson [92, 93], Soave-Redlich-Kwong [94], excess function (g^E) [95], group contribution [96], and lattice model equations of state (EOS) [97], have been shown to provide the most rigorous method for predicting phase equilibrium behavior.

The Chrastil correlation [91] for estimating lipid solubilities in SCFs takes the form:

$$\ln c = k \ln \rho + a / T + b \tag{5.2}$$

where ***c*** is the solute solubility, ***k*** is the association number representing the number of molecules in the solute-solvent complex, ρ is the pure solvent density, and ***a*** and ***b*** are empirical constants. Parameter ***a*** is dependent on the total heat of reaction (heat of solvation + heat of vaporization), and ***b*** is dependent on the association constant and solute and solvent molecular weights. Parameter ***k*** reflects the density dependence of solubility at constant temperature, and parameter ***a*** reflects the temperature dependence at constant density. Table 5.6 provides some Chrastil correlation parameters for the solubility of several fish oil lipid components in CO_2 [98–101].

Although, binary lipid/CO_2 systems have been studied extensively, multicomponent data are relatively scarce. In such multicomponent mixtures, complex intermolecular interactions may lead to significant deviations from pure component solubilities. However, pure component solubility information is still important, as it can be used to give a guide to the degree of separation possible between two or more classes of lipid at a given temperature and pressure. General solubility trends in binary systems contribute to our basic understanding of the principals of lipid solubility in SCFs. This information provides a sound basis on which to evaluate the solubility behavior of more complex multicomponent systems. However, researchers should exercise caution when making solubility measurements, interpreting the data and, ultimately, designing separation processes based on this information. Temelli and Güçlü-Üstündağ [102] report several discrepancies between reported solubility data for lipid/CO_2 systems from different laboratories. An example of such discrepancies is given in Figure 5.8 for some selected fatty acids in CO_2 [103–107]. Impurities, sample degradation, isomeric purity, and limitations to experimental methods are all contributing factors to the reported data variations.

Solute vapor pressures and solute-solvent and solute-solute intermolecular interactions govern solubility behavior. In binary systems of a particular homologous

TABLE 5.6
Chrastil Parameters for Some Selected Fish Oil Components (All Solubilities and Densities are in Units of $g \cdot L^{-1}$ Unless Otherwise Stated)

Lipid Component	k ± Standard Error	a ± Standard Error	b ± Standard Error	Range T/K; P/MPa	Ref.
		Fatty Acids			
Myristic acid, $C_{14:0}$	6.42 ± 0.33	−9300 ± 1727	−10.2 ± 5.9		98
Palmitic acid, $C_{16:0}$	7.00 ± 0.39	−12029 ± 1043	−7.0 ± 4.1		98
Stearic acid, $C_{18:0}$	5.81 ± 0.54	−15890 ± 741	12.0 ± 3.7		98
Oleic acid, $C_{18:1}$	7.92 ± 0.37	−3982 ± 691	−38.1 ± 2.3		98
Linoleic acid, $C_{18:2}$	9.71 ± 0.90	−5211 ± 1626	−46.3 ± 5.3		98
		Triglycerides			
Triolein	10.28 ± 0.66	−2057 ± 480	−61.5 ± 4.6		98
		Ethyl esters			
Stearic acid	5.80 ± 0.50	−2446 ± 857	−26.7 ± 3.9		98
Oleic acid	7.78 ± 0.34	−1947 ± 503	−40.9 ± 2.7		98
Linoleic acid	7.17 ± 0.63	−2193 ± 896	−36.2 ± 4.4		98
EPA	8.62 ± 0.17	2473 ± 262	−45.2 ± 1.2		98
DHA	7.76 ± 0.32	−1784 ± 529	−42.1 ± 2.5		98
		Hydrocarbons			
Squalene[a]	6.54 ± 0.06	−3936.6 ± 155	−28.24 ± 0.7	313–333; 10–30	99
		Minor Components			
Vitamin A[a]	5.07 ± 0.44	−3072 ± 339	−21.7 ± 2.12	313–353; 20–35	100[b]
Vitamin A palmitate[a]	7.66	0	−49.2	333; 12.5–30	101
β-carotene[a]	8.63 ± 0.61	−11576 ± 461	−23.3 ± 3.04	313–353; 20–35	100[b]
		Fish oils			
Cod liver oil[a]	10.91 ± 0.18	−4078 ± 122	−59.2 ± 0.98	313–333; 20–30	99[b]
Spiny dogfish liver oil[a]	9.97	0	−65.4	333; 20–30	101
Orange roughy oil[a]	7.79	0	−50.6	333; 20–30	101

[a] Solubilities are in $g \cdot kg^{-1}$ and densities are in $kg \cdot m^{-3}$; [b]Derived from data presented in this reference.

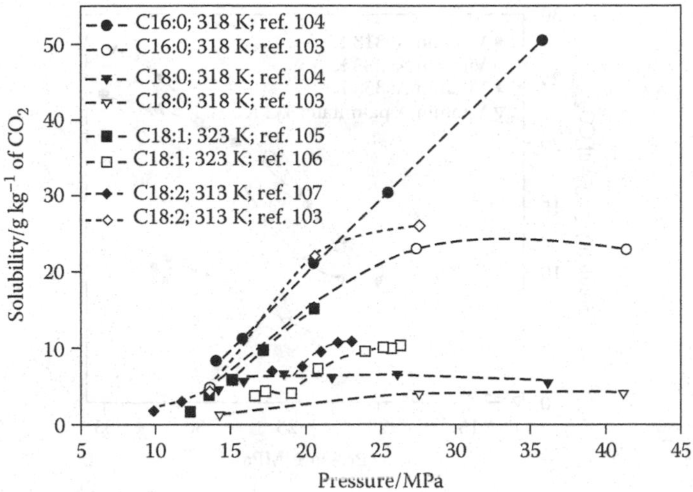

FIGURE 5.8 Comparison of fatty acid solubilities in CO_2 from various literature sources. (Lines have been drawn to aid the eye.)

series, where intermolecular interactions are similar, molecular weight and vapor pressures determine component solubilities. For example, in Figure 5.8, the solubility of fatty acids increases with decreasing molecular weight (chain length). Of the systems reported in the literature, fatty acid esters have the highest vapor pressures, followed by fatty acids. Vapor pressures of the glyceride lipid class follow the trend monoglycerides > diglycerides > triglycerides [98]. Esterification with a C_1 or C_2 alcohol substantially increases the solubility of fatty acids in CO_2 because the polar acid group is converted to a less polar ester group [108]. For fatty acids of the same chain length, melting points decrease with increasing degree of unsaturation (Figure 5.8). In this instance, solubility is affected by the physical state of the compound.

Johannsen and Brunner [100] measured the solubilities of fat-soluble vitamins in supercritical CO_2 in the temperature range of 313 to 353 K and pressure range of 20 to 35 MPa. The solubility for both β-carotene (provitamin A) and vitamin A increased with increasing pressure. Over the temperature range studied, vitamin A shows retrograde condensation behavior (solubility decreases with increasing temperature) up to around 30 MPa (Figure 5.9). At higher pressures, the solubility curves of vitamin A exhibit a crossover point and the system exhibits nonretrograde behavior. The solubility of vitamin A palmitate has been measured by Catchpole et al. [99] at 333 K and 12.5 to 30.0 MPa. The solubility is compared to that of the vitamin A free acid in Figure 5.9. The decrease in polarity of the ester over the free acid is counterbalanced by the large increase in molecular weight, leading to a modest decrease in solubility.

Mollerup et al. [109–111] carried out a series of phase equilibria measurements for fish oil fatty acid ethyl esters (FAEEs) of the sand eel with CO_2. Measurements were obtained using the crude fish oil FAEEs (283 to 343 K, 2 to 22 MPa), urea-fractionated fish oil FAEEs (313 to 343 K, 1.6 to 25 MPa), and ω3-rich fish oil FAEEs (313 to 343 K, 8 to 26 MPa). The initial FAEE oil compositions are given

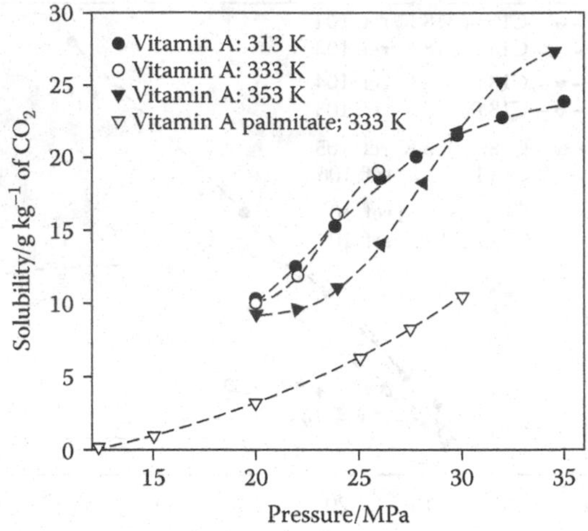

FIGURE 5.9 Solubility of vitamin A and vitamin A palmitate in CO_2.

in Table 5.7. The K-values, on a CO_2 free basis, for some selected ω3-FAEEs are given in Figure 5.10 at 313 K and 343 K as a function of pressure. The K-values depend strongly on temperature, pressure, and mixture composition. High selectivities (higher relative differences between K-values) and low solubilities were observed at low pressures, whereas at high pressures the selectivity was low but the solubility was high. The crude FAEE mixture was more soluble than the urea-fractionated and ω3-rich mixtures (which had similar solubilities) because the crude mixture contained a large amount of saturated and monounsaturated FAEEs of medium chain length (C_{14}–C_{18}). For the crude fish oil esters, the K-values were found to vary with chain length but not specifically the degree of unsaturation and position of double bonds. The authors reported that solubilities increase with increasing temperature, and Figure 5.10 shows that the selectivities at 343 K are greater than those at 313 K. The selectivities are largest in the ω3-rich system because the number of components and the amount of medium-chain-length material has decreased. Using urea-fractionation as a preconcentration step under the conditions investigated, the authors determined that optimum separation conditions in terms of selectivity and solubility were in the range of 16 to 18 MPa at 343 K (corresponding to CO_2 densities of 550 to 615 kg m^{-3}).

Catchpole and von Kamp [92] studied the phase equilibria of the system squalene/CO_2 and shark liver oil/CO_2 over the range 313 to 333 K and 10 to 25 MPa. The shark liver oil consisted of around 50% squalene and 50% of a mixture of TAGs and DAGEs. There was also 0.5% by mass pristane. The TAG/DAGE mixture has been assumed to be a single pseudocomponent, with a hypothetical carbon number of 54. The phase equilibria data for squalene/CO_2 at 313 K and 333 K are shown in Figure 5.11. The amount of oil dissolved in the vapor phase increases with increasing pressure and decreasing temperature, as does the amount of CO_2 dissolved in the oil phase. The experimental data for the binary systems squalene/CO_2 and C_{54}-TAG/CO_2 were modelled using the Peng-Robinson EOS with the usual mixing rules [92]. The Peng-Robinson EOS was also used to predict phase equilibrium and separation

TABLE 5.7
The Initial FAEE Fish Oil Compositions Used in the Studies of Mollerup et al. [109–111]

Component	Crude Fish Oil	Urea-fractionated	ω3-rich
	Weight (%) of FAEE		
10:0	0.4		
12:0	0.2		
14:0	7.5	0.8	
14:1 ω5	0.5	0.2	
15:0	0.5		
15:1 ω5	0.2	0.1	
16:0	18.5		
16:1 ω7	12.4	0.5	
16:2	1.4	2.3	
16:3 ω3	0.6	1.6	0.3
16:4 ω3	0.8	2.9	0.2
18:0	2.2	0.03	0.5
18:1 ω9	10.2	0.6	
18:1 ω7	2.3		
18:2 ω6	2.9	1.2	
18:3 ω6	0.4	0.8	
18:3 ω3	1.3	0.9	
18:4 ω3	3.8	12.7	2.3
20:0	0.2		
20:1 ω9	4.2	0.6	
20:2 ω6	0.3	0.5	
20:3 ω3	0.2	0.1	
20:4 ω3	0.7	1.0	2.9
20:5 ω3	10.0	35.9	52.5
22:1 ω11	6.5	0.3	
22:1 ω9	1.0	0.1	
21:5 ω3	0.4	1.5	
22:5 ω3	0.5	1.1	2.5
22:6 ω3	9.6	33.2	36.1

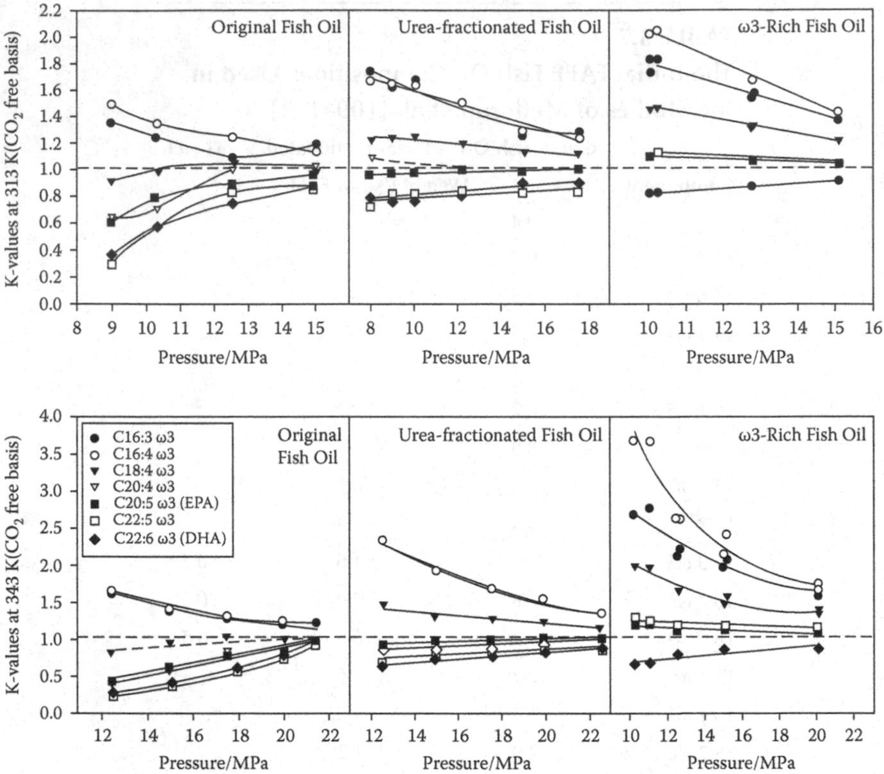

FIGURE 5.10 Partition coefficients (K-values) on a CO_2-free basis for whole sand eel oil, urea-fractionated sand eel oil, and ω3-rich sand eel oil [109–111]. (Reprinted from *Fluid Phase Equilibria*, 161, 169, ©1999. With permission from Elsevier.)

factors for the ternary system C_{54}-TAG/squalene/CO_2. The predicted vapor and liquid mole fractions of squalene in the ternary system are shown for selected temperatures and pressures in Figure 5.12. The liquid mole fractions equate to discrete mass fractions of squalene on a CO_2 free basis ranging from 0 to 1. It is interesting to note that the equilibrium relationship is almost linear, as shown by the regression lines in Figure 5.12. The mass fraction of CO_2 dissolved in the liquid phase at a given temperature and pressure stays almost constant even when the squalene mass fraction varies from 0 to 1 (on a CO_2-free basis). The predicted vapor and liquid-phase mass fractions of squalene for the same system are given on a CO_2-free basis in Figure 5.13. The K-values for squalene are also shown. The selectivity toward squalene is best at low pressure and low mass fraction of squalene. The solubility of TAGs decreases more sharply with decreasing pressure than squalene, and so the increase in selectivity is to be expected. The K-value decreases as the temperature and vapor-phase density increase, although it is still sufficiently high at 333 K and 25.0 MPa to enable separation of squalene and C_{54}-TAG.

Ruivo et al. [112] measured the phase equilibria of the ternary system methyl oleate/squalene/CO_2 over the range 313 to 343 K and 11 to 21 MPa. Four different

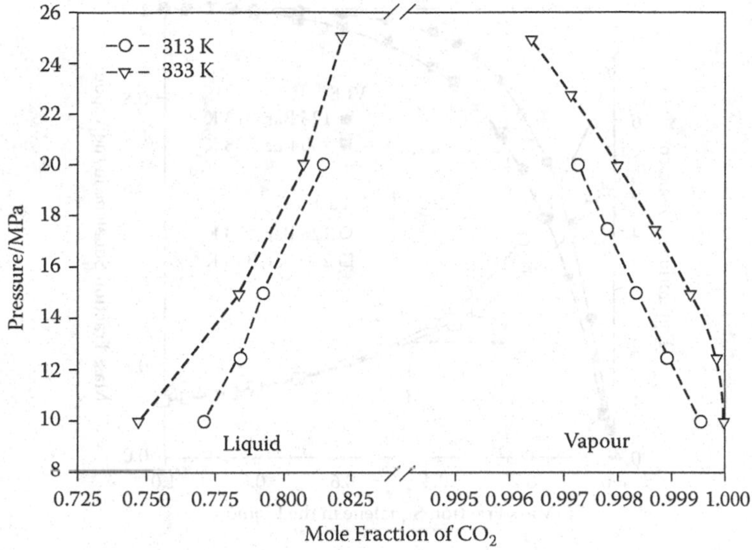

FIGURE 5.11 Liquid and vapor mole fractions for the squalene–CO_2 system [92].

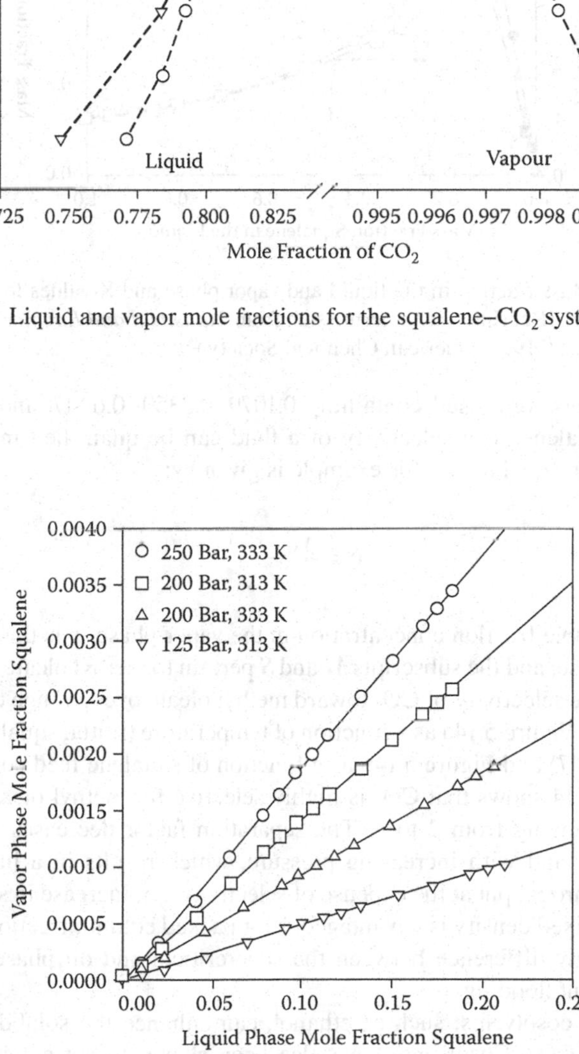

FIGURE 5.12 Liquid and vapor phase mole fraction of squalene at selected temperatures and pressures [92]. Points: Peng-Robinson equation of state predictions; lines: linear regressions. (Reprinted with permission from *Industrial and Engineering Chemistry Research*, 36, 3762. ©1997 American Chemical Society.)

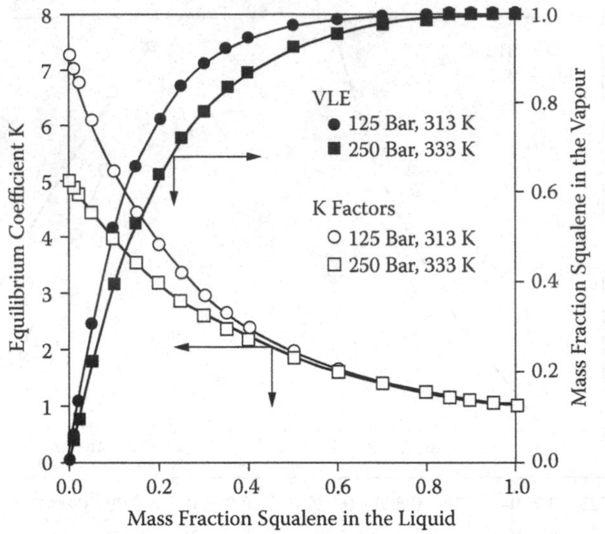

FIGURE 5.13 Mass fraction in the liquid and vapor phase and K-values for squalene on a CO_2-free basis [92]. (Reprinted with permission from *Industrial and Engineering Chemistry Research*, 36, 3762. ©1997 American Chemical Society.)

feed compositions were used containing 0.1079, 0.3350, 0.6447, and 0.8779 mole fractions of squalene. The selectivity of a fluid can be quantified in terms of the separation factor, α, which in this example is given by:

$$\alpha = \frac{y_M \cdot x_S}{x_M \cdot y_S} \tag{5.3}$$

where $\boldsymbol{y}$ is the mole fraction concentration in the vapor phase, $\boldsymbol{x}$ is the mole fraction in the liquid phase, and the subscripts $\boldsymbol{M}$ and $\boldsymbol{S}$ pertain to methyl oleate and squalene, respectively. The selectivity of CO_2 toward methyl oleate over a range of pressures is demonstrated in Figure 5.14a as a function of temperature (initial squalene feed mole fraction of 0.6447) and Figure 5.14b as a function of squalene feed concentration at 313 K. Figure 5.14 shows that CO_2 is highly selective for methyl oleate, with separation factors ranging from 2 to 8. The separation factor decreases with decreasing temperature and with increasing pressure, which results in a higher loading, giving greater throughput at the expense of selectivity. An increase in solubility with temperature at fixed density is advantageous for packed column fractionation, which requires a density difference between the supercritical and oil phases to be large enough to prevent flooding.

Addition of cosolvents, such as ethanol, can enhance the solubility of solutes in supercritical CO_2. Cosolvents have also been shown to act as entrainers. The entrainer effect has been defined as a phenomenon in which the solvent power of a fluid is increased by the addition of cosolvents, whilst the selectivity of that fluid is maintained or enhanced [113]. In many studies, the enhanced solubilities have been attributed to solute-cosolvent interactions, such as hydrogen bonding or dipole-dipole

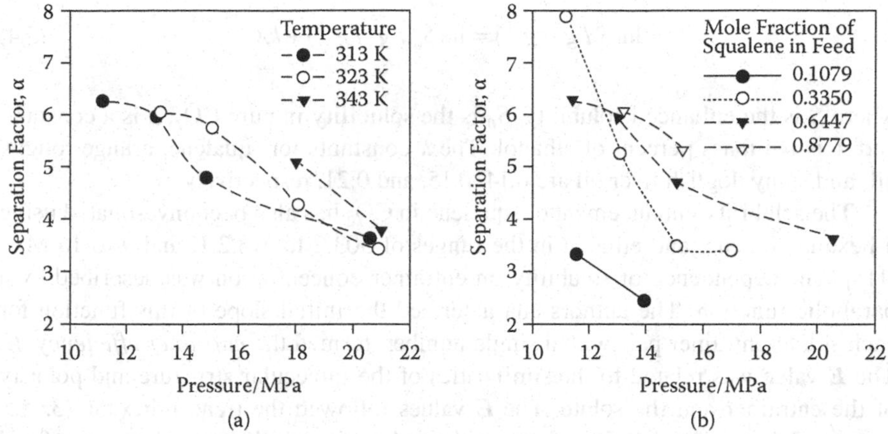

FIGURE 5.14 The separation factors for methyl oleate and squalene in CO_2 [112]. (Reprinted from *Journal of Supercritical Fluids*, 29, 77, ©2004. With permission from Elsevier.)

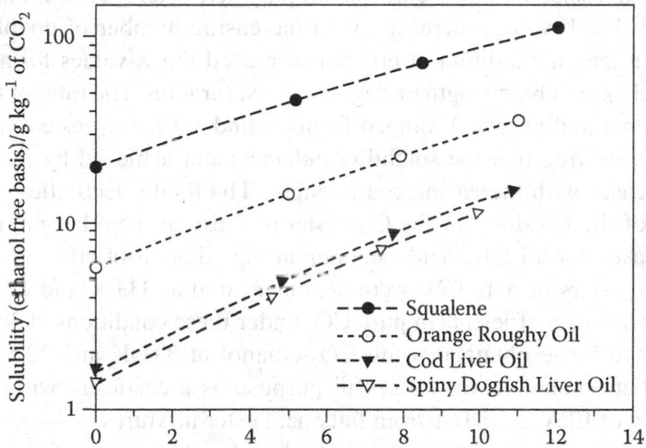

FIGURE 5.15 Enhancement of fish oil component solubilities as a function of ethanol cosolvent concentration at 333 K and 20 MPa [99]. (Reprinted with permission from *Journal of Chemical and Engineering Data*, 43, 1091. ©1998 American Chemical Society.)

interactions. Specific intermolecular interactions between cosolvents and solutes can enhance the solubility of those specific components, which can be particularly beneficial for improving separation selectivities.

Catchpole et al. [99] measured the solubility of squalene, orange roughy oil, cod liver oil, and spiny dogfish liver oil in supercritical CO_2 and CO_2/ethanol mixtures at 313 to 333 K and 20 to 30 MPa. Ethanol mass concentrations up to 12% (on a solute-free basis) were used. Catchpole and coworkers found that ethanol substantially increased the solubility of all fish oil components studied (Figure 5.15). At 333 K, the authors correlated the increase in solubility due to the addition of ethanol using the following equation:

$$\ln(S/g \cdot kg^{-1}) = \ln(S_0/g \cdot kg^{-1}) + kX \tag{5.4}$$

where ***S*** is the enhanced solubility, S_0 is the solubility in pure CO_2, ***k*** is a constant, and ***X*** is the mass percent of ethanol. The ***k*** constants for squalene, orange roughy oil, and spiny dogfish liver oil are 0.14, 0.15, and 0.21, respectively.

The solubility enhancement of squalene in CO_2 has also been investigated using n-hexane, toluene, and ethanol in the ranges of 303.2 to 313.2 K and 9 to 10 MPa [114]. The dependence of solubility on entrainer concentration was described by a parabolic function. The authors characterized the initial slope of this function for each solute-entrainer pair with a single number, termed the *entrainer efficiency, **E***. The ***E*** value was related to the similarities of the molecular structure and polarity of the entrainer and the solute. The ***E*** values followed the trend n-hexane (3.6) > toluene (2.4) > ethanol (1.5), indicating the order of solubility enhancement at fixed temperature, pressure, and entrainer concentrations.

Nilsson et al. [115] investigated the effects of adding 5% ethanol as a cosolvent on the K-values and selectivity of menhaden oil fatty acid esters at 333 K and 12.5 MPa. It was noted that K-values increased with increasing number of double bonds for a given chain length. Addition of ethanol increased the K-values for all fatty acid esters, regardless of chain length or degree of unsaturation. The ratio of the K-values for CO_2-ethanol and pure CO_2 ranged from around 1.5 for C_{14} esters to 3.1 for C_{22} esters, demonstrating that the solubility enhancement achieved by the addition of ethanol increases with increasing chain length. The fluid selectivities, defined here as the ratio of the K-values for the $C_{14:0}$ ester to those of the other mixture components, decreased for all fatty acid esters upon the addition of ethanol. K-values for the fatty acid esters in pure CO_2 were also measured at 333 K and 13.1 MPa. The measured partition coefficients in pure CO_2 under these conditions yielded K-values very similar to those obtained using CO_2-ethanol at 333 K and 12.5 MPa. They concluded that ethanol serves no useful purpose as a cosolvent with CO_2 for the concentration of EPA and DHA from fatty acid ester mixtures.

Although the use of cosolvents may be beneficial in terms of increasing solubility, the complex nature of fish oils means their application may be limited in terms of selectivity enhancement. Their use should be considered carefully because it can increase the complexity of process design. An increase in solvent loading may result in the coextraction of undesirable components. Also, using cosolvents can affect mass transfer, greatly increase processing costs, and can potentially induce degradation of the desired extract. For a more detailed discussion of binary/CO_2 and multicomponent/CO_2 lipid systems, the reader is directed to the critical and in-depth reviews of Temelli and Güçlü-Üstündağ [98, 102]. Several articles containing comprehensive tabulated data on lipid/CO_2 systems studied are also available in the literature [98, 116–118].

5.3.2.2 Polyunsaturated Fatty Acid Processing

The main focus of studies involving SCF processing of fish oils has been the isolation of ω3-PUFAs, and in particular, the isolation of EPA and DHA. It is well known that only modest enrichments using a triglyceride feedstock can be achieved unless the oil is already rich in DHA. The complex triglyceride structure, containing different chain length acids and degree of unsaturation, makes separations to achieve concentration of long-chain PUFAs impractical. For this reason, triglycerides are either saponified or converted to fatty acid alkyl esters prior to fractionation. The resultant fatty acids or esters have greater solubility in CO_2 by virtue of their lower molecular weight and greater volatility [119]. Some groups have investigated the fractionation of the free fatty acids but the reported degree of separation was low.

SCF extraction/fractionation processes involving ethyl esters are carried out in contacting devices comprising two fluid phases: 1) a more dense, ester-rich phase containing dissolved CO_2 and 2) a less dense, CO_2-rich phase in which some esters are dissolved. Separation processes exploit solubility differences of various feed components in the fluid-rich phase. Eisenbach [120] carried out the fractionation of cod liver oil fatty acid alkyl esters using CO_2 in a batch-continuous process. Here, a quantity of feed material is loaded into the base of the fractionation column, and CO_2 is continuously passed through the column until the feedstock is depleted. The low-molecular-weight components with short chain lengths are preferentially extracted. To enhance separation, a temperature gradient along the column was used to generate internal reflux in the column caused by a reduction in solubility. This caused the higher-molecular weight compounds to precipitate and individual chain-length fractions were collected in a separator as a function of time. In Eisenbach's study [120], the reflux was generated by the presence of a "hot finger" at 363 K in the top of the fractionation column. Extractions were carried out at 15.0 MPa using a CO_2 flow rate of 25 L hour^{-1}. The extractor and column temperatures were 298 and 323 K, respectively. The esters remaining in solution exited the top of the column and were collected in a separation vessel by pressure reduction. Using this method, Eisenbach was able to successfully separate the fatty acid esters according to carbon number. He reported that a fraction containing C_{20} esters with greater than 95% purity was obtained, consisting of around 13% of the feed material, which had an initial EPA content of 14.5%. The C_{20} fatty acid ester fraction had an EPA content of 48.2%.

In a later study by Nilsson and coworkers [121], a batch-continuous process for the fractionation of menhaden oil was carried out at 15.2 MPa using a temperature gradient along the height of the column. Temperatures in the column varied across four temperature zones from 293 K at the bottom to 373 K at the top. Fractionation was carried out with and without urea complexation as a preconcentration step. Two of the feed materials investigated are shown in Table 5.8. Fractionation of the whole esters (Feed 1) demonstrated that successful separation according to carbon number could be achieved. Fractions in excess of 60% purity with respect to C_{20} and 90% purity with respect to C_{22} esters were achieved with EPA and DHA contents of 51.9% (of the C_{20} fraction) and 59.5% (of the C_{22} fraction), respectively. The low purity of the C_{20} fraction emphasized the difficulty in separating the C_{18} and C_{20} esters. Separation using a urea-treated feed (Feed 2) produced fractions in excess of 80% purity with respect to C_{20} esters and greater than 95% purity with respect to C_{22} esters. More than 95% of the C_{20} fraction could be attributed to EPA and DHA accounted for around 90% of the total C_{22} fraction. The enhancement of the overall C_{20} purity was

TABLE 5.8
Composition of Feed Materials Used in Study by Nilsson et al. [121]

Fatty Acid	Feed 1[a]	Feed 2[b]
	Weight (%) of Fatty Acid Esters	
14:0	7.8	
16:0	15.6	
16:1 ω7	10.9	< 1.0
16:3 ω4	1.1	5.3
16:4 ω1	1.5	5.8
18:0	3.1	
18:1 ω9	7.6	
18:1 ω7	3.1	< 1.0
18:2 ω6	1.3	< 1.0
18:3 ω3	1.6	< 1.0
18:4 ω3	2.9	7.6
20:1 ω9	1.2	
20:4 ω6	1.0	1.4
20:4 ω3	1.5	< 1.0
20:5 ω3 (EPA)	16.5	48.6
21:5 ω3	< 1.0	1.3
22:5 ω3	2.5	< 1.0
22:6 ω3 (DHA)	10.9	22.2
Total by Carbon Number		
C_{14}	9.0	< 1.0
C_{16}	32.7	13.1
C_{18}	21.3	9.7
C_{20} (% EPA)	21.8 (75.7)	50.5 (96.2)
C_{22} (% DHA)	14.5 (75.2)	25.0 (88.8)

Source: Adapted with permission from *Journal of the American Oil Chemists' Society*, 65, 109. ©1988 American Oil Chemists' Society.

[a] Whole esters, [b]Urea fractionated esters.

attributed to the fact that most of the C_{18} fatty acid esters in typical menhaden oil are saturates and monounsaturates, which were preferentially removed during urea complexation. These maximum enrichments were achieved using a solvent-to-feed ratio (S/F) of around 475. The S/F has a significant impact on the economic viability of a process; high S/F values lead to lengthy fractionation times and/or high energy costs. A reduction in S/F can be achieved by increasing the operating pressure or decreasing the operating temperature, at the expense of separation performance. An appropriate balance between product yield and S/F should be investigated during the optimization process.

As an extension to this work, Nilsson et al. [122] employed an increasing pressure program in conjunction with a temperature gradient for fractionation of urea-crystallized fish oil ethyl esters. The feed sample had a composition very similar to that given for Feed 2 in Table 5.8. Using seven temperature zones along the height of the column, ranging from 313 K in the second zone to 353 K at the top (the bottom zone was at ambient temperature), and pressures ranging from 13.1 to 15.7 MPa (~ 6.9 MPa increments), they were able to recover 85% of EPA and 88% of DHA from the feed with a purity of 90%. The maximum enrichments in this study were achieved using a S/F of 340. The authors concluded that at a given pressure the S/F is primarily governed by the temperature of the uppermost zone and is insensitive to the lower zone temperatures and flow rate. Increasing the total number of temperature zones and reducing the temperature at the top of the column can maintain selectivity while lowering the S/F.

Batch-continuous processing is inappropriate for large-scale production because processing costs are adversely affected by operating parameters, such as high S/F values. Continuous processes are more efficient for production of large quantities of PUFAs. Krukonis [123] described a continuous countercurrent process for fractionation of fatty acid esters and large-scale production of EPA and DHA concentrates. In this process, the feed stream is continuously supplied to the top of a column, where it is contacted by the extracting solvent that is continuously flowing in the opposite direction. The direction of flow depends on the density of the two fluids. Generally, for CO_2 countercurrent extraction of fish oils, the heavier feed oil stream flows downward and the lighter CO_2 phase flows upward. Using the stage concept, Krukonis calculated that a total of 13 stages were required to obtain EPA of 90% purity in the top product (extract phase) and DHA of similar purity in the raffinate. The S/F was significantly decreased to around 30, which is much lower than the S/F for the batch-continuous processes described above. The cost of carrying out this process was estimated to be comparable to the production of ω3 fatty acid concentrates using conventional methods.

Riha and Brunner [124] investigated the continuous countercurrent fractionation of sardine oil FAEEs with supercritical CO_2 in the temperature and pressure ranges 313 to 353 K and 6.5 to 19.5 MPa, respectively. The experiments focused on separating low-molecular-weight components, with carbon numbers ranging from C_{14}–C_{18}, and high-molecular-weight components (HMCs), C_{20}–C_{22}. An economical operational point was determined and the process was deemed to be useful for the enrichment of HMCs, which could be further processed to obtain fractions rich in EPA and DHA. Using a S/F of 63 at 353 K and 19.5 MPa, in a 12m column with reflux (number of theoretical plates = 40), the HMC fractions were obtained in greater than 95% purity with greater than 95% yield.

The PUFA-rich filtrate obtained from a urea preconcentration step also contains uncomplexed urea. The fatty acids are usually recovered from the filtrate by solvent extraction with a mixture of a nonpolar organic solvent, such as hexane or isooctane, in which the urea is insoluble, and water in which the urea is soluble. The use of nonfood-grade organic solvents such as hexane in the extraction of PUFAs is

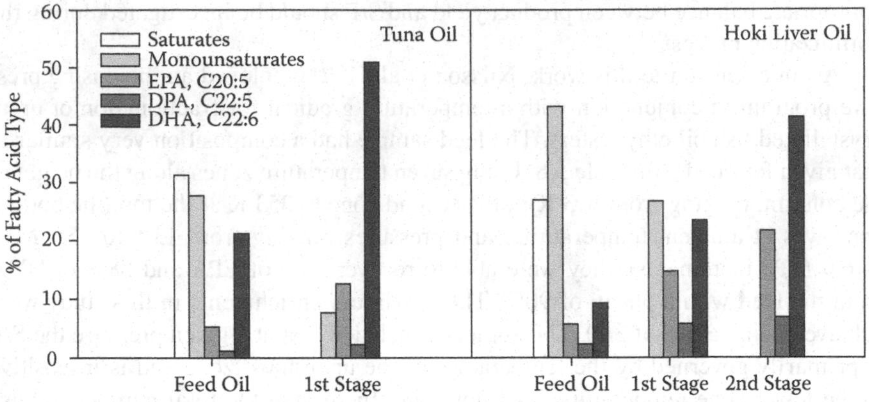

FIGURE 5.16 Fatty acid extract composition for single-stage urea fractionation of tuna oil and double-stage urea fractionation of hoki liver oil.

undesirable, particularly if the product is intended for dietary supplementation. Loss of PUFAs may occur during the reduction of organic solvent residues to regulatory levels. A patent by Catchpole et al. [125] describes a combined supercritical extraction-supercritical antisolvent process for extracting a wide range of lipophilic compounds from urea-containing solutions using near-critical fluids, such as polyunsaturated fatty acids from fish oils. The near-critical fluid is continuously contacted with the urea filtrate solution. The solvent properties of the near-critical fluid enable extraction of the fatty acid/alkyl esters from the solution (along with ethanol) while antisolvent properties result in urea (and water) being precipitated. The fatty acids and ethanol can then be recovered separately by consecutive pressure-reduction steps. Figure 5.16 [126] shows the results for a single-stage urea fractionation of tuna oil (initial EPA and DHA content of 5% and 22%, respectively), and a double-stage urea fractionation of hoki liver oil (initial EPA and DHA content of 6% and 9%, respectively). The supercritical antisolvent process is carried out after the final stage of urea processing and separates urea and oxidation products from the desired PUFAs. Typical antisolvent fractionation conditions are 333 K and 30 MPa. The final products contain around 40% to 50% DHA, and an ω3 content of 60% (tuna oil) and greater than 90% (hoki liver oil).

Kulas and Breivik [127] describe a process in which PUFA ethyl esters are extracted from solid urea/ethyl ester complexes. The complexes were obtained from the prior urea fractionation of fish oil ethyl esters. The authors note that the PUFAs are selectively extracted due to the low stability of the complex, while strongly bound monounsaturates are not extracted. They also note that some urea is extracted. It is also possible to use solid urea as a selective adsorbent for fatty acids or alkyl esters and CO_2 or CO_2 + ethanol as the mobile phase [125, 128]. The same processing considerations apply as for urea solutions: saturates are most strongly bound, and the strength of the bonding is temperature dependent. Figure 5.17 [129] shows the batch-continuous fractionation of hydrolyzed ling liver oil. A fatty acid/ethanol mixture was continuously mixed with CO_2 and then passed through two beds packed with finely ground solid urea. The resultant extract was around 50% DHA and

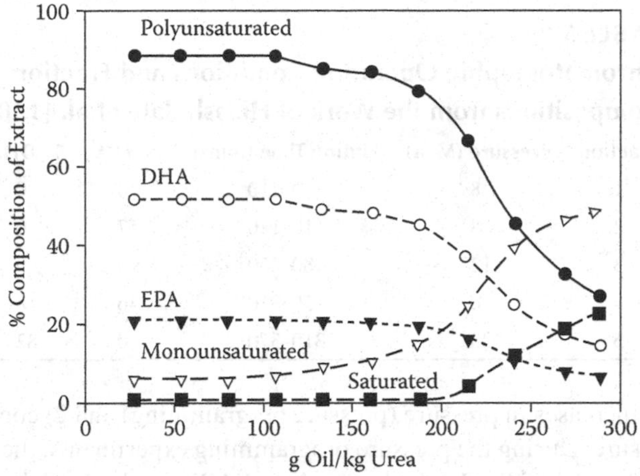

FIGURE 5.17 Extract composition versus fatty acid–urea ratio for batch-continuous fractionation of hydrolyzed ling liver oil using CO_2 and packed beds of finely ground solid urea at 290 K and 30 MPa.

greater than 90% ω3-PUFAs until a ratio of urea to fatty acid of approximately 5:1 was reached. The process was deemed not suitable for commercial scale operation because the solid urea complex could not be regenerated in-situ and became rock hard and very difficult to remove from extraction vessel baskets.

Supercritical fluid chromatography (SFC) can offer the same degree of fatty acid separation as that offered by SCF extraction of urea-concentrated feeds, but only if the stationary phase can specifically interact with double bonds. Higashidate et al. [130] enriched EPA and DHA from esterified sardine oil (initial EPA and DHA contents of 12% and 13%, respectively) by extracting the oil with supercritical CO_2 and directly introducing the resulting solution onto a silica gel column coated with silver nitrate. Extractions were carried out at 313 K and 8 MPa with a CO_2 flow rate of 9 g min^{-1}. Chromatographic separations were carried out at 313 K using pressure programming. Five fractions were collected sequentially. Using this method, they obtained EPA- and DHA-rich fractions with purities of 93% and 82%, respectively. The chromatographic operating conditions and fraction compositions are given in Table 5.9. The order of elution was in accordance with the interaction of carbon-carbon double bonds with silver ions; the interaction increases with an increasing degree of unsaturation, resulting in longer elution times.

Pettinello et al. [131] investigated the production of EPA-enriched mixtures from fish oil using SFC at both laboratory and pilot scale. Their starting mixture contained 67.6% EPA ethyl esters (EE) and 6.1% arachidonic acid (AA; 20:4 ω6)-EE, which were fractionated using a silica adsorption column and supercritical CO_2 as the eluting solvent. Samples were analyzed using capillary and packed column gas chromatography (GC). They used and compared several types of silica gel and investigated different process operating conditions. It was emphasized that sample loading should be optimized because this can strongly influence the yield obtained. Two modes of operation were carried out at the laboratory scale: 1) constant temperature

TABLE 5.9
Chromatographic Operating Conditions and Fraction Compositions from the Work of Higashidate et al. [130]

Fraction	Pressure (MPa)	Elution Time (mins)	% EPA	% DHA
1	8	0–110	0	0
2	8	110–180	57	0
3	12	180–250	93	0
4	20	250–310	46	18
5	20	310–370	0	82

with stepwise increases in pressure (pressure programming) and 2) constant temperature and pressure. During the pressure programming experiments, the pressure was raised during operation in order to change the solubility and selectivity of the FAEE in supercritical CO_2. The pressure was raised from 18.0 MPa to 22.0 or 24.0 MPa at 343 K during the chromatographic run. The lighter components were eluted during the early stages of the process, and the EPA-EE removal was enhanced at the higher pressures. The pressure program mode of operation gave larger yields of a given purity (EPA-EE + AA-EE purity of 90%, yield 49.0%), even if the constant pressure operation was carried out using a lower feed loading (EPA-EE + AA-EE purity of 90%, yield 10.0%). They reported that the selectivity obtained using pressure programming was not satisfactory and so the effect of temperature on selectivity was also investigated. The temperature was raised to 353 K and an operating pressure of 20.8 MPa was used to give the same CO_2 fluid density as that at 343 K and 18.0 MPa. A reduction in retention time and a sharpening of the GC peak was noted, indicating that silica gel has a weaker interaction with FAEE molecules at higher temperatures. A 95% EPA purity was achieved, with a yield of 11%. The purity of EPA dropped to 90% when the yield was increased to 43%.

In the pilot-scale operation, Pettinello and coworkers [131] used a system in which the CO_2 was recycled. Quantities of feed materials of the order of hundreds of grams were processed. The chromatographic fractionations were carried out using two types of silica gel (75 to 200 μm, 60 Å pore size and 75 to 200 μm, 40 Å pore size), and the researchers found that the behavior of both was similar in terms of separation performance. In terms of economics, for the process to be useful on an industrial scale, it is important that the silica adsorbent can be used several times. The researchers showed, using three successive runs with the same column of silica, that an 84% pure EPA fraction could be obtained with yields of 34.3%, 29.8%, and 28.2% for runs 1, 2, and 3, respectively. Pettinello et al. [131] stated that after the initial deactivation of silica during the first run, the adsorbent material maintained an acceptable level of activity. The effect of feed mass, pressure, and pressure programming were also investigated during pilot-scale operation. The EPA-EE purity was greatest at 93% with 24.6% yield when using a feed mass of 275.0 g, a temperature of 343 K, and a pressure programming method in the range 15.0 to 24.0 MPa.

A patent by Perrut et al. [132] describes using SFC for fish oil processing in a continuous mode of operation by applying simulated countercurrent moving bed

(SCMB) technology. The authors claimed that a 93% EPA-EE-rich fraction and an 85% DHA-EE-rich fraction were achieved using the combined SFC-SCMB method. In the first step, a starting mixture containing 32.8% EPA and 20.9% DHA was passed through a SFC column (300 × 60 mm) at 323 K and 16.0 MPa using RPC-18 (octadecyl silica gel, 12 to 45 μm) as a stationary phase and a CO_2 flow rate of 40 kg $hour^{-1}$. They obtained a 73.6% EPA-EE-rich fraction with a yield of 67.1% (this EPA-rich fraction contained 15.1% DHA). The DHA-EE-rich fraction was obtained with 56.3% purity with yield 76.1% (this DHA-rich fraction contained 35.4% EPA). In a second step, these fractions were reprocessed using eight RPC-18 columns (100 × 80 mm) at 323 K and 13.0 MPa with a CO_2 flow rate of 55 kg $hour^{-1}$. In the second step, the total yields for both EPA and DHA were 99% and the fraction purities increased to 93% and 85% for EPA-EE and DHA-EE, respectively.

Alkio and coworkers [133] studied the economic feasibility of producing large quantities of EPA and DHA from tuna oil using SFC. Using supercritical CO_2 as the mobile phase, they carried out a systematic study to find optimum process parameters for maximum production rate. At 338 K and 14.5 MPa, using octadecyl silane-type reversed-phase silica as the stationary phase, DHA and EPA could be produced simultaneously in one chromatographic step with purities of 80% to 95% and 50%, respectively. A process for producing 1,000 kg of DHA and 410 kg of EPA concentrate per year requires 160 kg of stationary phase and 2.6 tons $hour^{-1}$ of CO_2 recycle. They stated that the stationary phase would preferably be packed in four parallel 600 mm i.d. columns and the estimated cost of such a plant was around U.S. $2 million. Assuming that the stationary phase would have to be replaced once a year, the process operating costs were calculated to be U.S. $550 per kilogram of DHA and EPA concentrate (calculations were based on material and operating costs for the year 2000).

Enzymatic methods of enrichment have also been investigated under supercritical conditions. Lin et al. [134] studied the enrichment of ω3-PUFA content in TAGs of menhaden oil by lipase-catalyzed transesterification in supercritical CO_2. Prior to reaction, menhaden oil was treated by the urea inclusion method to produce an 80.1% ω3-PUFA concentrate, 71.2% of which was EPA + DHA. Using the immobilized 1,3-regiospecific lipase, IM60 from *Mucor miehei*, the authors studied the effects of several operating parameters on the reaction, including cosolvent concentration, reaction time, temperature, pressure, and substrate ratio (free ω3-PUFA:TAG). For all reactions, the enzyme concentration was kept at 10% w/w of the total substrates. Both water and ethanol were examined as cosolvents and both fluids exhibited a maximum for ω3-PUFA content in TAGs as a function of concentration. The reaction was studied with a water content ranging from 0% to 10% w/w and a maximum ω3-PUFA content of 46% was observed at 4% water content. A similar trend was observed for ethanol (studied in the range 0% to 15% w/w) with a maximum ω3-PUFA content of 56% being obtained at 10% w/w. Following this observation, all other reactions were carried out using 10% w/w ethanol as a cosolvent. For a process to be economically viable, optimum results must be obtained within the shortest timeframe. At 323 K and pressures ranging from 10.3 to 20.7 MPa, the total content of ω3-PUFAs in TAGs increased with time up to 5 hours, irrespective of pressure. The temperature effect on reaction was investigated in the range 313 to 333 K, and it was found that the ω3-PUFA content in TAGs increased with increasing temperature. However, it should be noted that elevated temperatures should be used with caution because most protein denaturation occurs at 318 to 323 K. Also, Kamet et al. [135] noted that below 313 K, CO_2 reacts readily with the free amino group on the enzyme surface to form a carbamate-enzyme complex, which reduces enzyme activity. The total ω3-PUFA content in TAGs decreased with increasing pressure. Two possibilities were offered to explain this phenomenon: 1) The suppression of the three-dimensional molecular structure of the active site on the enzyme leads to a reduction of enzyme activity; 2) At elevated pressures, a higher dissolution of CO_2 into the water on the surface of the enzyme causes a decrease in pH and induces the reverse reaction. The ω3-PUFA content in TAGs also increased as the free ω3-PUFA:TAG ratio increased. Under optimum conditions (10% ethanol, 5 hour reaction time, 10.3 MPa, 323 K, and a substrate ratio of 4:1), the authors were able to produce TAGs containing 56% w/w of ω3-PUFAs using this method.

5.3.2.3 Squalene and DAGE Processing

Catchpole and coworkers [101, 136, 137] investigated the continuous extraction of squalene from shark liver oil in laboratory scale (5 mL min^{-1} oil processing capability) pilot scale (30 mL min^{-1} oil) and production scale (1 L min^{-1} oil) packed column plant and a laboratory and pilot scale static mixer apparatus using supercritical CO_2. Separation performance was determined as a function of temperature, pressure, oil-to-CO_2 flow rate ratio, packed height, static mixer dimensions, packing type, and reflux ratio. The initial shark liver oil contained 50% by weight squalene and 0.1% by weight pristane ($C_{19}H_{40}$), with the balance being nonvolatile triglycerides and glyceryl ethers. The pilot scale packed column apparatus is shown in Figure 5.18. The basic apparatus consisted of a CO_2 compressor, a 2.5 m × 56 mm i.d. packed column, high pressure piston pumps for supply of the liquid shark liver oil and liquid reflux, and two jacketed separation vessels for the recovery of squalene, fish odors, and pristane. CO_2 was passed upward through the column at operating pressure. The shark liver oil was pumped into the top of the first (with no reflux) or second (with reflux) section of the column. The reflux liquid was pumped into the top of the first section. The raffinate, which is highly enriched in DAGEs and TAGs and stripped of squalene, was collected at regular time intervals from the bottom of the column. The CO_2 solution passed through a pressure reduction valve into the first separation vessel, where the bulk of the squalene was recovered. The fish odors and pristane were recovered in the second separation vessel by further pressure reduction, and CO_2 was recycled. The production scale plant (Figure 5.19) was operated in a similar manner [101]. The laboratory scale experiments also used the same methodology, although CO_2 was not recycled. The total oil loading was investigated as a function of oil-to-CO_2 ratio (the ratio of the total mass of top product from the column to the CO_2 mass that has passed through the column over a given time) over the temperatures and pressures 313 to 333 K and 12.5 to 25.0 MPa, respectively. Investigations were carried out without reflux at fixed temperature and with internal reflux using a temperature gradient over the height of the column in the range 313 to 333 K. The loading followed the pattern of squalene solubility, with the highest values obtained at 333 K and 25.0 MPa, and the lowest at

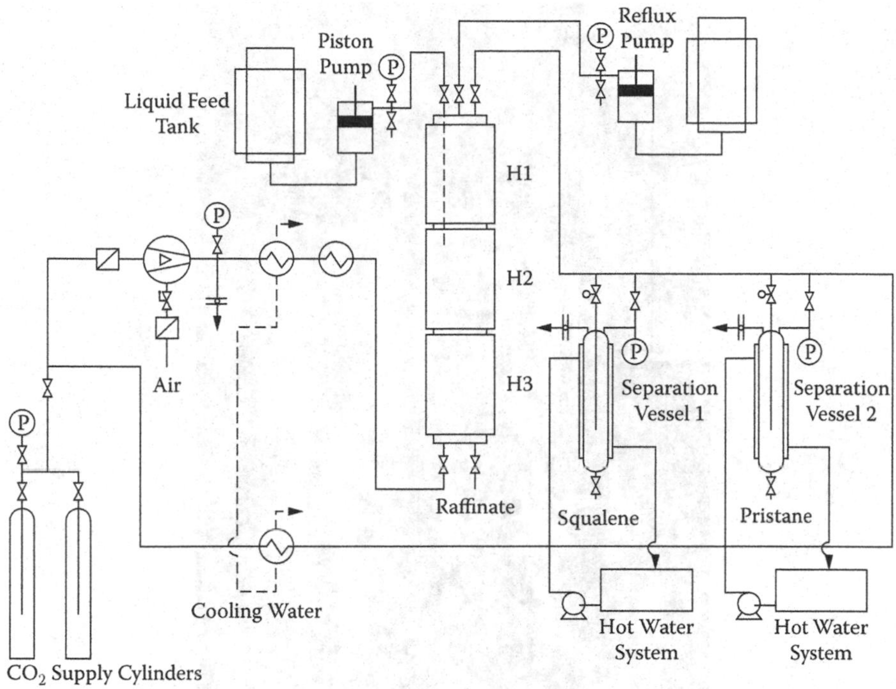

FIGURE 5.18 The pilot scale packed column apparatus used by Catchpole et al. [101, 136, 137] for the fractionation of fish oils. (Reprinted with permission from *Industrial and Engineering Chemistry Research*, 36, 4318. ©1997 American Chemical Society.)

313 K and 20.0 MPa. Pristane was the most soluble oil component and was virtually completely extracted from the feed oil, even at high oil-to-CO_2 mass ratios. Pristane is a skin irritant and is thus undesirable in squalene that is destined for use in cosmetic applications. Separator conditions were optimized to maximize the recovery of squalene in the first separator (313 K and 9.0 MPa), and maximizing pristane recovery in the second separator (313 K and 6.0 MPa).

The effect of packing type, scale of operation, and countercurrent versus cocurrent contacting on mass transfer efficiency was investigated. Stainless steel wool, Raschig rings, and Fenske helices were investigated as packings across a range of packed heights at a laboratory scale. No reliable results using Fenske helices were achieved due to excessive hold-up of liquid in the column. Raschig rings gave poorer mass transfer performance than stainless steel wool. Other researchers have also found that Raschig rings compared poorly with wire wool type packings for lipid/near-critical fluid packed column separations [138, 139]. The performance of wire wool also diminished significantly when only one packed section was used. The authors concluded that at least 0.8 m of packing was required to achieve a high level of separation of squalene from triglycerides at a laboratory scale, and 2.5 m at a pilot scale. The packed height available in the demonstration scale plant was insufficient to achieve mass transfer significantly better than a static mixer, which gives only one equilibrium stage (Figure 5.20). Under optimum conditions at pilot scale, a 92%

FIGURE 5.19 Production scale plant used for fractionation of squalene from shark liver oil [101].

squalene by mass product was obtained in separator 1 at column conditions of 333 K and 25.0 MPa. Using the 92% squalene fraction as a reflux feedstock, experiments were performed with a fixed ratio of feed oil to CO_2 using a range of reflux pump rates. Increasing the reflux to feed oil mass flow ratio caused both the top product stream loading and concentration of squalene in the top product to increase, as shown in Figure 5.21. The oil loading increased linearly toward the equilibrium loading of pure squalene in CO_2, with increasing reflux-to-feed ratio. Since the mass flow of the feed oil and CO_2 was fixed, there was also an increase of squalene concentration in the raffinate with increasing reflux ratio. To achieve optimum separation performance in terms of product purity and loss in raffinate, it is desirable to use reflux of the top product with a lowered feed rate of shark liver oil at fixed CO_2 flow rate. Pilot-scale operation under optimum process conditions with reflux yielded a 99% by mass squalene-rich fraction containing less than 0.5% pristane, with a loss of less than 5% by mass of squalene in the raffinate stream.

5.3.2.4 Vitamin A Processing

The processing of fish oils to recover fractions enriched in vitamin A was first carried out using near-critical propane [140]. A multicolumn process was used to

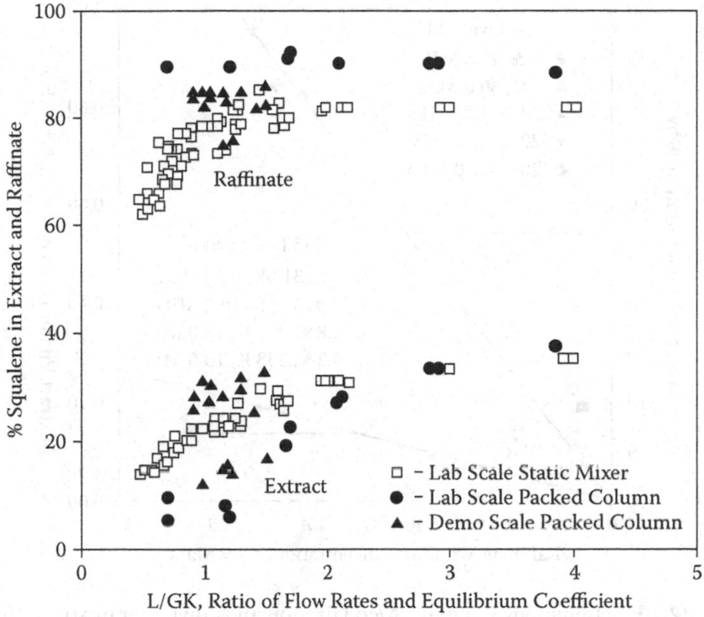

FIGURE 5.20 Extract and raffinate concentrations of squalene for static mixer, laboratory scale, and demonstration scale columns versus modified oil-to-CO_2 flow rate ratio.

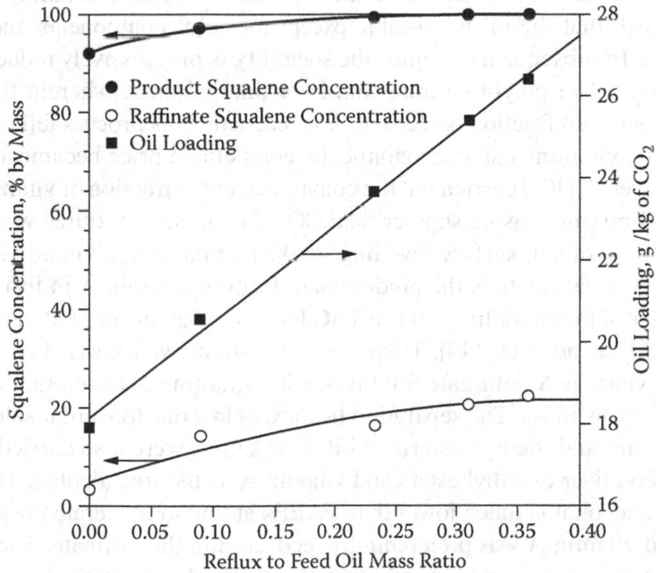

FIGURE 5.21 Squalene concentration in the top product and oil loading in CO_2 as a function of reflux to feed mass ratio [136]. (Reprinted with permission from *Industrial and Engineering Chemistry Research*, 36, 4318. ©1997 American Chemical Society.)

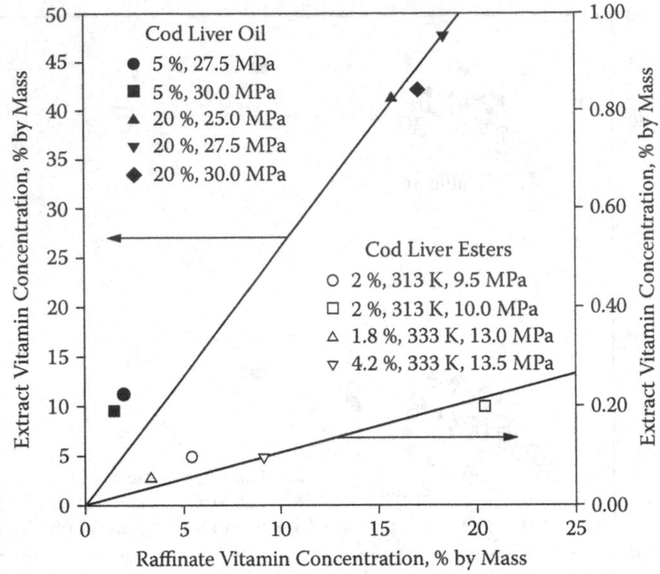

FIGURE 5.22 Fractionation of vitamin A/cod liver oil and ethyl ester mixtures [101]. Filled symbols: ●, ■, ▲, ▼, ◆ Extraction of vitamin A palmitate from cod liver oil; hollow symbols: ○, □, △, ▽ Extraction of vitamin A from fatty acid ethyl esters. (Reprinted from *Journal of Supercritical Fluids*, 19, 25, ©2000. With permission from Elsevier.)

fractionate menhaden and cod liver oils by utilizing differences in solubility of oil components near to, but below, the critical point. In the first column, conditions are chosen such that the oil is miscible except for color components and nonlipid contaminants. In subsequent columns, the solubility is progressively reduced to give fractions enriched in polyunsaturates until the final column, wherein the vitamin A-rich (most soluble) fraction is recovered, is reached. This process fell out of favor when synthetic vitamin A at an economically competitive price became available.

Catchpole et al. [101] carried out the countercurrent extraction of vitamin A from model fish oil mixtures using supercritical CO_2. The model mixtures were made to simulate the liver oils of surface dwelling sharks that have high vitamin A contents [141]. Vitamin A palmitate is the predominant form of vitamin A in fish oils [142]. The remainder of the shark liver oil is a TAG form that has similar fatty acid compositions to cod liver oil [143, 144]. Therefore, separations were carried out using cod liver oil and vitamin A palmitate mixtures, with vitamin A concentrations ranging from 1% to 20% by mass. The separation factor was low due to similar solubilities of the vitamin ester and the non-esterified oil. Extractions were also carried out using mixtures of cod liver oil ethyl esters and vitamin A as the free alcohol. The solubility of the free alcohol is much lower than FAEEs at low to moderate pressures (9 to 12 MPa), and vitamin A was preferentially recovered in the raffinate. The results of the fractionation experiments for cod liver oil/vitamin A palmitate and cod liver oil ethyl esters/vitamin A are shown in Figure 5.22. The concentration of vitamin A palmitate (cod liver oil) in the extract was enhanced over that of the raffinate and the enhancement was not strongly pressure dependent. High losses of the vitamin ester occurred in the raffinate. Increasing the extract to raffinate ratio can reduce this loss, but this results in lower extract concentrations. The separation was substantially improved when using cod liver oil ethyl esters. The majority of fatty acid esters were extracted from the feed to leave a raffinate enriched in vitamin A. The concentration of the vitamin in the extract did not exceed 0.5%.

5.3.2.5 Processing of Other Marine Oil Components

Wax ester oils were processed using the pilot scale supercritical pilot plant described earlier for the processing of squalene from shark liver oil [101]. The fish oil was partially degraded due to extended storage at room temperature and had high peroxide levels. CO_2 + ethanol was used as the solvent, which was countercurrently contacted with the wax ester oil. A high oil-to-solvent ratio could be used due to the high solubility of the oil mixture in the solvent phase. High-molecular-weight esters and astaxanthin were recovered in the raffinate, and medium molecular weight esters were recovered in the first separator, along with part of the ethanol. The overall extract had very high peroxide values. The extract mixture separated into two phases, a top wax ester-rich phase, and a bottom ethanol-rich phase, which also contained most of the peroxides and malodorous compounds present in the original oil. The final separator contained largely ethanol, with volatile odor compounds.

The extraction of green-lipped mussel oil, sold under the name Lyprinol™, using supercritical CO_2 has been carried out commercially for several years. The mussels are endemic to New Zealand. The oil has anti-inflammatory properties and has found application as an antiarthritic and antiasthmatic natural remedy. The extraction process and properties of the extract are described in a patent by Macrides and Kalafatis [145]. The oil is a complex mixture of free fatty acids, TAGs, sterols, and sterol esters [146].

5.4 SUMMARY

The conventional methods for isolation of high-value fish oil components include vacuum distillation, urea crystallization, hexane extraction, and conventional crystallization. These methods have the disadvantages of requiring high processing temperatures, resulting in the thermal degradation or decomposition of the thermally labile compounds or employing flammable or toxic solvents, which have adverse health effects. In this instance, separations employing SCF technologies offer new opportunities for resolving these separation problems.

REFERENCES

1. Gruger, E.H., Nelson, R.W. and Stansby, M.E., Fatty acid composition of oils from 21 species of marine fish, freshwater fish and shellfish, *J. Am. Oil Chem. Soc.*, 41, 662, 1964.
2. Burr, G.O. and Burr, M.M., A new deficiency disease produced by rigid exclusion of fats from the diet, *J. Biol. Chem.*, 82, 345, 1929.
3. Dyerberg, J., Linolenate-derived polyunsaturated fatty acids and prevention of atherosclerosis, *Nutr. Rev.*, 44, 125, 1986.

4. Mehta, J., Lopez, L.M. and Wargovich, T., Eicosapentaenoic acid: Its relevance in atherosclerosis and coronary artery disease, *Am. J. Cardiol.*, 59, 155, 1987.
5. Urakaze, M. et al., Infusion of emulsified treicosapentaenoyl-glycerol into rabbits: The effects on platelet aggregation, polymorphonuclear leukocyte adhesion, and fatty acid composition in plasma and platelet phospholipids, *Thromb. Res.*, 44, 673, 1986.
6. Kremer, J.M. et al., Effects of manipulation of dietary fatty acids on clinical manifestations of rheumatoid arthritis, *Lancet*, 1, 184, 1985.
7. Braden, L.M. and Carroll, K.K., Dietary polyunsaturated fat in relation to mammary carcinogenesis in rats, *Lipids*, 21, 285, 1986.
8. Reddy, B.S. and Maruyama, H., Effect of dietary fish oil on azoxymethane-induced colon carcinogenesis in male F344 rats, *Cancer Res.*, 46, 3367, 1986.
9. Svaneborg, N. et al., The acute effects of a single very high dose of n-3 fatty acids on plasma lipids and lipoproteins in healthy subjects, *Lipids*, 29, 145, 1994.
10. Frontini, M.G. et al., Distribution and cardiovascular risk correlates of serum triglycerides in young adults from a biracial community: The Bogalusa Heart Study, *Atherosclerosis*, 155, 201, 2001.
11. Morris, M.C., Sack, F. and Rosner, B., Does fish oil lower blood pressure? A meta-analysis of controlled trials, *Circulation*, 88, 523, 1993.
12. Weisinger, H.S. et al., Perinatal omega-3 fatty acid deficiency affects blood pressure later in life, *Nature Med.*, 7, 258, 2001.
13. Ziboh, V.A., Miller, C.C. and Cho, Y., Metabolism of polyunsaturated fatty acids by skin epidermal enzymes: Generation of antiinflammatory and antiproliferative metabolites, *Am. J. Clin. Nutr.*, 71, S361, 2000.
14. Brown, M.D. et al., Promotion of prostatic metastatic migration towards human bone marrow stoma by Omega 6 and its inhibition by Omega 3 PUFAs, *Br. J. Cancer*, 94, 842, 2006.
15. Shirota, T. et al., Apoptosis in human pancreatic cancer cells induced by eicosapentaenoic acid, *Nutrition*, 21, 1010, 2005.
16. Yonezawa, Y. et al., Inhibitory effect of conjugated eicosapentaenoic acid on mammalian DNA polymerase and topoisomerase activities and human cancer cell proliferation, *Biochem. Pharmacol.*, 70, 453, 2005.
17. Jho, D.H. et al., Eicosapentaenoic acid supplementation reduces tumor volume and attenuates cachexia in a rat model of progressive non-metastasizing malignancy, *J. Parenter. Enteral Nutr.*, 5, 291, 2002.
18. Adams, P.B. et al., Arachidonic acid to eicosapentaenoic acid ratio in blood correlates positively with clinical symptoms of depression, *Lipids*, 31, S157, 1996.
19. Assies, J. et al., Significantly reduced docosahexaenoic and docosapentaenoic acid concentrations in erythrocyte membranes from schizophrenic patients compared with a carefully matched control group, *Biol. Psychiatry*, 49, 510, 2001.
20. Ayton, A.K., Azaz, A. and Horrobin, D.F., A pilot open case series of ethyl-EPA supplementation in the treatment of anorexia nervosa, *Prostaglandins, Leukot. Essent. Fat. Acids*, 71, 205, 2004.
21. Mitchell, D.C. et al., Why is docosahexaenoic acid essential for nervous system function? *Biochem. Soc. Trans.*, 26, 365, 1998.
22. Buranudeen, F., Squalene, *Infofish Market. Digest*, 1, 42, 1986.
23. Gopakumar, K. and Thankappan, T.K., Squalene. Its sources, uses, and industrial applications, *Seafood Export J.*, 17, 1986.
24. McKenna, R.M., Wheatley, U.R. and Warmall, A., The composition of the surface skin fat (subum), *J. Invest. Dermatol.*, 15, 33, 1950.
25. Langdon, R.G. and Bloch, K., The utilization of squalene in the biosynthesis of cholesterol, *J. Biol. Chem.*, 200, 135, 1953.

26. Gloor, M. and Karenfeld, A., Effect of ultraviolet light therapy, given over a period of several weeks, on the amount and composition of the skin surface lipids, *Dermatologica*, 154, 5, 1977.
27. Kohno, Y. et al., UV-B exposure to skin surface: Photo-oxidation of surface lipid, *J. Jpn. Oil Chem. Soc. (Yukagaku)*, 42, 204, 1993.
28. Kohno, Y. et al., Kinetic study of quenching reaction of singlet oxygen and scavenging reaction of free radical by squalene in *n*-butanol, *Biochem. Biophys. Acta: Lipids and Lipid Metab.*, 1256, 52, 1995.
29. Storm, H.M. et al., Radioprotection of mice by dietary squalene, *Lipids*, 28, 55, 1993.
30. Strandberg, T.E., Tilvis, R.S. and Miettinen, T.A., Variations of hepatic cholesterol precursors during altered flows of endogenous and exogenous squalene in the rat, *Biochim. Biophys. Acta: Lipids and Lipid Metab.*, 1001, 150, 1989.
31. Strandberg, T.E., Tilvis, R.S. and Miettinen, T.A., Metabolic variables of cholesterol during squalene feeding in humans: Comparison with cholestyramine treatment, *J. Lipid Res.*, 31, 1637, 1990.
32. Reddy, B.S., Dietary fat and colon cancer: Animal model studies, *Lipids*, 27, 807, 1992.
33. Desai, K.N., Wei, H. and Lamartiniere, C.A., The preventive and therapeutic potential of the squalene-containing compound, Roidex, on tumor promotion and regression, *Cancer Lett.*, 101, 93, 1996.
34. Rao, C.V., Newmark, H.L. and Reddy, B.S., Chemopreventive effect of squalene on colon cancer, *Carcinogenesis*, 19, 287, 1998.
35. Das, B., *The Science Behind Squalene, the Human Antioxidant*, 1st ed., University of Toronto Press, Toronto, Canada, 2000.
36. Wetherbee, B.M. and Nichols, P.D., Lipid composition of the liver of deep-sea sharks from the Chatham Rise, New Zealand, *Comp. Biochem. Physiol. B*, 125, 511, 2000.
37. Bakes, M.J. and Nichols, P.D., Lipid, fatty acid, and squalene composition of liver oil from six species of deep-sea sharks collected in southern Australian waters, *Comp. Biochem. Physiol. B*, 110, 267, 1995.
38. Hallgren, B., Therapeutic effects of ether lipids, in *Ether Lipids: Biochemical and Biomedical Aspects*, Mangold, H.K. and Paltauf, F., Eds., Academic Press, New York, 1983, 261.
39. Saffioti, U. et al., Experimental cancer of the lung, *Cancer*, 20, 857, 1967.
40. Bershad, S., Developments in topical retinoid therapy for acne, *Semin. Cutan. Med. Surg.*, 20, 154, 2001.
41. Wolf, G., A history of vitamin A and retinoids, *FASEB J.*, 10, 1102, 1996.
42. Underwood, B.A., Vitamin A deficiency disorders: International efforts to control a preventable Pox, *J. Nutr.*, 134, 231, 2004.
43. Swern, D., Ed., *Baily's Industrial Oil and Fat Products*, Vol. 1, 4th Ed., John Wiley and Sons, New York, 1979, 81.
44. Buisson, D.H. et al., Oil from deep water fish species as a substitute for sperm whale and jojoba oils, *J. Am. Oil Chem. Soc.*, 59, 390, 1982.
45. Hagemann, J.W. and Rothfus, J.A., Oxidative stability of wax esters by thermogravimetric analysis, *J. Am. Oil Chem. Soc.*, 56, 629, 1979.
46. Nevenzel, J.C., Occurrence, function, and biosynthesis of wax esters in marine organisms, *Lipids*, 5, 308, 1970.
47. Sargent, J.R., Lee, R.F. and Nevenzel, J.C., Marine waxes, in *Chemistry and Biochemistry of Natural Waxes*, Kolattukudy, P.E., Ed., Elsevier Scientific Publishing, Amsterdam, 1976, 80.
48. Warth, A.H., *The Chemistry and Technology of Waxes*, Reinhold, New York, 1947, 89.

49. Body, D.R., Johnson, C.B. and Shaw, G.J., The monounsaturated acyl- and alkyl-moieties of wax esters and their distribution in commercial orange roughy (*Hoplostethus atlanticus*) oil, *Lipids*, 20, 680, 1985.
50. Fournier, V. et al., Thermal degradation of long-chain polyunsaturated fatty acids during deodorization of fish oil, *Eur. J. Lipid Sci. Technol.*, 108, 33, 2006.
51. Stout, V.F. et al., Fractionation of fish oils and their fatty acids, in *Fish Oils in Nutrition*, Stansby, M.E., Ed., Van Nostrand Reinhold, New York, 1990, 73.
52. Ackman, R.G., Ke, P.J. and Jangaard, P.M., Fractional vacuum distillation of herring oil methyl esters, *J. Am. Oil Chem. Soc.*, 50, 1, 1973.
53. Privett, O.S., Weber, R.P. and Nickell, E.C., Preparation and properties of methyl arachidonate from pork liver, *J. Am. Oil Chem. Soc.*, 36, 443, 1959.
54. Privett, O.S. and Nickell, E.C., Preparation of highly purified fatty acids via liquid-liquid partition chromatography, *J. Am. Oil Chem. Soc.*, 40, 189, 1963.
55. Weitkamp, A.W., Distillation, *J. Am. Oil Chem. Soc.*, 32, 640, 1955.
56. Chawla, P. and deMan, J.M., Measurement of the size distribution of fat crystals using a laser particle counter, *J. Am. Oil Chem. Soc.*, 76, 329, 1990.
57. Singleton, W.S., Solution properties, in *Fatty Acids*, Markley, K.S., Ed., Interscience Publishers, New York, 1960, 609.
58. Haraldsson, G.G., Separation of saturated/unsaturated fatty acids, *J. Am. Oil Chem. Soc.*, 61, 219, 1984.
59. Rubin, D. and Rubin, E.J., US Patent number - US4792418, 1989.
60. Brown, L.B. and Kolb, D.X., Application of low temperature crystallization in the separation of fatty acids and their compounds, *Prog. Chem. Fats Lipids*, 3, 57, 1955.
61. Bengen, M.F., German Patent 869 070, 1953.
62. Domart, C., Miyauchi, D.T. and Sumerwell, W.N., The fractionation of marine oil fatty acids with urea, *J. Am. Oil Chem. Soc.*, 32, 481, 1955.
63. Swern, D., Techniques of separation: E. urea complexes, in *Fatty Acids: Their Chemistry, Properties, Production, and Uses*, Part 3, 2nd Ed., Markley, K.S., Ed., Interscience Publishers, New York, 1964, 2309.
64. Abu-Nasr, A.M. and Holman, R.T., Highly unsaturated fatty acids. II. Fractionation by urea inclusion compounds, *J. Am. Oil Chem. Soc.*, 31, 16, 1954.
65. Robles Medina, A. et al., Concentration and purification of stearidonic, eicosapentaenoic, and docosahexaenoic acids from cod liver oil and the marine microalga *Isochrysis galbana*, *J. Am. Oil Chem. Soc.*, 72, 575, 1995.
66. Wille, H.J., Traitler, H. and Kelly, M., Production of polyenoic fish oil fatty acids by combined urea fractionation and industrial scale preparative high performance liquid chromatography, *Rev. Fr. Corps Gras*, 34, 69, 1987.
67. Ratnayake, W.M.N. et al., Preparation of omega-3 PUFA concentrates from fish oils via urea complexation, *Fat Sci. Technol.*, 90, 381, 1988.
68. Traitler, H., Wille, H. J. and Studer, A., Fractionation of blackcurrant seed oil, *J. Am. Oil Chem. Soc.*, 65, 755, 1988.
69. Wanasundara, U.N. and Shahidi, F., Concentration of omega-3 polyunsaturated fatty acids of seal blubber oil by urea complexation: Optimization of reaction conditions, *Food Chem.*, 65, 41, 1999.
70. Robles Medina, A. et al., Downstream processing of algal polyunsaturated fatty acids, *Biotechnol. Adv.*, 16, 517, 1998.
71. Haagsma, N. et al., Preparation of an omega-3 fatty acid concentrate from cod liver oil, *J. Am. Oil Chem. Soc.*, 59, 117, 1982.
72. Schlenk, H., *Progress in the Fats and Other Lipids*, Vol. 2, Pergamon Press, London, 1954, 243.
73. Smith, A.E., The crystal structure of the urea-hydrocarbon complexes, *Acta Cryst.*, 5, 224, 1952.

74. Beebe, L.M., Brown, P.R. and Turcotte, L.G., Preparative-scale-high-performance liquid chromatography of omega-3 polyunsaturated fatty acid esters derived from fish oil, *J. Chromatogr.*, 495, 369, 1988.
75. Adlof, R.O. and Emiken, E.A., The isolation of omega-3 polyunsaturated fatty acids and methyl esters of fish oils by silver resin chromatography, *J. Am. Oil Chem. Soc.*, 62, 1592, 1985.
76. Teshima, S., Kanazawa, A. and Tokiwa, S., Separation of polyunsaturated fatty acids by column chromatography on solver nitrate-impregnated silica gel, *Bull. J. Soc. Sci. Fish*, 44, 927, 1978.
77. Guil-Guerrero, J.L. and Belarbi, E-H., Purification process for cod liver oil polyunsaturated fatty acids, *J. Am. Oil Chem. Soc.*, 78, 477, 2001.
78. Perrut, M., Purification of polyunsaturated fatty acid (EPA and DHA) ethyl esters by preparative high performance liquid chromatography, *LC-GC*, 6, 914, 1988.
79. Shahidi, F. and Wanasundara, U.N., Omega-3 fatty acid concentrations: nutritional aspects and production technologies, *Trends in Food Sci. Technol.*, 9, 230, 1998.
80. El-Boustani, S. et al., Enteral absorption in man of eicosapentaenoic acid in different chemical forms, *Lipids*, 22, 711, 1987.
81. Lawson, L.D. and Hughes, B.G., Human absorption of fish oil fatty acids as triacylglycerols, free acids, or ethyl esters, *Biochem. Biophys. Res. Commun.*, 152, 328, 1988.
82. He, Y. and Shahidi, F., Enzymatic esterification of ω3-fatty acid concentrates from seal blubber oil with glycerol, *J. Am. Oil Chem. Soc.*, 74, 1133, 1997.
83. Bottino, N.R., Vandenberg, G.A. and Reiser, R., Resistance of certain long-chain polyunsaturated fatty acids of marine oils to pancreatic lipase hydrolysis, *Lipids*, 2, 489, 1967.
84. Garcia, H.S. et al., Synthesis of glycerides containing n-3 fatty acids and conjugated linoleic acid by solvent-free acidolysis of fish oil, *Biotechnol. Bioeng.*, 70, 587, 2000.
85. Hoshino, T., Yamane, T. and Shimizu, S., Selective hydrolysis of fish oil by lipase to concentrate n-3 polyunsaturated fatty acids, *Agri. Biol. Chem.*, 54, 1459, 1990.
86. Tanaka, Y., Hirano, J. and Funada, T., Concentration of docosahexaenoic acid in glyceride by hydrolysis of fish oil with *Candida cylindracea* lipase, *J. Am. Oil Chem. Soc.*, 69, 1210, 1992.
87. Gámez-Meza, N. et al., Concentration of eicosapentaenoic acid and docosahexaenoic acid from fish oil by hydrolysis and urea complexation, *Food Res. Int.*, 36, 721, 2003.
88. Schmitt-Rozieres, M., Deyris, V. and Comeau, L-C., Enrichment of polyunsaturated fatty acids from sardine cannery effluents by enzymatic selective esterification, *J. Am. Oil Chem. Soc.*, 77, 329, 2000.
89. Zuyi, L. and Ward, O.P., Lipase-catalyzed alcoholysis to concentrate the n-3 polyunsaturated fatty acid of cod liver oil, *Enzyme Microb. Technol.*, 15, 601, 1993.
90. Breivik, H., Haraldsson, G.G. and Kristinsson, B., Preparation of highly purified concentrates of eicosapentaenoic acid and docosahexaenoic acid, *J. Am. Oil Chem. Soc.*, 74, 1425, 1997.
91. Chrastil, J., Solubility of solids and liquids in supercritical gases, *J. Phys. Chem.*, 86, 3016, 1982.
92. Catchpole, O.J. and von Kamp, J-C., Phase equilibria for the extraction of squalene from shark liver oil using supercritical carbon dioxide, *Ind. Eng. Chem. Res.*, 36, 3762, 1997.
93. Jaubert, J-N. and Coniglio, L., From the correlation of binary systems involving supercritical CO_2 and fatty acid esters to the prediction of (CO_2-fish oil) phase behaviour, *Ind. Eng. Chem. Res.*, 38, 3162, 1999.
94. Coniglio, L., Knudsen, K. and Gani, R., Model prediction of supercritical fluid-liquid equilibria for carbon dioxide and fish oil related compounds, *Ind. Eng. Chem. Res.*, 34, 2473, 1995.

95. Coniglio, L., Knudsen, K. and Gani, R., Prediction of supercritical fluid-liquid equilibria for carbon dioxide and fish oil related compounds through the equation of state-excess function (EOS-g^E) approach, *Fluid Phase Equilib.*, 116, 510, 1996.
96. Espinosa, S., Diaz, S. and Brignole, E.A., Thermodynamic modelling and process optimization of supercritical fluid fractionation of fish oil fatty acid ethyl esters, *Ind. Eng. Chem. Res.*, 41, 1516, 2002.
97. Bamberger, T. et al., Measurement and model prediction of solubilities of pure fatty acids, pure triglycerides, and mixtures of triglycerides in supercritical carbon dioxide, *J. Chem. Eng. Data*, 33, 327, 1998.
98. Güçlü-Üstündağ, Ö. and Temelli, F., Correlating the solubility behaviour of fatty acids, mono-, di-, and triglycerides, and fatty acid esters in supercritical carbon dioxide, *Ind. Eng. Chem. Res.*, 39, 4756, 2000.
99. Catchpole, O.J., Grey, J.B. and Noermark, K.A., Solubility of fish oil components in supercritical CO_2 and CO_2 + ethanol mixtures, *J. Chem. Eng. Data*, 43, 1091, 1998.
100. Johannsen, M. and Brunner, G., Solubilities of the fat-soluble vitamins A, D, E, and K in supercritical carbon dioxide, *J. Chem. Eng. Data*, 42, 106, 1997.
101. Catchpole, O.J., Grey, J.B. and Noermark, K.A., Fractionation of fish oils using supercritical CO_2 and CO_2 + ethanol mixtures, *J. Supercrit. Fluids*, 19, 25, 2000.
102. Temelli, F. and Güçlü-Üstündağ, Ö., Supercritical technologies for further processing of edible oils, in *Bailey's Industrial Oil & Fat Products*, Vol. 5, 6th Ed., Shahidi, F., Ed., John Wiley & Sons, New Jersey, 2005, 397.
103. Maheshwari, P. et al., Solubility of fatty acids in supercritical carbon dioxide, *J. Am. Oil Chem. Soc.*, 69, 1069, 1992.
104. Kramer, A. and Thodos, G., Solubility of 1-octadecanol and stearic acid in supercritical carbon dioxide, *J. Chem. Eng. Data*, 34, 184, 1989.
105. Nilsson, W.B., Gauglitz, E.J., Jr. and Hudson, J.K., Solubilities of methyl oleate, oleic acid, oleyl glycerols, and oleyl glycerol mixtures in supercritical carbon dioxide, *J. Am. Oil Chem. Soc.*, 68, 87, 1991.
106. Škerget, M., Knez, Ž. and Habulin, M., Solubility of α-carotene and oleic acid in dense CO_2 and data correlation by a density based model, *Fluid Phase Equilib.*, 109, 131, 1995.
107. Chen, C-C., Chang, C-M. J. and Yang, P-W., Vapor-liquid equilibrium of carbon dioxide with linoleic acid, a-tocopherol and triolein at elevated pressures, *Fluid Phase Equilib.* 175, 107, 2000.
108. Dandge, D.K., Heller, J.P. and Wilson, K.V., Structure solubility correlations: Organic compounds and dense carbon dioxide binary systems, *Ind. Eng. Chem. Prod. Res. Dev.*, 24, 162, 1985.
109. Staby, A., Forskov, T. and Mollerup, J., Phase equilibria of fish oil fatty acid ethyl esters and sub- and supercritical CO_2, *Fluid Phase Equilib.*, 87, 309, 1993.
110. Borch-Jensen, C., Staby, A. and Mollerup, J.M., Phase equilibria of urea-fractionated fish oil fatty acid ethyl esters and supercritical carbon dioxide, *Ind. Eng. Chem. Res.*, 33, 1574, 1994.
111. Borch-Jensen, C. and Mollerup, J., Phase equilibria of long-chain polyunsaturated fish oil fatty acid ethyl esters and carbon dioxide, ethane, or ethylene are reduced gas temperatures of 1.03 and 1.13, *Fluid Phase Equilib.*, 161, 169, 1999.
112. Ruivo, R., Paiva, A. and Simoes, P., Phase equilibria of the ternary system methyl oleate/squalene/carbon dioxide at high pressure conditions, *J. Supercrit. Fluids*, 29, 77, 2004.
113. Walsh, J. M., Ikonomou, G.D. and Donohue, M.D., Supercritical phase behaviour: The entrainer effect, *Fluid Phase Equilib.*, 33, 295, 1987.
114. Sovova, H. et al., Solubility of squalene and dinonyl phthalate in CO_2 with entrainers, *J. Supercrit. Fluids*, 14, 145, 1999.

115. Nilsson, W.B., Seaborn, G.T. and Hudson, J.K., Partition coefficients for fatty acid esters in supercritical fluid CO_2 with and without ethanol, *J. Am. Oil Chem. Soc.*, 69, 305, 1992.
116. Staby, A. and Mollerup, J., Separation of constituents of fish oil using supercritical fluids: A review of experimental solubility, extraction, and chromatographic data, *Fluid Phase Equilib.*, 91, 349, 1993.
117. Brunner, G. and Dohrn, R., High-pressure fluid phase equilibria: Experimental methods and systems investigated (1988–1993), *Fluid Phase Equilib.*, 106, 213, 1995.
118. Christov, M. and Dohrn, R., High-pressure fluid phase equilibria: Experimental methods and systems investigated (1994–1999), *Fluid Phase Equilib.*, 202, 153, 2002.
119. Krukonis, V.J., Supercritical fluid fractionation of fish oils. Concentrations of eicosapentaenoic acid, *J. Am. Oil Chem. Soc.*, 61, 698, 1984.
120. Eisenbach, W., Supercritical fluid extraction: A film demonstration, *Ber. Bunsenges. Phys. Chem.*, 88, 882, 1984.
121. Nilsson, W.B. et al., Fractionation of menhaden oil ethyl esters using supercritical fluid CO_2, *J. Am. Oil Chem. Soc.*, 65, 109, 1988.
122. Nilsson, W.B., Gauglitz, E.J., Jr. and Hudson, J.K., Supercritical fluid fractionation of fish oil esters using incremental pressure programming and a temperature gradient, *J. Am. Oil Chem. Soc.*, 66, 1596, 1989.
123. Krukonis, V.J., Processing with supercritical fluids: Overview and applications, in *Supercritical Fluid Extraction and Chromatography: Techniques and Applications*, ACS Symposium Series number 366, Charpentier, B.A. and Sevenants, M.R., Eds., American Chemical Society, Chicago, 1988, 26.
124. Riha, V. and Brunner, G., Separation of fish oil ethyl esters with supercritical carbon dioxide, *J. Supercrit. Fluids*, 17, 55, 2000.
125. Catchpole, O.J., MacKenzie, A.D. and Grey, J.B., Patent WO 03/089399, US2006035350, 2003.
126. Catchpole, O.J. et al., Unpublished data.
127. Kulas, E. and Breivik, H., Patent WO 01/10809, US6528669, 2001.
128. Hiroshi, U. and Hiroshi, S., Patent JP 60214757, 1985.
129. Catchpole, O.J. et al., Unpublished data.
130. Higashidate, S., Yamauchi, Y. and Saito, M., Enrichment of eicosapentaenoic acid and docosahexaenoic acid esters from esterified fish oil by programmed extraction-elution with supercritical carbon dioxide, *J. Chromatogr.*, 515, 295, 1990.
131. Pettinello, G. et al., Production of EPA enriched mixtures by supercritical fluid chromatography: from the laboratory scale to the pilot plant, *J. Supercrit. Fluids*, 19, 51, 2000.
132. Perrut, M., Nicoud, R.M. and Breivik, H., U.S. Patent 5719302, 1998.
133. Alkio, M. et al., Purification of polyunsaturated fatty acid esters from tuna oil with supercritical fluid chromatography, *J. Am. Oil Chem. Soc.*, 77, 315, 2000.
134. Lin, T-J., Chen, S-W. and Chang, A-C., Enrichment of n-3 PUFA contents on triglycerides of fish oil by lipase-catalyzed trans-esterification under supercritical conditions, *Biochem. Eng. J.*, 29, 27, 2006.
135. Kamet, S. et al., Biocatalytic synthesis of acrylates in organic solvents and supercritical fluids. III. Does carbon dioxide covalently modify enzymes?, *Biotechnol. Bioeng.*, 46, 610, 1995.
136. Catchpole, O.J., von Kamp, J-C. and Grey, J.B., Extraction of squalene from shark liver oil in a packed column using supercritical carbon dioxide, *Ind. Eng. Chem. Res.*, 36, 4318, 1997.
137. Catchpole, O.J. et al., Fractionation of lipids in a static mixer and packed column using supercritical carbon dioxide, *Ind. Eng. Chem. Res.*, 39, 4820, 2000.

138. Czech, B. and Peter, S., Efficiency of different packings in counter-current near-critical fluid extraction, in *Pre-prints of High Pressure Chemical Engineering,* 2nd International Symposium, DECHEMA-GVC, 1990, 419.
139. Böhm, F. et al., Design, construction, and operation of a multipurpose plant for commercial supercritical gas extraction, in *ACS Symposium Series*, Vol. 406, Johnston, K.P. and Penninger, J.M.L., Eds., American Chemical Society, Washington D.C., 1989, 502.
140. Passino, H.J., The solexol process, *Ind. Eng. Chem.*, 41, 280, 1949.
141. Shortland, F.B., The aquatic animal oil resources of New Zealand, *N.Z. J. Sci. Technol.*, 2, 30, 1950.
142. Winholz, M., *The Merck Index*, Vol. 9, Merck, New York, 1976, 1133.
143. Hilditch, T.P. and Williams, P.N., *The Chemical Composition of Natural Fats*, Chapman Hall, London, 1964.
144. Vlieg, P. and Body, D.R., Lipid contents and fatty acid composition of some New Zealand freshwater finfish and marine finfish, shellfish, and roes, *N.Z. J. Marine Freshwater Res.*, 22, 151, 1988.
145. Macrides, T. and Kalafatis, N., U.S. Patent 6,083,536, 2000.
146. Wolyniak, C.J. et al., Gas chromatography-chemical ionization-mass spectrometric fatty acid analysis of a commercial supercritical carbon dioxide lipid extract from New Zealand green-lipped mussel (*Perna canaliculus*), *Lipids*, 40, 355 2005.

6 Supercritical Fluid Extraction of Active Compounds from Algae

Rui L. Mendes

CONTENTS

6.1 Introduction 189
6.2 Supercritical Fluid Extraction from Algae 191
6.2.1 *Botryococcus Braunii* 191
6.2.2 *Chlorella Vulgaris* 193
6.2.3 *Dunaliella* 196
6.2.4 *Haematococcus pluvialis* 198
6.2.5 *Hypnea charoides* 201
6.2.6 *Nannochloropsis* 202
6.2.7 *Spirulina (Arthrospira)* 205
6.2.7.1 *Spirulina maxima* 205
6.2.7.2 *Spirulina platensis* 207
6.3 Conclusion 209
References 209

6.1 INTRODUCTION

Microalgae are all the eukaryotic photosynthetic microorganisms, which present a great genetic variety. Some authors include in this definition prokaryotic photosynthetic organisms [1], such as the cyanobacteria. Phytoplankton, which comprise the autotrophic prokaryotic and eukaryotic microorganisms suspended near the water surface, are the base of the aquatic life food chain.

More than 50,000 species of microalgae are supposed to exist, but only about 50 have been well studied and a much lower number has been cultivated in large scale [2]. Microalgae produce a great variety of secondary metabolites, which are synthesized at the end of growth phase and at the stationary one. The unlimited structural diversity of these compounds can still be enlarged applying techniques of combinatorial chemistry [3]. The bioactive molecules produced by these microorganisms can be beneficial or harmful, but even phycotoxins and related products may serve as materials for useful drugs [4]. On the other hand, until now, microalgae had not been used much for the production of chemicals and biochemicals, although some, such as some polyunsaturated fatty acids (PUFAs), are the greatest reserve of the biosphere [5].

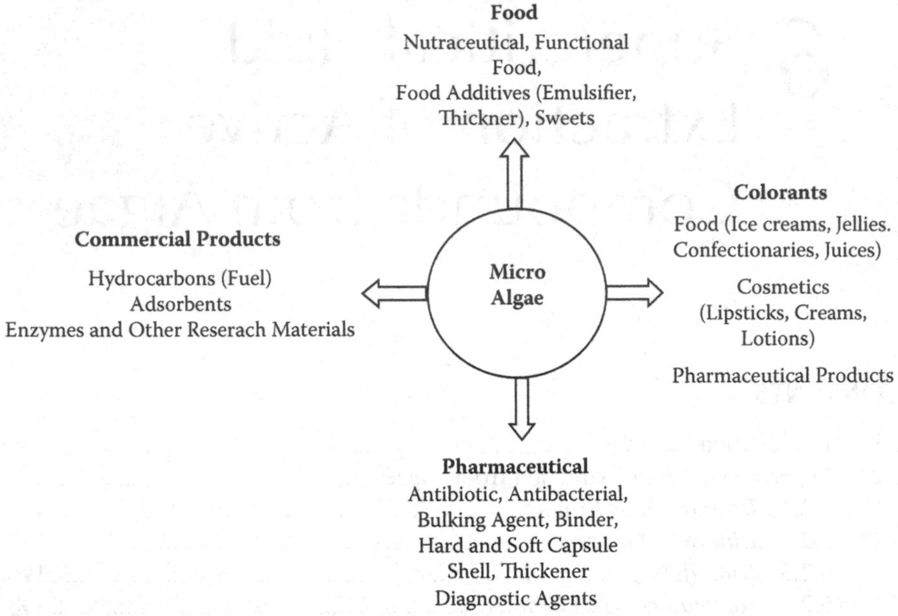

FIGURE 6.1 Microalgae applications in various fields. (*Source*: Adapted from Dufossé et al., *Trends in Food Science & Technology*, 16, 389, 2005. With permission.)

In fact, microalgae can produce, in significant amounts, lipids similar to vegetable oils, fuels, proteins, and essential fatty acids, with dietary applications, such as linoleic, g-linolenic, eicosapentaenoic, docosahexaenoic, and arachidonic acids; vitamins (β-carotene, B12, and E); pigments (carotenoids, phycobiliproteins), waxes, biosurfactants, sterols, and other chemical specialties [6–8]. Reviews focusing on commercial aspects of microalgae biotechnology have recently appeared [9, 10]. Figure 6.1 summarizes the main applications of microalgae [11].

The marine macroalgae, also known as seaweed, have been used as human food and fertilizers and to obtain agar, alginic acid, and carrageenan, among other uses. In this field, the discovery of metabolites with biological activity has increased significantly in the last three decades [12]. The pharmaceutical industry has primarily focused its attention on the following substances: sulphated polysaccharides as antiviral substances, halogenated furanones as antifouling compounds, and kahalalide F as a possible treatment of lung cancer, tumors, and acquired immunodeficiency syndrome [12].

The lipid content of microalgae can reach 85% (dry weight basis), although values between 20% and 40% are more typical [7]. The production yield of these compounds relies on the conditions of culture, with the amount of nitrogen being one of the main factors to control lipid content [5]. The final yield also depends on the extraction method, degree of crushing, and type of solvents used.

The most important feature of algae is their photosynthetic capacity. Therefore, the culture of microalgae is usually made in open ponds, but this can lead to their contamination and to diseases, restricting this method to species that are more resistant, such as *Dunaliella*, *Spirulina,* and *Chlorella* [13]. *Dunaliella salina* is cultivated in ponds with high levels of salt, whereas *Spirulina* can be cultivated in highly alkaline water. On the other hand, the costs associated with harvesting the algae (which can include microscreen, centrifugation, or flocculation) can prevent the economic success of this process.

Genetic engineering offers the possibility of converting photosynthetic microalgae able to produce autotrophically chemical specialties into ones able to produce them heterotrophically [14]. The use of fermenters and photobioreactors (which can be tanks provided with a light source, polyethylene bags, plastic or glass tubes, etc.) for the selective production of particular compounds is a promising method for the culture of microalgae. The future of this technology also passes by the use of mixotrophic cultures [15]. Genetic engineering has also been used to create overproducing strains, a condition that helps some species be more competitive, as well as to modify them in order to obtain specific compounds [16, 17].

Supercritical fluid extraction (SFE) of this type of compound has some advantages over the conventional methods. These compounds can be obtained without thermal degradation and without solvents, therefore there are neither issues related to the toxicity of the organic solvents used nor the legal restrictions of their use. Moreover, with SFE it is possible to obtain a high efficiency of extraction, a shortened extraction time, and, in some cases, a higher yield. On the other hand, the selectivity for certain compounds is much higher with SFE than with organic extraction. Generally all algae has a remaining biomass after lipid extraction that consists mainly of protein (about 50%) and carbohydrates. The organic solvent extraction can lead to denaturation of the proteins, unlike the SFE, what would be detrimental to its use in food or feed applications [18, 19].

Some of the more interesting early studies of supercritical carbon dioxide (CO_2) extraction of compounds from algae were carried out using *Scenedesmus obliquus* [20], *Dilophus ligulatus* (a brown macroalga) [21], *Dunaliella salina* [22], *Skeletonema costatum* [46], and *Ochronomas danica* [46], with the objective of obtaining fatty acids, several secondary metabolites, eicosapentaenoic acid (EPA), and β-carotene, respectively. Also, some reviews on this field have appeared [23–25]. Table 6.1 shows a collection of research literature on several macroalgae and microalgae (including also other microorganisms, such as fungi and yeasts), and the respective target compounds, which have been the object of SFE.

The aim of this chapter is to review in some detail the studies of SFE to obtain several active compounds from some of the most important microalgae and also some seaweed species, focusing on the types of compounds extracted, comparison with the conventional methods to obtain them, and the most relevant aspects of their SFE.

6.2 SUPERCRITICAL FLUID EXTRACTION FROM ALGAE

6.2.1 *Botryococcus Braunii*

Botryococcus braunii is a species of green microalgae that live in large natural colonies, either in salt water or freshwater, in temperate and tropical regions. It presents a unique particularity among the known photosynthetic microorganisms

TABLE 6.1
Typical Compounds Extracted from Algae and Other Related Organisms

Organism	Target Compounds	Refs.
Arthrospira (Spirulina)	Lipids, γ-linolenic acid, carotenoids, phycocyanin	26, 27, 28, 29, 30, 31, 32
Botryococcus braunii (race A)	Linear alkadienes (C25, C27, C29, C31), triene (C29)	33
Chlorella vulgaris	Astaxanthin, canthaxanthin, lipids	34, 35
Dilophus ligulatus	Lipids, secondary metabolites	21
Dunaliella bardawill	*Trans*-β-carotene, *cis*-β-carotene	36
Dunaliella salina	*Trans*-β-carotene, *cis*-β-carotene	37, 38
Isochrisis galbana	Lipids, EPA, PUFAs	39
Haematococus pluvialis	Astaxanthin, astaxanthin esthers	30, 40, 41
Hypnea charoides	ω3 fatty acids (EPA, DHA)	42
Mortierella (fungus)	Lipids, GLA	43
Nannochloropsis gaditana	Chlorophyll a, β-carotene, vaucheriaxanthine	44
Nannochloropsis sp.	Lipids, EPA	45
Ochronomas danica	Lipids, EPA, PUFAs	46
Phaffia rodozyma (yeast)	Astaxanthin	47
Pilayella littoralis	Nonhumic compounds	48
Saprolegnia parasitica (fungus)	Lipids, EPA	49
Scenedesmus obliquus	Lipids, proteins	20
Skeletonema costatum	Lipids, EPA, PUFAs	46
Torulaspora delbrueckii (yeast)	Squalene	50

because its content in hydrocarbons can reach 85% of the dry biomass. These algae have already been proposed as a future renewable source of fuel [51]. The localization of the hydrocarbons is extracellular. Each cell is surrounded by several external walls, where most of the hydrocarbons accumulate in globular formations. The remaining hydrocarbons (about 5%) are located in the cytoplasm [52].

Race A of these algae produces and accumulates linear alkadienes, C_{25}–C_{31}, with an odd number of carbon atoms and the triene $C_{29}H_{54}$ [53]. These compounds, after hydrogenation or functionalization, can be used as substitutes for paraffinic and natural waxes, with cosmetic and pharmaceutical applications [54].

Supercritical CO_2 studies were carried out at 40°C and pressures of 125, 200, and 300 bar [33]. Figure 6.2 shows the cumulative curves of the hydrocarbon extraction yield versus the hexane extraction. The maximum extraction yields of hydrocarbons obtained were 76 g/kg (dry algae basis) and 72 g/kg for hexane and SFE extractions, respectively.

About 95% of the extracellular hydrocarbons were rapidly extracted at 300 bar. Moreover, the supercritical extracts were limpid and golden due to the nonextraction of chlorophyll, and they contained about 60% hydrocarbons, unlike what happened with hexane extracts, which contained only 37% of these compounds. When hydrocarbons were depleted from the algae, the extracts became more viscous. On the

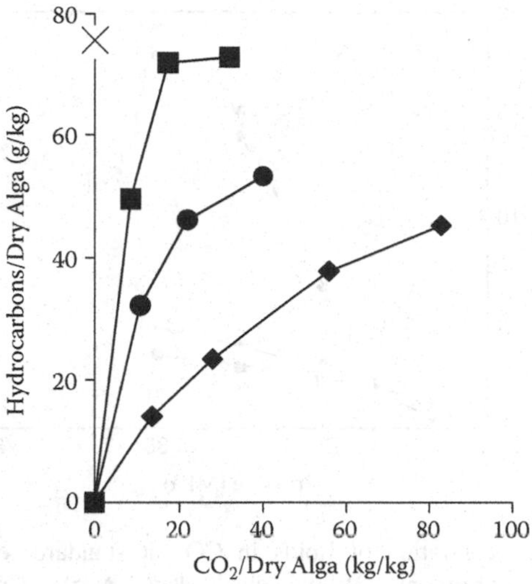

FIGURE 6.2 Hydrocarbons supercritical extraction yield, as a function of solvent-algae ratio, from the microalgae *Botryococcus braunii* in supercritical CO_2 at 313.1 K. ◆ 12.5 MPa, ● 20.0 MPa, ■ 30.0 MPa, × hexane extraction. (*Source*: Mendes et al., *Inorganica Chimica Acta*, 356, 328, 2003. With permission.)

other hand, C_{27} and C_{29} in the extracts decreased proportionately during the extraction progress, while C_{31} alkadiene proportion increased [35]. Hydrocarbon fraction increased with pressure: the initial concentration of hydrocarbons increased more steeply than that of intracellular lipids, leading to extracts more rich in hydrocarbons at higher pressures [55].

6.2.2 *Chlorella Vulgaris*

Chlorella vulgaris microalgae are carotenoid producers (mainly of canthaxanthin and astaxanthin) [56]. Carotenogenesis can be induced through saline, luminous, or nutritional stress. On the other hand, the content in carotenoids can be tailored through the duration of the process and the intensity of the imposed stresses.

Carotenoids belong to a hydrocarbon class (carotenes) and their oxygenated derivatives (xanthophylls). Their basic structure, reflecting its synthesis path, consists of eight isoprenoid units, which are assembled in such way that two methyl groups near the molecule center are in position 1,6, while the other methyl groups stay in position 1,5 [57]. The set of conjugated double bonds (eleven to thirteen) constitutes the chromophore responsible for the color of these compounds. These colors range from yellow to red and are influenced by the presence of more double bonds, functional groups, and the type of molecular conformation.

Canthaxanthin (β-β-carotene-4,4’-dione), $C_{40}H_{52}O_2$, a red pigment, together with astaxanthin, is one of the most important keto carotenoids [13]. It is used as colorant to improve the color of poultry meats and egg yolks as well as in aquaculture to give a pink tonality to salmon and trout flesh. Although lacking pro-vitamin A activity, canthaxanthin has anticarcinogenic capacity [58].

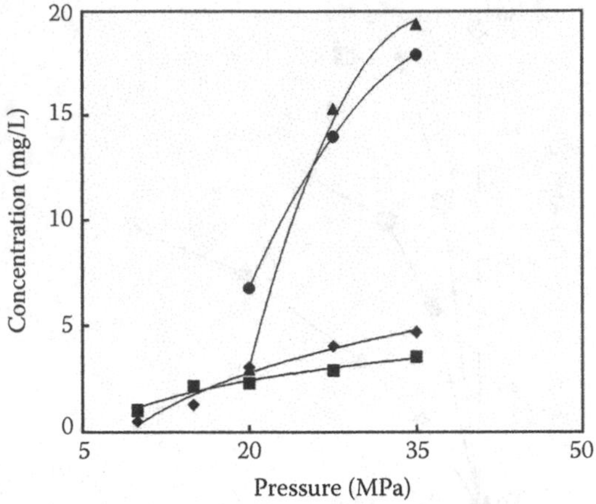

FIGURE 6.3 Initial concentration of lipids in CO_2 at standard temperature and pressure (STP), as a function of pressure. Whole cells, ■ 40°C, ● 55°C. Crushed cells, ◆ 40°C, ▲ 55°C. (*Source*: Mendes, R.L. and Palavra, A.F., in *Chemistry, Energy, and the Environment*, Sequeira, C.A.C. and Moffat, J.B., Eds., Royal Society of Chemistry, Cambridge, 51, 1998. With permission.)

These microalgae were submitted to supercritical CO_2 at pressures between 10.0 and 35.0 MPa and temperatures of 40°C and 55°C. The extractions were carried out on 5 g of freeze-dried *Chlorella* at several physical conditions of the microalgae: not crushed (whole), partially crushed, and totally crushed cells [34, 35].

The initial concentration of lipids (as determined by the slope of the cumulative curve of the extraction at origin) in supercritical fluid increased with pressure, for both temperatures, either with whole or crushed cells, but with the latter the increase was more significant (Figure 6.3). Above 15.0 MPa, there was an initial concentration increase with temperature for whole cells, but with the crushed algae, that value was around 25.0 MPa. For crushed cells, the highest value for this concentration was obtained at 35.0 MPa/55°C (19 mg/L CO_2) and the lowest (3 mg/L) at 20.0 MPa/55°C. This pressure was the lowest used for crushed cells. For whole cells, the highest concentration obtained was 5 mg/L CO_2 at 35.0 MPa/55°C. This behavior can be related to the different amount and type of lipids available in the supercritical CO_2, according to the physical condition of the algae.

For the conditions of pressure and temperature studied, using whole cells, the global yield of lipids obtained increased either with the pressure at constant temperature or with temperature at constant pressure (20.0 and 35.0 MPa) [34]. With crushed cells at 20.0 MPa, the yield decreased with temperature, whereas at 35.0 MPa, it increased.

The highest yield of lipids obtained by SFE was 13.3% (dry weight, partial crushed cells) at 35.0 MPa/55°C; this value dropped to 5% at the same conditions using the whole algae. The yield of organic solvent extraction for crushed cells using acetone and hexane were 16.8% and 18.5%, respectively.

The extraction of carotenoids showed a similar behavior to that of lipids for pressure and temperature variations. However, the yield of carotenoids was higher with supercritical CO_2 at 35.0 MPa (50 mg/100 g dry weight algae) than that obtained

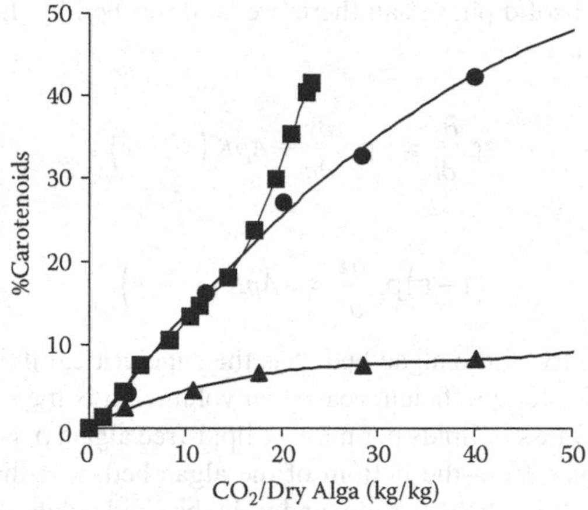

FIGURE 6.4 Recovery of carotenoids from *Chlorella vulgaris*, as a function of solvent-algae ratio, at 35.0 MPa and 328.1 K. ▲ Whole cells, ● Slightly crushed cells, ■ Well crushed cells. (*Source*: Mendes et al., *Inorganica Chimica Acta*, 356, 328, 2003. With permission.)

either with acetone (40 mg/100 g dry weight algae) or hexane (30 mg/100 g dry weight algae). When the degree of crushing increased, the yields of lipids and carotenoids also increased [35].

The fraction of carotenoids in the extracted oils increased with the pressure. For whole cells, this fraction ranged from 18 mg/100 g oil (at 20.0 MPa/55°C) to 171 mg/100 g of oil (at 35.0 MPa/55°C) when 100 g of CO_2 on 5 g algae were used [34]. For crushed cells, the fractions obtained were higher: 100 mg/100 g oil at 20.0 MPa/55°C and 275 mg/100 g of oil at 35.0 MPa/55°C [59].

When a completely crushed *Chlorella* is used, with a carotenoids content of 3 mg/g of dry algae, the yield of carotenoids extraction increased steeply when about 20% of these compounds were extracted (Figure 6.4). This increase can be attributed to a higher accessibility of the supercritical fluid to the carotenoids bound to the cell fragments after the lipids' outside particles are extracted or to a competitive effect between the lipids and carotenoids for the supercritical solvent [35].

Astaxanthin and canthaxanthin account for about two-thirds of the carotenoids in *C. vulgaris*. The ratio of astaxanthin to canthaxanthin in the supercritical extracts (0.8) is the same as that found in acetone extracts when completely crushed cells are used. However, when partially crushed or whole cells are used, the ratio of astaxanthin to canthaxanthin in the supercritical extracts drops considerably, from 0.6 (with acetone) to 0.2. Possible explanations for this behavior are a stronger bond between astaxanthin and the cellular matrix or a different localization of the two carotenoids.

An unsteady model to describe the SFE of lipids from the microalgae *C. vulgaris* was based on the model used to describe the extraction of lipids from fungi (*Saprolegnia parasitica*) using supercritical CO_2 and supercritical CO_2 + cosolvent (10% ethanol) [49]. The model assumes that the axial and radial dispersions are negligible, the properties of the fluid and algal bed remain constant, and the complex mixture of lipids is considered as a single component [60]. The mass balance in fluid and solid phase can therefore be described by the following set of differential equations:

$$\varepsilon\rho\frac{\partial y}{\partial t} = -\rho u\frac{\partial y}{\partial h} + ApK\left(y^{*} - y\right) \tag{6.1}$$

$$\left(1-\varepsilon\right)\rho_s\frac{\partial x}{\partial t} = -ApK\left(y^{*} - y\right) \tag{6.2}$$

where ε is the porosity of the algae bed, ρ is the supercritical fluid density, ApK is the overall mass transfer coefficient (based on volume), u is the superficial velocity of the fluid, x is the mass of lipids per mass of lipid-free algae, ρ_s is the algae density, h is the axial distance from the bottom of the algae bed, y is the lipid concentration in supercritical fluid, t is the time, and y^* is the lipid solubility. The initial and boundary conditions are x = xo at t = 0 for any h, and y = 0 at h = 0 and t ≥ 0.

To have a complete description of the extraction curve, the overall mass transfer coefficient (which aggregates the external and the intraparticle resistances) must vary throughout the whole extraction, according to an empirical expression (3) [49, 60], found by a trial-and-error method:

$$ApK = ApKo\exp\left[\ln\left(0.01\right)\left(x_o - x\right)/\left(x_o - x_{\text{shift}}\right)\right] \tag{6.3}$$

where x_{shift} is the concentration of lipids (in the algae) at which the diffusion-controlled regime starts and x_o is the initial concentration of lipids. Mass transfer coefficients were determined using a least squares regression of the experimental extraction curves.

The model was applied to different physical conditions of *C. vulgaris* at a pressure of 35.0 MPa and 55°C, using a CO_2 flow rate of 0.8 g/min. For algae with whole cells, the shift for the diffusion-controlled regime occurred when 4% of the lipids were obtained, in the case of a *C. vulgaris* with a thicker cellular wall, and 10% when an alga with a thinner cellular wall was used. The thickness of this wall seems related to the carotenoid content of *C. vulgaris* [61]. With a partial crushing, the shift occurred when about 25% of the lipids were extracted, whereas with a total crushing (almost all the cells disrupted), the shift occurred when 55% of the lipids were extracted. The last value is similar to those obtained with SFE of lipids from fungi and of oil from rape seed, although in this case, the shift occurred because the remaining oil was in other form of binding to the matrix [62]. The calculated mass transfer coefficients (ApKo) ranged between 0.171 kg/m^{3}s and 0.531 kg/m^{3}s for the extraction using whole cells and 0.7 kg/m^{3}s and 2.2 kg/m^{3}s using crushed cells [60, 61].

6.2.3 *Dunaliella*

The microalgae *Dunaliella salina* can produce β-carotene up to 14% (dry weight basis) [2], with this compound being mainly a mixture of two geometric isomers (all-*trans* and 9-*cis*). The synthetic compound is the crystalline all-*trans* isomer,

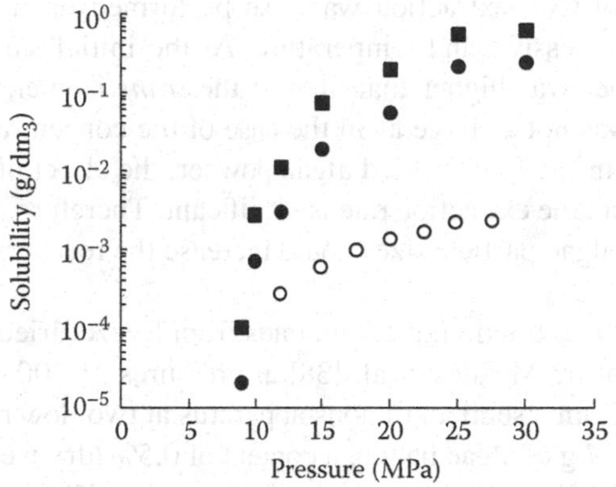

FIGURE 6.5 Solubility of β-carotene isomers in supercritical CO_2 (40°C), ● all-*trans*-β-carotene (mixture), ■ *cis*-β-carotene (mixture), ○ synthetic *trans*-β-carotene. (*Source*: Adapted from Mendes et al., *Inorganica Chimica Acta*, 356, 328, 2003. With permission.)

which is not really fat-soluble; however, natural β-carotene (from algae) is not crystalline and is highly fat-soluble.

Separating the two isomers is advantageous because the *cis* is more easily absorbed in human tissues [63] and has more antioxidant activity [63]. The conventional techniques of isomer separation involve large amounts of organic solvents and are very time-consuming. On the other hand, for medical and food applications, it is important that this separation is done without the use of toxic solvents.

The separation of isomers can be carried out through SFE if there is a significant difference in solubility of the compounds. The *cis* isomer is about three times more soluble than the all-*trans*, and both carotenes present higher solubility than the synthetic all-*trans* isomer.

The concentration in supercritical CO_2 of the *cis*-β-carotene and all-*trans*-β-carotene from a solid mixture of carotenoids from *Dunaliella salina* is shown in Figure 6.5. This mixture was obtained by acetone extraction from wet *Dunaliella* [37]. The organic extract was saponified in order to remove the lipids and chlorophyll, and the separation of phases was carried out with diethyl ether. The carotenoids contained in the ether phase were recovered by evaporation of this solvent with a stream of nitrogen.

Gamlieli-Bonshtein et al. [36] used a different approach to compare the solubility of the β-carotene isomers. Due to nonavailability of the pure 9-*cis* isomer to carry out solubility measurements, the solubility of the compound was calculated indirectly. Supercritical CO_2 extraction of β-carotene from *Dunaliella bardawil* was performed at 448 bar and 40°C from a concentrate extract obtained with a mixture ethanol/hexane/water. These conditions of pressure and temperature were previously found to be the optimal ones for the supercritical extraction of β-carotene [36]. Entering with the ratio of the initial rates of the extraction (from the concentrate) of the two isomers and knowing the solubility of the *trans* isomer, which was previously measured, the authors calculated the solubility of the 9-*cis* β-carotene as being, at 448 bar /40°C, 7,64 × 10^{-5} g isomer/g CO_2, nearly four times the value of the all-*trans* isomer. Supercritical CO_2 extraction was also performed on raw algae under the same conditions of pressure and temperature. At the initial stages, the extraction rate of the *cis* isomer was higher than that of the *trans* isomer, but the difference between the rates was not as large as in the case of the concentrate. This suggested to the authors that, in the freeze-dried algae powder, the effect of internal diffusion limitations on β-carotene extraction rate is significant. Therefore, the results indicate that decreasing the algae particle size should increase the recovery and selectivity of isomer separation.

Supercritical CO_2 extraction of compounds from freeze-dried *Dunaliella salina* was also carried out by Mendes et al. [38] at pressures of 200 and 300 bar and a temperature of 40°C, in a semicontinuous apparatus at two flow rates, of 18.9 g/min and 10.8 g/min, on 27 g of algae having a content of 0.5% (dry weight) of β-carotene and 10% (dry weight) lipids. The yield of the extraction of lipids increased with the pressure, but no increase in the extraction yield of β-carotene was observed with the pressure. When the supercritical extraction of lipids and β-carotene were compared for the two different flow rates at 300 bar, it was verified that the extraction yield was higher for the lower flow rate. Recoveries of 25% and 80% were reached for lipids and β-carotene, respectively, at the flow rate of 10.8 g/min. It was also verified that it was possible to obtain values of the *cis-trans* ratio well above of the initial one in the algae (1.3). For instance, when the extraction was carried out at 300 bar and at a flow rate of about 18.9 g CO_2/min, the *cis-trans* ratio reached the highest value (3.6) at its beginning and decreased exponentially along the extraction. For a lower flow rate of CO_2 (10.8 g/min), the ratio decreased to 2.2. These results showed that the *cis*-β-carotene was quickly depleted at the surface of the algae particles. This behavior can be explained by the higher solubility of the *cis* isomer in CO_2 and by the presence of a higher amount of this compound in the outer part of the algae particles (where the *trans* isomer was more easily isomerized), as the capacity of *Dunaliella* to produce the *cis* isomer is related to light exposition [65]. The values of the *cis-trans* ratio were higher (about 4) when β-carotene concentrates of *Dunaliella salina* [37] or *Dunaliella bardawil* [36] were used.

In an attempt to increase the selectivity of the separation of β-carotene isomers, Nobre et al. used a combination of supercritical CO_2 and silica gel [66]. The experiments were carried out at a pressure of 200 bar and a temperature of 40°C in a flow apparatus. The extractor was a 32-cm^3 vessel filled with 20 g of glass beads, in which about 100 mg β-carotene (obtained from *Dunaliella*) had been previously precipitated. A second 5-cm^3 vessel in series contained the silica gel. Several loads of silica were used: 1.5 g, 0.5 g, and 0.15 g. The *cis-trans* ratio increased when the amount of silica gel decreased, being always higher than the one obtained by simple SFE, reaching a value of 7.4 for the lowest amount of silica gel used. This value of the ratio was about three times the one obtained by SFE at the same conditions of pressure and temperature.

6.2.4 *Haematococcus pluvialis*

Haematococcus pluvialis is a freshwater flagellate that accumulates astaxanthin in its aplanospores. It can produce astaxanthin in relatively high yields (up to

45 mg/g dry weight algae) in laboratory conditions, but due to heavy contamination, the use of open ponds is not satisfactory to obtain the high yields necessary for algal astaxanthin to compete with the synthetic one. Other methods for growing *H. pluvialis* have been tried to overcome that drawback and also to increase yields [13]. Olaizola [67] explored the strategy used by a pharmaceutical company to make *Haematococcus* astaxanthin more acceptable to consumers and to make it more competitive. This strategy includes the use of SFE.

Astaxanthin (3,3′-dihydroxy-β,β-carotene-4,4′-dione), $C_{40}H_{52}O_2$, is one of the most important xanthophylls either from a commercial or a biotechnological point of view, being the most abundant carotenoid and pigment found in certain aquatic animals, such as salmon, trout, shrimp, and lobster. Several isomers are found in astaxanthin present in nature, with 3S, 3S' being the main one found in *H. pluvialis* [68]. The presence of the hydroxyl and keto groups in the ionone ring explains why most of the astaxanthin appears in *Haematococcus* in esterified forms (mono and diester) and also why it is very prone to oxidation. The activity of the compound is several times higher than that of β-carotene and vitamin E. Guerin et al. [68] reviewed the main applications for human health and nutrition of *Haematococcus* astaxanthin, namely its uses against ultraviolet-light photooxidation, inflammation, cancer, and aging and age-related diseases and in the promotion of immune response, liver function, and heart, eye, and prostate health. On the other hand, a single high dose (100 mg) was administered to humans in a study of bioavailability, and it was verified that astaxanthin was readily absorbed and incorporated in human plasma lipoprotein at a convenable degree [69].

Valderrama et al. [30] and Machmudah et al. [41] carried out supercritical CO_2 extraction of astaxanthin from *H. pluvialis*. Another study of the same type that focused on astaxanthin, as well as on the other carotenoids present in these microalgae, was carried out by Nobre et al. [40]. Pressured fluid extraction (PFE), which uses conventional solvents at controlled temperatures and pressures, has also been performed to extract carotenoids from *H. pluvialis* [70].

In the studies of Valderrama et al. [30], the semicontinuous supercritical apparatus was provided with an extraction vessel of 450 ml, and the experiments were carried out at 300 bar and 60°C. Three types of runs were performed: run 1, in which the dried samples of microalgae were crushed by cutting mills prior to extraction; run 2, in which the samples were crushed like in run 1, followed by manual grinding using dry ice (solid CO_2); and run 3, in which the samples were treated as in run 2, but the extraction was performed using supercritical CO_2 containing ethanol as a cosolvent (9.4% weight).

The astaxanthin yield depended strongly on sample preparation and on the use of ethanol as a cosolvent. A more efficient grinding led to higher yield. With pure CO_2, astaxanthin is recovered only partially. With *H. pluvialis* well crushed and using the cosolvent, an astaxanthin recovery of 97% was reached. This recovery was determined from the initial and final content (residue of the extraction) of the compound in the microalgae. Valderrama et al. [30] did not provide the method by which those contents were determined and no mention of esterified astaxanthin, the usual form of astaxanthin in these microalgae, was made. The supercritical extraction (yield

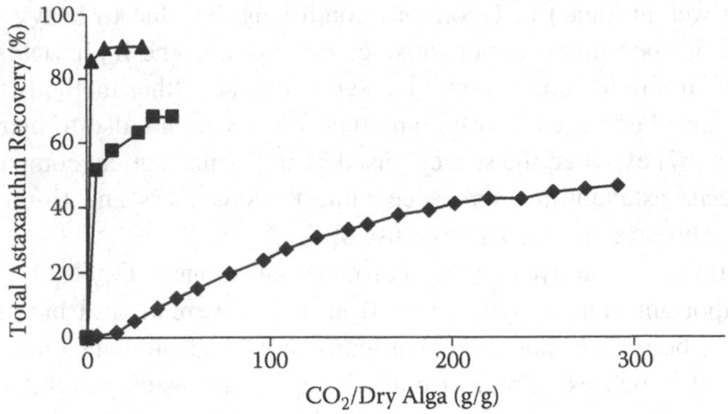

FIGURE 6.6 Recovery of total astaxanthin (free plus esters) as a function of CO_2 amount with (■) and without (▲) ethanol as a cosolvent using slightly crushed algae, and CO_2 with ethanol using well-crushed algae (◆), at 60°C and 300 bar. (*Source*: Nobre et al., *Eur. Food Res. Tecnol.*, 223, 787, 2006. With permission.)

versus solvent-feed ratio) was modeled with the same empirical model used in the supercritical extraction of lipids from *Spirulina* (see section 6.2.7.1).

Nobre et al. [40] carried out a more detailed study of supercritical CO_2 extraction of carotenoids from *H. pluvialis* using freeze-dried samples ground with a ball mill. The effects of pressure (200 and 300 bar), temperature (40 and 60°C), degree of crushing (related to milling time), and the use of ethanol (10%) as cosolvent on the extraction yield were determined. The extraction with acetone was carried out using glass beads mixed with the algae until total absence of color was present in the biomass.

Besides free astaxanthin (which represents about 2% of the carotenoids), the yield of esterified astaxanthin (73% of the carotenoids present in *H. pluvialis*), β-carotene (7% of the carotenoids), canthaxanthin, and lutein were assessed and the recoveries obtained by SFE were compared with the total content of carotenoids from the microalgae. Figure 6.6 shows the total astaxanthin recovery as a function of solvent-feed ratio, showing the beneficial effect of ethanol as a cosolvent as well as the degree of crushing of the microalgae cells. A recovery of about 90% was obtained. The improvement in yield due to ethanol was ascribed to the increase of astaxanthin solubility in supercritical CO_2 due to its polar character and, on the other hand, to the swelling of the microalgae particle pores, which eased the release of the compounds. The recovery improvement due to the crushing of the cells can be attributed to the increase in the number of disrupted cells and the degree of disruption, increasing the amount of extractable carotenoids. In this work, the researchers also verified an increase in the recoveries of carotenoids when the pressure increased from 200 to 300 bar, but only a slight improvement was obtained when the temperature, at the same pressures, increased from 40°C to 60°C.

All the carotenoids identified in *H. pluvialis* (esterified and free astaxanthin, β-carotene, lutein, canthaxanthin) were recovered with values near or higher than 90% at a pressure of 300 bar and a temperature of 60°C, using ethanol as a cosolvent.

Machmudah et al. [41] studied the supercritical CO_2 extraction of astaxanthin from *H. pluvialis* using a flow apparatus provided with a 50-cm^3 extraction vessel containing 7 g of algae in each experiment. The content of astaxanthin was 3.33% (dry basis) as determined by Soxhlet using dichloromethane. The conditions tested were temperature (313 to 353 K), pressure (20 to 55 MPa), CO_2 flow rate (2 to 4 cm^3/min), and ethanol concentration (ethanol-solvent = 1.67–7.5%). The algae were not crushed, unlike in the aforementioned studies. Without ethanol entrainer, the amount of total extract, the amount of astaxanthin extract, and the astaxanthin content in the extract increased with an increase in temperature. This behavior was attributed to the increase of vapor pressure of astaxanthin and also to the decomposition of the cell wall with the temperature, which contributed to the increase of the extractable compounds. The increase of pressure also led to an increase of those yields. The amount of the total extract and the amount of astaxanthin extract, but not the astaxanthin content in the extract, slightly increased with CO_2 flow rate.

The highest astaxanthin extracted (recovery) and astaxanthin content (in the total extract) were 77.9% and 12.3%, respectively, which were obtained at 55 MPa, at 343 K, and at a CO_2 flow rate of 3 cm^3/min.

Using ethanol as an entrainer, a higher amount of astaxanthin (80.6%) could be extracted at a more moderate pressure (40 MPa). On the other hand, the increase in entrainer concentration, up to 5% (v/v) ethanol, increased the amount of astaxanthin. With the use of entrainer, the CO_2 flow rate considerably affected the astaxanthin extracted and a higher amount of astaxanthin was extracted as CO_2 flow rate decreased. This last point suggested to the authors the importance of the internal mass diffusion in the extraction of astaxanthin from *H. pluvialis*. This is in accordance with the previous studies of Valderrama [30] and Mendes [40], which demonstrated that the crushing of the algae contributed highly to the success of the extraction.

In studies of PFE using acetone as solvent, higher or equal amounts of carotenoids were extracted than were extracted with the traditional organic solvent method [70]. The time of extraction was 20 minutes with PFE, whereas 90 minutes were necessary to perform the traditional acetone extraction.

6.2.5 *Hypnea charoides*

Hypnea charoides is a subtropical red seaweed. Among the seaweeds, the red ones are known for their high omega 3 (w3) fatty acid contents [71]. Among the fatty acid constituents of algal lipids, the most important are the polyunsaturated fatty acids (PUFAs). They range from C12 (twelve carbon atoms) to C22 and can contain up to six allylic bonds separated by a carbon atom. Within this group, the essential fatty acids (EFAs) for humans are linoleic acid, g-linolenic acid (GLA), dihomo-g-linoleic acid, arachidonic acid, and EPA. There are two types of PUFAs: w3 and omega 6 (w6), known officially by n-3 or n-6 fatty acids, respectively [72]. These numbers correspond to the position of the last double bond counted from the last methyl group. For most eukaryotic algae, which contain predominantly mono and saturated fatty acids, the triglycerides constitute up to 80% of the lipid fraction [73]. In a general way, the marine algae are rich in w3 fatty acids, namely EPA (C20:5, w3) and docosahexaenoic acid (DHA, C22:6, w3). The main source for human consumption of these PUFAs is marine fish, whose feeding basis is constituted by microalgae. With the depletion of fish stocks, microalgae and macroalgae (seaweeds) could be alternative sources of w3 fatty acids [42].

A large number of epidemiological, animal, and clinical studies have shown that intake of w3 fatty acids is beneficial for preventing a certain number of diseases (e.g., cardiovascular diseases, some forms of cancer, and diseases with immunoinflammatory components) and can also play a role in brain and nerve development of growing fetuses and infants [72]. The therapeutic effects of these fatty acids have also been shown [74].

Cheung [42] carried out supercritical CO_2 extraction of n-3 fatty acids from these macroalgae, with emphasis on the effects of temperature, pressure on the yield, and composition of the extracts. The experiments were performed on 2 g of dried and ground (1 mm sieve) algae, at temperatures of 40°C and 50°C and pressures of 24.1, 31.0, and 37.9 MPa at a CO_2 flow rate of 2 mg/min. The extraction cell was provided with a 10-cm^3 stainless cartridge.

Cheung found that lipid extraction increased with the pressure at constant temperature. At higher pressures (31.0 and 37.9 MPa), an increase in temperature led to an increase in the yield and also in the rate of extraction (in the initial period). However, at the lowest pressure used, the yield decreased when the temperature passed from 40°C to 50°C. These results are similar to published data of supercritical CO_2 extraction of lipids from *Chlorella vulgaris* [34]. This behavior is explained in terms of the combined effect of the pressure and temperature on the density of the solvent and vapor pressure of the solutes.

The maximum yield of lipids from *H. charoides* at the conditions studied by Cheung was 67.1 mg/g (dry weight algae) at 37.9 MPa/50°C and the lowest was 33.7 mg/g at 24.1 MPa/50°C. Six w3 fatty acids were found in the extracts, representing at the conditions of maximum yield (37.9 MPa/50°C), 24% and 39% of the lipids and total fatty acids, respectively. EPA (20:5 w3) represented about 60% of the total w3 fatty acids and α-linolenic acid (18:3 w3) 16% of these fatty acids. DPA (22:5 w3) and DHA (22:6 w3) were also found to represent 8% and 12%, respectively, of the total w3 fatty acids.

With the exception of DPA and DHA, the yield of all the w3 fatty acids had a similar behavior to that of the total lipids, with the variation of pressure and temperature. Even at higher pressures, the C22 w3 fatty acids showed a lower yield at higher temperatures. This behavior could be attributed to the fact that chain length is a more important factor in supercritical CO_2 solubility than the degree of saturation.

The author concludes that algal lipids of *H. charoides* could be a nonconventional alternative to w3 fatty acids.

6.2.6 *Nannochloropsis*

Nannochloropsis species are marine microalgae able to produce a high content of both lipids and eicosapentaenoic acid (EPA). This compound is the major fatty acid of those found in *Nannochloropsis* (ranging from 29% to 33%), with the content of n-3 PUFAs being about 40% of the total fatty acids [45]. It is also a source of valuable pigments, such as chlorophyll a, astaxanthin, zeaxanthin, and canthaxanthin [75].

Andrich et al. [45] verified that in supercritical extraction studies of algae little information about the kinetics of the process and the influence of the operating conditions on the composition of the lipid extracts were available. Therefore, they focused their studies at examining these points.

Supercritical CO_2 extraction of bioactive lipids from the microalgae *Nannochloropsis* sp. was carried out in a pilot plant apparatus at pressures of 40, 55, and 70 MPa and temperatures of 40°C and 55°C. An amount of 180 g microalgae, mixed with 100 g of 3-mm glass spheres, was used for each run. The supercritical CO_2 flow rate was 10 kg/h. The authors also performed extraction by percolation with hexane (Soxhlet).

The following equation is used to describe the evolution of extracted oil over time (t), for both supercritical extraction and hexane one:

$$Oe = H^*\left[Os\right]\left(1 - e^{-kt}\right) \tag{6.4}$$

where Oe is the amount (g) of oil extracted at a time (t) per gram of algal biomass, H^* is a constant ranging from 0 to 1, [Os] is the amount (g) of oil present in 1 g of starting material, and k is the kinetic constant. The product H^* [Os] represents the asymptotic value of the extraction curve. In a highly efficient process, H^* tends to 1, meaning that the amount of extractable oil per gram of biomass coincides with the concentration of oil in the starting material.

The maximum extraction rate (R_{max}) is given by:

$$R_{max} = kH^+\left[Os\right] \tag{6.5}$$

The authors interpreted the cumulative extraction curves as a function of time in terms of the kinetic parameters. In terms of oil extractable, all the processes (supercritical CO_2 extractions at all the conditions and hexane extraction) are substantially equivalent (about 250 mg lipids/g dry algae). The value of k increases at constant temperature with the increase of pressure and also increases for a given pressure when the temperature increases, although in this case more moderately. The combined effect of pressure-temperature (P-T) affects k more than P and T alone, increasing its value three times from the extraction at 40°C and 400 bar to the extraction at 55°C and 700 bar. SFE was clearly faster than the extraction by hexane (R_{max} was several times higher).

R_{max} was expressed in terms of the concentration of lipids in the supercritical fluid at the beginning of the extraction, R^*_{max} (g/l), and related to the density, r (g/l), and temperature, T (K), of the solvent through the equation due to Chrastill [76]:

$$R^*_{max} = \rho^a e^{(b/T+C)} \tag{6.6}$$

The values of the parameters, obtained from the several extraction conditions studied, were a = 10.92 ± 2.57; b = 3506.57 ± 1225.65; and c = 62.68 ± 16.18. The authors claim that this kinetic model seems suitable to evaluate the economic feasibility of the process, although more information is needed for its generalization.

The percentage of EPA in the total fatty acids ranged from 29.4% to 33% for the conditions of extraction studied. However, these values seem lower than the 37% claimed for the commercial product. The authors attributed this to the fact that the use of supercritical CO_2 caused some losses in the most polar fractions, where EPA is possibly primarily located. On the other hand, no particular differences were found in the fatty acid profiles of extracts. However, a slight increase in EPA and DPA (C22:5*n*-3) seems perceivable when passing from the mildest to the hardest SFE conditions.

Macías-Sánchez et al. [44] carried out supercritical CO_2 extraction of pigments from *Nannochloropsis gaditana* in a microscale apparatus, which is provided with two extractors of 10 ml each. In one of the extractors was inserted a cartridge containing 0.2 g of freeze-dried microalgae and, after 15 minutes of static extraction, the runs were performed at a flow rate of 4.5 mmol/min for 3 h. A total of 15 experiments were carried out in a random way, in order to fulfill a multilevel factorial design to determine the effect of temperature and pressure on the extraction of the carotenoids and chlorophyll a.

Organic solvent extraction, using methanol, was performed by sonication on 0.2 g of *N. gaditana,* with 5 ml of solvent for each of the 14 cycles necessary to obtain the methanol without any coloration. The algae pellet remained greenish after the solvent extraction. The yields obtained were 0.8 mg/mg and 18.5 mg/mg (dry weight basis) for carotenoids and chlorophyll a, respectively.

SFE was carried out at temperatures of 40°C, 50°C, and 60°C and pressures of 100, 200, 300, 400, and 500 bar. The analysis of the experimental design shows that the temperature, pressure, and interaction of both variables significantly influence the process (p-value < 0.05). The yield increases with pressure at a given temperature from 100 to 400 bar, where the maximum yield is reached at 60°C (400 bar) for both carotenoids and chlorophyll a (0.343 mg/mg and 2.238 mg/mg, respectively). At 100 bar, no pigments were extracted and at each pressure the yield increased with temperature for pressures above 200 bar. As usual, this behavior is explained in terms of balance between solvent density and vapor pressure. On the other hand, at 500 bar, the yield decreased. The fact of the maximum yield be obtained at an intermediate pressure (400 bar) is explained in terms of the decrease of the diffusion coefficient due to the increase of the CO_2 density, which reduces the penetration capacity of the fluid and the yield at the highest pressure used. This also already had been suggested for the extraction of carotenoids from *Spirulina* [28].

The following empirical correlations for carotenoids (6.7) and chlorophyll a (6.8) were obtained [44]:

$$R = -0.233163 + 0.00492577\,T + 0.00121779\,P - 0.0000923077\,T^2 \\ +0.00002705\,PT - 0.00000353013 - P^2 \tag{6.7}$$

$$R = 3.43203 - 0.00140362\,P - 0.14499\,T - 0.00000725952\,P^2 \\ +0.0001963\,PT + 0.001144\,T^2 \tag{6.8}$$

where R is the yield of pigment (mg/mg), T is the temperature (°C), and P is pressure (bar).

Furthermore, the ratio of carotenoids to chlorophylls (carot/chlor) in the extracts is always higher than that obtained with methanol (carot/chlor = 0.043) and decreases with pressure, with the highest ratio obtained at 200 bar and 60°C (carot/chlor = 1.389). This leads the authors to conclude that supercritical CO_2 is more selective for the extraction of carotenoids in the presence of more polar pigments like chlorophyll a.

The SC extraction of chlorophyll in this work contrasts previous work from other authors, in which the nonextraction of this type of compound by supercritical CO_2 from *Botryoccocus braunii* [33], *Skeletonema costatum* [46], and *Ochronomas danica* [46] was reported. However, in the study of supercritical CO_2 extraction of oil from *Scenedesmus obliquus* [20], the extraction of chlorophyll a is mentioned. This matter is discussed by Balaban et al. [23], who suggested that entrained particles of algae might have caused the discrepancy between this result and previous reports of insolubility of chlorophyll in supercritical CO_2.

6.2.7 *Spirulina* (*Arthrospira*)

Spirulina is one of the most promising microalgae. It is rich in the essential fatty acid all-*cis*-6, 9, 12-octadecatrienoic acid (GLA); pigments (phycocyanin, myxoxanthophyl, zeaxanthin, and β-carotene); proteins; and sulfolipids [77]. Like other algae, such as *Chlorella, Spirulina* has also been used as functional food (food derived from natural sources whose consumption is beneficial to health of the human body). In this case, algae are provided either as supplements or complete food [78]. Research has also shown the therapeutic value of *Spirulina* and its extracts in a large number of diseases [79].

The GLA from *Spirulina* actually cannot compete economically with the GLA from higher plants (e.g., evening primrose, blackcurrant, and borage), but there are some advantages to using *Spirulina* microalgae as a GLA producer. This compound is found mainly in the glycolipid fraction of the lipids, which eases its purification as a pharmaceutical commodity and, on the other hand, unlike the GLA of the higher plants, it is not associated with undesirable fatty acids [77].

GLA presents higher EFA activity and antithrombotic and hypolipidemic effects than linoleic acid (18:2w6) [80]. It also has been used in several medical applications, such as the treatment of schizophrenia, multiple sclerosis, atopic eczema, premenstrual syndrome, diabetes, and rheumatoid arthritis [80, 81]. It can also play an important role in the synthesis of a kind of prostaglandin, a hormone involved in essential tasks of the human body, namely the control of the arterial tension, cholesterol, and inflammation.

The antioxidant activity of extracts from *Spirulina platensis* obtained with pressurized liquid extraction was studied and attributed to the presence of carotenoids, phenolic compounds, and degradation products of chlorophyll [82].

6.2.7.1 *Spirulina maxima*

Supercritical CO_2 extraction of lipids from *S. maxima* was studied by Mendes et al. [32], Canela et al. [29], and Valderrama et al. [30]. The first authors focused their work on the extraction of GLA, the second ones on the extraction of fatty acids and carotenoids, and the third on the residue of the extraction in order to obtain the pigment phycocyanin.

The objective of the work of Mendes et al. [32] was to carry out the supercritical CO_2 extraction of lipids, focusing on the GLA content of these lipids and to assess the influence of pressure, temperature, and the use of ethanol as entrainer on the extraction yields and selectivity of the several fatty acids and classes of lipids. The extraction of the *Spirulina* lipids was also performed with organic solvents (ethanol, hexane, acetone, and a mixture of water, chloroform, and methanol), having in view the comparison of the two types of extraction (SFE and organic). The supercritical experiments were carried out in a flow apparatus on 3.6 g of ash-free, freeze-dried *Arthrospira* (*Spirulina*) *maxima* at a flow rate of 2 g/min at pressures of 250 bar and 300 bar and temperatures of 50°C and 60°C. At 250 bar and 50°C with pure CO_2, a yield of 0.05% (GLA/dry biomass wt%) was obtained when 1.4 kg CO_2 was used, but above a solvent-feed ratio of 300 g/g, the concentration of lipids in was very low. At the same conditions, using supercritical CO_2 plus ethanol (10 mol%), the yield improved to 0.17%, but higher yields are possible because the concentration of the lipids in the solvent grew steadily even for high solvent-feed ratios. Ethanol can have an entrainment effect on the extraction of the lipids, which are mainly polar, increasing its solubility and, on the other hand, can counterbalance the hydrogen bonds and ionic forces between the membrane-associated lipids and proteins [83], allowing the lipids to be available for extraction by the supercritical fluid. The increase of temperature at 250 bar led to an increase in GLA yield, but the increase in pressure to 350 bar at 60°C led to the highest GLA yield obtained (0.44%). The GLA yields reached with ethanol, acetone, and hexane were 0.68%, 0.63%, and 0.01%, respectively. In another study [31], in which ethanol was mixed with *Spirulina*, GLA yield increased with the amount of ethanol and, for equal masses of ethanol and microalgae, the yield was more than 10 times that obtained with supercritical CO_2 extraction from non-pretreated *Spirulina*.

Valderrama et al. [30] extracted the CO_2-soluble material from *Spirulina maxima*, with the aim of concentrating the pigment phycocyanin in the extraction residue. The first extraction was carried out using pure CO_2 at 300 bar and 60°C, which led to a lipid yield of 1.1%. In a second set of experiments, the co-solvent ethanol (10%) was used, with an increased yield of 3%. The content in phycocyanin in the samples of *Spirulina maxima* increased from 6.17 wt% to 8.3% when pure CO_2 was used and from 6.91% to 8.4% when the solvent was CO_2 plus 10% ethanol.

An empirical model for the concentration of phycocyanin correlates well the lipid extraction yield (Y) with the solvent-feed ratio (X):

$$Y = \alpha\left(1 - e^{-\beta X}\right) \tag{6.9}$$

In the model, α and β are empirical constants obtained from the experimental yield. Y = 0 for X = 0, and the yield has as a limit the empirical maximum, which is represented by the parameter α.

Canela et al. [29] studied the supercritical CO_2 extraction of fatty acids and carotenoids from *S. maxima*. The experiments were conducted at temperatures between 20°C and 70°C and with pressures up to 180 bar. These authors previously determined *Spirulina* composition (3.1% lipids, 54.5% proteins, 9.1% humidity, and 12.2% ashes). The total amount of CO_2-soluble material was determined, using a 3.5-cm^3 extraction at 180 bar (30°C and 40°C) and 150 bar (30°C, 40°C, 60°C, and 70°C). The highest amount, 0.93% (dry weight basis of *Spirulina*), was obtained at 150 bar and 60°C.

Kinetic experiment runs of supercritical CO_2 extraction of lipids from *S. maxima* were performed using a standard extraction unit provided with an extraction cell of about 368 cm^3. For each run, an amount of 175 g microalgae, mixed with equal amount of glass beads, was used. The solvent flow rate was maintained at 0.12 kg/h. The effects of temperature and pressure were quantified using a factorial experimental design. The optimal extraction conditions were 150 bar and 60°C. But taking into account the target components (fatty acids and carotenoids), which are prone to degradation at high temperatures, a temperature of 50°C or lower is suggested. Canela et al. [29] reported the composition of the supercritical extracts obtained in terms of total carotenoids. The yields are very low when compared with the carotenoid content of the microalgae reported by Cohen [77]. This behavior can be attributed in part to the low pressures used, as the carotenoids present a very low solubility at these pressures [84].

Supercritical extractions curves were modeled according to the Goto et al. model [85], which was developed for the supercritical extraction of essential oil from peppermint. This model treats the solid substrate as a porous matrix. The solute is extracted after its desorption from the solid. Diffusion occurs inside the particle pores, and there is a mass-transfer resistance in the film surrounding the particles. The model gave a good representation of the experimental data, and a partition coefficient and a combined mass-transfer coefficient were obtained from the fitting of the experimental results to the main equation of the model.

6.2.7.2 *Spirulina platensis*

Santos et al. [26] performed experiments of SFE in samples of *S. platensis* using CO_2 at a temperature of 40°C and a pressure of 200 bar. Beds of 10 g algae (and sometimes 20 g) were placed in a half-liter extraction vessel.

To evaluate the effect of the moisture content of the algae, three experiments were carried out. The first one, directly on the microalgae supplied (moisture content of 3.24%), led to a lipid yield of about 1.1% (dry basis wt); the second test was performed on a sample that had been dried in P_2O_5 desiccator for several hours; the third test was on the residue from the second one, which had been damped to give a moisture content of 7%. The drying of the microalgae produced a substantial fall in the lipid yield (0.3% in a dry weight basis). Further damping of the residue enabled more lipidic material to be extracted from the residue, but still a low yield was obtained (0.5%).

In this work, the fatty acids from the glyceride fraction obtained in the experiments were identified. A typical percentage of the obtained fatty acids is similar to this one: palmitic (33%), palmitoleic (8%), stearic (2%), oleic (4%), linoleic (21%), and g-linolenic (32%).

In order to model the SFE, the authors conducted experiments at two superficial velocities. When the amount of extracted lipids is plotted against the CO_2 mass used, the curves for the higher flow rate fall below that corresponding to the lower flow rate, but when yields are plotted against time, the experimental points fall in a common curve. This is consistent with a mass transfer mechanism within the algae particles (by diffusion), which is so slow that the extractant phase composition changes very little along the bed and never approaches close to the equilibrium value. The external film resistance is negligible. Based on these assumptions, Santos et al. [26] applied the well-known single sphere extraction model and found that the best fit for De/r^2, where De is the effective diffusivity (of the lipids) and the r is the algae particle radius, led to the following value: $De\pi^2/r^2 = 0.00120$, for the data resulting from the SFE of lipids from *S. platensis* at a pressure of 200 bar and temperature of 40°C.

Qiuhui [27] used supercritical CO_2 extraction of lipids from these microalgae, with the goal of removing its bad smell, an obstacle to the marketing and acceptance of *Spirulina*, which is sold and exported as a powder. In fact, this author pretended to separate and purify the active components of the algae. The SFE experiments were carried out in a semicontinuous apparatus, provided with recycling of CO_2, at a flow rate of 24 kg/h; pressures of 30, 35, and 40 MPa; extraction time of 2, 3, and 4 h, and temperature of 40°C.

The highest lipid yield (7.2%) was obtained at 35 MPa and 4 h of operation, a value near that obtained at 40 MPa (7%) for the same extraction time. However, for an extraction time of 2 h, the yield obtained at 40 MPa (6.5%) was higher than the one obtained at 35 MPa (5.9%). The yield of GLA increased with pressure, having obtained 0.12% at 20 MPa, while at 40 MPa 0.29% was reached. Both yield values are higher than the one (0.05%) obtained by Mendes et al. [32] at 25 MPa/50°C from *S. maxima*.

The removing of the deleterious smell was also reached. Moreover, after treatment with supercritical CO_2, the protein content was practically not altered, with only a slight reduction of about 1% in essential amino acids having been detected, thereby preserving the nutritive value of *Spirulina*.

Careri et al. [28] carried out studies of SFE on the strain pacifica (a carotenoid-rich dietary product) of the microalgae *S. platensis*. The target compounds for the study were the carotenoids β-carotene, β-cryptoxanthin, and zeaxanthin, and the authors proposed to find the best experimental conditions for their recovery through an experimental design procedure. Four parameters were investigated: pressure (150, 250, and 350 bar), temperature (40, 60, and 80°C), dynamic extraction time (40, 70, and 100 min), and percentage of ethanol (in volume) added to the CO_2 (5, 10, and 15%). Other experimental conditions were 0.5 g of dry powdered algae (50 mm particle size) in a 7-cm^3 extractor. The extraction by organic solvent was performed using tetrahydrofuran and petroleum ether [86].

For all three carotenoids, the highest pressure (350 bar) led to the highest recoveries. The temperatures that maximized the recoveries were 80°C, 70°C, and 60°C for zeaxanthin, β-cryptoxanthin, and β-carotene, respectively. The effects of the various factors were analyzed and, for all the compounds, the temperature was found not to be significant as the main effect. On the contrary, the pressure of the supercritical fluid plays an important role, appearing to be significant for all the carotenoids.

At constant temperature, a pressure increase causes an increase in fluid density and thus could have a double effect: increased solvent power of CO_2 and reduced interaction between the fluid and the solid matrix, having as a consequence the decreasing of the diffusion coefficient at higher densities.

The amount of ethanol was also significant for all the carotenoids. The increase in the percentage of the entrainer led to an increase of the extraction yields. This effect is not only related to the modification of the polarity of the supercritical fluid but also to the interaction of the ethanol with the solid matrix because the carotenoids with different polarities show better recoveries when ethanol is added to the fluid.

Careri et al. [28] also compared the SFE obtained at the best conditions with that obtained from organic solvent extraction. In terms of yield, organic solvent extraction showed a slight advantage, but because it is time-consuming (with multiple extraction and purification steps) and very expensive, in the end SFE, proved to be a more effective procedure.

6.3 CONCLUSION

The use of algae to obtain useful biochemicals for medical, pharmaceutical, and dietary applications shows a large potential in the near future. However, many algae also produce harmful compounds. Although some of these can be directed to obtain useful drugs, they must be screened before human consumption occurs.

In many cases, SFE shows advantages over the use of organic solvents to obtain compounds from algae, namely a shorter time of extraction and better or similar yields, avoiding the use of expensive and polluting solvents. Moreover, SFE can be used with supercritical fluid fractionation and supercritical fluid chromatography to separate or purify the extracted compounds. Because it is an expensive procedure, SFE is only reserved for high-value compounds in commercial applications.

REFERENCES

1. Borowitzka, M.A. and Borowitzka, L.J., Eds., *Micro-algal Biotechnology*, Cambridge University Press, Cambridge, 1988.
2. Borowitzka, M.A., in Vitamins and fine chemicals from micro-algae, *Micro-algal Biotechnology*, Borowitzka, M.A. and Borowitzka, L.J., Eds., Cambridge University Press, Cambridge, 1998, 153.
3. Burja, A.M. et al., Marine cyanobacteria-a prolific source of natural products, *Tetrahedron*, 57, 9347, 2001.
4. Skulberg, O.M., Microalgae as a source of bioactive molecules: Experience from cyanophyte research, *J. of App. Phycol.*, 12, 341, 2000.
5. Kyle, D.J., Specialty oils from microalgae, *New Perspectives in Biotechnology of Plant Fats and Oils*, Ratray, J., Ed., American Oil Chemists' Society Press, Champaign, Illinois, 1991, 130.
6. Cohen, Z., *Chemicals from Microalgae*, Taylor & Francis, London, 1999.
7. Borowitzka, M.A., Fats, oils and hydrocarbons *Micro-algal Biotechnology*, Borowitzka, M.A. and Borowitzka, L.J., Eds., Cambridge University Press, Cambridge, 1998, 257.

8. Yamaguchi, K., Recent advances in microalgal bioscience in Japan, with special reference to utilization of biomass and metabolites: A review, *J. of Appl. Phycol.*, 8, 487, 1997.
9. Apt, K.E. and Behrens, P.W., Commercial developments in microalgal biotechnology, *J. Phycol.*, 35, 215, 1999.
10. Pulz, O. and Gross, W., Valuable products from biotechnology of microalgae, *App. Microbiol. Biotechnol.*, 65, 635, 2004.
11. Dufossé, L. et al., Microorganisms and microalgae as sources of pigments for food use: A scientific oddity or an industrial reality?, *Trends Food Sci. Technol.*, 16, 389, 2005.
12. Smit, A.J., Medicinal and pharmaceutical uses of seaweed natural products: A review, *J. of Appl. Phycol.*, 16, 245, 2004.
13. Margalith, P.Z., Production of ketocarotenoids by microalgae, *App. Microbiol. Biotechnol.*, 51, 431, 1999.
14. Kyle, D.J., The future development of single cell oils, in *Single Cell Oils*, Cohen, Z. and Ratledge, C., Eds., American Oil Chemists' Society Press, Champaign, Illinois, 2005, 239.
15. Richmond, A., Microalgal biotechnology at the turn of the millennium: A personal view, *J. of Appl. Phycol.*, 12, 441, 2000.
16. Craig, R., Reichelt, B.Y. and Reichelt, J.L., Genetic engineering of micro-algae, in *Micro-algal Biotechnology*, Borowitzka, M.A. and Borowitzka, L.J., Eds., Cambridge University Press, Cambridge, 1998, 415.
17. Borowitzka, M.A., Carotenoid production using microorganisms, in *Single Cell Oils*, Cohen, Z. and Ratledge, C., Eds., American Oil Chemists' Society Press, Champaign, Illinois, 2005, 124.
18. Stahl, E., Quirin, K.W. and Blagrove, R. J., Extraction of seed oils with supercritical carbon dioxide: Effect on residual proteins, *J. Agric. Food Chem.*, 32, 938, 1984.
19. Nakhost, Z., Karol, M. and Krukonis, V.J., Non-conventional approaches to food processing in CELSS. I-Algal proteins, characterization, and process optimization, *Adv. Space Res.*, 7, 29, 1987.
20. Choi, K.J. et al., Supercritical fluid extraction and characterization of lipids from algae *Scenedesmus obliquus, Food Biotechnol.*, 1, 263, 1987.
21. Subra, P. and Boissinot, P., Supercritical fluid extraction from a brown alga by stage-wise pressure increase, *J. Chromatogr.*, 543, 413, 1991.
22. Erazo, S. et al., Estudio de la biomassa y de los pigmentos carotenoides contenidos en una especie nativa de la microalga *Dunaliella salina* sp., *Rev. Agroquim. Tecnol. Aliment.*, 29, 538, 1989.
23. Balaban, M.O., O' Keefe, S. and Polak, J.T., Supercritical extraction of algae, in *Supercritical Fluid Technology in Oil and Lipid Chemistry*, King, J.W. and List, G.R., Eds., American Oil Chemists' Society Press, Champaign, Illinois, 1995, 247.
24. Wisniak, J. and Korin, E., Supercritical fluid extraction of lipids and other materials from algae, in *Single Cell Oils*, Cohen, Z. and Ratledge, C., Eds., American Oil Chemists' Society Press, Champaign, Illinois, 2005, 220.
25. Herrero, M., Cifuentes, A. and Ibanez, E., Sub- and supercritical fluid extraction of functional ingredients from different natural sources: Plants, food-by-products, algae, and microalgae, a review, *Food Chem.*, 98, 136, 2006.
26. Santos, R. et al., Extraction of valuable components from micro-algae *Spirulina platensis* with compressed CO_2, in *Proceedings of the Fourth Italian Conference on Supercritical Fluids and Their Applications*, Reverchon, E., Ed., Capri (Napoli)-Italy, 1997, 217.
27. Qiuhui, H., Supercritical carbon dioxide extraction of *Spirulina platensis* component and removing the stench, *J. Agric. Food Chem.*, 47, 2705, 1999.

28. Careri, M. et al., Supercritical fluid extraction for liquid chromatographic determination of carotenoids in Spirulina Pacifica algae: a chemometric approach, *J. Chromatogr.*, 912, 61, 2001.
29. Canela, A.P.R.F. et al., Supercritical fluid extraction of fatty acids and carotenoids from the microalgae *Spirulina* maxima, *Ind. Eng. Chem. Res.*, 41, 3012, 2002.
30. Valderrama, J.O., Perrut, M. and Majewski, W., Extraction of astaxanthin and phycocianin from microalgae with supercritical carbon dioxide, *J. Chem. Eng. Data*, 48, 827, 2003.
31. Mendes, R.L. et al., Supercritical CO_2 extraction of gamma-linolenic acid (GLA) from cyanobacterium *Arthrospira (Spirulina) maxima*: Experiments and modelling, *Chem. Eng. J.*, 105, 147, 2005.
32. Mendes, R.L., Reis, A.D. and Palavra, A.F., Supercritical CO_2 extraction of γ-linolenic acid and other lipids from *Arthrospira (Spirulina) maxima*: Comparison with organic solvent extraction, *Food Chem.*, 99, 57, 2006.
33. Mendes, R.L. et al., Supercritical carbon dioxide extraction of hydrocarbons from the microalga *Botryococcus braunii*, *J. Appl. Phyc.*, 6, 289, 1994.
34. Mendes, R.L. et al., Supercritical CO_2 extraction of carotenoids and other lipids from *Chlorella vulgaris*, *Food Chem.*, 53, 99, 1995.
35. Mendes, R.L. et al., Supercritical carbon dioxide extraction of compounds with pharmaceutical importance from microalgae, *Inorganica Chimica Acta*, 356, 328, 2003.
36. Gamlieli-Bornshtein, I., Korin, E. and Cohen, S., Selective separation of *cis-transi* geometrical isomers of β-carotene via CO_2 supercritical fluid extraction, *Biotechnol. Bioeng.*, 80, 169, 2002.
37. Mendes, R.L. et al., Solubility of synthetic and natural β-carotene in supercritical carbon dioxide, in *Proceedings of the Fourth Italian Conference on Supercritical Fluids and Their Applications*, Reverchon, E., Ed., Capri (Napoli)-Italy, 1997, 423.
38. Mendes, R.L. et al., Supercritical CO_2 extraction of β-carotene from algae, in *Proceedings of the Sixth Conference on Supercritical Fluids and Their Applications*, Reverchon, E., Ed., Maiori, Italy, 2001, 187.
39. Perretti, G. et al., Extraction of PUFA rich oils from algae with supercritical carbon dioxide, in *Proceedings of the Sixth International Symposium on Supercritical* Fluids, Brunner, G., Kikic, I. and Perrut, M., Eds., Versailles, France, 2003, 29.
40. Nobre, B. et al., Supercritical carbon dioxide extraction of astaxanthin and other carotenoids from the microalga *Haematococcus pluvialis*, *Eur. Food Res. Tecnol.*, 223, 787, 2006.
41. Machmudah, S. et al., Extraction of astaxanthin from *Haematococcus pluvialis* using supercritical CO_2 and ethanol as entrainer, *Ind. Eng. Chem. Res.*, 45, 3652, 2006.
42. Cheung, P.C.K., Temperature and pressure effects on supercritical carbon dioxide extraction of n-3 fatty acids from red seaweed, *Food Chem.*, 65, 399, 1999.
43. Sakaki, K. et al., Supercritical fluid extraction of fungal oil using CO_2, N_2O, CHF_3, and SF_6, *J. Am. Oil Chem. Soc.*, 67, 553, 1990.
44. Macías-Sánchez, M.D. et al., Supercritical fluid extraction of carotenoids and chlorophyll a from *Nannochloropsis gaditana*, *J. Food Engin.*, 66, 245, 2005.
45. Andrici, G. et al., Supercritical fluid extraction of bioactive lipids from the microalga *Nannochloropsis* sp., *Eur. J. Lipid Sci. Technol.*, 107, 381, 2005.
46. Polak, J.T., Balaban, M.B. and Philips, A.J., Supercritical carbon dioxide extraction of lipids from algae, *ACS Symposium Series*, 406, 449, 1989.
47. Lim, G-B. et al., Separation of astaxanthin from red yeast *Phaffia rhdozyma* by supercritical carbon dioxide, *Biochem. Engin. J.*, 11, 181, 2002.
48. Radwan, A. et al., Supercritical fluid CO_2 extraction accelerates isolation of humic acid from live *Pilayella littoralis* (Phaeophyta), *J. of Appl. Phycol.*, 8, 545, 1997.

49. Cygnarowicz-Provost, M. et al., Supercritical fluid extraction of fungal lipids using mixed solvents: Experiment and modeling, *J. Supercritical Fluids*, 5, 24, 1992.
50. Bhattacharjee, P. and Singhal, R.S., Extraction of squalene from yeast by supercritical carbon dioxide, *World J. Microbiol. Biotechnol.*, 19, 605, 2003.
51. Calvin, M. and Taylor, S.E., Fuels from algae, in *Algal and Cyanobacterial Biotechnology*, Cresswell, R.C., Rees, T.A.V. and Shah, N., Eds., Longman Scientific & Technical, London, 1989, 137.
52. Templier, J., Largeau, C. and Casadevall, E., Mechanism of non-isoprenoid hydrocarbon biosynthesis in *Botryococcus braunii*, *Phytochemistry*, 23, 1017, 1984.
53. Yamaguchi, K. et al., Lipid composition of a green alga *Botryococcus braunii*, *Agric. Biol. Chem.*, 51, 493, 1987.
54. Casadevall, E., Production d'hydrocarbures par l'algue *Botryococcus braunii*, *Biomasse Actualités*, 3, 64, 1983.
55. Mendes, R.L. and Palavra, A.F., Supercritical extraction of compounds from microalgae, in *Chemistry, Energy, and the Environment,* Sequeira, C.A.C and Moffat, J.B., Eds., The Royal Society of Chemistry, Cambridge, 51, 1998.
56. Gouveia, L. et al., Evolution of the pigments in *Chlorella vulgaris* during carotenogenesis, *Bioresources Technol.*, 57, 157, 1996.
57. Davies, B.H., Carotenoids, in *Chemistry and Biochemistry of Plant Pigments*, Goodwin, T.W., Ed., Academic Press, London, 2, 38, 1976.
58. Bendich, A., Non-provitamin A activity of carotenoids: Immunoenhancement, *Trends in Food Sci. Technol.*, 2, 127, 1991.
59. Mendes, R.L., PhD Thesis, Extracção supercrítica de lípidos de microalgas, Universidade Técnica de Lisboa, Portugal, 1995.
60. Mendes, R.L. et al., Supercritical CO_2 extraction of lipids from microalgae, in *Proceedings of the Third International Symposium on Supercritical Fluids*, Brunner, G. and Perrut, M., Eds., Strasbourg, 1994, 2, 477.
61. Mendes, R.L. et al., Modelisation de l ´extraction supercritique de lipides d ´algues, in *Proceedings 3eme Colloque sur les Fluides Supercritiques, Applications aux Produits Naturels*, Pellerin, P. and Perrut, M., Eds., Grasse (France), 1996, 141.
62. King, M.B. et al., Equilibrium and rate data for the extraction of lipids using compressed carbon dioxide, *Separation Sci. Technol.*, 22, 1103, 1987.
63. Hebuterne, X. et al., Intestinal absorption and metabolism of 9-cis-beta-carotene in vivo: Biosynthesis of 9-cis-retinoic acid, *J. Lipid Res.*, 36, 1264, 1995.
64. Levin, G. and Mokady, S., Antioxidant activity of 9-cis compared to all-trans beta-carotene in vitro, *Free Radic. Biol. Med.*, 17, 77, 1994.
65. Ben-Amotz, A., Lers, A. and Avron, M., Stereoisomers of beta-carotene and phytoene in the alga *Dunaliella bardawil*, *Plant Physiol.*, 86, 1286, 1988.
66. Nobre, B.P. et al., Separation of the isomers of β-carotene with supercritical CO_2 and silica gel, in *Proceedings of the Seventh Meeting on Supercritical Fluids*, Perrut, M. and Reverchon, E., Eds., Antibes, France, 2000, 2, 781.
67. Olaizola, M., Commercial development of microalgal biotechnology: From the test tube to the marketplace, *Biomol. Eng.*, 20, 459, 2003.
68. Guerin, M., Huntley, M.E. and Olaizola, M., *Haematococcus* astaxanthin: Applications for human health and nutrition, *Trends Biotechnol.*, 21, 210, 2003.
69. Østerlie, M., Bjerkeng, B. and Liaaen-Jensen, S., Plasma appearance and distribution of astaxanthin *E/Z* and *R/S* isomers in plasma lipoproteins of men after single dose administration of astaxanthin, *J. Nutr. Biochem.*, 11, 482, 2000.
70. Denery, J.R. et al., Pressurized fluid extraction of carotenoids from *Haematococcus pluvialis* and *Dunaliella salina* and kavalactones from *Piper methysticum*, *Analytica Chimica Acta*, 501, 175, 2004.

71. Ackman, R.G., Algae as a source of edible oils, in *New Sources of Fats and Oils*, Pryde, E.H., Princen, L.H. and Mukherjee, K.D., Eds., American Oil Chemists' Society Press, Champaign, Illinois, 1981, 189.
72. Trautwein, E.A., n-3 fatty acids: Physiological and technical aspects for their use in food, *Eur. J. Lipid Sci. Technol.*, 103, 45, 2001.
73. Becker, E.W., *Microalgae: Biotechnology and Microbiology*, Cambridge University Press, 1994, 177.
74. Peet, M. et al., Two double-blind placebo-controlled pilot studies of eicosapentaenoic acid in the treatment of schizophrenia, *Schizophr. Res.*, 49, 243, 2001.
75. Lubían, L.M. et al., *Nannochloropsis* (*Eutigmatophyceae*) as source of commercially valuable pigments, *J. Appl. Phycol.*, 12, 249, 2000.
76. Chrastill, J., Solubility of solids and liquids in supercritical gases, *J. Phys. Chem.*, 86, 3016, 1982.
77. Cohen, Z., The chemicals of *Spirulina*, in *Spirulina platensis* (*Arthrospira*): *Physiology, Cell-Biology, and Biotechnology*, Vonshak, A., Ed., Taylor & Francis, London, 1997, 175.
78. Otles, S. and Pire, R., Fatty acid composition of *Chlorella* and *Spirulina* microalgae species, *J. AOAC Int.*, 84, 1708, 2001.
79. Khan, Z., Bhadouria, P. and Bisen, P.S., Nutritional and therapeutic potential of *Spirulina*, *Current Pharma. Biotechnol.*, 6, 373, 2005.
80. Nakahara, T. et al., Gamma-linolenic acid from genus *Mortierella*, in *Industrial Applications of Single Cell Oils*, Kyle, D.J. and Ratledge, C., Eds., American Oil Chemists' Society Press, Champaign, IL, 1992, 61.
81. Kennedy, M.J., Reader, S.L. and Davies, R.J., Fatty acid production characteristics of fungi with particular emphasis on gamma linolenic acid production, *Biotechno. Bioengin.*, 42, 625, 1993.
82. Jaime, L. et al., Separation and characterization of antioxidants from *Spirulina platensis* microalga combining pressurized liquid extraction, TLC, and HPLC-DAD, *J. Sep. Sci.*, 28, 2111, 2005.
83. Certik, M., Andrası, P. and Sajbidor, J., Effect of extraction methods on lipid yield and fatty acid composition of lipid classes containing g-linolenic acid extracted from fungi, *J. Am. Oil Chem. Soc.*, 73, 357, 1996.
84. Mendes, R.L. et al., Solubility of β-carotene in supercritical carbon dioxide and ethane, *J. Supercritical Fluids*, 16, 99, 1999.
85. Goto, M., Sato, M. and Hirose, T., Extraction of peppermint oil by supercritical carbon dioxide, *J. Chem. Eng. Jpn.*, 26, 401, 1993.
86. Hart, D.J. and Scott, K.J., Development and evaluation of an HPLC method for the analysis of carotenoids in foods, and the measurement of the carotenoid content of vegetables and fruits commonly consumed in the U.K., *Food Chem.*, 54, 101, 1995.

7 Application of Supercritical Fluids in Traditional Chinese Medicines and Natural Products

Shufen Li

CONTENTS

7.1 Introduction ... 216
7.2 Special Features of SFE Technique in Processing TCM and Natural Products ... 217
7.3 Status of SFE in Processing TCM and Natural Products in China ... 219
7.3.1 National Symposiums on SCF Technology ... 219
7.3.2 SCF Equipment Made in China ... 219
7.3.3 Summary of Applying SFE in Processing TCM and Natural Products ... 220
7.3.3.1 SFE with Pure Supercritical Carbon Dioxide ... 220
7.3.3.2 SFE with CO_2 in Presence of Cosolvent ... 220
7.3.3.3 SFE with CO_2 in Presence of Surfactant ... 221
7.3.3.4 Combing SFE with Ultrasound-Enhanced Extraction Method ... 223
7.3.3.5 Combining SFE with Enhanced Separation Methods ... 223
7.3.3.6 Combining SFE with Other Techniques to Make Full Use of Herbal Materials ... 224
7.4 Select Examples of SFE of TCM and Natural Products ... 225
7.4.1 Extraction of Essential Oil from Clove Bud with SC-CO_2 ... 225
7.4.2 Extraction of Medical Ingredients from the Mixture of *Angelica sinensis* and *Ligusticum chuanxiong* Hort with SC-CO_2 ... 228
7.4.3 Extraction of Edible and Medicinal Ingredients from Grape Seeds with SC-CO_2 ... 230
7.4.4 Isolation of Organochlorine Pesticide from Ginseng with SC-CO_2 ... 233
7.5 Summary and Prospect ... 236
References ... 237

7.1 INTRODUCTION

Traditional Chinese medicine (TCM) is a scientific summary of rich experiences of the Chinese nation's struggle against disease for thousands of years. It is one of the oldest and strongest traditional medical systems in the history of the world. Classical Chinese herbal medicines encompass a large number of herbal formulations with known mild pharmaceutical effects and minimum side effects that are used for the treatment of a wide variety of difficult-to-treat diseases. Over the course of many centuries, TCM has greatly formed a unique theoretical system and diagnosing and treating techniques that have made an indelible and substantial contribution to both the health and prosperity of the Chinese people [1, 2]. TCM has not only enjoyed an excellent reputation in China but also in the rest of the world. It will play a more and more important role in contributing to the health and longevity of mankind.

In China, more than 11,000 plants are considered to be medicine herbs. Almost 2,000 Chinese traditional patent medicines are listed in the official Chinese Pharmacopoeia. These medicines are widely used as TCM in China, even in Southeast Asia [1–4]. Most of them are processed with many kinds of medicine plants according to the theory of prescription composition. However, some are composed of only a single plant. The efficacy of Chinese herbal medicines is considered a synergism of many effective components, including not only the small molecule compounds such as volatile oils, alkaloids, flavonoids, and saponins but also biological macromolecules such as polysaccharides, proteins, and peptides.

The most traditional method for processing herbs involves boiling them in water for hours so that most of the ingredients are dissolved. Another method involves the use of conventional organic solvents for extraction instead of boiling water; the most commonly used organic solvents are ethanol, ether, chloroform, and methanol. When the traditional extraction methods for processing herbs are used, the extracts consist of various compounds, including some undesired substances that dissolve with the desired products. Therefore, further purification steps are necessary to remove the coextracted impurities. In these processes, long processing times of 2 to 7 days are generally needed. High boiling or extraction temperatures often lead to degradation of heat-sensitive compounds. Moreover, traces of toxic solvents are hardly removed from the extracts, which directly influences the quality of the products. Hydrodistillation (steam distillation) is generally used for obtaining volatile oils from plants. Its high processing temperature can also lead to degradation of heat-sensitive compounds. Therefore, alternative extraction techniques with better selectivity and efficiency are highly desirable.

In recent years, the catchphrase "modernization and internationalization of Traditional Chinese medicine" has often been presented in Chinese papers, magazines, and symposiums, and this concept has become a very hot topic [5, 6]. For this reason, the pharmaceutical study of natural products has become one of the most interesting and active research areas in China. Some new chemical separation technologies, such as supercritical fluid extraction (SFE), membrane separation, ultrasonic-assist extraction, molecular distillation (MD), and polymeric adsorbent technology, have been considered to help improve the productive process of TCM

[7, 8]. These efforts have been successful. Among these new technologies, SFE is presently considered as one of the most clean and highly effective technologies for processing TCM.

The high solvent power of supercritical fluid (SCF) was first reported over a century ago. Demonstration of SFE technology for industrial applications was reported by Zosel in 1970 [9]. Since then, the fundamental and applied aspects of SCF and processes with applications cover a wide range of topics in energy, environment, medicine, chemical industries, and analytical field. It has been also rapidly extended to other fields, such as chemical reaction, supercritical fluid chromatography (SFC), and particle formation in material processing with SCF [10, 11].

This chapter focuses on introducing the SFE techniques used in processing TCM and natural products. The special features of SFE techniques and the status of SFE in processing TCM and natural products in China are briefly reviewed. Four typical examples of SCF application in TCM and natural products from our laboratory research have been selected to make further description.

7.2 SPECIAL FEATURES OF SFE TECHNIQUE IN PROCESSING TCM AND NATURAL PRODUCTS

A gas, when compressed isothermally to pressures greater than its critical pressure, exhibits enhanced solvent power in the vicinity of its critical temperature. Such fluid is called *supercritical fluid*. SCFs possess desirable specific characteristics that make them attractive as solvents. Liquid-like densities and gas-like viscosities, coupled with diffusion coefficients that are at least an order of magnitude higher than those of liquids, contribute to the enhancement of mass transfer. In particular, adjusting pressure and temperature can control the solvent density and hence solvent power, because the solvent power of a SCF relates to the solvent density in the critical region [12–14].

Among SCFs, supercritical CO_2 (SC-CO_2) remains the most commonly used fluid for SFE application because of its mild critical properties (Tc = 31.1°C, Pc = 7.38 MPa), nontoxicity, chemical inertness, and availability in high purity at low cost. These excellent properties lead CO_2 to be considered an "environmentally friendly" solvent for extraction of natural products, such as coffee, tea, hops, and selected spices [12–13].

As is known, the dipole moment of CO_2 is zero and its polarizability is only 26.5×10^{-25} cm^{-3}, which is less than that of all of hydrocarbons except methane [14]. Therefore, SC-CO_2 is only a good solvent for extraction of nonpolar compounds, such as hydrocarbons, while its large quadripole moment also enables it to dissolve some moderately polar compounds, such as alcohols, esters, aldehydes, and ketones [9]. When pure SC-CO_2 is employed as a solvent for processing natural products, mixtures containing both nonpolar and moderately polar substances are generally extracted.

Both the properties of the solute and the solvent can affect the extractability of natural products. Vapor pressure, polarity, and molecular weight of solutes are the most important factors affecting the solubility of solutes in SCF. Raising the temperature raises the vapor pressure or sublimation pressure of solutes and hence increases the solubility of the solute. However, increasing the temperature also causes a simultaneous decrease in the density of CO_2, which tends to decrease solubility. Two competing factors need to be considered to find the suitable temperature. Raising the pressure increases the density of the solvent of SC-CO_2 and hence the solvent power of the solutes. However, the benefit of this increase is often limited by manufacturing ability of high-pressure equipment and capital costs. Selecting a suitable cosolvent or entrainer that can maintain or improve selectivity and increase solubility may be one of the keys to expanding the application of SFE. The addition of a cosolvent can not only shift the critical properties from the pure solvent critical properties and hence affect the properties of the SCF but can also induce cosolvent-solute interactions or associations, such as acid-base interactions, depending on the properties of the solute, solvent, and cosolvent, any of which may enhance the solubility [15–19].

In most cases of solvent extraction from botanical substances, four steps of mass transport occur:

1. Diffusion of solvent into the botanical substance
2. Solvation of solute
3. Diffusion of solute into bulk fluid phase
4. Transport of solute and the bulk fluid phase from the extraction zone

Usually, the diffusion of the solutes out of the matrix is the limit step. In order to reduce the diffusion distance of solutes through the botanical substrate and further rupturing of the cell wall, hence eliminating some diffusion barriers, it is necessary to ground the natural raw material into the optimum size because particle size has some effect on yield and rate of recovery [20–25].

SFE processes need to consider both extraction and separation. Three basic operation models can be used to separate solutes from SCF solvents: pressure reduction, temperature variation, and adsorption. Each operational model has its advantages and limitations. SFE with extract separation by varying the temperature is operated in isobaric state. Extract separation by adsorption allows the SFE process to run isobarically and isothermally, and so SCF can be circulated without recompression. These two modes require less energy consumption. However, pressure reduction is the most used mode in processing of TCM because the operation can be made more stable by effectively controlling the liquid CO_2 level when reducing the separation pressure and temperature to below the critical values.

For successful SFE, some factors must be taken into consideration prior to the experiments. These factors include the type of raw materials, method of feed preparation, type of fluid, choice of cosolvents, method of feeding cosolvents, and extraction and separation conditions, including pressure, temperature, flow rate, and extraction time. To optimize SFE conditions, a statistical experimental design based on orthogonal experiments is commonly used and reported, where the yield and the content of the active compound in the extracts are often considered as the target index.

TABLE 7.1
Presentations in the Five Chinese Symposiums and VIP on SCF Technology

	1st, 1996	2nd, 1998	3rd, 2000	4th, 2002	5th, 2004	VIP	Total
SFE	23	31(4)	42(1)	58	68	129	351
SCR	13(6)	16(4)	18(4)	17(6)	21	76	155
Particle formation	3	5	11	10	24	38	91
SFC	2	1	2	1	2	8	16
Others	1	2	7	8	13	21	52
Theoretical study	12	13	19	18	16	37	125
Papers in each symposium	48	68	99	112	144	309	790

7.3 STATUS OF SFE IN PROCESSING TCM AND NATURAL PRODUCTS IN CHINA

7.3.1 National Symposiums on SCF Technology

As one of the green chemistry and engineering technologies, SCF science and technology calls great attention from the government. Many national projects relating to SCF science and technology were supported by the government. Many enterprises also carry out some research and development together with the researchers of universities and institutes. In addition, national symposiums on SCF science and technology have been held every 2 years since 1996. Up to five symposium proceedings have now been published in China [26–30].

Table 7.1 summarizes the presentations from the five Chinese symposiums and articles appearing in the Chinese core journals in VIP Chinese Databank, which was the authoritative professional database in China through 1998. It can be seen from Table 7.1 that, although the application of SCF has also been rapidly extended to supercritical chemical reaction, particle formation, SFC, and other fields, the earliest and most active research field centers on SFE, especially its applications in extracting active components from Chinese herbs.

7.3.2 SCF Equipment Made in China

As supercritical states of fluids are at high pressures, the industrialization design and scale-up plant are the core issues for most enterprises and research institutions. Currently, at least seven sets of large-scale industrial SFE equipment were introduced from Europe, of which the largest one was made in Germany (Uhde High Pressure Technologies GmbH) and the extraction vessels' configuration for each of the three vessels are 3500 L, respectively. Aside from for importing large-size instruments, the domestic instruments for SCF technology especially in SFE are being developed. There are more than 30 sets of homemade SFE instruments with extractor sizes over 100 L, and the largest size is 2000 L. Additionally, there are more than 150 SFE laboratory units with exactor volumes less than 25 L located in the many provinces of China [6].

There exist some obstacles to developing homemade SFE instruments. For example, the auto-control system is still relatively backward and some manufacture levels for high-pressure parts cannot reach international standards.

7.3.3 Summary of Applying SFE in Processing TCM and Natural Products

At least 150 kinds of Chinese traditional herbal plants were selected as raw materials to investigate with SFE technology in China. The research methods used can be classified into six different cases:

1. SFE with pure SC-CO_2
2. SFE with CO_2 in presence of cosolvent
3. SFE with CO_2 in presence of surfactant
4. Combing SFE with ultrasound-enhanced extraction method
5. Combing SFE with other separation methods
6. Combining SFE with other techniques to make full use of herbal materials.

7.3.3.1 SFE with Pure Supercritical Carbon Dioxide

An overview of recent publications on applications of pure SC-CO_2 in TCM and natural products is given in Table 7.2. The target extracts are most commonly volatile essential oils, which are a mixture of nonpolar components and moderately polar substances. The extraction temperatures are generally from 30°C to 60°C, and the investigated pressures were from 8 to 40 MPa, depending on the properties of both the raw material used and the desired extracts. These kinds of applications can fully show the advantages of SFE over traditional solvent extraction, hydrodistillation, and steam distillation, such as higher yield with better quality, less hydrocarbon pollution, greater safety, lower production cost, and no degradation of heat-sensitive compounds. As we know, traditional processing methods for these kinds of herbal materials are mostly hydrodistillation and steam distillation or solvent extraction, in which high boiling temperatures often lead to degradation of heat-sensitive compounds. Furthermore, traces of toxic solvents are hardly removed from the extracts when solvents are used, which directly influences the quality of the products.

7.3.3.2 SFE with CO_2 in Presence of Cosolvent

Some examples of using SC-CO_2 in the presence of a cosolvent are listed in Table 7.3. Most of the extracts obtained are middle-polar substances, such as alkaloids, saponins, and flavonoids. The most commonly used cosolvents are ethanol and different concentrations of aqueous ethanol solutions. As is known, compared with methanol and chloroform, ethanol is less toxic and more acceptable for processing TCM. Water is the most acceptable and cheapest solvent. By regulating the ratio of water and ethanol, one can readily manipulate the properties of the fluids. Usually, addition of

TABLE 7.2
Overview on the Extraction of Active Compounds from Chinese Herbals with SC-CO_2

Raw Materials	Conditions	Extracts	Yield (%)	References
Arnebia euchroma (Royle) Johnst	35°C, 27 MPa	Naphthaquinonic compounds	4.1–4.6	[31]
Atractylodes macrocephala Koidz	50°C, 28 MPa	Volatile components	4.27	[32]
Bee pollen	55°C, 30 MPa	Lipophilic components	5.0	[33]
Cortex albiziae	35°C, 30 MPa	Lipophilic components	5.4	[34]
Curcuma kwangsiensis	60°C, 26 MPa	β-elemene	0.0271	[35]
Ear of Schizonepeta tenifolia Briq.	50°C, 20 MPa	Essential oil	6.31	[36]
Fig residues	45°C, 30 MPa	Anticancer components	2.53	[37]
Glycyrrhiza uralensis Fisch.	50°C, 25 MPa	Essential oil	1.69	[38]
Leaves of Artemisiae argyi	32°C, 15 MPa	Essential oil	2.71	[39]
Lilium brownii	50°C, 18 MPa	Essential oil	2.92	[40]
Ocimum basilicum L.	45°C, 16 MPa	Essential oil	4.96	[41]
Orris	55°C, 26 MPa	Orris oil	12.71	[42]
Perilla frutescens (L.) Britton	50°C, 15 MPa	Essential oil	2.5	[43]
Radix Angelicae dahuricae	35°C, 25 MPa	Essential oil	3.6	[44]
Radix Litseae Cubebae	55°C, 30 MPa	Lipophilic components	2.6	[45]
Salvia castanea Diels f. tomentosa Stib.	65°C, 35 MPa	Tanshinones	2.9	[46]
Saposhnikovia divaricata (Turcz) Schischk	35°C, 22 MPa	Lipophilic components	4–4.5	[47]
Schisandra Chinensis (Turcz) Baill	50°C, 25 MPa	Schizandrin	Not available	[48]
Stroma of Cordyceps kyushuensis	50°C, 20 MPa	Essential oil	9.72	[49]
Wheat plumule	35°C, 20 MPa	Wheat plumule oil	Not available	[50]
Zanthoxylum seed	35°C, 40 MPa	Zanthoxylum seed oil	10.32	[51]

a small amount of a liquid cosolvent can significantly enhance the extraction efficiency and consequently reduce the extraction time or pressure.

7.3.3.3 SFE with CO_2 in Presence of Surfactant

The research and progress of SC-CO_2 microemulsion verify the possibility of extracting polar compounds with SC-CO_2. When an appropriate surfactant is added into the SC-CO_2, a reverse microemulsion can form, which facilitates the dissolution of hydrophilic molecules in SC-CO_2. The formation of SC-CO_2 microemulsion needs

TABLE 7.3
Extraction of Active Compounds from Chinese Herbals by SC-CO_2 in the Presence of Cosolvent

Raw Materials	Condition	Cosolvent	Extracts	Yield (%)	References
Aplinia Oxyphylla Miquel seeds	35°C, 25 MPa	Ethanol	Volatile oil	3.21	[52]
Astragalus root	45°C, 40 MPa	95% ethanol	Astragaloside IV	0.27	[53]
Branches and needles of Taxus yunnanensis	40°C, 34 MPa	Methanol	Taxol	0.0057	[54]
Cornus officinalis Sieb. et Zucc.	45°C, 35 MPa	Ethanol	Urosolic acid	0.239	[55]
Corydalis yanhusuo W. T. Wang	40°C, 15 MPa	95% ethanol	Tetrahydro-palmatine	0.039	[56]
Curcuma longa	55°C, 25 MPa	Ethanol	Curcumin	0.0024	[57]
Iris tectorum	50°C, 25 MPa	Chloroform	Irone	Not available	[58]
Ligusticum chuanxiong Hort.	45–65°C, 30–50 MPa	Ethanol	Femlic acid	0.735	[59]
Polygonum cuspidatum	50°C, 25 MPa	95% ethanol	Resveratrol	Not available	[60]
Polygonum multiflorum Thunb	50°C, 30 MPa	Chloroform + methanol	Phospholip	3.11	[61]
Pricklyash Peel	35°C, 20 MPa	Ethanol	Essential oil	13	[62]
Propolis	40°C, 35 MPa	95% ethanol	Flavonoids	34.9	[63]
Pteris semipinnata L.	60°C, 25 MPa	Ethanol	Diterpenoids	0.117	[64]
Rhizome of Coptis chinensis Franch	60°C, 50 MPa	1,2-propanediol	Berberine	7.53	[65]
Salvia miltiorrhiza bunge	60°C, 25 MPa	Methanol	Tanshinone IIA	0.038	[66]
Sinomenium acutum (Thumb) Rehd et Wils	60°C, 30 MPa	Methanol	Sinomenine	0.747	[67]
Taxus mairei bark	45–50°C, 30–35 MPa	Ethanol	Taxoids	Not available	[68]

surfactant with CO_2-philic groups, such as siloxane, fluoroalkane, fluoroether, tertiary amine, and alkyonol [69].

SC-CO_2 extraction in the presence of surfactant and cosolvent has not been widely used yet in the field of TCM. However, some researchers in China have explored this kind of extraction. For examples, Chen et al. [70] observed the effect of surfactant on enhancing the efficiency of SC-CO_2 extraction of ephedrine from ephedra. Dioctyl sodium sulfosuccinate (DSS), sodium dodecyl sulfate (SDS), 1-heptanesulfonate (SHS), and carboxymethylcellulose sodium (CMC-Na) were used as the surfactants, and their influences on extraction of ephedrine from ephedra by SFE were studied. The results indicate that DSS, SDS, SHS, and CMS-Na enhanced the efficiency of SFE by 246.8%, 123.4%, 83.0%, and 53.2%, respectively, which was due to their molecular constitutions. The more liposoluble parts the surfactants have, the higher the efficiency. The application of surfactants offered a valuable way for SFE of alkaloid. Ge et al. [71] studied the use of Tween-80 and Span-80 in the extraction of matrines from Kuh-seng. And they found that the yield is 1.8–2.2 times more than that of the method without the surfactant. Satisfactory results were also achieved when Wang et al. [72] use nonionic surfactant, Span-80, and Tween-80, together with water and ethanol, in a certain proportion as modifiers to extract lactones from atractylodes macrocephala Koidz. The content of the lactones in the extractive can reach to 87.78% at extraction temperature of 15°C and pressure of 30 MPa.

7.3.3.4 Combing SFE with Ultrasound-Enhanced Extraction Method

Most TCMs are solid materials and the mass transfer rate of the solid materials is limited by the diffusion inside the particles. That makes the mass transfer rate slow in the SCF and leads to a longer extraction time as a result. This bottleneck of SFE, however, can be solved by introducing ultrasound into the SCF. A number of physical effects (turbulence, particle agglomeration, and biological cell rupture) as well as chemical effects (free radical formation) possessed by ultrasound facilitate the mass transfer in SCF.

Ding et al. [73] used double-frequency ultrasounds alternately to enhance SFE of flavonoids from Toona sinensis. Ethanol was also used as a cosolvent. Ding et al. demonstrated in their research that the successive order of different effect factors on the yield is cosolvent amount > ultrasonic frequency > extraction temperature > extraction pressure > ultrasonic power, and the optimum conditions of the extraction include temperature 50°C, pressure 20 MPa, cosolvent amount 2 mL/g, ultrasonic frequency 20 kHz, and power 150W. When extracting oil and coixenolide from adlay seeds, Hu et al. [74] found energy savings after the introduction of ultrasound because the extraction temperature, pressure, and rate of CO_2 could be decreased and extraction time could also be shortened.

7.3.3.5 Combining SFE with Enhanced Separation Methods

7.3.3.5.1 SFE Combined with Pressured Fractional Distillation

SFE coupled with pressured fractional distillation has reportedly been used to concentrate w-3 fatty acids from fish oils [75]. It is now used in separation and purification of TCM and natural products. For example, Liu et al. [76] used SFE coupled with distillation to extract and concentrate vitamin E from soybean oil deodorizer. Li [77] has successfully used this combined technology for extracting the lipid fractions from adlay seeds. The industrial scale-up of this process has led to replacement of the original solvent extraction method and has obtained approval from China's Food and Drug Administration for application in the pharmaceutical industry for manufacturing TCM.

The lipid fractions of adlay seeds are the active pharmaceutical ingredients of Kanglaite Injection, a parenteral drug approved in China and Russia for treatment of advanced non-small-cell lung cancer. Extraction of adlay seed oil in the extraction tank was firstly carried out with SC-CO_2 at temperatures of 30°C to 45°C and pressures of 22 MPa. The CO_2 with dissolved crude adlay seed oil then in turn entered to the separation column and two separation tanks to remove impurities, including fatty acids, moisture, and pigments. When SFE is combined with pressured fractional distillation, the change of temperature and pressure can result in a change of relative separation factor. As a result, the quality of adlay seed oil extracted with SC-CO_2 extraction has attained the standard of refined oil in terms of its quality specifications.

7.3.3.5.2 SFE Combined with MD or HSCCC Techniques

SFE extracts are generally not a single compound but rather a complex mixture of effective components, including some impurities. Sometimes, in order to get a component of high purity, it is necessary to use other separation techniques to further treatment extracts of SFE. MD, high-speed countercurrent chromatography (HSCCC), silica gel column separation, solid phase extraction, and other processes are presently being investigated for this purpose, with MD and HSCCC being the most widely investigated processes in China [78–88].

MD is a liquid-liquid extraction technique in a high-vacuum condition, which has the feature of low distillation temperature, short heating time, and high selectivity. The combined technique of SFE and MD is mostly used to extract the essential components of TCM. After MD processing, the components of low molecular weight in the SFE extracts can be concentrated. Zhang et al. [78–83] used this combined technique to extract the effective components of rhizoma atractylodis macrocephala, garlic, forsythia suspense, radix angelicae pubescentis, Spirulina, and ligusticum wallichii Franch, and all the results were satisfactory. For example, when garlic was extracted with SC-CO_2, 16 compounds in the extractive were identified, whereas only 4 active compounds (diallyl disulfide, 3-ethenyl-1,2-dithia-cyclohex-5-ene, 2-ethenyl-1,3-dithia-cyloohex-5-ene, and diallyl trisulfide) were obtained by molecular distillation of the extractive [80].

HSCCC is a unique liquid-liquid partition chromatography technique that uses no solid support matrix. It eliminates the irreversible adsorptive loss of samples onto the solid support matrix that occurs with use of the conventional chromatographic column [84]. Cao et al. [85–86] used HSCCC to purify the catechins and free fatty acids extracted by SFE from cratoxylum prunifolium Dyer and grape seeds and obtained purities of 98% and 99%, respectively. Wang et al. [87] got psoralen and isopsoralen purities of above 99% when they combined SFE with HSCCC. Similarly, Peng et al. [88] got flavonoids of 97.6~99.2% purity from Patrinia villosa Juss. The literature mentions [84] that countercurrent chromatography could be used in a little larger scale, which suggests a bright future for combined SFE and HSCCC techniques.

7.3.3.6 Combining SFE with Other Techniques to Make Full Use of Herbal Materials

Sometimes, medical plants have more than one kind of effective component. In order to make full use of the plant, researchers try to isolate different effective components with different methods. SFE has the advantage of extracting nonpolar and moderately polar substances, so it is usually used to extract the lipophilic compounds.

Zhang et al. [89] obtained lipophilic components of tanshinone and hydrophilic components of danshensu and protocatechualdehyde at one time by combining SFE with water boiling. Our lab also did some work to make full use of the raw materials. Ye et al. [90] used SC-CO_2 to extract oil from grape seeds, the residues of which were extracted with hot water and then deposited with appropriate alcohol to get proanthocyanidin. Xiao et al. [91] combined SC-CO_2 extraction with solvent extraction to obtain both essential oil and alkaloids from Nelumbo Nucifera Gaertn.

Another case is that the active compounds with stronger polar in herbs are desired, but the herbs also contain a certain amount of lipophilic substances, which were once removed as impurities with traditional solvent extraction methods. With the feature of convenient operation, high safety, high removal ratio, and easy isolation of solvent, SFE is now adopted to remove the lipophilic compounds before extraction of effective components with other techniques.

In order to extract polysaccharide from mongolia mushroom, Wang et al. [92] first investigated the effect of pretreatment to degrease and decolar mongolia mushroom by SC-CO_2 or by solvent extraction. The experimental results indicated that, when suitable extraction conditions were used with CO_2, the effect of degreasing and decolar was excellent. Moreover, the pretreatment process favored the extraction of polysaccharide. The extraction yield of the polysaccharide with pretreatment by SFE is 1.8 fold that with solvent pretreatment and 4.2 times that without pretreatment.

7.4 SELECT EXAMPLES OF SFE OF TCM AND NATURAL PRODUCTS

In our laboratory, more than 20 kinds of herbal plants were selected as raw materials to investigate SFE processing. Four typical examples were briefly reported, which indicate that different SFE processes and parameters can be developed depending on processing purpose and the properties of the raw materials.

7.4.1 Extraction of Essential Oil from Clove Bud with SC-CO_2

Clove (*Eugenia caryophyllata* Thunb.) is widely cultivated in the south of China. Clove bud oils contain high contents of eugenol, which give it strong biological activity and antimicrobial activity. Beside eugenol, clove bud oils also contain some amount of other active compounds of eugenol acetate and β-caryophyllene. Clove bud oil has several therapeutic effects, including antiphlogistic, antivomiting, analgesic, antispasmodic, carminative, kidney reinforcement, and antiseptic effects. It also is used as a flavoring agent and antimicrobial material in food [93–95].

Extraction of clove oils from clove bud with SC-CO_2 was investigated [96]. The herbal materials of clove bud were ground by a FW80 Sample Mill machine in different periods to get different particle distribution, which was measured by mechanical sieving after extraction and calculated by weight of different size of clove bud particle. Grades of particle size were classified on the following scale: 1 = < 10 mesh; 2 = 10~20 mesh; 3 = 20~40 mesh; 4 = 40~60 mesh; 5 = 60~80 mesh; 6 = 80~100 mesh; 7 = 100~120 mesh; 8 = > 120 mesh. Particle size index was calculated by the following formula:

TABLE 7.4
Three-Level Orthogonal Design and Experimental Results for Extraction of Clove Oil with SC-CO_2 [96]

Run No.		Factor A (T/°C)	Factor B (P/MPa)	Factor C (particle size/#)	Yield (kg extract/kg feed)	Eugenol Content (%)
1		1(30)	1(10)	1(1#)	0.2056	53.69
2		1(30)	2(20)	2(2#)	0.1943	54.22
3		1(30)	3(30)	3(3#)	0.1830	55.64
4		2(40)	1(10)	3(3#)	0.1910	56.20
5		2(40)	2(20)	1(1#)	0.2224	54.52
6		2(40)	3(30)	2(2#)	0.2043	55.30
7		3(50)	1(10)	2(2#)	0.1956	58.77
8		3(50)	2(20)	3(3#)	0.1827	57.83
9		3(50)	3(30)	1(1#)	0.2395	56.97
	K1	0.5830	0.5923	0.6676		
	K2	0.6178	0.5995	0.5942		
	K3	0.6179	0.6269	0.5568		
Yield (%)	K1/3	0.1943	0.1974	0.2225		Σ = 1.8186
	K2/3	0.2059	0.1998	0.1981		
	K3/3	0.2060	0.2090	0.1856		
	R	0.0117	0.0116	0.0369		
	K1	163.55	168.66	165.18		
	K2	166.02	166.57	168.29		
	K3	173.57	167.91	169.67		
Eugenol content (%)	K1/3	54.52	56.22	55.06		Σ = 503.14
	K2/3	55.34	55.52	56.10		
	K3/3	57.86	55.97	56.56		
	R	3.34	0.25	1.04		

$$\text{Particle size index} = \sum \frac{\text{weight of each grade} \times \text{grade}}{\text{total weight} \times \text{highest grade}} \quad (7.1)$$

The particle size indexes of material in this experiment were 0.7944, 0.6430, and 0.5223, named as 1#, 2#, and 3# respectively.

The following parameters were used: temperature, 30°C, 40°C, and 50°C; pressure, 10 MPa, 20 MPa, and 30 MPa; and particle size, 1#, 2#, and 3#. All the selected factors were examined using a three-level orthogonal array design with an OA9 (3^3) matrix, as shown in Table 7.4. It can be seen from the order of the maximum differences that particle size had the most influence on the oil yield, then temperature and

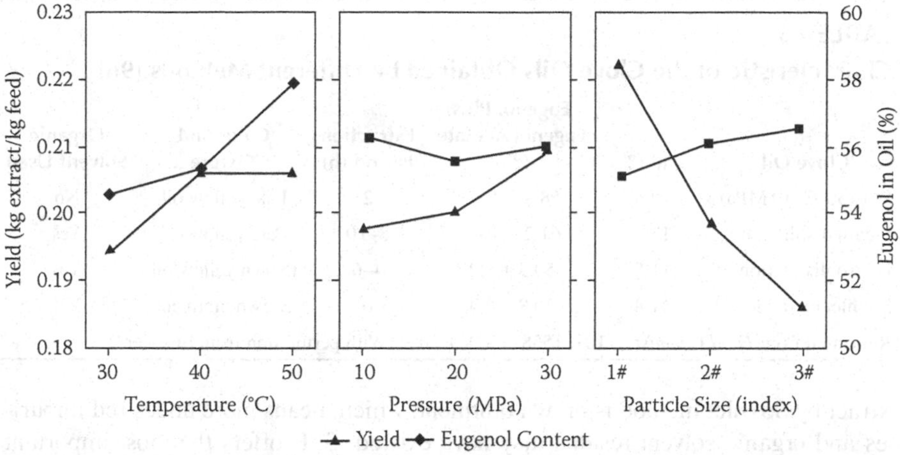

FIGURE 7.1 Effect of temperature, pressure, and particle size on yield and eugenol content of clove oil extracted by SC-CO_2. (Reprinted from *Food Chemistry,* 101, 1558–1564, ©2007. With permission from Elsevier.)

pressure. However, the sequence of the influences of the parameters on the eugenol content in the oils was temperature, particle size, and then pressure. The factor of temperature shows the maximum influence on the eugenol content in the oils.

Based on the above data from the three-level orthogonal array design, Figure 7.1 further shows the effect of temperature, pressure, and particle size on the yield and eugenol content of clove oil extracted by SC-CO_2. It can be observed that increase of temperature from 30°C to 40°C results in an increase of the extraction yield and high eugenol content in the oils, while the increase of temperature from 40°C to 50°C does not result in an increase of the oil yield, and there is an increase of eugenol content in clove oil. The extraction yield enhanced significantly with increase of pressure due to the increase of the solubility of the oil components. This increase is attributed to the increase of the CO_2 density, which results in an increase of its dissolving ability. However, because the high-molecular-weight compounds in clove buds (fatty acids, fatty acids methyl esters, sterols, etc.) were also coextracted with increase of pressure, the eugenol content of the clove oil does not obviously change. The extraction yield increases by decreasing the particle size of the comminuted clove buds due to the higher amount of oil released when the bud cells are destroyed by milling, and this amount of oil is easily extracted for direct exposure to the SC-CO_2. However, the eugenol content in the clove oil increases as particle size increases. Therefore, the particle size should not be too small in order to avoid coextraction of more compounds with high-weight molecules.

Gas chromatography with Mass spectrometry (GC/MS) analysis was used to identify the compounds in the clove oils extracted with SC-CO_2. Twenty-three compounds were identified. Comprehensive comparison of the clove oils obtained by different methods is listed in Table 7.5. The content of the main biological ingredients of eugenol plus eugenol acetate in the clove oil by Soxhlet extraction is lowest, although its yield of clove oil is highest among the four extraction methods. Furthermore, the

TABLE 7.5
Characteristic of the Clove Oils Obtained by Different Methods [96]

Clove Oil	Yield (%)	Eugenol Plus Eugenol Acetate (%)	Extraction Period (h)	Color and Texture	Organic Solvent Used
SFE(50°C, 10 MPa)	19.6	58.8 + 19.6	2	Pale yellow oil	No
Steam distillation	10.1	61.2 + 10.2	8–10	Pale yellow oil	Yes
Hydro distillation	11.5	50.3 + 3.2	4–6	Brown yellow oil	Yes
Soxhlet extraction	41.8	30.8 + 9.3	6	Brown ointment	Yes

Reprinted from *Food Chemistry*, 101, 1558–1564, ©2007. With permission from Elsevier.

extract by Soxhlet method is brown ointment, which means more undesired impurities and organic solvent residue may have existed. SFE offers the most important advantages over other methods. Extraction yield of SFE was about two times as high as that obtained by steam and hydrodistillation. The highest content of eugenol plus eugenol acetate in the extracted oil was also obtained. Pale yellow oil is desired and shortest extraction time is needed for SFE compared with the other three extraction methods.

7.4.2 Extraction of Medical Ingredients from the Mixture of *Angelica sinensis* and *Ligusticum chuanxiong* Hort with SC-CO_2

Angelica sinensis (Oliv.) Diels and *Ligusticum chuanxiong* hort have been widely used as TCM to treat pathological conditions such as atherosclerosis and hypertension. Their phytochemical profiles analyses suggest that *Angelica sinensis* and *Ligusticum chuanxiong* hort contain similar substances, such as ferulic acid and essential oil. Ferulic acid is one of the most important medical components in *Angelica sinensis* and *Ligusticum chuanxiong* because it possesses antioxidative properties by virtue of the phenolic hydroxyl group in its structure. Studies have shown that ferulic acid could inhibit malondialdehyde (MDA) production from platelets, inhibit erythrocyte lyses induced by MDA and hydroxyl radical, and inhibit lipid peroxidation induced by H_2O_2 and O_2 [97–99].

Extraction of ferulic acid from a mixture of similar portions of *Angelica sinensis* and *Ligusticum chuanxiong* hort was firstly carried out with SC-CO_2. As ferulic acid was considered to be the active component for preventing heart disease, its content in extracts was analyzed by high-performance liquid chromatography and the analyzing results were used as the quality control index for medical efficiency.

The effects of extraction temperature, extraction pressure, particle size, and material sources on the extract yield (E) and the content of ferulic acid in extracts (C) using pure CO_2 was first experimentally investigated, as listed in Table 7.6. As shown in Table 7.6, the extract yields increased from 2.95% to 3.95% and the content of ferulic acid in extracts increased from 0.25% to 0.28% when pressure increased from 20 to 50 MPa at constant temperature of 65°C. These changes occurred because increasing the pressure at constant temperature increases the density of SC-CO_2, which further increases the solvation power of the SCF. When temperatures increased from

TABLE 7.6
Experimental Data of Extraction of Ferulic Acid from the Mixture with Pure CO_2

Temperature (°C)	Pressure (MPa)	Particle Size (mesh)	E(%)	C(%)
65	20	20–40	2.95	0.25
65	30	20–40	3.63	0.26
65	40	20–40	3.79	0.28
65	50	20–40	3.95	0.28
35	30	20–40	3.18	0.18
45	30	20–40	3.32	0.23
55	30	20–40	3.50	0.24
65	30	20–40	3.63	0.26
65	30	40–60	4.40	0.22
65	30	60–80	5.52	0.21

TABLE 7.7
Experimental Results of Different Cosolvents

Cosolvent	T (°C)	P(MPa)	E (%)	C (%)
Pure CO_2	65	30	3.63	0.26
Ethanol	65	30	5.16	0.58
And n-butyl alcohol	65	30	5.13	0.52
Ethyl acetate	65	30	4.69	0.31

35°C to 65°C at a pressure of 30 MPa, the extract yields increased from 3.18% to 3.63% and the content of ferulic acid in extracts increased from 0.18% to 0.26%.

The effect of particle size on the extract yield is also shown in Table 7.6. Extract yields increased from 3.63% to 5.52% with particle size decreasing from 20–40 meshes to 60–80 meshes. Therefore, it is necessary to ground the natural raw materials into the optimum size in order to reduce the diffusion distance and to improve extraction efficiency. However, the content of ferulic acid in the extracts decreased from 0.26% to 0.21%, indicating other components may also be extracted.

Although SC-CO_2 has been widely investigated, it is a poor extractant for polar substances. In order to increase the power of solvent for extracting polar ferulic acid, three cosolvents were employed: ethanol, ethyl acetate, and n-butyl alcohol. Different ratios of cosolvents to raw material (w/w) were also studied. Cosolvents were directly added into the raw materials and soaked for 4 hours before carrying out SC-CO_2 extraction under the conditions of 65°C and 30 MPa. After the SFE process, the extracts were vaporized with an evaporator in a vacuum to remove the solvent. The experimental results are listed in Table 7.7.

Table 7.7 shows that all of three cosolvents not only greatly enhanced the contents of ferulic acid in the extracts but also increased the extract yield greatly

TABLE 7.8
Comparison of SFE with Percolation Method

Extraction Method	Temperature (°C)	Pressure (MPa)	E (%)	C (%)
SFE with pure CO_2	65	30	3.63	0.26
Percolation			4.54	0.61
SFE with ethanol	65	30	7.12	0.83

compared with pure CO_2 extraction. Because ethanol is one of the few accepted organic solvents in the food and medicine industries, it was employed to study the ratio of cosolvent. The experimental results show that the extract yield of 7.12% and the content of ferulic acid of 0.83% in extractive were obtained when the ratio of the ethanol to raw materials was 1.6.

The percolation method described in the Pharmacopoeia of People's Republic of China (2000 year) was employed. Ethanol with concentration of 95% was used as a solvent. Before percolation, the powered mixture was soaked with solvent for 24 hours. The percolation flow rate was 15 drops/min and the percolation time was about 8 hours. After percolation, the extracts were vaporized with a vacuum rotatory evaporator to remove the solvent. The material was ground using a mixer-grinder. Comparisons of SFE with percolation method are listed in Table 7.8.

It can be seen that both the extract yields and the content of ferulic acid in extracts by pure CO_2 are the lowest among the three processing methods. Adding a suitable cosolvent, such as ethanol in this study, could greatly increase the content of ferulic acid in extracts, which is superior to the traditional percolation method of extracting polar ferulic acid from the mixture of *Angelica sinensis* and *Ligusticum chuanxiong* hort. This method may be one key way to make use of a suitable cosolvent for increasing the solvent power of CO_2 and to expand the application of SFE. However, SFE extracts from the herbs of the mixture of *Angelica sinensis and Ligusticum chuanxiong* hort are generally a complex mixture of the components, which may have some differences in both composition and contents compared with the extractive obtained by the original patented traditional methods. Therefore, further research on pharmacology and medicine efficiency is needed for the safe and effective use of this method for TCM.

7.4.3 Extraction of Edible and Medicinal Ingredients from Grape Seeds with SC-CO_2

In our lab, we also investigated the use of SC-CO_2 in the extraction of active compounds from grape seeds. Grape seeds contain seed oil and procyanidins, which are generally named plant polyphenol. The weight proportion of grape oil in the total grape seed is about 10–15% and that oil is rich in linoleic acid, which belongs to unsaturated fatty acid. In the fields of food, cosmetics, and medicine, it is considered to be beneficial to use oils high in linoleic acid. Grape procyanidins have been increasingly paid much attention as one of the ten most popular herbal medicines in the world and can be used for medicine, hygienic food, and cosmetics due to their

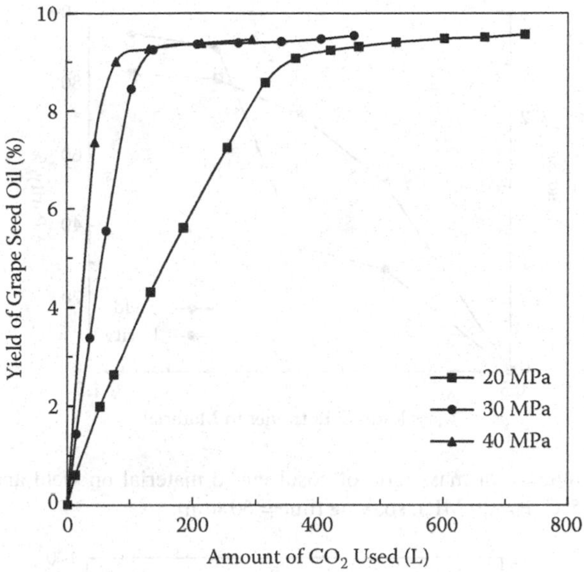

FIGURE 7.2 Plot of extraction yield of grape seed oil vs. SC-CO_2 amount (45°C).

biological and pharmacological actives, such as anti-oxidation and anti-mutation [101–103].

The influences of extraction temperature and pressure on the extraction yield of the seed oil were investigated [104]. Figure 7.2 illustrates the effect of extraction pressure on the extraction yield of seed oil at 45°C. With pure CO_2, we obtained the qualified grape oils. The results show that the yield of oilseed was up to 9.5% at 45°C and 30 MPa. However, when using seeds supplied from another source area, the result shows that the yield of oilseed was up to 13.51% at 55°C and 30 MPa. GC-MS analysis shows that the unsaturated fatty acid in the extracted oil constituent was up to 90.1%. Therefore, it is important to investigate source area of the raw materials and make some necessary analysis for the active components.

After extracting oil from the grape seeds with SC-CO_2, the extraction of procyanidins was further studied with SC-CO_2 in the presence of the cosolvent of ethanol. Three kinds of methods for adding cosolvent were investigated in order to enhance the solvent power of CO_2 for increased yield and purity of procyanidins. The three methods included adding cosolvent to raw material in static mode, adding cosolvent to SC-CO_2 in flowing mode, and a combination of the two modes. The effects of extraction temperature, pressure, the concentration and dosage of cosolvents, and soaking time on the extraction yield and purity of procyanidins were studied. The experimental results show that, when the mass ratio of cosolvent added to the raw material was 1.2:1(w:w) and soaked for 60 minutes before the SC-CO_2 extraction was carried out. At temperature of 55°C and 30 MPa and when a concentration of 60% cosolvent in CO_2 was applied in flowing mode, a yield of 10.9% with a purity of 95.9% of procyanidins could be obtained (Figure 7.3). When a concentration of 60% cosolvent in CO_2 was used as extraction solvents in flowing mode to make SFE at extraction

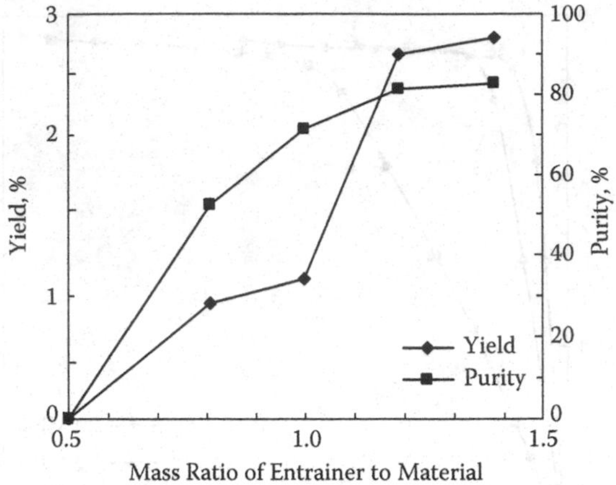

FIGURE 7.3 Influence of mass ratio of cosolvent to material on yield and purity of procyanidins (T = 55°C, P = 30 MPa, soaking time = 60 min).

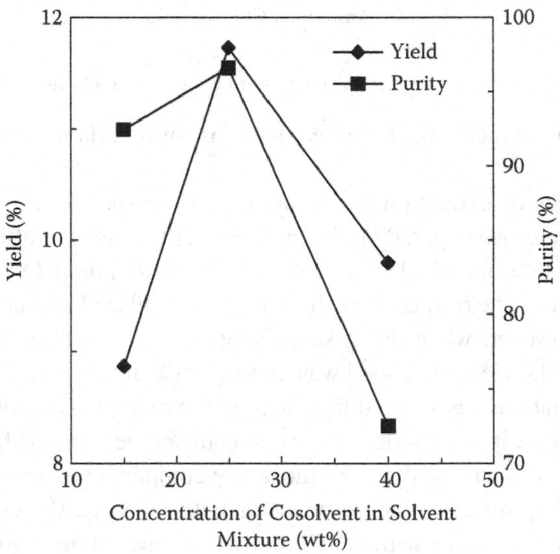

FIGURE 7.4 Effect of dosage of cosolvent flowing on yield and purity of procyanidins (T = 55°C, P = 35 MPa, mass ratio of cosolvent resting to material = 1.2:1, soaking time = 90 min).

temperature of 55°C and extraction pressure of 35 MPa, a yield of 10.9% with a purity of 95.9% of procyanidins could be obtained. When the mass ratio of 1.2:1(w:w) for cosolvent added to the raw material was used and soaked for 60 minutes, and then 25% concentration of cosolvent in SC-CO_2 in flowing mode was used as extraction solvents for SFE at the extraction temperature of 55°C and extraction pressure of 35 MPa, the highest yield of 11.73%, with a purity of 96.6% of procyanidins, could be obtained (Figure 7.4). Therefore, adding cosolvent to raw material in static mode combined with adding a suitable concentration of cosolvent to SC-CO_2 in flowing mode may be the best method among the three for adding cosolvents.

For comparison, another integrated technology was also investigated for obtaining procyanidins from grape seeds by using the combined method of SFE and macroporous resin adsorption technology [105]. SFE with SC-CO_2 was first used to remove the grape seed oil. Then macroporous resin adsorption technology was used to purify the crude procyanidins extracted by hot water extraction with alcohol deposition. The experimental results show that a yield of 4.88% having a purity of 95% of procyanidins could be obtained, which is far less than the yield obtained by adding cosolvent to raw material in static mode combined with adding a suitable concentration of cosolvent to SC-CO_2 in flowing mode. It reveals that extracting natural products with SFE has obvious predominance.

7.4.4 Isolation of Organochlorine Pesticide from Ginseng with SC-CO_2

Many Chinese traditional and herbal drugs are being exported abroad as food additives or plant drugs. Radix ginseng is a rare Chinese traditional medicine material which has therapeutic effects that can be used to treat many diseases and can also be used as a health food. However, a high content of residues of prohibitory organochlorine pesticides, such as hexachlorocyclohexanes (BHC), exists in radix ginseng, which exceeds the limited level greatly according to the international standard regulation [106–108].

The safety issue of Chinese herbal medicines is a subject of scientific interest. SCFs as "environmentally friendly" alternatives to liquid solvents for sample preparation in analytical chemistry have received much attention in the past few years. SFE has been shown to be an efficient and rapid method for the isolation of organochlorine pesticides from vegetables [109]. However, no studies report on the process of removal of BHC from radix ginseng with SFE. The feasibility of removing BHC pesticide residues from radix ginseng with SC-CO_2 was explored in our lab [110–113].

The roots with hairs of radix ginseng were powdered and sifted out prior to extraction, in which radix ginseng with sizes of 550 to 1120 μm was selected. Extractions were performed with a Spe-ed SFE instrument (Applied Separations Inc., Allentown, PA).

For the determination of BHC, Gas Chromatograph with Electrical Conductivity Detector (GC-ECD) analysis was carried out using an Agilent 6890 plus (U.S.) gas chromatograph equipped with ^{63}Ni electron-capture detector, using a BPX608 capillary column (25 m × 0.32 mm). The chromatographic conditions were as follows: injector temperature, 280°C; detector temperature, 320°C; and nitrogen flow-rates, 10.0 mL/min (carrier gas). The column temperature was programmed as follows: increased at a rate of 25°C/min from initial temperature of 50°C to 150°C, retained for 1 min, and then increased at a rate 6°C·min^{-1} to 240°C.

Determination of BHC in radix ginseng samples was carried out. The BHC content was 2.380 mg·kg^{-1} feeds. According to the international standard regulation for food and drugs, the content of the pesticide residues of BHC in radix ginseng should be lower than 0.1 mg·kg^{-1}. Therefore, at least 95.2% of BHC should be removed from the radix ginseng samples in this work.

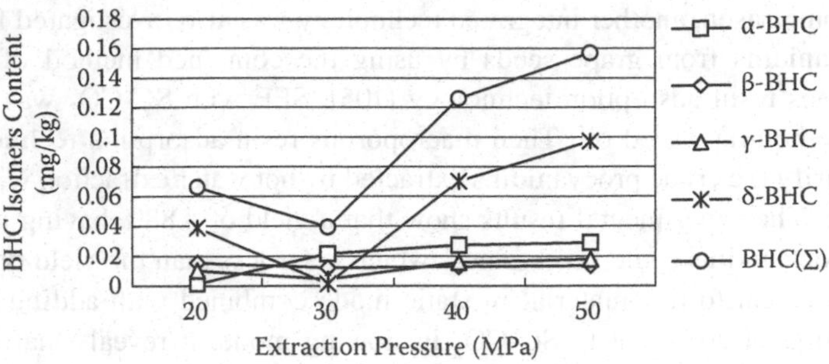

FIGURE 7.5 Effect of pressure on removal of BHC with SC-CO_2 in the presence of water at 60°C. (From Li, S.F. and Quan, C., *Chinese Journal of Chemical Engineering,* 13, 433, 2005. With permission.)

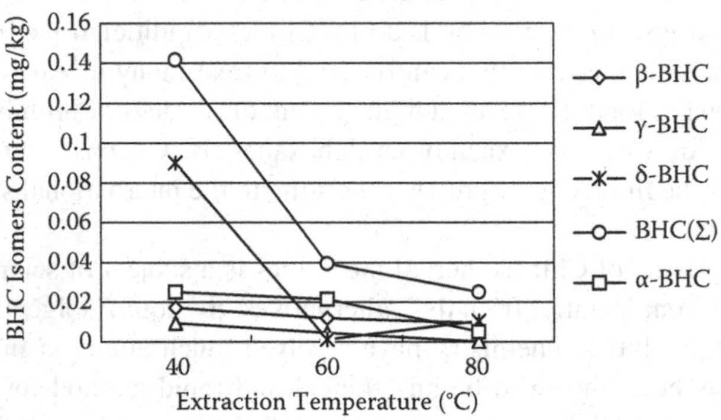

FIGURE 7.6 Effect of temperature on removal of BHC with SC-CO_2 in presence of water at 30 MPa. (From Li, S.F. and Quan, C., *Chinese Journal of Chemical Engineering,* 13, 433, 2005. With permission.)

Extraction of BHC from radix ginseng was first investigated with pure SC-CO_2 in temperatures ranging from 60°C to 80°C and pressure ranging from 25 MPa to 50 MPa. However, the results of GC-ECD analyses indicated that pure CO_2 could not reduce BHC content to the level of 0.1 mg·kg^{-1} to meet the BHC permission limit, so that a certain modifier was necessary.

Removal of BHC residues from radix ginseng with CO_2 in the presence of cosolvent was investigated using three kinds of solvents: water, ethanol, and hexane. When a suitable amount of water is added into the feedstock before extraction, BHC content in radix ginseng could be reduced to 0.0394 mg·kg^{-1} at 60°C and 30 MPa, while addition of the same amount of ethanol or hexane as cosolvents did not serve such a purpose.

The effect of extraction pressure on removal of BHC from radix ginseng by SC-CO_2 in the presence of cosolvent water at 60°C is shown in Figure 7.5. The BHC content in radix ginseng was reduced to levels lower than 0.1 mg·kg^{-1}, which were 0.08 mg·kg^{-1} and 0.04 mg·kg^{-1} separately in the pressures of 20 MPa and 30 MPa.

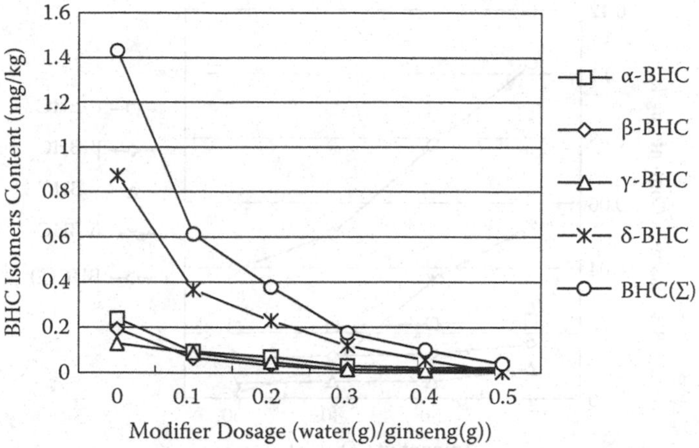

FIGURE 7.7 Influences of dosage of water on removal of BHC from radix ginseng at a temperature of 60°C and pressure of 30 MPa. (From Li, S.F. and Quan, C., *Chinese Journal of Chemical Engineering,* 13, 433, 2005. With permission.)

However, further increase in pressure increased the BHC residues, so the optimal pressure obtained in this study was 30 MPa.

The effect of extraction temperature was studied at a constant pressure of 30 MPa with water as cosolvent at temperatures ranging from 40°C to 80°C (Figure 7.6). Increasing temperature resulted in decreasing content of BHC in radix ginseng. At a temperature of 40°C, BHC residues in radix ginseng were higher than 0.1 mg·kg^{-1}. However, the BHC content was reduced to 0.04 mg·kg^{-1} when temperature increased to 60°C. However, in light of the thermo-sensitive properties of the natural plant, the optimal temperature is 60°C.

To seek the minimum cosolvent dosage, influences of dosage of water on SFE of BHC in radix ginseng were studied at a temperature of 60°C and pressure of 30 MPa. The content of BHC in extracted radix ginseng decreased rapidly from 1.43 mg·kg^{-1} to 0.04 mg·kg^{-1} (Figure 7.7) when the dosage of water increased from 0 to 0.5 [water (g)/ginseng (g)], so dosage of cosolvent had significant effect on removal of BHC with SFE. When the dosage of water was 0.4, content of BHC was 0.11 mg·kg^{-1}, and when it was 0.5, content of BHC was 0.04 mg·kg^{-1}. Therefore, the dosage of water must be more than 0.4. However, an excessive amount of water does not have a positive function because it cannot be absorbed by the powdered ginseng, so the suitable dosage of water was about 0.5 g per gram of ginseng.

At 60°C, 30 MPa, and a dosage of water of 0.5 g, influences of the amount of CO_2 on removal of BHC from radix ginseng were investigated. As shown in Figure 7.8, α-BHC, β-BHC, and γ-BHC change less while δ-BHC decreased obviously with increasing amount of CO_2 for certain ginseng material. As the total results, about 30 to 50 standard liter CO_2 per gram ginseng could match the total BHC residue permission limit—that is, 150 L or more of CO_2 could reduce the content of BHC to less than 0.1 mg·kg^{-1} for 5 g ginseng.

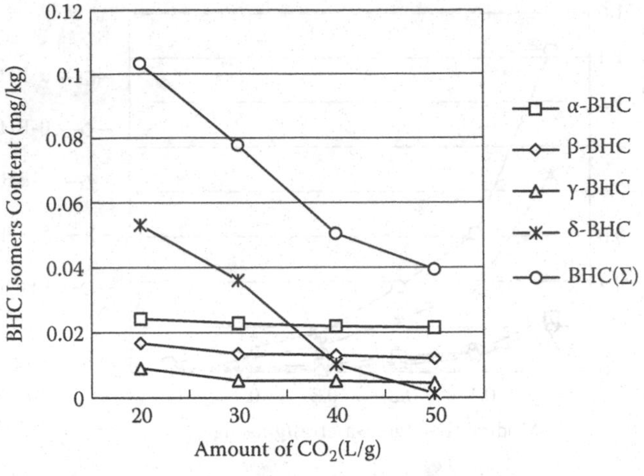

FIGURE 7.8 Impact on the amount of CO_2 on BHC content. (From Li, S.F. and Quan, C., *Chinese Journal of Chemical Engineering*, 13, 433, 2005. With permission.)

7.5 SUMMARY AND PROSPECT

From the above-stated examples, we find that SFE has sufficiently advanced technology and superior efficiency when extracting nonpolar or low-polar compounds. With the modification of cosolvent and surfactant, the limitation of SFE in extracting moderately polar and intensively polar compounds is somewhat improved. In addition, the combination of SFE with other techniques widens the application of SFE in the fields of TCM and natural products. As a clean and green separation technique, SFE has a promising future in its application in the fields of TCM and natural products. However, we also learned that SFE extracts from herbs are generally not a single compound but a complex mixture of components, which commonly have some differences in composition or in contents compared with the extractive obtained by other traditional methods. Therefore, most of the research results stated in this chapter are generally in the laboratory level, few of them have scaled up. Further research on pharmacology and medicine efficiency is necessary to ensure the safety and efficacy of combined extraction processes for TCM.

Additionally, some factors that limited the rapid development of SFE should not be overlooked. As lack of a deeper understanding to SCF state itself, much experimental work is needed to determine the process conditions that cannot be predicted. Theoretical research of the thermodynamic and dynamic control mechanisms of the SFE process also need to be investigated because this research can provide important guidance for optimizing the industrial production of SC-CO_2 extraction. As operating SFE process is at high pressure which demands equipment to guarantee safety, how to increase the quality of the equipments and to reduce the operating costs also need the manufacturers to take their efforts.

SCF technology is a promising technology with exciting commercial potential. It is replacing older solvent technologies and creating new technologies for processing TCM and natural products because it is safe, environmentally benign, and cost-effective. It fascinates many researchers and deserves further investigation to expand its application.

REFERENCES

1. Chinese Pharmacopoeia Commission, *Chinese Pharmacopoeia (Chinese)*, Vol. 1, Chemical Industry Press, Beijing, 2000.
2. Zheng, H.Z., Dong, Z.H. and She, J., *Modern Study and Application of Chinese Medicine (Chinese)*, Vol. 2, Xueyuan Press, Beijing, 1997.
3. Huang, T.K., *Traditional Chinese Patent Pharmacology (Chinese)*, 3rd ed., China Press of Traditional Chinese Medicine, Beijing, 1996, 610.
4. Xiao, C.H., *Chinese Medicine Chemistry (Chinese)*, Shanghai Scientific and Technical Publisher, Shanghai, 1994.
5. Gan, S.J. et al., *Development Strategy in Modernization of Traditional Chinese Medicines (Chinese)*, Scientific and Technical Documents Publishing House, Beijing, 1998, 110.
6. Luo, G.A. et al., Modernization of traditional Chinese medicine and materia medica. *World Sci. Technol.*, 5, 11, 1999.
7. Li, S.F., Supercritical fluid extraction technology, in *Critical Technologies in Modernization of Traditional Chinese Medicines (Chinese)*, Yuan, Y.J., Liu, M.Y. and Dong, A.J., Eds., Chemical Industry Press, Beijing, 2002, 11.
8. Zhu, Z.Q., Introduction, *Supercritical Fluids Technologies-Principles and Applications (Chinese)*, Chemical Industry Press, Beijing, 2000, chapt. 1, p. 2.
9. Zosel, K. (Studiengesellschaft Kohle), DBP 2005293 *The process for the decaffeination of green coffee beans*, 1970.
10. Han, B.X., Extraction and separation technology of SCF *Science and Technology of Supercritical Fluid (Chinese)*, China Petrochemical Press, Beijing, 2005, chapt. 8, p. 219.
11. McHugh, M.A. and Krukonis, V.J., Introduction, *Supercritical Fluid Extraction: Principles and Practice*, 2nd ed., Butterworth, Boston, 1993, chapt. 1, p. 5.
12. Bruno, T.J. and Ely, J.F., Supercritical fluid technology, in *Reviews in Modern Theory and Applications*. Boston, MA: CRC Press, 1991.
13. King, J.W., Fundamentals and application of applications of supercritical fluid extraction in chromatography science, *J. Chromatogr.*, 27, 355, 1989.
14. Zhang, J.C., The application of SFE in the research and development of traditional Chinese medicines, *Supercritical Fluid Extraction Technology (Chinese)*, Chemical Industry Press, Beijing, 2002, chapt. 6, p. 105.
15. Walsh, J.M., Ikonomou, G. D. and Donohue, M.D., Supercritical phase behavior: The entrainer effect, *Fluid Phase Equil.*, 33, 295, 1987.
16. Ariel, A.C., Solute-solute and solute-solvent correlations in dilute near-critical ternary-mixtures: Mixed-solute and entrainers effects, *J. Phys. Chem. B*, 97, 2740, 1993.
17. Mijeong, L.J. and David, J.C., Investigation of modifier effects in supercritical CO_2 extraction from various solid matrixes, *J. Supercrit. Fluids*, 16, 33, 1999.
18. Kerry, M.D., Chien-Ping, K. and Robert, P.G., The use of entrainers in the supercritical extraction of soils contaminated with hazardous organics, *Ind. Eng. Chem. Res.*, 26, 2058, 1987.
19. John, J.L. et al., Role of modifiers for analytical-scale supercritical fluid extraction of environmental samples, *Anal. Chem.*, 66, 909, 1994.
20. Marentis, R.T., *In Supercritical Fluid Extraction and Chromatography*, Charpentier, B.A. and Sevenants, M.R., Eds., ACS, Washington, 1988, 127.
21. Motonobu, G., Bhupesh, C.R. and Tsutomu, H., Shrinking-core leaching model for supercritical fluid extraction. *J. Supercrit. Fluids*, 9, 128, 1996.

22. Tan, C. and Liou, D., Modeling of desorption at supercritical conditions, *AICHE J.*, 35, 1029, 1989.
23. Sovová, H., Rate of the vegetable oil extraction with supercritical CO_2, I Modelling of extraction curves, *Chem. Eng. Sci.*, 49, 409, 1994.
24. Özkal, S.G., Yener, M.E. and Bayïndïrlï, L., Mass transfer modeling of apricot kernel oil extraction with supercritical carbon dioxide, *J. Supercrit. Fluids*, 35, 119, 2005.
25. Reverchon, E. and Marrone, C., Modeling and simulation of the supercritical CO_2 extraction of vegetable oils, *J. of Supercritical Fluids*, 19, 161, 2001.
26. Shen, Z.Y. et al., *The First National Symposium on Supercritical Fluids (Chinese)*, Shi-Ja-Zhuang, China, 1996.
27. Zhang, J.C. et al., *The Second National Symposium on Supercritical Fluids (Chinese)*, Guang Zhou, China, 1998.
28. Chen, K. X. et al., *The Third National Symposium on Supercritical Fluids (Chinese)*, Xi-An, China, 2000.
29. Yu, D.S. et al., *The Fourth National Symposium on Supercritical Fluids (Chinese)*, GuiYang, China, 2002.
30. Han, Y.Q. et al., *The Fifth National Symposium on Supercritical Fluids (Chinese)*, Qing Dao, China, 2004.
31. Liang, R.H., Xie, M. Y. and Liu, W., Response surface analysis study on naphthaquinonic compounds yield by supercritical CO_2 extraction of *Amebia euchroma* (Royle) John, *Food Sci.*, 25, 76, 2004.
32. Wu, S.X., Lv, G.Y. and Li, W.L., Supercritical CO_2 extraction process of atractylodes macrocephala Koidz and the determination of the extracts, *Chin. Trad. Pat. Med. (Chinese)*, 27, 885, 2005.
33. Lei, H.P. et al., Supercritical CO_2 extraction of fatty oils from bee pollen and its GC-MS analysis, *J. Chin. Med. Mater. (Chinese)*, 27, 177, 2004.
34. Wu, G. et al., GC-MS analysis of lipophilic components of cortex albiziae extracted by supercritical CO_2, *Chin. Trad. Herbal Drugs (Chinese)*, 36, 832, 2005.
35. Wu, L.H., Du, X. and Liu, H.M., β-Elemene from volatile oil of curcuma kwangsiensis by supercritical fluid extraction, *Chin. Trad. Herbal Drugs (Chinese)*, 37, 368, 2006.
36. Chen, Y., Jiang, Z.H. and Tian, J.K., GC-MS analysis of volatile oil from the ear of *Schizonepeta tenifolia* Briq., *J. Chin. Medic. Mater. (Chinese)*, 29, 140, 2006.
37. Wang, Z.B. and Ma, H.L., Study on anti-cancer components of fig residues with supercritical fluid CO_2 extracting technique, *Chin. J. Chin. Mater. Med. (Chinese)*, 36, 1443, 2005.
38. Fu, Y.J., Wang, W. and Zu, Y.J., Study on volatile components of seed oil from *Glycyrrhiza uralensis* fisch. by supercritical carbon dioxide extraction-gas chromatography-mass spectrometry, *Chin. J. Anal. Chem. (Chinese)*, 33, 498, 2005.
39. Zeng, H.Y. and Li, J.H., Technology study on the volatile oils from the leaves of *Artemisiae argyi* extracted by supercritical CO_2 or microwave, *Food Sci.*, 25, 124, 2004.
40. Liu, C.M., Huang, G.S. and Liu, X.H., Extraction of volatile oil from *Lilium brownii* by supercritical CO_2, *Nat. Prod. Res. Develop. (Chinese)*, 17, 485, 2005.
41. Yan, G.H. and Lin, Z.Y., The volatile oils from *Ocimum basilicum* L. by supercritical CO_2 extraction, *Res. Pract. Chin. Med. (Chinese)*, 19, 56, 2005.
42. Li, C.H. et al., Study of supercritical CO_2 extraction of orris oil, *Nat. Prod. Res. Develop. (Chinese)*, 17, 773, 2005.
43. Qiu, Q. et al., GC-MS analysis of chemical constituents of the essential oil from *Perilla frutescens* (L.) britton. by different extraction methods, *Chin. J. Pharm. Anal.*, 26, 114, 2006.
44. Mi, H. et al., Analysis of volatile components in essential oil of radix angelicae dahuricae of supercritical fluid extraction by gas chromatography mass spectrometry, *Chin. J. Anal. Chem. (Chinese)*, 33, 366, 2005.

45. Liu, J.Z. and Zhong, Z.J., GC-MS analysis of lipophilic components of radix litseae cubebae extracted by supercritical CO_2, *J. Chin. Med. Mater. (Chinese)*, 29, 142, 2006.
46. Mo, S.Z. et al., Extraction of lipophilic components from salvia castanea diels f. tomentosa stib. by supercritical CO_2, *J. Chin. Med. Mater. (Chinese)*, 27, 735, 2004.
47. Gao, Y. et al., Hemostatic effects of the extracts of saposhnikovia divaricata (Turcz.) schischk by supercritical extraction, *Chin. Trad. Herbal Drugs (Chinese)*, 36, 254, 2005.
48. Zhang, Y., Extraction of schisandra chinensis (Turcz) by supercritical carbon dioxide, *Chin. Tradit. Patent Med. (Chinese)*, 27, 880, 2005.
49. Zhang, G.Y., Lin, J.Y. and Qiu, Q., Supercritical carbon dioxide extraction and GC-MS of essential oil from the stroma of cordyceps kyushuensis, *Chin. J. Pharm. Anal.*, 26, 191, 2006.
50. Zhang, X.W. et al., Supercritical carbon dioxide extraction of wheat plumule oil, *J. Food Engin.*, 37, 103, 1998.
51. Wang, J.P. et al., Supercritical CO_2 extraction of zanthoxylum seed oil, *Chin. Trad. Herbal Drugs (Chinese)*, 35, 997, 2004.
52. Liu, H. et al., Extraction of aplinia oxyphylla miquel seed essence via supercritical carbon dioxide and antioxidant activity of the extracts, *J. South Chin. Univ. Technol. (Natural Science Ed.) (Chinese)*, 34, 54, 2006.
53. Sun, H.Y., Guan, S. and Huang, M., Study on new extraction technology of astragaloside IV, *J. Chin. Med. Mater. (Chinese)*, 28, 705, 2005.
54. Liu, L. et al., Supercritical-CO_2 fluid extraction of taxol from branches and needles of taxus yunnanensis, *Chin. Tradit. Patent Med. (Chinese)*, 28, 480, 2006.
55. Han, Z.H. et al., Supercritical fluid CO_2 extraction of urosolic acid from cornus officinalis, *Chin. Trad. Herbal Drugs (Chinese)*, 36, 1159, 2005.
56. Chen, F., Guo, L.W. and Jin, S.L., Supercritical CO_2 fluid extraction of tetrahydropalmatine from corydalis yanhusuo W. T. Wang by orthogonal design, *Tradit. Chin. Drug Res. Clin. Pharm. (Chinese)*, 16, 137, 2005.
57. Xiu, S.L., Wu, Q.N. and Zhen, O.Y., Study on the SFE condition for curcumin in curcuma longa, *Chin. J. Chin. Mater. Med. (Chinese)*, 29, 857, 2004.
58. Guo, T. and Chen, J., The process study on the extraction of irone by supercritical fluids extraction, *J. Chin. Med. Mater. (Chinese)*, 27, 768, 2004.
59. Sun, Y.Y. et al., Extraction of medical components from *Ligusticum chuanxiong* hort. with supercritical CO_2, *Chem. Engin. (Chinese)*, 34, 60, 2006.
60. Zhou, J.K., Li, J.H. and Ge, F.H., Extraction of resveratrol from *Polygonum cuspidatum* by supercritical CO_2, *J. Chin. Med. Mater. (Chinese)*, 27, 675, 2004.
61. Geng, Y.L. et al., Study on supercritical CO_2 extraction of phospholipid from *Polygonum multiflorum* thumb, *Food Sci.*, 27, 140, 2006.
62. Huo, W.L., Study on technology of supercritical CO_2 extraction of pricklyash peel volatile oils, *Food Sci.*, 26, 153, 2005.
63. Gu, Y.H. et al., Extraction of flavonoids from propolis by supercritical CO_2, *Chin. Traditi. Herbal Drugs (Chinese)*, 37, 380, 2006.
64. Deng, Y.F. and Liang, N.C., Extraction of diterpenoids from *Pteris semipinnata* by supercritical CO_2 fluid and their analysis with HPLC-MS, *Chin. Trad. Herbal Drugs (Chinese)*, 35, 145, 2004.
65. Liu, B. et al., Extraction of berberine from rhizome of coptis chinensis franch using supercritical fluid extraction, *J. Pharm. Biomed. Anal.*, 41, 1056, 2006.
66. Dean, J.R., Liu, B. and Price, R., Extraction of tanshinone IIA from salvia miltiorrhiza bunge using supercritical fluid extraction and a new extraction technique, phytosol solvent extraction, *J. Chromatogr. A*, 799, 343, 1998.

67. Liu, B. et al., Supercritical fluid extraction of sinomenine from *Sinomenium acutum* (thumb) Rehd et Wils, *J. Chromatogr. A*, 1075, 213, 2005.
68. Cui, H.W. and Ge, F.H., Studies on constituents from *Taxus mairei* bark, *J. Chin. Med. Mater. (Chinese)*, 27, 566, 2004.
69. Johnston, K.P. et al., Water-in-carbon dioxide microemulsions: An environment for hydrophiles including proteins, *Science*, 271, 624, 1996.
70. Chen, B. et al., Role of surfactant in supercritical fluid extraction, *Acad. J. Second Mil. Med. Univ. (Chinese)*, 21, 463, 2000.
71. Ge, F.H. et al., Effect of non-ionogenic surfactants on the extraction of matrines from kuh-seng, *J. Chin. Med. Mater.*, 26, 426, 2003.
72. Wang, F., Li, X.Y. and Zhen, X.H., The affection of modifier in extraction and isolation of sesquiterpenesolide from the essential oil of atractylodes macrocephala Koidz, *J. Xian Instit. Technol. (Chinese)*, 25, 465, 2005.
73. Ding, C.M., Qiu, T.Q. and Lu, H.Q., Double-frequency ultrasounds alternately enhanced supercritical fluid extraction of effective components of plants, *Chem. Eng.*, 33, 67, 2005.
74. Hu, A.J. et al., Ultrasound assisted supercritical fluid extraction of oil and coixenolide from adlay seed, *Ultrasonics Sonochemistry*, 14, 219, 2007.
75. Eisenbach, W., *Ber. Bunsenges Phys. Chem.*, 88, 882, 1984.
76. Liu, Y., Ding, X.L., and Zhu, D.H., Extraction and concentration natural vitamin E using supercritical CO_2, *Chem. Eng. (China)*, 34, 59, 2006.
77. Li, D.P., Chinese Patent ZL02 1 37312.4, 2004.
78. Zhang, Z.Y. et al., Analysis of essential oil from rhizoma atractylodis macrocephalae by GC-MS with supercritical CO_2 extraction and molecular distillation, *J. Instrum. Anal. (Chinese)*, 22, 61, 2003.
79. Zhang, Z.Y., Lei, Z.J. and Wang, P., Studies on chemical composition of forsythia suspensa by supercritical fluid extraction and molecular distillation, *J. Instrum. Anal. (Chinese)*, 22, 60, 2002.
80. Wang, P., Zhang, Z.Y. and Wu, H.Q., Studies on volatile oil of garlic by supercritical fluid extraction and molecular distillation, *Chin. Hosp. Phatm. J.*, 22, 253, 2002.
81. Gu, W.X. et al., Extraction of active principles of radix angelicae pubescentis by CO_2 SFE-MD, *Acad. J. Guangdong Coll. Pharm. (Chinese)*, 18, 85, 2002.
82. Shi, Y., Gu, W.X. and Zhang, Z.Y., Extraction of active principles of spirulina by CO_2 SFE-MD, *Guangdong Pharm. J. (Chinese)*, 13, 10, 2003.
83. Zhou, B.J., Zhang, Z.Y. and Shi, Y., Analysis with gas chromatography and mass spectrography of volatile components of *Ligusticum wallichii* franch extracted by CO_2 supercritical fluid extraction in combination with molecular distillation, *J. First Milit. Med. Univ. (Chinese)*, 22, 652, 2002.
84. Li, Z.C., Development and future of countercurrent chromatograph, *Chem. Ind. Eng. Prog.*, 24, 816, 2005.
85. Cao, X.L. et al., Supercritical fluid extraction of catechins from cratoxylum prunifolium dyer and subsequent purification by high-speed counter-current chromatography, *J. Chromatogr. A*, 898, 75, 2000.
86. Cao, X.L. and Yoichiro, I., Supercritical fluid extraction of grape seed oil and subsequent separation of free fatty acids by high-speed counter-current chromatography, *J. Chromatogr. A*, 1021, 117, 2003.
87. Wang, X. et al., An efficient new method for extraction, separation, and purification of psoralen and isopsoralen from *Fructus psoraleae* by supercritical fluid extraction and high-speed counter-current chromatography, *J. Chromatogr. A*, 1055, 135, 2004.
88. Peng, J.Y. et al., Efficient new method for extraction and isolation of three flavonoids from patrinia villosa juss. by supercritical fluid extraction and high-speed counter-current chromatography, *J. Chromatogr. A*, 1102, 44, 2006.

89. Zhang, H. et al., Extraction of effective components in salvia miltiorrhiza bge. by supercritical extraction combined with water decoction, *Chin. Trad. Herbal Drugs (Chinese)*, 35, 1360, 2004.
90. Ye, C.H., Li, S.F. and Tang, S.K., Extraction and purification of procyanidins from grape seed with supercritical CO_2 and macroporous adsorption resin, *Chem. Ind. Eng. (Chinese)*, 23, 220, 2006.
91. Xiao, L. et al., Combining supercritical CO_2 extraction with solvent extraction for obtaining active ingredients from nelumbo nucifera gaertn, presented at the Third International Symposium on Supercritical Fluid Technology for Energy and Environment Applications, China, Oct. 24–26, 2004, 74.
92. Wang, D.W., Shan, Y.L. and Tolgor, B., Effect of extraction rate of supercritical CO_2 extraction on mongolia mushroom polysaccharide, *Food Sci.*, 27, 107, 2006.
93. Deans, S.G. and Ritchie, G., Antibacterial properties of plant essential oils, *Int. J. Food Microbiol.*, 5, 165, 1987.
94. Kim, H. M., Lee, E.H. and Hong, S.H., Effect of syzygium aromaticum extract on immediate hypersensitivity in rats, *J. Ethnopharmacol.*, 60, 125, 1998.
95. Della Porta, G. et al., Isolation of clove bud and star anise essential oil by supercritical CO_2 extraction, *Lebensm.-Wiss. u.-Technol.*, 31, 454, 1998.
96. Guan, W.Q. et al., Comparison of essential oils of clove buds extracted with supercritical carbon dioxide and other three traditional extraction methods, *Food Chem.*, 101, 1558–1564, 2007.
97. He, D.P. and Song, G.S., The development report on the extraction of *Ligusticum chuangxiong* Hort. as solvent, *Chin. Oil (Chinese)*, 21, 10, 1996.
98. Shi, L. F., Deng, Y. Z. and Wu, B. S., Studies on chemical constituents and their stability of the essential oil from *Ligusticum chuangxiong* Hort., *Chin. J. Pharm. Anal. (Chinese)*, 15, 26, 1995.
99. Sun, Y.Y. and Li, S.F., Solubility of ferulic acid and tetramethylpyrazine in supercritical carbon dioxide, *J. Chem. Eng. Data (Chinese)*, 50, 1125, 2005.
100. Daniel, S., *Bailey's Industrial Oil and Fat Products*, A Wiley-Interscience Publication, Wiley J., New York, 1979, 449.
101. Bagchi, D. et al., Free radicals and grape seed proanthocyanidin extract: Importance in human health and disease prevention, *Toxicology*, 148, 187, 2000.
102. Castillo, J. et al., Antioxidant activity and radioprotective effect against chromosomal damage induced in vivo by X-rays of flavan-3-ols (procyanidins) from grape seeds (vitis vinifera): Comparative study versus other phenolic and organic compounds, *J. Agric. Food Chem.*, 48, 1738, 2000.
103. Gabor, M., Engi, E. and Sonkodi, S., Effect of flavone derivatives on the arterial wall and its resistance in rats with spontaneous hypertension, *Kiserl Orvostud*, 39, 425, 1987.
104. Tang, S.K. et al., Extraction of grape seed oil from grape seed with supercritical CO_2, *J. Chem. Eng. Chin. Univ. (Chinese)*, 18, 23, 2004.
105. Ye, C.H., Li, S.F. and Tang, S.K., Extraction and purification of procyanidins from grape seed with supercritical CO_2 and macroporous adsorption resin, *Chem. Indust. Eng.*, 23, 220, 2006.
106. Liu, F. and He, L.C., Studies on the qualitative and quantitative methods of quality control of Panax ginseng, *Medicament Analysis (Chinese)*, 22, 173, 2002.
107. Wang, Y.H. and Rong, H., Determination of organic chloride pesticide residue in ginseng with capillary gas chromatography, *Chin. J. Anal. Chem.*, 22, 931, 1994.
108. Yang, S.L. et al., Determination of the benzene hexachloride residue in *Ligusticum chuanxiong* hort. and other 6 Chinese herbal drugs, *Chin. Trad. Herbal Drugs (Chinese)*, 33, 423, 2002.

109. Jian, H.W., Qiang, X. and Kui, J., Supercritical fluid extraction and off-line clean-up for the analysis of organochlorine pesticide residues in garlic, *J. Chromtogr. A*, 818, 138, 1998.
110. Quan, C. et al., Determination of organochlorine pesticides residue in ginseng root by orthogonal array design soxhlet extraction and gas chromatography, *Chromatographia*, 59, 89, 2004.
111. Quan, C. et al., Supercritical fluid extraction and clean-up of organochlorine pesticides in ginseng, *J. Supercrit. Fluids*, 31, 149, 2004.
112. Li, S.F. and Quan, C., Isolation of organochlorine pesticide from Ginseng with supercritical CO_2, *Chin. J. Chem. Eng.*, 13, 433, 2005.
113. Li, S.F. et al., Determination and clean-up of the pesticides residue in the Chinese medicinal materials, *Chin. Trad. Herbal Drugs (Chinese)*, 35, 232, 2004.

8 Extraction of Bioactive Compounds from Latin American Plants

M. Angela A. Meireles

CONTENTS

8.1 Introduction: Examples of Latin American Bioactive Compounds........... 243
8.1.1 Examples of SFE from Native Latin American Plants........................ 244
8.2 Extracting Bioactive Compounds by SFE.. 252
8.2.1 Relevant Process Information for COM Estimation........................ 254
8.2.2 Selecting Parameters Intended for COM Estimation 254
8.2.2.1 Pressure and Temperature Processes 255
8.2.2.2 The Kinetic Parameters... 257
8.2.3 The COM for Latin American Plants 260
8.3 Conclusions 262
References 262

8.1 INTRODUCTION: EXAMPLES OF LATIN AMERICAN BIOACTIVE COMPOUNDS

In this chapter, a brief review of supercritical fluid extraction (SFE) of bioactive compounds from solid substratum is presented. The state of the art of SFE in Latin America is described. Examples of research in development, embodying experimental and modeling of mass transfer and thermodynamics, for several systems are discussed. For preliminary studies of technical and economical feasibility, a very simple empirical model can be used to describe the mass transfer in the extractor cell. To calculate the cost of manufacturing (COM), no solubility data are required, and considering the flash separator ideal, the required information is the global yield in extract at a given condition of temperature and pressure along with an estimate time interval for an extraction cycle. COM estimated this way is provided for some Latin American plants.

Latin America (LA) is formed by 33 countries: Antigua and Barbuda, Argentina (AR), Bahamas, Barbados, Belize, Bolivia (BO), Brazil (BR), Chile (CH), Colombia (CO), Commonwealth of Dominica, Costa Rica, Cuba, Dominican Republic, Ecuador, El Salvador, Granada, Guatemala, Guyana, Haiti, Honduras, Jamaica, Mexico, Nicaragua, Panama, Paraguay (PA), Peru (PE), Saint Kitts, Saint Vincent and the Grenadines, Santa Lucia, Suriname, Trinidad and Tobago, Uruguay, and

Venezuela. Few of these countries are in the Amazonian region; those that are include Brazil, Bolivia, Colombia, Ecuador, Guyana, Peru, Suriname, and Venezuela. The richness of the Amazonian biodiversity may help the region development as well as its devastation. Presently, the governments of the Amazonian countries have demonstrated their concern with the devastation of their natural resources. Many initiatives of sustained development are available from governmental agencies, nongovernmental organizations, and private companies. These initiatives include sustained harvesting of native plants by local communities. Adding value to the raw material by processing it has also been stimulated.

Besides that, LA countries are producers of condiments, aromatic herbs, roots, and tropical fruits used by the food, pharmaceutical, and cosmetic industries. Some of these products are used locally, and others are exported. Among the exported products are black pepper, clove buds, and ginger. Essential oils and oleoresins of vetivergrass, eucalyptus, cinnamon, mint, and other plants are also exported. Brazil and Paraguay are large producers of stevia, a plant whose aqueous extract has been used for years as a sucrose substitute in special diets [1]. In addition, several other plants possess lipids, starches, and cellulose that can potentially be economically explored. Examples of these are turmeric, saffron, and bacuri. In addition, Chile cultivates certain microalgae, such as *Spirulina maxima,* which maybe used as source of fatty acids [2–3].

Questions related to the use of techniques that avoid or minimize damages to the environment are currently being debated. Consumers' demands indicate that, in the near future, products of better quality will be requested more and more. This tendency can be explored thoroughly by the LA countries. To take advantage of their potential, these countries need to develop or adapt technologies that are economically viable and ecologically responsible. Products obtained by SFE are free from toxic residues and generally possess higher quality than products obtained by conventional techniques.

Therefore, raw materials from LA countries represent a business opportunity for producers of vegetable extracts, moreover if these extracts are prepared by SFE. Combining this rich biodiversity with an ecologically correct technology would represent the ideal marriage!

8.1.1 Examples of SFE from Native Latin American Plants

Compilations of literature on SFE were done recently by Meireles [5], Rosa and Meireles [6], Diáz-Reinoso et al. [7], and del Valle et al. [8]. Therefore, the information presented in this chapter represents an update of the previous works. In spite of that, the compilation of literature data was not meant to be exhaustive; instead it focused on including information that was not easily accessible. Table 8.1 to Table 8.3 list Latin-American plants (spontaneous or cultivated) studied; the SFE studies on the microalgae *Spirulina maxima* were also included [2,3]. The common names were confirmed in the U.S. Department of Agriculture Plant Database [9], the Rainforest Database [10], w^{3}TROPICOS of The Missouri Botanical Garden [11], Searchable World Wide Web Multilingual Multiscript Plant Name Database [12], and CropINDEX [13]. The scientific name spellings were confirmed using the same databases and the Flora brasiliensis [14]. The regions of occurrence (spontaneous or

TABLE 8.1
Bioactive Compounds from Latin America Plants (Spontaneous and Introduced): Volatile Oil, Oleoresin, and Other Aroma Compounds

Common Name	Scientific Name	Part Used	Bioactive Compound(s)	Region of Occurrence (Spontaneous or Cultivation)	SFE Conditions/ MPa/K/Cosolvent	Yield (%)	Reference
Aguaribay	*Schinus mollis*	Fractionation of steam distillation volatile oil	β-Pinene, α-pinene, and limonene	LA	9/323	—	[15]
Annatto	*Bixa orellana*	Seeds	Volatile oil and oleoresin: bixin and nor bixin	BR-N	20–30/313–333/EtOH	1–45	[16, 17]
Bamboo piper or pimenta-longa	*Piper aduncum*	Leaves	α-humulene, asaricin, β-caryophyllene	SA	10–30/303–313	1.4–1.8	[18]
Basil (sweet)	*Ocimum gratissimum*	Leaves	Eugenol	BR-SE	10–30/313	1–1.8	[19]
Black pepper	*Piper nigrum* L.	Seeds	β-caryophyllene, limonene, 3-δ-carene, sabinene	BR, PA	15–30/303–323	0.5–2.1	[20–22]
Bushy lippia	*Lippia alba*	Leaves	Carvone and limonene	BR-SE/NE	8–12/313–323	1.5–5	[23]
Chamomile	*Chamomilla recutita* [L] R.	Flowers	Azulene and chamazulene	BR-S	10–20/303–313	0.82–4.3	[24]
Citronella	*Cymbopogon winterianus*, Jowitt	Leaves	Citronellal, citronellol geraniol	BR-SE	7–16/289–298	0.45–1	[25]
Clove buds	*Eugenia caryophyllus*	Fruit	Eugenol, β-caryophyllene, and α-humulene	BR-NE	6.7–10/283–308	~ 14	[26, 27]

continued

TABLE 8.1 (continued)
Bioactive Compounds from Latin America Plants (Spontaneous and Introduced): Volatile Oil, Oleoresin, and Other Aroma Compounds

Common Name	Scientific Name	Part Used	Bioactive Compound(s)	Region of Occurrence (Spontaneous or Cultivation)	SFE Conditions/ MPa/K/Cosolvent	Yield (%)	Reference
Coffee	*Coffea arabica*	Fruit	Aroma compounds (pyrazines, pyridines, and furan derivatives)	LA	24–31/315–371	—	[28]
Coriander	*Coriandrum sativum* L.	Seeds	Volatile oil, phenolic compounds	LA	20–30/298–331	0.8–2.3	[29, 30]
Croton	*Croton zehntneri* Pax et Hoff	Leaves	(E)-Anethole	BR-NE/S	6.7–7.9/283–301	2.1–3.8	[31]
Erva baleeira or wild sage	*Cordia verbenacea*	Leaves	β-caryophyllene and α-humulene	BR-SE	7.8–30/299–323	0.11–5.5	[32]
Eucalyptus	*Eucalyptus citriodora*, Hook	Leaves	Citronellal and citronellol	BR-SE	7–16/289–298	0.31–0.68	[25]
Eucalyptus	*Eucalyptus tereticornis*	Leaves	Aromadendrene, 1,8-cineol and globulol	BR-NE	6.7–7.9/283–298	0.45–1.13	[33]
Fennel	*Foeniculum vulgare*	Seeds	Anethole, fenchone, and fatty acids	BR-SE	10–30/313	3–12.5	[34]
Ginger	*Zingiber officinalis* R.	Rhyzome	β-pinene, m-diethyl-benzene, o-diethyl-benzene, ar-curcumene, α-zingiberene, β-sesquiphellandrene	BR-SE	15–30/293–313	2–3	[35–37]
Green pepper basil	*Ocimum selloi*	Leaves	Volatile oil	BR-SE	10–30/303–323	0.71–2.2	[38]
Horsetail (giant)	*Equisetum giganteum* L.	Aerial parts	Oleoresin	BR-S	12–30/303–313	1.44	[39]
Khoa	*Satureja boliviana*	Leaves	Pulegone, isomenthone, tymol	BO	6.5–7/289–294/EtOH	2–4.6	[40]

Lemon verbena	*Aloysia triphylla*	Leaves	Neral (or Z-citral) and geranial (or E-citral), and spathulenol	BR-SE	10–35/308–318	0.6–1.5	[41]
Lemongrass	*Cymbopogon citratus*	Aerial parts	Neral and geranial	BR (NE, SE, and S)	6.9–7.4/288–297	0.21–42	[42]
Lippia sidoides	*Lippia sidoides* C.	Leaves	Timol	BR-NE	6.7–7.9/283–298	2.2–3.3	[43]
Macela	*Achyrocline satureioides* and *A. alata*	Leaves	α-humulene, β-caryophyllene, quercetin	BR-SE	10–30/303–313	1.2–4.2	[44]
Marigold	*Calendula officinalis*	Flowers	Oleoresin	BR-S	12–20/293–313 K	2–2.8	[45]
Mastranto	*Hyptis suaveolens*	Leaves	Spatulene, Germacrene B, Caryophyllene	VE	8–9/308–318	0.1–0.3	[46]
Orange (sweet)	*Citrus sinensis* (L.)	Shells	Volatile oil	SA	20/313	0.6–0.15	[47, 48]
Oregano	*Origanum vulgare* L.	Leaves	Cis-sabinene hydrate, thymol, carvacrol	LA	10–20/293–313	0.4–1.3	[49]
Palmarosa	*Cymbopogon martini* Roxb.	Leaves	Geraniol, linalool	BR-AM	7–16/289–298	0.07–0.2	[25]
Piprioca	*Cyperus sesquiflorus*	Rhyzomes	Spathulenol, trans-β-guaiene, germacrene D	BR-AM	10–12/333–353	0.3–0.4	[50]
Rosemary	*Rosmarinus officinalis*	Leaves	Camphor, carnosic and rosmarinic acids, phenolic diterpenes	BR-SE	10–30/303–313	1–5	[51, 52]
Stevia	*Stevia rebaudiana* B.	Leaves	Austroinulin, n-tetracosane, n-pentacosane	BR/ PA	25/303	1.4–1.6	[53,54]
Vetivergrass	*Vetiveria zizanioides* (L.) Nash	Roots	Khuzymol	BR-NE/SE	20/313	3.2	[50, 55, 56]
Xylopia aromatica	*Xylopia aromatica*	Fruit	β-Phellandrene, β-myrcene, α-pinene	LA	7.5/318	1.5	[57]

LA: Latin America; SA: South America; BR-AM: Amazonian region of Brazil; BR-NE: Northeast of Brazil; BR-S: South of Brazil; BR-SE: Southeast of Brazil; AR: Argentina; BO: Bolivia; PA: Paraguay; VE: Venezuela

TABLE 8.2
Bioactive Compounds from Latin America Plants (Spontaneous and Introduced): Lipids and Lipid-Soluble Compounds

Common Name	Scientific Name	Part Used	Bioactive Compounds	Region	SFE Conditions/ MPa/K/Cosolvent	Yield (%)	Reference
Bacuri	*Platonia insignis*	Shell	Free fatty acids	BR-AM	6.3–7.0/289–294	0.15–0.47	[58]
Buriti	*Mauritia flexuosa* L.	Fruit	Carotenoids, tocopherols, and lipids: fatty acids, etc.	BR-AM	20–30/ 313–328	4.7–7.8	[59]
Cupuassu	*Theobroma grandiflorum*	Seeds	Lipids: fatty acids, etc.	BR-AM	24.8–35.2/323–353 Solvents: CO_2/Ethane	2–6/ 5–6	[60]
Jojoba	*Simmondsia chinensis*	Seeds	Fatty acids	AR	40/313–353	0.34–0.4	[61]
Olive husk	—	Husk	Vegetable oil	CH	30/313	7.5–12.5	[62]
Palm	*Elaeis guineensis* L.	Pressed palm fibers or kernel	Carotenoids, tocopherols, fatty acids, etc.	BR-AM	15–30/318–328	1.8–4.9	[63–66]
Paprika powder	*Capsicum annuum*	—	Carotenoids	AR	30/333	0.9	[67]
Passion fruit	*Passiflora edulis*	Seeds	Lipids, fatty acids, etc.	BR-AM/SE	20–30/317–343	13.7–27.7	[68]
Pejibaye or pupunha	*Guilielma speciosa* or *Bactris gasipaes*	Fruit	Fatty acids	SA-AM	8–30/293–323	9–13	[69, 70]
Rapeseed	*Brassica napus*	Seeds	Vegetable oil	CH	30/313	6–12	[62]
Rice bran	*Oryza sativa*	Parboiled rice bran	Tocotrienol and tocopherols	BR-S/SE	15–30/298–333	8	[71, 72]
Rosehip	*Rosa canina* L.	Seeds	Carotenoids and fatty acids	CH	10/301	4–6.5	[62]
Tucuman	*Astrocaryum vulgare*	Seeds	Fatty acids	BR-AM	20–30/313–343	31	[73]
Ucuuba	*Virola surinamensis*	Seeds	Trimirystin	SA	20–25/323	44	[74]

SA: South America; SA-AM: Amazonian region of South America; BR-AM: Amazonian region of Brazil; BR-SE: Southeast of Brazil; AR: Argentina; CH: Chile

TABLE 8.3
Bioactive Compounds from Latin America Plants (Spontaneous and Introduced): Miscellaneous

Common name	Scientific name	Part used	Bioactive compound(s)	Region	SFE conditions/ MPa/K/Cosolvent	Yields/%	Reference
Andreadoxa	*Andreadoxa flava*	Leaves	8-methoxy-N-methyl-flindersine	BR-NE	20.4–30.6/313	—	[75]
Arnica	*Solidago chilensis*	Leaves	Isoquercetin, rutin, vitexin	BR	10/333/EtOH	12.5	[76]
Aroeira	*Schinus therebinthifolius*	Leaves, flowers and stems	Anacardic acids	BR-NE	13.6–27.2/313–323	—	[77]
Arruda da serra or arruda brava	*Poiretia bahiana* C. Muller	Leaves	Waxes, isoflavones, rotenones and monoterpenoid	BR-NE	20.4–23.8/313	—	[78]
Artemisia or sweet sagewort	*Artemisia annua*	Leaves	Artemisinin	BR-SE	15–35/303–323	5–6.5	[79]
Avocado	*Persea americana*	Leaves	Quercetin, isoquercetin, rutin, vitexin	LA	10/333/EtOH	14	[76]
Baccharia or vassoura	*Baccharis dracunculifolia* DC	Leaves	(E)-nerolidol and spathulenol	SA	9–12/313–333	0.4	[80]
Basil	*Ocimum basilicum*	Leaves	Antioxicant compounds	BR-SE	10–35/303–323 K/H_2O: 1/10/20%	1–2/6–11/12–24	[82]
Black wattle or black acacia	*Acacia mearnsii*	Bark	Tannin	BR	15–20/313–353	4.8–23.7 of tannin	[81]
Brazilian ginseng	*Pfaffia glomerata*	Rhizome	Ecdysterone	BR	20/303	0.6	[83]
Boldo	*Peumus boldus* M.	Leaves	Boldine	CH	6–15/303–333/H_2O	0.5–3.4	[84]
Cacao	*Theobroma cacao*	Fruit	Caffeine, theobromin, fatty acids, and metyilxantines	LA	24.8/323	3–13	[85]

continued

TABLE 8.3 (continued)
Bioactive Compounds from Latin America Plants (Spontaneous and Introduced): Miscellaneous

Common name	Scientific name	Part used	Bioactive compound(s)	Region	SFE conditions/ MPa/K/Cosolvent	Yields/%	Reference
Cashew	*Anacardium occidentale* L.	Nuts	Cardanol, anacardic acid, catequin	BR-NE	9.8–30/313–333	2–22	[86, 87]
Chilean hop	*Humulus lupulus*	Leaves	Alpha-acids	CH	20/313	6–14	[88]
Coca	*Erythroxylum coca* Lam.	Leaves	Cocaine	CO	17–22/313/MeOH + H_2O	0.17–0.60	[89]
Copaiba	*Copaifera* sp.	Leaves	Phenolic compounds	BR-AM	10–25/323–333	0.5–4	[90]
Grape	*Vitis vinifera*	Fruit skin	Resveratrol	AR, CH	15/313/ EtOH	Recovery of 20–100% of resveratrol	[91]
Grapefruit	*Citrus paradisi* L.	Peel	Naringin	AR	9.5/331.8/15% EtOH	Recovery of 14.4% of naringin	[92]
Guaco	*Mikania glomerata*	Leaves	Rutin	BR	10/333/ EtOH	8.8	[76]
Guarana	*Paullinia cupana*	Seeds	Caffeine	BR-N	10–40/313–343	Recovery of 98% of caffeine	[93, 94]
Jackfruit	*Artocarpus heterophyllus*	Leaves	Isoquercetin, vitexin	LA	10/333/EtOH	6	[76]
Jalapeno pepper or red pepper	*Capsicum annuum* L.	Fruit	Capsaicin	CH	32/313	14	[95–98]
Mango	*Mangifera indica*	Leaves	Phenolic compounds	SA	25/318	1	[100]
Marigold (pot)	*Calendula officinalis*	Flowers	Faradiol-3-O-laurate, palmitate and myristate	LA	50/323	1.2–2.5	[45, 101]

Mate	*Ilex paraguariensis* A.St.-Hil.	Leaves	Caffeine and methyl-xanthines	SA-S	40/343	2–4	[102]
Passion flower or maypops	*Passiflora incarnate*	Leaves	Quercetin, vitexin	LA	10/333/EtOH	9.2	[76]
Pink trumpet tree	*Tabebuia avellanedae*	Wood	Lapachol	AR, BR	9–20	Recovery of lapachol 0.2–1.9%	[103]
Pitanga or Surinam cherry	*Eugenia uniflora*	Leaves	Isoquercetin, rutin	LA	10/333/EtOH	10.6	[76]
Stevia	*Stevia rebaudiana* B.	Leaves	Diterpenes glycosides: stevioside and rebaudioside A	BR, PA	12–25/283–318/H_2O, EtOH	0.3–1.2	[53, 54]
Tabernaemontana	*Tabernaemontana catharinensis*	Bark and leaves	Alkaloids: Coronaridine, voacangine, etc.	BR-SE	20–300/308–328/ EtOH, IsoC3/H_2O	0.4–15	[105, 106]
Turmeric	*Curcuma longa* L.	Roots	Curcuminoids and terpenoids	BR-C	20–30/303–318/ EtOH + IsoC3	5–22	[51, 107–111]
Vinca or Madagascar periwinkle	*Catharanthus roseus*	Leaves	Isoquercetin, rutin, vitexin	LA	10/333/EtOH	15.3	[76]

LA: Latin America; SA: South America; SA-S: South of South America; BR: Brazil; BR-C: Central region of Brazil; BR-NE: Northeast of Brazil; BR-SE: Southeast of Brazil; AR: Argentina; CH: Chile; CO: Colombia cultivation places) were indicated; this information was gathered from the publications or the databases previously mentioned. Plants were classified as producers of volatile oils and/or oleoresins (Table 8.1) [15–57], lipids, and lipid-soluble substances (carotenoids, tocopherols, etc.) (Table 8.2) [58–74]. Plants that produce miscellaneous compounds such as phenolic compounds, isoflavones, and so on are grouped in Table 8.3 [75–111]. The experimental data in Table 8.1 to Table 8.3 were obtained focused in the SFE process, thus, yields, kinetic behavior, and chemical composition of the extracts (or the content of the target component) and, in few cases, the biological activities (antioxidant, anticancer, antimycobacterium) of the various systems were determined. Some plants, in spite of their economical importance for the LA countries (such as the orange for Brazil) have received little attention; nonetheless, an interesting study on the fractionation of the oily-fraction of concentrated-frozen orange juice oil was done by Marques [112].

Table 8.4 summarizes the phase equilibrium data [113–133] measured for some of the systems in Table 8.1 to Table 8.3; some entirely predictive studies have also been included [122, 125–127]. Phase equilibrium measurement and modeling were mostly done for lipid systems [5, 118, 132]. Alkaloids + CO_2 phase equilibrium data were measured for caffeine in CO_2 and CO_2 + cosolvent [115] and predicted for purine alkaloids [130]. Phase equilibrium data for artemisinin in CO_2 were measured and fitted to density based model and cubic equation of state [113]. The phase equilibrium of quercetin + CO_2 + ethanol was measured and the data were fitted to group contribution and equation of state (EOS) models [131]. The solubilities of oleoresin compounds such as boldine and capsaicin in CO_2 were measured and the data were fitted to density-based models [114, 117]. The phase equilibria of SFE extracts of clove and fennel in CO_2 [119, 120, 133] showed liquid-vapor and liquid-liquid-vapor phase split; the phase equilibrium of vetivergrass SFE extract + CO_2 showed liquid-vapor split. The phase equilibria of camphor + CO_2, camphor + propane, and camphor + CO_2 + propane were measured and fitted to the Peng-Robinson EOS (PR-EOS) [116]. These highly asymmetrical systems show liquid-vapor phase split as well as liquid-liquid-vapor that were quantitatively described by the PR-EOS. The experimental phase equilibrium of limonene oxidation products + CO_2 was also well described by the PR-EOS [123, 124].

Because of the importance of SFE as an analytical tool, Table 8.5 is given here to present some analytical applications in development in LA [134–146]; these studies are concentrated in Brazil and Chile.

Other applications of supercritical fluids (SCFs) are related to the use of curcuminoids to impregnate Polyethylene Terephthlate (PET) films [156], to hydrolyze starchy matrices such as ginger and turmeric [111, 157], to hydrolyze cellulosic matrices [99, 104], and to fraction volatile oils that couple SFE and membranes [158].

8.2 EXTRACTING BIOACTIVE COMPOUNDS BY SFE

SFE from vegetable matrices is complex by nature. Therefore, hardly any characteristic of the system can be described by a simple model. Nonetheless, a very simplified model can be extremely useful for COM estimation. In this simplified model, the vegetable material described has been formed by a cellulosic structure (CS) and

TABLE 8.4
Phase Equilibrium (or Solubility) for Bioactive Compounds at High Pressures

Pure Component or Mixture	Solvent	Type of Equilibrium/MPa/K	Reference
Artemisinin	CO_2	Solubility/10–25/308.2–328.2	[113]
Boldine (Hydro-alcoholic bold leaf extracts)	CO_2	Solubility/8–40/298–333	[114]
Caffeine	CO_2 + EtOH and IsoC3	Solubility/15–30/323–343	[115]
Camphor	CO_2/propane	LV/3.2–13.6/304–354	[116]
Capsaicin	CO_2	Solubility/6–40/298–318	[117]
Castor oil and their fatty acid ethyl esters	CO_2	LV/1.7–25.4/313–343	[118]
Clove SFE extract	CO_2	LV, L_1L_2V/303–328/5.8–13.2	[119]
Fennel SFE extract	CO_2	LV, L_1L_2V/4.7–22/303–333	[120]
Fish oil	CO_2	Solubility/14.7–29.4/301–323	[121]
L-dopa	CO_2	*	[122]
Limone oxidation	CO_2	LV/4.9–14/313–343	[123, 124]
Orange peel oil	CO_2	LV/4–11/313–333	[125–127]
Palm fatty acid distillate	CO_2	Solubility/20–35/313–363	[128, 129]
Purine alkaloids	CO_2	*	[130]
Quercetin	CO_2 + EtOH	Solubility/8–12/313	[131]
Soybean oil and its fatty acid ethyl esters	CO_2	LV/1.3–26.4/313–343	[118]
Triglycerides	CO_2	LV/3.3–14/278.3–368. 5	[132]
Vetivergrass SFE extract	CO_2	LV/7.7–30.8/303–333	[133]

EtOH: ethanol; IsoC3: isopropyl alcohol; LV: Liquid-vapor equilibrium; L_1L_2V: Liquid-Liquid-Vapor equilibrium; Sol: Solubility data
*No experimental data are available.

a solute mixture. The CS contains all insoluble materials, including proteins, carbohydrates, and salts, and is insoluble in the solvent but strongly interacts with the solute mixture. The solute is formed by a multicomponent mixture containing compounds from a variety of chemical functions from low-molecular-mass substances such as, for instance, ethanol (this substance occurs naturally in orange oil from certain varieties cultivated in Brazil [112]), terpenoids, and high-molecular-mass substances such as stevia glycosides [53, 54] and curcuminoids [51, 107–109]. In this definition, the presence of cosolvent modifies the composition of the solute mixture as well as that of the CS. Yet, the solid matrix can be viewed as previously described by incorporating into the solute the substances that are now soluble due to the presence of the cosolvent. Analogously, in the CS phase remains only the solvent (SCF + cosolvent) insoluble material. Therefore, in the extractor vessel, the mixture CS + solute + solvent can be treated as a pseudoternary system. The solute is a

TABLE 8.5
Analytical Applications

Substratum	Bioactive Compound	Reference
Honey	Pesticides	[134]
Fish	Mercury	[135]
Lacteal matrices	Fat soluble vitamins	[136]
Organotin	Organotin	[137]
Sausages	N-nitrosamines	[138]
Soil samples	Butyltin	[139]
Soil samples	Pesticides	[140–142]
Vegetable oils	Polycyclic aromatichydrocarbons	[143]
Human hair	Cadmium	[144]
Urine	Chromium	[145]
Urine	Nitrofurantoin	[145]

multicomponent mixture whose nature depends on the vegetable material used. The CS is also a multicomponent mixture and, thus, a pseudocomponent that is entirely inert to the action of the solvent (or solvent mixture); nonetheless, it does interact with the solute mixture. The system in a very simplified conception can be considered as a two-phase system. The light, or solvent, phase constitutes of solute mixture + solvent, and the heavy phase contains the cellulosic structure + solute mixture.

8.2.1 Relevant Process Information for COM Estimation

The huge amount of data on SFE from several different solid matrices published [5–8] indicates that SFE has been proven to be technically feasible for virtually any solid substratum. Nonetheless, in spite of the recent development of new industrial plants all over the world, LA has none. One of the reasons is the restraint imposed by the fixed cost of investment of an SFE unit. Therefore, to fulfill the current pursuit of clean technology, one must show investors that in addition to being technically viable, SFE is indeed an attractive choice for an extraction process. In order to do so, we must build a benchmark for the COM of SFE to be compared with the COM of conventional processes. So, a simple and yet reliable method to estimate COM for preliminary analysis or business plan analysis is needed. For this, a simple procedure that employs minimum experimental information would be adequate. Rosa and Meireles [147] have demonstrated such methodology based on the method described by Turton et al. [148]; these authors estimated COM for clove bud oil and ginger oleoresin. In the next section, we describe in detail the required information to employ the procedure adopted by Rosa and Meireles [147].

8.2.2 Selecting Parameters Intended for COM Estimation

For a business plan, that is, at the very early stages of process development, the following questions must be answered for each vegetable matrix:

1. What is the best process for obtaining the desired extract? (At this point, the alternative extraction processes should be considered, including SFE, low-pressure solvent extraction [LPSE] with a variety of solvents, and steam distillation [for volatile oil only].)
2. For a given extract with specified functional properties, would SFE be a good choice?
3. If SFE is an alternative, what are the pressure and temperature of extraction?
4. At this pressure and temperature, what is the process yield?
5. How long does it take to obtain such yield?

In order to answer these questions, two types of experimental data must be available:

1. Global yield, or total amount of soluble substances present in the vegetable matrix for a given condition of temperature and pressure.
2. SFE kinetics for the system under consideration.

To develop mass transfer and phase equilibrium models, several types of information are required [36, 61, 97, 149–151]. For instance, the characteristics of the solid matrix, such as humidity, content of soluble material, structure, particle size, and distribution, are required for evaluation of the mass transfer parameters from different models. To choose a thermodynamic model suitable for describing phase equilibrium, experimental data must be available. These parameters must be measured, estimated, or both, preferably using standard procedures. Nonetheless, for the business plan, the needed information is less. For instance, with the knowledge of the bed apparent density, the required bed volume for a given production can be estimated. Or, if extractors of a given volume are available, it is possible using the bed apparent density to calculate the raw material demand to be processed. Adding to this simple information, the global yield and the time for an extraction cycle is enough for COM estimation [147].

Additionally, the composition of the extract in terms of its major compounds and one functional property would help the decision makers. To obtain the required information for process design, identification of the solute mixture is necessary; therefore, the chemical composition of SFE extracts must be determined by appropriate methods, such as gas chromatography with flame ionization detector (GC-FID), Gas Chromatography-Mass Spectrometry (GC-MS), high-performance liquid chromatography, or ultraviolet spectrophotometry. For extracts that will be used as nutraceuticals, biological activity must also be monitored. This can be done using a simple technique to access the antioxidant activity of the SFE extract. One such method is that of Hammerschmidt and Prat [152], which has been adapted to be used for SFE extracts [51]. Another key issue is the optimization of the separation step that can be simulated using a simple cubic EOS, such as the PR-EOS.

8.2.2.1 Pressure and Temperature Processes

The selection of the process conditions, such as temperature, pressure, solvent flow rate, cosolvent (if required), solid matrix preparation, and so on, is required for process optimization. Nonetheless, before selecting the parameters related to mass transfer (i.e., to the kinetics that will ultimately be used for process optimization), it is interesting to choose the pressure and temperature of extraction. This can be done considering the thermodynamics of the system as well as the composition of the extract at a given condition of temperature and pressure. In order to select the pressure and temperature of process, the phase equilibrium or solubility of the system solute + SCF, the solubility of the pseudoternary system CS + solute + SF, or the global yield can be used. At this point, it is important to remember that in spite of the CS being inert to the solvent, it strongly interacts with the solute; therefore, the interaction of the solute with the solvent, as well as with the CS, must be considered. Thus, the solubility of the solute in solvent measured in the pseudobinary system formed by solute and solvent, and the solubility of the solute measured for the solid matrix-solvent system will be quite different. In the second case, as reported by Brunner [4], the solubility can be an order of magnitude smaller than the solubility measured for the first case. As the CS strongly interacts with the solute mixture, it can be expected that the various compounds that form the solute mixture will have different affinities for the CS; therefore, the choice of process temperature and pressure will be better done considering parameters measured for the pseudoternary system. The phase equilibrium of several pseudobinary systems (solute/solvent) has been systematically reported in literature [5]; these data are very helpful in optimizing the SFE separation step. Other authors have measured the solubility (Y^*) of the pseudoternary system CS + solute + SF using the dynamic method. As discussed by Rodrigues et al. [27], to obtain correct values of Y^* would require a tedious and costly work to determine the solvent flow rate that can be safely used to measure this parameter. The solubility has also been reported by several authors as the initial slope of an overall extraction curve (OEC) in terms of total yield as a function of the ratio of solvent mass (S) to the feed mass (F); this parameter is referred as $Y^*_{S/F}$. The difference between Y^* and $Y^*_{S/F}$ can be understood by recalling that in order to measure Y^* true equilibrium for the pseudoternary system is expected to be obtained while $Y^*_{S/F}$ is measured at a given ratio of solvent mass to feed mass. The results of Moura et al. [34] for fennel + CO_2 have shown that $Y^*_{S/F}$ can be a function of the fixed bed geometry, i.e., the ratio of the bed height (H_B) to the bed diameter (D_B). These authors obtained increasing values of $Y^*_{S/F}$ as the ratio H_B/D_B increased from 2.21 to 8.84. On the other hand, similar experiments done by Carvalho et al. [52] for rosemary + CO_2 showed that $Y^*_{S/F}$ varied in a narrow range as H_B/D_B increased from 0.67 to 8.4. Even so, as reported by Moura et al. [34] and Carvalho [52], these results were dependent on the solvent flow rate. More precisely, the interstitial velocity in the fixed bed plays an important role in the process, thus affecting the measurement of $Y^*_{S/F}$. Additionally, to obtain reliable results of either Y^* or $Y^*_{S/F}$, the experimental assays must be performed in SFE units containing extractor vessels with volumes of at least 50 mL, since an OEC must be built. Alternatively, the choice of the operating temperature and pressure can be done considering the global yields isotherms (GYI) as reported by Rodrigues et al. [154] and Moura et al. [34], among others. The total or global yield (X_o) can be obtained throughout exhaustive extraction in a SFE unit. There is no need to build an OEC; therefore, extractor vessels of small volumes (V_E < 50 mL) and small amounts of feed are required. This is a very convenient choice when only small amounts of the vegetable material are available, which is often the case for spontaneous crops. A trick question is what should be the ratio of solvent mass to feed mass? Excess solvent must be used in order to obtain the true value of X_0. Considering that the total yield should be an intensive property as discussed elsewhere [43, 154], it should depend only on temperature and pressure, therefore, it is enough to establish a suitable value for the ratio S/F to measure the yield in such a way as to guaranty its usability for selecting the operating pressure and temperature. Since the global yield was determined at a selected ratio of S/F it should be denoted as $X_{0,S/F}$. Based on our results [100], values of S/F greater than 15 are good enough for the selection of the operating temperature and pressure. Considering that an extraction experiment to build an OEC for some vegetable matrices can take several hours while global yields assays are shorter, the usage of the GYI to select the operating temperature and pressure instead can save hours of experimental work. If OECs are available and global yield information is missing, then the global yield can be estimated using the spline fitting of the OEC, as will be discussed next.

8.2.2.2 The Kinetic Parameters

For the production of essential oils, oleoresins, vegetable oils from exotic vegetable seeds, sweeteners, and so on, in general, a multipurpose plant containing at least two fixed-bed extractors will be employed. At the laboratory level, the analysis of such a process can be done considering the OEC. The effects of the process variables pressure, temperature, and solvent flow rate on the total yield as well as on the chemical profile of the extract are not easily seen from the OEC. Therefore, for first approximations, it would be interesting to establish a simple procedure to analyze the effects of the process variables. Afterward, the conditions can be optimized considering the global process.

An OEC is obtained considering the amount extracted (mass of extract or yield) as a function of time. The information provided by an OEC is the time required for an extraction batch. A typical OEC can be described by three steps:

1. A constant extraction rate period (CER)
2. A falling extraction rate period (FER), which represents the step for which both convection and diffusion in the solid substratum controls the process
3. A diffusion-controlled rate period (DC)

Prior to SFE, the solid substratum requires preprocessing that at least includes comminution. In order to avoid channeling, the particle size used in SFE will depend on the ratio of bed diameter to particle diameter, which has been reported to be between 50 and 250 [5]. Because of the strong interaction between water and CS, solute dehydration is required if the water content is more than 20%, wet basis. Therefore, the preprocessing of the solid substratum promotes the rupture of cell walls; thus, the solid matrix subjected to SFE will contain ruptured as well as unruptured cells. Even so, for certain solid matrices, the severe treatment suffered during the pretreatment results in about 100% of ruptured cells, as for instance, for the recovery of carotenoids from pressed palm oil fibers [63, 64]. However, for special solid matrices for which the solute is located very superficially, no comminution is needed. Indeed, Silva [17] has shown that extraction of bixin and norbixin from urucum (*Bixa orellana*) seeds is more effective using whole seeds than milled seeds.

The CER period is characterized by the extraction of the solute contained in the surface of the solid substratum particles or in cells that were broken during preprocessing. Sovová [149] called the solute removed during the CER period *easily accessible solute*. The mass transfer in the external film near the particle's surface is controlled by convection. The CER period is characterized by the following kinetic parameters: (1) the mass transfer rate (M_{CER}), (2) the duration of the CER period (t_{CER}), (3) the yield during the CER period (R_{CER}), and (4) mass ratio of solute in the fluid phase at the extractor vessel outlet (Y_{CER}). About 70% to as much as 90% of the soluble material can be extracted from the substratum during the CER period if careful pretreatment is used [20, 26]. In the FER period, a considerable portion of the solid particles is no longer coated with solute or the number of broken cells is no longer uniform. Thus, the mass transfer rate diminishes as a result of the decrease in the effective mass transfer area as well as the increase in importance of the diffusional mechanism. In the DC period, the solute coating of the solid particles has been completely removed and, thus, the extraction process is controlled by the diffusion of the solvent to the inner parts of the particles followed by the diffusion of the solute-solvent mixture to the surface of the particles.

8.2.2.2.1 Describing the OEC by a Spline

An OEC can be described by a family of straight lines. The mass of extract (or the yield) can be obtained from the following equations.

For N lines:

$$m_{Ext} = \left(b_0 - \sum_{i=1}^{i=N} C_i b_{i+1} \right) + \sum_{i=1}^{i=N} b_i t \tag{8.1}$$

For two straight lines:

$$m_{Ext} = \left(b_0 - C_1 b_2 \right) + \left(b_1 + b_2 \right) t \tag{8.2}$$

For three straight lines:

$$m_{Ext} = \left(b_0 - C_1 b_2 - c_3 b_3 \right) + \left(b_1 + b_2 + b_3 \right) t \tag{8.3}$$

where b_i for i = 0,1,2 are the linear coefficients of lines 1, 2… and C_i for i = 1,2 are the intercepts of these lines (for instance, C_1 is the intercept of the first and second lines, and C_2 is the intercept of the second and third lines), m_{Ext} is the mass of extract (or the yield), and t is time.

Figure 8.1 shows that two straight lines can quantitatively describe the OEC for ginger + CO_2 [36], whereas Figure 8.2 shows that three lines are required for chamomile + CO_2 [24]. The first line represents the entire CER plus the beginning of the FER period. The slope of the line represents the mass-transfer rate of the CER period, M_{CER}. The time corresponding to the interception of the two lines is denoted

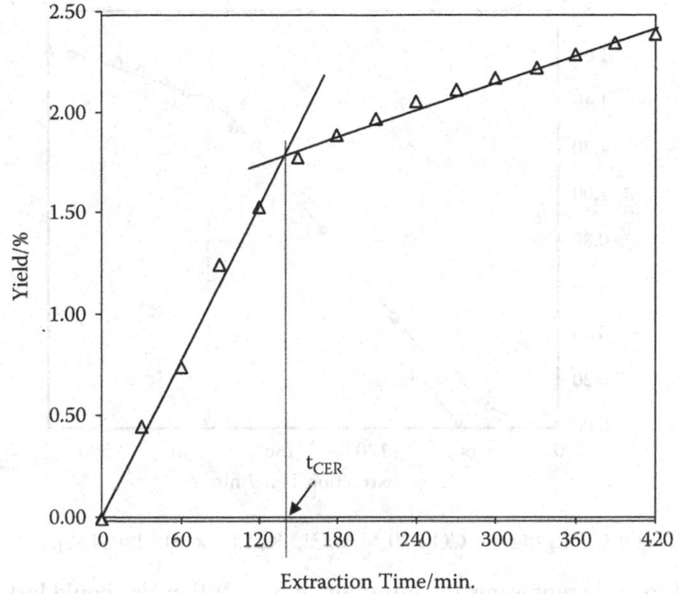

FIGURE 8.1 OEC for ginger + CO_2: 20 MPa, 308 K, 5.91 × 10^{-5} kg/s [36].

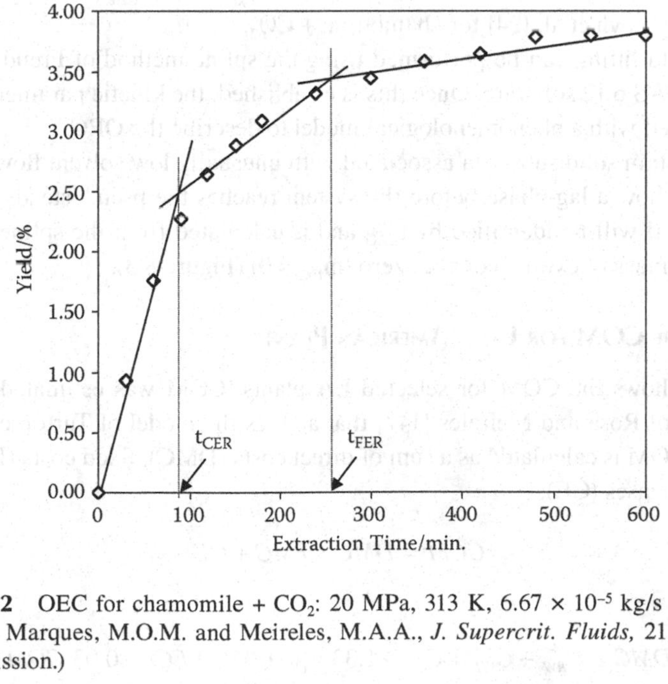

FIGURE 8.2 OEC for chamomile + CO_2: 20 MPa, 313 K, 6.67 × 10^{-5} kg/s [24]. (From Povh, N.P., Marques, M.O.M. and Meireles, M.A.A., *J. Supercrit. Fluids,* 21, 245, 2001. With permission.)

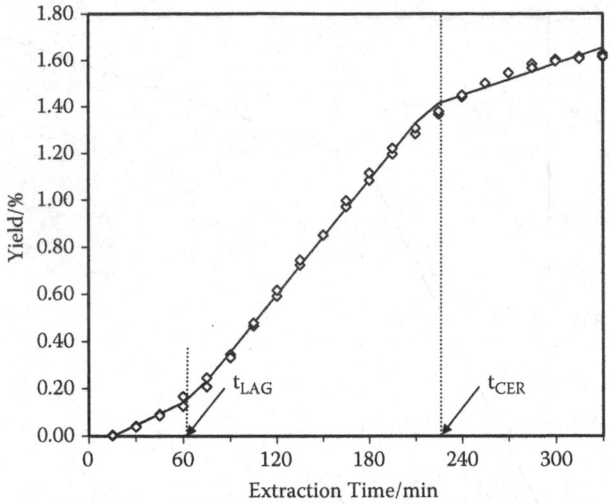

FIGURE 8.3 OEC for ginger + CO_2: 20 MPa, 313 K, 1.60 × 10^{-5} kg/s [36].

by t_{CER} and roughly represents the minimum time a SFE cycle should last. The mass ratio of solute in the supercritical phase at the bed outlet (Y_{CER}) is obtained by dividing M_{CER} by the mean solvent flow rate for the CER period. The yield relative to the CER period is R_{CER}, or minimum yield expected from SFE process at a given temperature, pressure, solvent flow rate, and solid substratum preprocessing. If the GYI is missing but an OEC is available, then the global yield can be estimated using Equation 8.1 or Equation 8.2 by calculating the m_{Ext} at $t = 3\ t_{CER}$; this approximation was used by Povh et al. [24] for chamomile + CO_2.

The data fitting can be performed using the spline method of Freud and Little [153] and SAS 6.12 software. Once this is established, the kinetic parameters should be associated with a phenomenological model to describe the OEC.

For certain solid substrata associated with unusually low solvent flow rates, the OEC will show a lag-phase before the system reaches the pseudosteady state; this time interval will be identified by t_{LAG} and is calculated from the spline model by setting the mass of extract equal to zero ($m_{Ext} = 0$) (Figure 8.3).

8.2.3 The COM for Latin American Plants

Table 8.6 shows the COM for selected LA plants. COM was estimated using the procedure of Rosa and Meireles [147] that applies the model of Turton et al. [148], in which COM is calculated as a sum of direct costs (DMC), fixed costs (FMC), and general expenses (GE):

$$COM = DMC + FMC + GE \tag{8.4}$$

$$DMC = C_{RW} + C_{WT} + C_{UT} + 1.33\ C_{OL} + 0.069\ FCI + 0.03\ COM \tag{8.5}$$

TABLE 8.6
COM for SFE Extracts from Selected Plants

Raw Material	Botanic Name	Target Component	SFE Conditions MPa/K/Cosolvent	Yield(%)	$^*t_{Ext}$(min.)	COM (US $/kg)	Reference
Anise	*Pimpinella anisum*	Anethole	10/303	7.9	100	21.21	[160]
Brazilian ginseng	*Pfaffia glomerata*	Ecdysterone	20/303	0.6	140	1,648.00	[83]
Clove	*Eugenia caryophyllus*	Volatile oil	10/288	12.9	70	9.18	[147]
				13.5	90	9.88	
				14.1	120	10.97	
Fennel	*Foeniculum vulgare*	Anethole	25/303	12.5	80	8.81	[160]
Ginger	*Zingiber officinalis*	Oleoresin	20/313	2.7	150	99.80	[147]
Rosemary	*Rosmarinus officinalis*	Volatile oil	30/313	5	100	42.69	[160]
Tabernaemontana	*Tabernaemontana catharinensis*	Coronaridine and voacangine	35/308–318/EtOH	1.4	90	440.31	[161]

$^*t_{Ext}$: extraction time

$$FMC = 0.708\ C_{OL} + 0.168\ FCI \tag{8.6}$$

$$GE = 0.177\ C_{OL} + 0.009\ FCI + 0.16\ COM \tag{8.7}$$

where C_{RW} is the cost raw material, C_{WT} is the cost of waste treatment, C_{OL} is the cost of operational labor, and FCI is a fraction of the investment.

COM estimation was done using the software Tecanalysis v. 2.0 developed in LASEFI–DEA/FEA - UNICAMP. In calculating the cost, the yields and the minimum time of the SFE cycles were estimated as discussed in Section 8.2.2.2.1. The SFE unit chosen as the benchmark contains two extractor vessels of 400 L (approximate cost U.S. $2 million). The total annual operating time was assumed to be 7920 h, which corresponds to 330 days per year based on a 24-h shift. The cost of operational labor was estimated to be U.S. $3.00/h. The SFE unit is multipurpose; therefore, it should be operating the assumed 7920 h per year regardless of the raw material used. The cost of waste treatment was assumed to equal zero. Table 8.7 and Table 8.8 show the Tecanalysis report for the input data and results, respectively, for COM estimation of clove volatile oil. The extraction times (t_{Ext}) were assumed equal or 1.10 to 1.90t_{CER}.

8.3 CONCLUSIONS

SFE can be a true alternative for obtaining high quality extracts from several LA plants. Comparing the COM estimated with the selling prices, it is clear that there is a business opportunity.

REFERENCES

1. Meireles, M.A.A., Wang, G.-M., Hao, Z.-B., Shima, K. and Teixeira da Silva, J., Stevia (Stevia rebaudiana Bertoni): Futuristic view of the sweeter side of life, in *Floriculture, Ornamental and Plant Biotechnology: Advances and Topical Issues,* Teixeira da Silva, J., Ed., GSB, 2006, chap. 46, pp. 415–425.
2. Canela, A.P.R.F., Rosa, P.T.V., Marques, M.O.M. and Meireles, M.A.A., Supercritical fluid extraction of fatty acids and carotenoids from the microalgae *Spirulina maxima, Ind. Eng. Chem. Res.,* 41, 3012–3018, 2002.
3. Valderrama, J.O., Perrut, M. and Majewski, W., Extraction of astaxantine and phycocyanine from microalgae with supercritical carbon dioxide, *J. Chem. Eng. Data,* 48, 827–830, 2003.
4. Brunner, G., *Gas Extraction: An Introduction to Fundamentals of Supercritical Fluids and the Application to Separation Processes,* Springer, New York, 1994.
5. Meireles, M.A.A., Supercritical extraction from solid: Process design data (2001–2003), *Curr. Opin. Solid St. M. Sci.,* 7, 321–330, 2003.
6. Rosa, P.T.V. and Meireles, M.A.A., Supercritical technology in Brazil: System investigated (1994–2003), *J. Supercrit. Fluids,* 34, 109–117, 2005.
7. Diáz-Reinoso, B., Moure, A., Dominguéz, H. and Parajoá, J.C., Supercritical CO_2 extraction and purification of compounds with antioxidant activity, *J. Agric. Food Chem.,* 54, 2441–2469, 2006.

TABLE 8.7
Technical-Economical Analysis Report: Input Data

Raw Material: Clove Buds	
Initial Investment	
Price (US$): SFE unit with two extractors	2,000,000.00
Extractor volume (m^3)	0.40
Total annual operation time (h)	7920
Operation labor cost (US$/h)	3
Raw material cost (US$/MT)	505
Initial humidity (%)	10
Final humidity (%)	10
Grinding and drying cost (US$/MT)	30
CO_2 cost (US$/kg)	0.1
Loss of CO_2 (% of total used in a cycle)	2
Electrical power cost (US$/Mcal)	0.0703
Cooling water cost (US$/Mcal):	0.0837
Saturated steam (5 barg) Cost (US$/Mcal)	0.0133
Depreciation (%/Year)	10
Sea freight cost (US$/MT·km)	0.01
Sea freight distance (km)	0
Total road freight cost (US$/MT·km)	0
Operational Data	
Extraction time (min)	120
Extraction temperature (°C)	15
Extraction pressure (MPa)	10
Flash tank pressure (MPa)	4
CO_2 flow rate (kg/h)	90
Bed density (kg/m^3)	520
Scale-Up Model	
Yield (kg extract/kg feed)	0.1414
Waste Treatment Cost	
Solid waste (US$)	0
Liquid waste (US$)	0
Gas waste (US$)	0

TABLE 8.8
Technical-Economical Analysis Report: Results

Raw Material: Clove Buds

Fraction of Investment

Total investment (US$) - IT	2,000,000.00
Column volume (m^3) - Cv	0.40

Operational Labor Cost

Equipment	Hmo/Hop	Total Hmo (h)	Cost (US$)
Extractor	1	7920	23,760.00
Flash distillation	0.1	792	2,376.00
Condenser	0.1	792	2,376.00
CO_2 tank	0.5	3960	11,880.00
Pump	0.05	396	1,188.00
Heat exchanger	0.1	792	2,376.00
		Total	43,956.00

Hmo/Hop: Man-labor hours per equipment per hour of operation of the system; Hop: Annual operating hours of the equipment

Raw Material Cost

Solid matter cost (US$)	415,958.40
CO_2 used in process (kg)	712,800.00
Loss of CO_2 (%)	2.00
CO_2-specific cost (US$/kg)	0.10
CO_2 cost (US$)	1,425.60
Preprocessing cost (US$)	24,710.40
Sea cargo cost (US$)	0.00
Road cargo cost (US$)	0.00
Raw material cost (US$)	442,094.40

Utility Cost

Equipment	Energy (Mcal)	Specific Cost (US$/MCal)	Cost (US$)
Flash distillation	33,854.79	0.0133	450.27
Condenser	−36,385.97	0.0837	3,045.51
Pump	1,013.96	0.0703	71.28
Heat exchanger	1,922.80	0.0133	25.57
		Total	3,592.63

Waste Treatment Cost

Solid waste (US$)	0.00
Liquid waste (US$)	0.00

TABLE 8.8 (continued)
Technical-Economical Analysis Report: Results

Waste Treatment Cost (continued)	
Gas waste (US$)	0.00
Total (US$)	0.00

Cost of Manufacturing

Variable	Value (US$)	Value in COM (US$)	% of COM
Investment (US$) - IT	2,000,000.00	607,407.40	47.53
Raw material (US$) - CRM	442,094.40	545,795.55	42.71
Operation labor (US$) - COL	43,956.00	120,200.67	9.41
Utilities (US$) - CUT	3,592.63	4,435.34	0.35
Waste treatment (US$) - CWT	0.00	0.00	0.00
Cost of manufacturing (US$) - COM	1,277,838.95		
Mass of extract (kg)	116,468.30		
Specific cost (US$/kg)	10.97		

8. del Valle, J.M., de la Fuente, J.C. and Cardarelli, D.A., Contributions to supercritical extraction of vegetable substrates in Latin America, *J. Food Eng.*, 67, 35–57, 2005.
9. U.S. Department of Agriculture Natural Resources Conservation Service. Plant Database, *http://plants.usda.gov/*, accessed June 2006.
10. Raintree Nutrition, Rainforest Plant Database, *http://www.rain-tree.com/*, accessed June 2006.
11. Missouri Botanical Garden, w^{3}TROPICOS, *http://mobot.mobot.org/W3T/Search/vast.html*, accessed June 2006.
12. Searchable World Wide Web Multilingual Multiscript Plant Name Database (M.M.P.N.D.) *http://www.plantnames.unimelb.edu.au/Sorting/Frontpage.html*, accessed July 2006.
13. Center for New Crops and Plant Products, CropINDEX, Purdue University, *http://www.hort.purdue.edu/newcrop/morton/pejibaye.html*, accessed July 2006.
14. Flora brasiliensis, *http://florabrasiliensis.cria.org.br/*, accessed June 2006.
15. Daghero, J. et al., Fractionation of *Schinus molle* L. (aguaribay) essential oil with supercritical CO_2, in *V Brazillian Meeting on Supercritical Fluids*, Ferreira, S.R.S., Ed., UFSC, Florianópolis, Brazil, 2004, CD-ROM.
16. Nobre, B.P. et al., Supercritical carbon dioxide extraction of pigments from *Bixa orellana* seeds (experiments and modeling), *Braz. J. Chem. Eng.*, 23, 251–258, 2006.
17. Silva, G.F., Extração de corantes do urucum com dióxido de darbono Supercrítico, PhD thesis, State University of Campinas, Campinas, Brazil, 1999.
18. Martinez, J. et al., Extraction of volatile oil from *Piper aduncum L.* leaves with supercritical carbon dioxide, in *Proceedings of the Sixth International Symposium on Supercritical Fluids,* Brunner, G., Kikic, I. and Perrut, M., Eds., ISASF, Versailles, France, 65–70, 2003.

19. Leal, P.F. et al., Global yields, chemical compositions, and antioxidant activities of clove basil (*Ocimum gratissimum* L.) extracts obtained by supercritical fluid extraction, *J. Food Proc. Eng.*, 29, 547–559, 2006.
20. Ferreira, S.R.S., Meireles, M.A.A. and Cabral, F.A., Extraction of essential oil of black pepper with liquid carbon dioxide, *J. Food. Eng.*, 20, 121–133, 1993.
21. Ferreira, S.R.S. et al., Supercritical fluid extraction of black pepper (*Piper nigrum* L.) essential oil, *J. Supercrit. Fluids,* 14, 235–245, 1999.
22. Perakis, C., Louli, V., and Magoulas, K., Supercritical fluid extraction of black pepper oil, *J. Food Eng.*, 71, 386–393, 2005.
23. Braga, M.E.M., Ehlert, P.A.D., Ming, L.C. and Meireles, M.A.A., Supercritical fluid extraction from *Lippia alba:* Global yields, kinetic data, and extract chemical composition, *J. Supercrit. Fluids,* 34, 149–156, 2005.
24. Povh, N.P., Marques, M.O.M. and Meireles, M.A.A., Supercritical CO_2 extraction of essential oil and oleoresin from chamomile (*Chamomilla recutita* [L.] Rauschert), *J. Supercrit. Fluids,* 21, 245–256, 2001.
25. Muller, C., et al., Supercritical extraction of essential oils of *Cymbopogon martini, Cymbopogon winterianus,* and *Eycalyptus citriodora* with carbon dioxide, Chem. Preprint Archive Volume 2001, Issue 5, May 2001, pp. 158–177. Available at http://www.sciencedirect.com/preprintarchive
26. Meireles, M.A.A. and Nikolov, Z.L., Extraction and fractionation of essential oils with liquid carbon dioxide, in *Spices, Herbs and Edible Fungi,* Charalambous, G., Ed., Elsevier Science, Amsterdam, 1994, chap. 6, pp. 171–199.
27. Rodrigues, V.M. et al., Determination of the solubility of extracts from vegetable raw material in pressurized CO_2: A pseudo-ternary mixture formed by cellulosic structure + solute +solvent, *J. Supercrit. Fluids,* 22, 21–36, 2002.
28. Oliveira, A.L., Eberlin, M.N. and Cabral, F.A., Aromatic oil from Brazilian roasted coffee extracted by supercritical carbon dioxide: Composition analysis by GC/MS, in *V Brazillian Meeting on Supercritical Fluids,* Ferreira, S.R.S., Ed., UFSC, Florianópolis, Brazil, 2004, CD-ROM.
29. Yepez, B., Espinosa, M. and Bolaños, G., Producing antioxidant fractions from herbaceous matrices by supercritical fluid extraction, *Fluid Phase Equilibria,* 194–197, 879–884, 2002.
30. Kraut, S., Braga, M.E.M. and Meireles, M.A.A., Extraction of coriander (*Coriandrum sativum* L.) leaves oil by SFE: Total phenolic content, in *Proceedings of the Eighth Conference on Supercritical Fluids Application,* Reverchon, E., Ed., ISASF, Ischia, Italy, 2006, 143–146.
31. Sousa, E.M.B.D. et al., Extraction of volatile oil from *Croton zehntneri* Pax et Hoff with pressurized CO_2: Solubility, composition and kinetics, *J. Food Eng.*, 69, 325–333, 2005.
32. Quispe-Condori, S., Rosa, P.T.V., Foglio, M.A. and Meireles, M.A.A., Supercritical extraction of essential oil from *Cordia curassavica* (Jacq.) Roemer and Schultes, in *Proceedings of the Sixth International Symposium on Supercritical Fluids,* Brunner, G., Kikic, I. and Perrut, M., Eds. ISASF, Versailles, France, 2003, 285–290.
33. Sousa, E.M.B.D., Construção e utilização de um dispositivo de extração com fluido pressurizado, aplicado a produtos naturais, Ph.D. thesis, Federal University of Rio Grande do Norte (UFRN), Natal, Brazil.
34. Moura, L.S. et al., Supercritical fluid extraction from fennel (*Foeniculum vulgare*): Global yield, composition and kinetic data, *J. Supercrit. Fluids,* 35, 212–219, 2005.
35. Martinez, J. et al., Multicomponent model to describe extraction of ginger oleoresin with supercritical carbon dioxide, *Ind. Eng. Chem. Res.*, 42, 1057–1063, 2003.

36. Monteiro, A.R., Extração do óleo essencial/oleoresina de gengibre (*Zingiber officinale* Roscoe) com CO_2 supercrítico: uma avaliação do pré-tratamento e das variáveis de processo, Ph.D. thesis, State University of Campinas, Campinas, Brazil, 1999.
37. Zancan, K.C., Marques, M.O.M., Petenate, A.J. and Meireles, M.A.A., Extraction of ginger (*Zingiber officinale* Roscoe) oleoresin with CO_2 and co-solvents: A study of the antioxidant action of the extracts, *J. Supercrit. Fluids,* 24, 57–76, 2002.
38. Quispe-Condori, S., Determinação de parâmetros de processo nas diferentes etapas do processo de extração supercrítica de produtos naturais: Artemisia annua, Cordia verbenacea, Ocimum selloi e Foeniculum vulgare, Ph.D. thesis, State University of Campinas, Campinas, Brazil, 2005.
39. Michielin, E.M.Z. et al., Composition profile of horsetail (*Equisetum giganteum* L.) oleoresin: Comparing SFE and organic solvents extraction, *J. Supercrit. Fluids,* 33, 131–138, 2005.
40. Portillo, R., Extração do óleo essencial de khoa (*Satureja boliviana* Benth Briq) por diferentes processos: destilação por arraste a vapor, solventes orgânicos e dióxido de carbono pressurizado, Ph.D. thesis, State University of Campinas, Campinas, Brazil, 1999.
41. Pereira, C.G. and Meireles, M.A.A., Evaluation of global yield, composition, antioxidant activity, cost of manufacturing of extracts from lemon verbena (*Aloysia triphylla* L'Herit Britton) and mango (*Mangifera indica* L.) leaves, *J. Food Proc. Eng.,* 30, 150–173, 2007.
42. Ferrua, F.Q., Marques, M.O.M. and Meireles, M.A.A., Óleo essencial de capim-limão obtido por extração com dióxido de carbono líquido, *Ciênc. Tecnol. Aliment.,* 14, 83–92, 1994.
43. Sousa, E.M.B.D. et al., Experimental results for the extraction of essential oil from *Lippia sidoides* Cham. using pressurized carbon dioxide. *Braz. J. Chem. Eng.,* 19, 229–241, 2002.
44. Leal, P.F. et al., Global yields, chemical compositions, and antioxidant activities of extracts from *Achyrocline alata* and *Achyrocline satureioides*, *Phcog. Mag.,* 2, 153–159, 2006.
45. Campos, L.M.A.S., Michielin, E.M.Z., Danielski, L. and Ferreira, S.R.S., Experimental data and modeling the supercritical fluid extraction of marigold (*Calendula officinalis*) oleoresin, *J. Supercrit. Fluids,* 34, 163–170, 2005.
46. Chacin, J., Marquina-Chidsey, G. and Figueroa, Y., Extraction of mastranto (*Hyptis suaveolens*) essential oil using supercritical carbon dioxide, in *V Braz. Meet. Supercritical Fluids,* Ferreira, S.R.S., Ed., UFSC, Florianópolis, Brazil, 2004, CD-ROM.
47. Berna, A., Tarrega, A., Blasco, M. and Subirats, S., Supercritical CO_2 extraction of essential oil from orange peel: Effect of the height of the bed, *J. Supercrit. Fluids,* 18, 227–237, 2000.
48. Fernandes, J.B. et al., Citrus seed oil extractions and their activity against leaf cutting ant Atta sexdens and its symbiotic fungus, *Quim. Nova,* 25, 1091–1095, 2002.
49. Rodrigues, M.R.A. et al., Chemical composition and extraction yield of the extract of *Origanum vulgare* obtained from sub- and supercritical CO_2, *J. Agri. Food Chem.,* 52, 3042–3047, 2004.
50. Costa, T.S., Corrêa, N.C.F., Machado, N.T. and França, L.F., Extração de óleos essenciais de vetiver (*Vetiveria zizaniodes*) e piprioca (*Cyperus sesquiflorus*), *V Braz. Meet. Supercritical Fluids,* Ferreira, S.R.S., Ed., UFSC, Florianópolis, Brazil, 2004, CD-ROM.
51. Leal, P.F. et al., Functional properties of spice extracts obtained via supercritical fluid extraction, *J. Agri. Food Chem.,* 51, 2520–2525, 2003.

52. Carvalho, R.N., Moura, L.S., Rosa, P.T.V. and Meireles, M.A.A., Supercritical fluid extraction from rosemary (*Rosmarinus officinalis*): Kinetic data, extract's global yield, composition, and antioxidant activity, *J. Supercrit. Fluids,* 35, 197–204, 2005.
53. Yoda, S.K., Marques, M.O.M., Petenate, A.J. and Meireles, M.A.A., Supercritical fluid extraction from *Stevia rebaudiana* Bertoni using CO_2 and CO_2 plus water: Extraction kinetics and identification of extracted components, *J. Food Eng.,* 57, 125–134, 2003.
54. Pasquel, A., Meireles, M.A.A., Marques, M.O.M. and Petenate, A.J., Extraction of stevia glycosides with CO_2 + water, CO_2 + ethanol, and CO_2 + water + ethanol, *Braz. J. Chem. Eng.,* 17, 271–282, 2000.
55. Martinez, J. et al., Valorization of Brazilian vetiver (*Vetiveria zizanioides* (L.) Nash ex Small) oil, *J. Agri. Food Chem.,* 52, 6578–6584, 2004.
56. Talansier, E. et al., Supercritical fluid extraction from vetiver roots: A study of SFE kinetics, *10th European Meeting on Supercritical Fluids,* Perrut, M., Ed., ISASF, Colmar, France, 2005, CD-ROM.
57. Stashenko, E.E., Jaramillo, B.E. and Martinez, J.R., Analysis of volatile secondary metabolites from Colombian *Xylopia aromatica* (Lamarck) by different extraction and headspace methods and gas chromatography, *J. Chromatog. A,* 1025, 105–113, 2004.
58. Monteiro, A.R., Meireles, M.A.A., Marques, M.O.M. and Petenate, A.J., Extraction of the soluble material from the shells of the bacuri fruit (*Platonia insignis* Mart) with pressurized CO_2 and other solvents, *J. Supercrit. Fluids,* 11, 91–102, 1997.
59. França, L.F. et al., Supercritical extraction of carotenoids and lipids from buriti (*Mauritia flexuosa*), a fruit from the Amazon region, *J. Supercrit. Fluids,* 14, 247–256, 1999.
60. Azevedo, A.B.A., Kopcak, U. and Mohamed, R.S., Extraction of fat from fermented cupuaçu seeds with supercritical solvents, *J. Supercrit. Fluids,* 27, 223–237, 2003.
61. Tobares, L. et al., Chemical characteristics of jojoba wax obtained by supercritical fluid extraction, in *V Braz. Meet. Supercritical Fluids,* Ferreira, S.R.S., Ed., UFSC, Florianópolis, Brazil, 2004, CD-ROM.
62. del Valle, J.M. et al., Microstructural effects on internal mass transfer of lipids in prepressed and flaked vegetable substrates, *J. Supercrit. Fluids,* 37, 178–190, 2006.
63. França, L.F. and Meireles, M.A.A., Extraction of oil from pressed palm oil (*Elaes guineensis*) fibers using supercritical CO_2, *Ciênc. Tecnol. Aliment.,* 17, 384–388, 1997.
64. Franca, L.F. and Meireles, M.A.A., Modeling the extraction of carotene and lipids from pressed palm oil (*Elaes guineensis*) fibers using supercritical CO_2, *J. Supercrit. Fluids,* 18, 35–47, 2000.
65. de Deus, G.A., Corrêa, N.C.F., Machado, N.T. and França, L.F., Effect of the solvent flow rate on the supercritical extraction of lipids and vitamins from pressed palm oil (*Elaes guineensis*) fibers, in *V Braz. Meet. Supercritical Fluids,* Ferreira, S.R.S., Ed., UFSC, Florianópolis, Brazil, 2004, CD-ROM.
66. Corrêa, N.C.F., Araújo, M.E., Machado, N.T. and França, L.F., Supercritical fluid extraction of palm kernel oil from dendê (*Elaeis guineensis*) with CO_2, in *ENPROMER'99–II Congresso de Engenharia de Processos do MERCOSUL,* CD-ROM.
67. Ambrogi, A., Cardarelli, D.A. and Eggers, R., Separation of natural colorants using a combined high pressure extraction-adsorption process, *Lat. Am. App. Res.,* 33, 3, 323–326, 2003.
68. Corrêa, N.C.F., Meireles, M.A.A., França, L.F. and Araújo, M.E., Extração de óleo da semente de maracujá (*Passiflora edulis*) com CO_2 supercrítico, *Ciênc. Tecnol. Aliment.,* 14, 29–37, 1994.
69. Araújo, M.E., Machado, N.T., França, L.F. and Meireles, M.A.A., Supercritical extraction of pupunha (*Guilielma speciosa*) oil in a fixed bed using carbon dioxide, *Braz. J. Chem. Eng.,* 17, 297–306, 2000.

70. Pasquel, A., Castillo, A. and Sotero, V., Oil extraction from shell of *Bactris gasipaes* HBK fruits using compressed carbon dioxide, in *Proceedings of the Sixth International Symposium on Supercritical Fluids,* Brunner, G., Kikic, I. and Perrut, M., Eds., ISASF, Versailles, France, 2003, 77–82.
71. Sarmento, C.M.P., Ferreira, S.R.S. and Hense, H., Supercritical fluid extraction (SFE) of rice bran oil to obtain fractions enriched with tocopherols and tocotrienols, *Braz. J. Chem. Eng.,* 23, 2, 243–249, 2006.
72. Danielski, L., Zetzl, C., Hense, H. and Brunner, G., A process line for the production of raffinated rice oil from rice bran, *J. Supercrit. Fluids,* 34, 133–141, 2005.
73. Christensen, T., Machado, N.T., França, L.F. and Brunner, G., Extração de vitaminas e lipídeos do tucumã (*Astrocaryum vulgare,* Mart.) em leito fixo usando CO_2 supercrítico, in *Proceedings of the 13th Brazilian Congress of Chemical Engineering,* Águas de São Pedro, Brazil, 2000, CD-ROM.
74. Morais, J.L.C., Machado, N.T. and Bayma, J.C., Extraction of trimirystin from ucuúba (*Virola surinamensis,* Miristicaceae) in fixed bed with supercritical CO_2, in *Proceedings of the Fifth International Symposium on Supercritical Fluids,* ISSF, Atlanta, 2000.
75. Santana, L.L.B. et al., Selectivity in the extraction of 2-quinolones alkaloids with supercritical CO_2, in *V Braz. Meet. Supercritical Fluids,* Ferreira, S.R.S., Ed., UFSC, Florianópolis, Brazil, 2004, CD-ROM.
76. Takeuchi, T.M. et al., in *Proceedings of the Eighth Conference on Supercritical Fluids Application,* Reverchon, E., Ed., ISASF, Ischia, Italy, 2006, 147–150.
77. Cardoso, L.A. et al., Preliminary study of *Schinus therebinthifolius* extraction by supercritical carbon dioxide, in *V Braz. Meet. Supercritical Fluids,* Ferreira, S.R.S., Ed., UFSC, Florianópolis, Brazil, 2004, CD-ROM.
78. Cardoso, L.A. et al., Supercritical extraction of isoflavones and monoterpenoid of Poiretia bahiana, in *V Braz. Meet. Supercritical Fluids,* Ferreira, S.R.S., Ed., UFSC, Florianópolis, Brazil, 2004, CD-ROM.
79. Quispe-Condori, S. et al., Global yield isotherms and kinetic of artemisinin extraction from *Artemisia annua* L leaves using supercritical carbon dioxide, *J. Supercrit. Fluids,* 36, 40–48, 2005.
80. Cassel, E. et al., Extraction of baccharis oil by supercritical CO_2, *Ind. Eng. Chem. Res.,* 39, 4803–4805, 2000.
81. Pansera, M.R. et al., Extraction of tannin by *Acacia mearnsii* with supercritical fluids, *Braz. Arch. Biol. Technol.,* 47, 995–998, 2004.
82. Leal, P.F., Maia, N.B., Carmello, Q.A.C. and Meireles, M.A.A., Supercritical extraction from *Ocimum basilicum* using CO_2 and $CO_2 + H_2O$: Global yields and antioxidant activity of the extracts, in *V Braz. Meet. Supercritical Fluids,* Ferreira, S.R.S., Ed., UFSC, Florianópolis, Brazil, 2004, CD-ROM.
83. Leal, P.F., Alexandre, F.C., Kfouri, M.B. and Meireles, M.A.A., in *Proceedings of the Eighth Conference on Supercritical Fluids Application,* Reverchon, E., Ed., ISASF, Ischia, Italy, 2006, 157–160.
84. del Valle, J.M., Rogalinski, T., Zetzl, C. and Brunner, G., Extraction of boldo (*Peumus boldus* M.) leaves with supercritical CO_2 and hot pressurized water, *Food Res. Int.,* 38, 203–213, 2005.
85. Saldaña, M.D.A., Mohamed, R.S. and Mazzafera, P., Extraction of cocoa butter from Brazilian cocoa beans using supercritical CO_2 and ethane, *Fluid Phase Equilibria,* 194–197, 885–894, 2002.
86. Patel, R.N., Bandyopadhyay, S. and Ganesh, A., Extraction of cashew (*Anacardium occidentale*) nut shell liquid using supercritical carbon dioxide, *Bioresour. Technol.,* 97, 847–853, 2006.
87. Smith, R.L., Jr. et al., Separation of cashew (*Anacardium occidentale* L.) nut shell liquid with supercritical carbon dioxide, *Bioresour. Technol.,* 88, 1–7, 2003.

88. del Valle, J.M., Rivera, O., Teuber, O. and Palma, M.T., Supercritical CO_2 extraction of Chilean hop (*Humulus lupulus*) ecotypes, *J. Sci. Food Agric.*, 83, 1349–1356, 2003.
89. Brachet, A. et al., Experimental design in supercritical fluid extraction of cocaine from coca leaves, *J. Biochem. Biophys. Methods,* 43, 353–366, 2000.
90. Carvalho, R.N., Jr., Braga, M.E.M. and Meireles, M.A.A., Determination of the global yields isotherms for the system copaiba (Copaifera sp.) + CO_2, in *Proceedings of the Eighth Conference on Supercritical Fluids Application,* Reverchon, E., Ed., ISASF, Ischia, Italy, 2006, 589–592.
91. Pascual-Martý, M.C., Salvador, A., Chafer, A. and Berna, A., Supercritical fluid extraction of resveratrol from grape skin of *Vitis vinifera* and determination by HPLC, *Talanta,* 54, 735–740, 2001.
92. Giannuzzo, A.N., Boggetti, H.J., Nazareno, M.A. and Mishima, H.T., Supercritical fluid extraction of naringin from the peel of Citrus paradise, *Phytochem. Anal.,* 14, 221–223, 2003.
93. Saldaña, M.D.A., Zetzl, C., Mohamed, R.S. and Brunner, G., Extraction of methylxanthines from guaraná seeds, mate leaves, and cocoa beans using supercritical carbon dioxide and ethanol, *J. Agr. Food Chem.*, 50, 4820–4826, 2002.
94. Saldaña, M.D.A., Zetzl, C., Mohamed, R.S. and Brunner, G., Decaffeination of guaraná seeds in a microextraction column using water saturated CO_2, *J. Supercrit. Fluids,* 22, 119–127, 2002.
95. del Valle, J.M. et al., Supercritical carbon dioxide extraction of pelletized jalapeno peppers, *J. Sci. Food Agri.*, 83, 550–556, 2003.
96. del Valle, J.M., Jimenez, M. and de la Fuente, J.C., Extraction kinetics of pre-pelletized jalapeno peppers with supercritical CO_2, *J. Supercrit. Fluids,* 25, 33–44, 2003.
97. Uquiche, E., del Valle, J.M. and Ihl, M., Microstructure-extractability relationships in the extraction of prepelletized jalapeno peppers with supercritical carbon dioxide, *J. Food Sci.,* 70, E379–E386, 2005.
98. Uquiche, E., del Valle, J.M. and Ortiz, J., Supercritical carbon dioxide extraction of red pepper (*Capsicum annuum* L.) oleoresin. *J. Food Eng.,* 65, 55–66, 2004.
99. Pasquini, D., Pimenta, M.T.B., Ferreira, L.H. and Curvelo, A.A.D., Extraction of lignin from sugar cane bagasse and Pinus taeda wood chips using ethanol-water mixtures and carbon dioxide at high pressures, *J. Supercrit. Fluids,* 36, 31–39, 2005.
100. Pereira, C.G., Obtenção de extratos de leiteira de dois irmãos (*Tabernaemontana catharinensis* A. DC.), cidrão (*Aloysia triphylla* L'Herit. Britton) e manga (*Mangifera indica* L.) por extração supercrítica: Estudo dos parâmetros de processo, caracterização e atividade antioxidante dos extratos. PhD thesis, State University of Campinas, Campinas, Brazil, 2005.
101. Hamburger, M. et al., Preparative purification of the major anti-inflammatory triterpenoid esters from marigold (*Calendula officinalis*), *Fitoterapia*, 74, 328–338, 2003.
102. Saldana, M.D.A., Zetzl, C., Mohamed, R.S. and Brunner, G., Extraction of methylxanthines from guaraná seeds, maté leaves, and cocoa beans using supercritical carbon dioxide and ethanol, *J. Agric. Food Chem.*, 50, 4820–4826, 2002.
103. Viana, L.M., Freitas, M.R., Rodrigues, S.V. and Baumann, W., Extraction of lapachol from *Tabebuia avellanedae* wood with supercritical CO_2: An alternative to Soxhlet extraction? *Braz. J. Chem. Eng.*, 20, 317–325, 2003.
104. Pasquini, D., Pimenta, M.T.B., Ferreira, L.H. and Curvelo, A.A.S., Sugar cane bagasse pulping using supercritical CO_2 associated with co-solvent 1-butanol/water, *J. Supercrit. Fluids,* 34, 125–131, 2005.
105. Pereira, C.G. et al., Extraction of indole alkaloids from *Tabernaemontana catharinensis* using supercritical CO_2 + ethanol: An evaluation of the process variables and the raw material origin, *J. Supercrit. Fluids,* 30, 51–61, 2004.

106. Pereira, C.G., Leal, P.F., Sato, D.N. and Meireles, M.A.A., Antioxidant and antimycobacterial activities of *Tabernaemontana catharinensis* extracts obtained by supercritical CO_2 + cosolvent. *J. Med. Food,* 8, 533–538, 2005.
107. Chassagnez, A.L.M., Corrêa, N.C. and Meireles, M.A.A., Extração de oleoresina de cúrcuma (*Curcuma longa* L.) com CO_2 supercrítico, *Ciênc. Tecnol. Aliment.,* 17, 399–404, 1997.
108. Chassagnez-Méndez, A.L. et al., Supercritical CO_2 extraction of curcuminoids and essential oil from rhizomes of turmeric (*Curcuma longa* L.), *Ind. Eng. Chem. Res.,* 39, 4729–4733, 2000.
109. Braga, M.E.M., Leal, P.F., Carvalho, J.E. and Meireles, M.A.A., Comparison of yield, composition, and antioxidant activity of turmeric (*Curcuma longa* L.) extracts obtained using various techniques, *J. Agric. Food Chem.,* 51, 6604–6611, 2003.
110. Braga, M.E.M., Moreschi, S.R.M. and Meireles, M.A.A., Effects of supercritical fluid extraction on *Curcuma longa* L. and *Zingiber officinale* R. starches, *Carbohydrate Polymers,* 63, 340–346, 2006.
111. Moreschi, S.R.M., Leal, J.C., Braga, M.E.M. and Meireles, M.A.A., Ginger and turmeric starches hydrolysis using subcritical water + CO_2: The effect of the SFE pretreatment, *Braz. J. Chem. Eng.,* 23, 235–242, 2006.
112. Marques, D.S., Desterpenação de Óleo Essencial de Laranja por Cromatografia Preparativa de Fluido Supercrítico, MSc thesis, State University of Campinas, Campinas, Brazil, 1996.
113. Coimbra, P. et al., Experimental determination and correlation of artemisinin's solubility in supercritical carbon dioxide, *J. Chem.Eng. Data,* 51, 1097–1104, 2006.
114. de la Fuente, J.C., Quezada, N. and del Valle, J.M., Solubility of boldo leaf antioxidant components (Boldine) in high-pressure carbon dioxide, *Fluid Phase Equilibria,* 235, 196–200, 2005.
115. Kopcak, U. and Mohamed, R.S., Caffeine solubility in supercritical carbon dioxide/co-solvent mixtures, *J. Supercrit. Fluids,* 34, 209–214, 2005.
116. Carvalho, R.N., Jr., Corazza, M.L., Cardozo-Filho, L. and Meireles, M.A.A., Phase equilibrium for (camphor + CO_2), (camphor + propane), and (camphor + CO_2 + propane), *J. Chem. Eng. Data,* 51, 997–1000, 2006.
117. de la Fuente, J.C., Valderrama, J.O., Bottini, S.B. and del Valle, J.M., Measurement and modeling of solubilities of capsaicin in high-pressure CO_2, *J. Supercrit. Fluids,* 34, 195–201, 2005.
118. Ndiaye, P.M. et al., Phase behavior of soybean oil, castor oil and their fatty acid ethyl esters in carbon dioxide at high pressures, *J. Supercrit. Fluids,* 37, 29–37, 2006.
119. Souza, A.T. et al., Phase equilibrium measurements for the system clove (*Eugenia caryophyllus*) oil + CO_2. *J. Chem. Eng. Data,* 49, 352–356, 2004.
120. Moura, L.S., Corazza, M.L., Cardozo-Filho, L. and Meireles, M.A.A., Phase equilibrium measurements for the system fennel (*Foeniculum vulgare*) extract + CO_2. *J. Chem. Eng. Data,* 50, 1657–1661, 2005.
121. Corrêa, A.P.A., Gonçalves, L.A.G. and Cabral, F.A., Fractionation of fish oil with supercritical carbon dioxide, in *V Braz. Meet. Supercritical Fluids,* Ferreira, S.R.S., Ed., UFSC, Florianópolis, Brazil, 2004, CD-ROM.
122. Vieira de Melo, S.A.B., Melo, R.L.F.V., Costa, G.M.N. and Alves, T.L.M., Solubility of L-dopa in supercritical carbon dioxide: Prediction using a cubic equation of state, *J. Supercrit. Fluids,* 34, 231–236, 2005.
123. Corazza, M.L., Cardozo, L., Antunes, O.A.C. and Dariva, C., Phase behavior of the reaction medium of limonene oxidation in supercritical carbon dioxide, *Ind. Eng. Chem. Res.,* 42, 3150–3155, 2003.

124. Corazza, M.L., Filho, L.C., Antunes, O.A.C. and Dariva, C., High pressure phase equilibria of the related substances in the limonene oxidation in supercritical CO_2, *J. Chem. Eng. Data,* 48, 354–358, 2003.
125. Diaz, M.S., Espinosa, S. and Brignole, E.A., Optimal solvent cycle design in supercritical fluid processes, *Lat. Am. Appl. Res.,* 33, 161–165, 2003.
126. Diaz, S., Espinosa, S. and Brignole, E.A., Citrus peel oil deterpenation with supercritical fluids: Optimal process and solvent cycle design, *J. Supercrit. Fluids,* 35, 49–61, 2005.
127. Espinosa, S., Diaz, M.S. and Brignole, E.A., Process optimization for supercritical concentration of orange peel oil, *Latin Am. Appl. Res.,* 35, 321–326, 2005.
128. Saito, A.M., Peixoto, C.A. and Cabral, F.A., Phase equilibria modelling of palm fatty acid distillate (PFAD) and supercritical carbon dioxide, in *V Braz. Meet. Supercritical Fluids,* Ferreira, S.R.S., Ed., UFSC, Florianópolis, Brazil, 2004, CD-ROM.
129. Peixoto, C.A., França, L.F. and Cabral, F.A., Phase equilibria of palm fatty acid distillate (PFAD) and supercritical carbon dioxide, in *V Braz. Meet. Supercritical Fluids,* Ferreira, S.R.S., Ed., UFSC, Florianópolis, Brazil, 2004, CD-ROM.
130. Favero, F.W. and Skaf, M.S., Solvation of purine alkaloids in supercritical CO_2 by molecular dynamics simulations, *J. Supercrit. Fluids,* 34, 237–241, 2005.
131. Chafer, A., Fornari, T., Berna, A. and Stateva, R.P., Solubility of quercetin in supercritical CO_2 plus ethanol as a modifier: Measurements and thermodynamic modelling, *J. Supercrit. Fluids,* 32, 89–96, 2004.
132. Florusse, L.J., Fornari, I., Bottini, S.B. and Peters, C.J., Phase behavior of carbon dioxide–low-molecular weight triglycerides binary systems: Measurements and thermodynamic modeling. *J. Supercrit. Fluids,* 31, 123–132, 2004.
133. Favareto, R. et al., Supercritical fluid extraction from vetiver (*Vetiveria zizanioides*) roots: Kinetics and phase equilibrium, in *Proc. VII Iberoamerican Conference on Phase Equilibria and Fluid Properties for Process Design - EQUIFASE,* Galicia-Luna, L.A., Ed., Morelia, Mexico, 2006, CD-ROM.
134. Rissato, S.R., Galhiane, M.S., Knoll, F.R.N. and Apon, B.M., Supercritical fluid extraction for pesticide multiresidue analysis in honey: Determination by gas chromatography with electron-capture and mass spectrometry detection, *J. Chromotogr. A,* 1048, 153–159, 2004.
135. Grinberg, P., Campos, R.C., Mester, Z. and Sturgeon, R.E., Solid phase microextraction capillary gas chromatography combined with furnace atomization plasma emission spectrometry for speciation of mercury in fish tissues, *Spectrochim. Acta B,* 58, 427–441, 2003.
136. Paixao, J.A. and Campos, J.M., Determination of fat soluble vitamins by reversed-phase HPLC coupled with UV detection: A guide to the explanation of intrinsic variability, *J. Liq. Chromatogr.,* 26, 641–663, 2003.
137. Godoi, A.F.L., Favoreto, R. and Santiago-Silva, M., Environmental contamination for organotin compounds, *Quim. Nova,* 26, 708–716, 2003.
138. Andrade, R., Reyes, F.G.R. and Rath, S., A method for the determination of volatile N-nitrosamines in food by HS-SPME-GC-TEA, *Food Chem.,* 91, 173–179, 2005.
139. Godoi, A.F.L., Montone, R.C. and Santiago-Silva, M., Determination of butyltin compounds in surface sediments from the Sao Paulo State coast (Brazil) by gas chromatography-pulsed flame photometric detection, *J. Chromatogr. A,* 985, 205–210, 2003.
140. Rissato, S.R., Galhiane, M.S., Apon, B.M. and Arruda, M.S.P., Multiresidue analysis of pesticides in soil by supercritical fluid extraction/gas chromatography with electron-capture detection and confirmation by gas chromatography-mass spectrometry, *J. Agr. Food Chem.,* 53, 62–69, 2005.

141. Rissato, S.R., Galhiane, M.S., de Souza, A.G. and Apon, B.M., Development of a supercritical fluid extraction method for simultaneous determination of organophosphorus, organohalogen, organonitrogen and pyretroids pesticides in fruit and vegetables and its comparison with a conventional method by GCECD and GCMS, *J. Braz. Chem. Soc.*, 16, 1038–1047, 2005.
142. Richter, P. et al., Screening and determination of pesticides in soil using continuous subcritical water extraction and gas chromatography-mass spectrometry, *J. Chromatogr. A*, 994, 169–177, 2003.
143. Zougagh, M., Redigolo, H., Rios, A. and Valcarcel, M., Screening and confirmation of PAHs in vegetable oil samples by use of supercritical fluid extraction in conjunction with liquid chromatography and fluorimetric detection, *Anal. Chimica Acta,* 525, 265–271, 2004.
144. Arancibia, V. et al., Quantitative extraction of sulfonamides in meats by supercritical methanol-modified carbon dioxide: A foray into real-world sampling, *J. Sep. Sci.,* 26, 1710–1716, 2003.
145. Arancibia, V., Valderrama, M., Silva, K. and Tapia, T., Determination of chromium in urine samples by complexation-supercritical fluid extraction and liquid or gas chromatography, *J. Chromatogr. B,* 785, 303–309, 2003.
146. Arancibia, V. et al., Extraction of nitrofurantoin and its toxic metabolite from urine by supercritical fluids. Quantitation by high performance liquid chromatography with UV detection, *Talanta,* 61, 377–383, 2003.
147. Rosa, P.T.V. and Meireles, M.A.A., Rapid estimation of the manufacturing cost of extracts obtained by supercritical fluid extraction, *J. Food Eng.*, 67, 235–240, 2005.
148. Turton, R.C. Bailie, Whiting, W.B. and Shaeiwitz, J.A., *Analysis, Synthesis, and Design of Chemical Process,* Prentice Hall, Upper Saddle River, 1998.
149. Sovová, H., Rate of the vegetable oil extraction with supercritical CO2 I. Modeling of extraction curves. *Chem. Eng. Sci.,* 49, 409–414, 1994.
150. del Valle, J.M. and de la Fuente, J.C., Supercritical CO_2 extraction of oilseeds: Review of kinetic and equilibrium models, *Crit. Rev. Food Sci. Nutrit.,* 46, 131–160, 2006.
151. Ruetsch, L., Daghero, J. and Mattea, M., Supercritical extraction of solid matrices. Model formulation and experiments, *Latin Ame. Appl. Res.,* 33, 103–107, 2003.
152. Hammerschmidt, P.A. and Pratt, D.E., Phenolic antioxidants of dried soybeans, *J. Food Sci.,* 43, 556–559, 1978.
153. Freud, R.J. and Little, R.C., *SAS System for Regression, SAS Series in Statistical Applications, 2nd Ed.*, SAS Institute, Cary, NC, 1995, 211.
154. Rodrigues, V.M. et al., Supercritical extraction of essential oil from aniseed (*Pimpinella anisum* L) using CO_2: Solubility, kinetics, and composition data, *J. Agric. Food Chem.,* 51, 1518–1523, 2003.
155. Germain, J.C., del Valle, J.M. and de la Fuente, J.C., Natural convection retards supercritical CO_2 extraction of essential oils and lipids from vegetable substrates, *Ind. Eng. Chem. Res.,* 44, 2879–2886, 2005.
156. Herek, L.C.S., Oliveira, R.C., Rubira, A.F. and Pinheiro, N., Impregnation of PET films and PHB granules with curcumin in supercritical CO_2, *Braz. J. Chem. Eng.,* 23, 227–234, 2006.
157. Moreschi, S.R.M., Petenate, A.J. and Meireles, M.A.A., Hydrolysis of ginger bagasse starch in subcritical water and carbon dioxide, *J. Agr. Food Chem.,* 52, 1753–1758, 2004.
158. Castelan, L.H. et al., Extraction of lemongrass essential oil with dense carbon dioxide, *J. Supercrit. Fluids,* 21, 33–39, 2001.

159. Pokrywiecki, J.C. et al., Separation of active principles from the essential oil of medicinal plants with supercritical carbon dioxide and reverse osmosis membrane, in *V Braz. Meet. Supercritical Fluids,* Ferreira, S.R.S., Ed., UFSC, Florianópolis, Brazil, 2004, CD-ROM.
160. Pereira, C.G. and Meireles, M.A.A., Manufacturing cost of essential oils obtained by supercritical fluid extraction, in *Proceedings of the Eighth Conference on Supercritical Fluids Application,* Reverchon, E., Ed., ISASF, Ischia, Italy, 2006, 77–82.
161. Pereira, C.G., Rosa, P.T.V. and Meireles, M.A.A., Extraction and isolation of indole alkaloids from *Tabernaemontana catharinensis* A.DC: Technical and economical analysis, *J. Supercrit. Fluids,* 40, 232–238, 2006.

9 Antioxidant Extraction by Supercritical Fluids

Beatriz Díaz-Reinoso, Andrés Moure, Herminia Domínguez, and Juan Carlos Parajó

CONTENTS

9.1 Introduction ... 275
9.2 Types of Antioxidants and Regulation Aspects ... 276
9.3 Natural Antioxidants and Sources ... 277
9.3.1 Phenolic Compounds ... 280
9.3.2 Terpenoids ... 280
9.3.3 Carotenoids ... 281
9.3.4 Vitamin E ... 281
9.3.5 Other Natural Antioxidants ... 282
9.4 Biological Properties of Antioxidant Compounds ... 282
9.4.1 Phenolic Compounds ... 282
9.4.2 Terpenoids ... 283
9.4.3 Carotenoids ... 284
9.4.4 Vitamin E ... 284
9.4.5 Antioxidant Properties of SC-CO_2 Extracts ... 284
9.5 Determination of Antioxidant Activity ... 285
9.6 Supercritical-CO_2 Extraction of Antioxidants ... 286
9.6.1 Processing Schemes ... 287
9.6.2 Effects of the Most Influential Operational Variables ... 288
9.6.2.1 Pressure and Temperature ... 289
9.6.2.2 Modifier ... 290
9.6.3 SC-CO_2 Extracts versus Conventional Solvent Extracts ... 292
References ... 293

9.1 INTRODUCTION

According to a widely used definition, an *antioxidant* is any substance that, when present at lower concentrations than those of an oxidizable substrate (such as lipids, proteins, deoxyribonucleic acid [DNA] or carbohydrates), significantly delays or prevents oxidation of that substrate [1, 2]. Neither this definition nor other definitions [3] restrict antioxidant activity to a specific group of compounds or to any particular mechanism of action. Natural antioxidants play a decisive role in different systems: *i*) in plants, they act as protecting agents against radiation or microbial infections, *ii*) in foods, they delay or inhibit the formation of toxic lipid oxidation products, maintaining nutritional quality and increasing shelf life, and *iii*) in biological systems, along with endogenous defenses (enzymes, vitamins, proteins, and others), dietary antioxidants may help prevent or slow the oxidative stress induced by free radicals [4]. Since considerable evidence indicates that oxidative damage may contribute to the development of age-related and degenerative diseases, the protective effects of beneficial compounds have been ascribed to their antioxidant activity, although many antioxidants *in vivo* probably act by other mechanisms than *in vitro* assays or are unlikely to have such effects at the concentrations available in plasma [5, 6].

Due to an increasing consumer demand to replace controversial synthetic antioxidants, such as butylated hydroxytoluene (BHT), butylated hydroxyanisole (BHA), tertiary butyl hydroquinone (TBHQ), and gallates, the preservation of foods is a promising application of natural antioxidants, which could confer additional biological activities to the products. Although natural antioxidants are assumed to be safe and innocuous, their lack of toxicity should be confirmed.

Great effort is being devoted to the search for alternative and cheap sources of natural antioxidants, as well as to the development of efficient and selective extraction techniques. Extraction with conventional solvents is sometimes characterized by poor selectivity and requires high temperatures, which could result in degradation of the desired compounds. Supercritical fluid extraction (SFE) is more selective than conventional extraction and is optimal when products free from residual solvents are required (for example, for food, cosmetic, and pharmaceutical purposes). Carbon dioxide (CO_2) is the most suited solvent for SFE of thermolabile compounds, owing to its nontoxic and nonflammable character and high availability at low cost and high purity, allowing an optimal reproduction of the physicochemical, biological, and therapeutic properties of the target compounds. Supercritical CO_2 (SC-CO_2) extracts are regarded as "natural"; are free from pathogenic and spoilage microorganisms, spores, and enzymes; the absence of light and oxygen prevents oxidation reactions. Future developments in extraction of antioxidants will probably be related to SFE [7], which is well positioned with respect to increasingly restrictive environmental, toxicological, and health regulations.

Theoretical and practical aspects of the SFE of compounds with recognized antioxidant activity have been revised [8–11] and particularized for the extraction of antioxidants [12–14].

9.2 TYPES OF ANTIOXIDANTS AND REGULATION ASPECTS

Lipid oxidation is important in food deterioration—for example when oxygen reacts with lipids in a series of free radical chain reactions [15] or in the oxidative modification of low-density lipoproteins (LDLs). According to the free radical theory of aging, various oxidative reactions occurring in the organism (mainly in mitochondria) generate free radicals as by-products, which damage nucleic acids, proteins, and lipids and result in aging and age-associated pathologies. The stages of the classical nonenzymatic free radical–mediated chain reactions are: 1) initiation (by heat, light, ionizing radiation, metal ions, or metalloproteins), 2) propagation,

3) branching, and 4) termination. The main features of the mechanisms of lipid oxidation and antioxidant action have been detailed in the literature [3, 4, 15–19].

Antioxidants have traditionally been divided into two groups: primary and secondary. Primary antioxidants (such as phenolic compounds or vitamin E) are destroyed during the induction period, when they delay or inhibit the initiation step by reacting with radicals. Secondary, or preventative, antioxidants slow the oxidation rate, removing substrate by binding oxygen from air, complexing with transition metal ions (acetates, citrates, tartrates, and phosphates), quenching singlet oxygen, binding certain proteins with prooxidant effects, absorbing ultraviolet (UV) radiation or (in the case of phospholipids) creating a protective layer between oil and air surface. Antioxidants can act according to several mechanisms, and synergism among different oxidation inhibitors can occur [15, 17].

Other nonmechanistic classifications have been established for antioxidants. For example, according to their origin, they can be classified as natural products, natural identical (α-tocopherol), or artificial. Depending on their chemical structure, antioxidants have been grouped into phenolics (BHA, BHT, TBHQ, gallates), quinones (hydroquinone, tocopherols, hydroxychromanes, hydroxycoumarins), organic acids (ascorbic, citric, tartaric, and lactic acids and their salts and ethylenediaminetet raacetic acid and its salts), sulfur compounds (inorganic: sulfites, bisulfites, and metasulfites; organic: methionine, cisteine), and enzymes (catalases, peroxidases, superoxide dismutase). The natural antioxidants found within biological systems include four general groups: enzymes, large molecules (albumin, ceruloplasmin, ferritin, other proteins), small molecules (ascorbic acid, glutathione, uric acid, tocopherol, carotenoids, polyphenols), and some hormones (estrogen, angiotensin, melatonin) [20].

Technological requirements for food antioxidants include low volatility and stability (to avoid losses during processing and storage), ability to protect from oxidation at low concentrations, solubility and compatibility with other components of the oxidizable substrate, nontoxic and nonirritant character at the effective concentration, and ability to not confer color, odor, or taste to the final product. Food utilization of synthetic antioxidants such as BHT, BHA, and gallates is not permitted in the European Union (EU) for some special foods, such as those for infants and young children [21, 22], and it is generally restricted to levels that depend on the considered application. However, antioxidants from natural origins (such as spices) do not need to be declared and are allowed at higher doses [12, 17]. Table 9.1 summarizes data on the major food antioxidants according to the EU and U.S. regulations, as well as maximum levels of acceptable daily intake (ADI) established by the *Codex Alimentarius*.

9.3 NATURAL ANTIOXIDANTS AND SOURCES

The most studied antioxidants extractable from vegetal biomass by SFE with CO_2 are phenolics, terpenoids, carotenoids, and tocopherols. As the available data show a comparatively higher antioxidant activity for phenolics, a more detailed discussion is provided for these compounds.

TABLE 9.1
Major Food Antioxidants and Their Maximum Level in Different Products According to European and U.S. Regulations

		EU REGULATION		U.S. REGULATION	
Name	**ADI (mg/kg)**	**Food**	**Maximum Level**	**Standardized Food**	**Maximum Level (mg/kg)**
Tocopherols (Toc)	0.15–0.2		*Quantum satis*	Animal fat or animal and vegetable fat mixture	300
Tocopherol rich–extract, α-tocopherol, γ-tocopherol, β-tocopherol		Nonemulsified oils and fat (except virgin oils and olive oils)	*Quantum satis*	Sausages and related products, meat products	300
		Refined live oil, except olive pomace oil	200 mg/l (α-toc)	Frozen raw breaded shrimp	200
		Nonemulsified oils and fat (except virgin oils and olive oils)	*Quantum satis*	Bacon (pump-cured)	500 (α-tocopherol)
				Poultry and poultry products	300 (200 + others except TBHQ)
BHT	0.125	Fats and oils (for heat treated foodstuffs)	200 mg/kg (G's and BHA singly or combined)	Dehydrated potato shreds	50 (BHA, BHT)
BHA	0.5			Active dry yeast	1000 (BHA)
				Dry mixes for beverages and desserts	2 (BHA)
Gallates (G's)					
- Propyl gallate (PG)	2.5	Oil and fat to fry (except olive pomace oil); lard; fish oil; beef, poultry and sheep fat	100 mg/kg (BHT)	Dry breakfast cereals	50 (BHA, BHT)
- Octyl gallate (OG)	—			Dry diced glazed fruit	32 (BHA)
- Dodecyl gallate (DG)	—			Dry mixes for beverages and desserts	90 (BHA)

Cake mixes, cereal-based snacks, milk powder, dehydrated meat, seasonings, processed nuts, precooked cereals	200 mg/kg (G's and BHA singly or combined)	Shortenings emulsion stabilizers	200 (BHA, BHT)
		Potato and sweet potato flakes	50 (BHA, BHT)
		Potato granules	10 (BHA, BHT)
		Sweet potato flakes	50 (BHA, BHT)
		Chewing gum	1000 (BHA, BHT, PG)
Dehydrated potatoes	25 mg/kg (G's and BHA, singly or combined)	Animal fat or animal and vegetable fat mixture	100 (BHA, BHT, PG) or 200
		Margarine	200
		Frozen raw breaded shrimp	200 (BHA, BHT)
		Dried meats	100 (BHA, BHT, PG)
		Dry sausage	30 singly or 60 (BHA, BHT, PG)
		Fresh pork, sausages and sausage products, meat products	100 singly or 200 (BHA, BHT, PG)
		Various poultry products	100 (BHA,BHT) or 200
		Flavoring substances	5000 (BHA)

9.3.1 Phenolic Compounds

Phenolic compounds are one of the main classes of secondary metabolites in plants, responsible for color development, pollination, and protection against UV radiation and pathogens. In foods, these compounds contribute to sensory properties (color, astringency). Phenolics refer to monomeric, oligomeric, or polymeric compounds with an aromatic ring bearing one or more hydroxyl substituents and functional derivatives (esters, methyl ethers, glycosides, etc.).

Phenolics include simple phenols, coumarins, flavonoids, stilbenes, lignans, and hydrolyzable and condensed tannins. Flavonoids (a large and complex group of compounds containing a three-ring structure with two aromatic centers and a central oxygenated heterocycle) are common antioxidants. The six major subclasses of flavonoids are flavones, flavonols, flavanones, catechins or flavanols, anthocyanidins, and isoflavones. Most flavonoids present in plants are conjugated with sugars, although occasionally they are found as aglycons [23]. More than 4,000 different naturally occurring flavonoids have been discovered, and more than 36,000 different flavone structures are possible.

Phenolic compounds have powerful antioxidant activities *in vitro* [24], based on their structure, hydrogen-donating potential, and ability to chelate metal ions. They may show higher efficacy than endogenous or synthetic antioxidants [25]. Their antioxidant activity [26–28] and their structure-activity relationships have been examined [17, 29–32].

The most-studied sources of phenolic antioxidants are fruits and vegetables [33–36], grains and cereals [37], and teas [38, 39]. Agricultural and industrial wastes are renewable, cheap, and highly available sources of phenolic antioxidants.

9.3.2 Terpenoids

Terpenoids, also known as *isoprenoids,* are secondary plant metabolites accounting for the largest family of natural compounds, widespread in plants and lower invertebrates. The isoprenoid biosynthetic pathway generates primary and secondary metabolites of ecological relevance to plant growth and survival. These compounds are involved in interactions between plants, between plants and microorganisms, and between plants and insects, acting as allelopathic agents and attractants or repellants in plants [40]. They are involved in the defense, wound sealing, and thermotolerance of the plants as well as in the pollination of seed crops, the flavor of fruits, and the fragrance of flowers, determining the quality of agricultural products. Some terpenoids or their precursors act as scavengers for external aggressive molecules in the gaseous phase (i.e., ozone). The term *terpenes* is used for a group of compounds with the basic C_5 isoprene unit. According to the number of these units (1 to 6), terpenoids are classified into hemiterpenoids, monoterpenoids (C_{10}) (limonene, carvone, carveol); sesquiterpenoids (C_{15}); diterpenoids (C_{20}) (retinoids); sesterterpenoids (C_{25}); tri- (C_{30}); and tetraterpenoids (carotenoids), having eight isoprenoid C_5 residues.

Terpenoid compounds (monoterpenes, sesquiterpenes, and diterpenes) are the main components of essential oils, which also contain oxygenated derivatives and other compounds (including aldehydes, ketones, phenolic, acetates, and oxides).

The antioxidant activity of different essential oils in different model systems is well known [41, 42], and synergistic effects with phenolics have been reported [40, 43]. Essential oils are the commercial sources of terpenoids, whereas enzymes and extracts from bacteria, cyanobacteria, yeasts, microalgae, fungi, plants, and animal cells have also been used for the production and bioconversion of terpenes. Their biotechnological transformations appear especially promising because applications such as fragrances and flavors in cosmetics and foods depend on the absolute configuration (different enantiomers present different properties).

9.3.3 Carotenoids

Carotenoids are a group of more than 600 different compounds, with isoprenoid (tetraterpenoid) structure, synthesized by plants, photosynthetic organisms, and some nonphotosynthetic bacteria, yeasts, and molds. They can be found as pigments in fruits, flowers, and animal species (birds, insects, fish, and crustaceans) and play an important role in the protection against photooxidative damage. Most carotenoids are composed of a central carbon chain of alternating single and double bonds (3 to 15 conjugated double bonds) with different cyclic or acyclic end groups. They are classified as carotenes (α- and β-carotene, lycopene), composed only of carbon and hydrogen atoms, or xanthophylls (zeaxanthin, lutein, α- and β-cryptoxanthin, canthaxanthin, astaxanthin), with at least one oxygen atom. Carotenoids predominantly occur in their all-*trans* configuration, although *cis*-isomers can be formed during food processing [44]. Lycopene exhibits the highest antioxidant activity, and its plasma level is slightly higher than that of β-carotene [45]. The results reported for the antioxidant activity of β-carotene differ widely due to the various test systems and the experimental conditions used [46]. The conjugated double-bond system is responsible for the antioxidant properties of carotenoids, which can act by quenching singlet oxygen formed due to the effects of UV light, scavenging peroxyl radicals, hydrogen transfer, or electron transfer [47–49].

Major sources of lycopene include tomatoes, rosehip, apricots, guavas, watermelons, papayas, and pink grapefruits; α-carotene is found in carrots, tomatoes, and green vegetables; β-carotene is present in the same materials as α-carotene as well as in paprika and sweet potatoes; β-cryptoxanthin is present in mangos, papaya, peaches, paprika, oranges, lutein in bananas, egg yolks, spinach, parsley, and marigold flowers; zeaxanthin in paprika; astaxanthin in salmon, the yeast *Phaffia rhodozyma,* and the algae *Haematococcus pluvialis*; and canthaxanthin in carrots.

9.3.4 Vitamin E

Vitamin E includes a family of tocopherols (having a phytyl tail attached to their chromanol nuclei), tocotrienols (with an unsaturated tail), and some of their ester derivatives (such as succinate and acetate). Vitamin E effectively inhibits the peroxidation of lipids because it can scavenge the peroxyl radicals. The radical-scavenging capacity of α-tocopherol and α-tocotrienol is similar in hexane, but α-tocotrienol is more active in membrane systems and α-tocopherol shows higher bioactivity. The major sources of vitamin E are plant species, and its content varies between tissues, with preferential accumulation in seeds. Due to their amphipathic nature, tocopherols are associated with membrane lipids or lipid storage structures. Vitamin E is the most important natural antioxidant in vegetable oil–derived foods, found in rice bran, palm oil, and wheat germ [50]. The richest source is a by-product of soybean processing (the oil deodorizer distillate).

9.3.5 Other Natural Antioxidants

Other compounds with antioxidant activity suitable as food additives are peptides and proteins [51, 52], Maillard products [53, 54], oligosaccharides, sugars and polyols [55], and microbial metabolites [56].

9.4 BIOLOGICAL PROPERTIES OF ANTIOXIDANT COMPOUNDS

9.4.1 Phenolic Compounds

A variety of biological effects have been reported for phenolic acids, including alleviation of hyperuricemia and protection against LDL oxidation, anti-inflammatory, antitumor, and autoimmune-related effects [57–61]. Caffeic and ferulic acids provide protection against carcinomas [62], ferulic acid esters protect against UV radiation [63], and trans-cinnamic acid can be used in the prevention or treatment of diabetes [64].

Research in flavonoids has increased since the discovery of the low cardiovascular mortality rate in Mediterranean populations that is associated with red wine consumption and high dietary saturated fat intake ("French paradox"). Their strong antioxidant power makes flavonoids able to quench free radicals and to act against the oxidation of LDLs, attenuating the development of atherosclerosis, reducing thrombosis, and promoting normal endothelial function [65–68]. Flavonoids are excellent candidates as health-promoting, disease-preventing, and chemopreventive agents because they are extremely safe and associated with low toxicity [69, 70]. Protective action has been postulated for chronic diseases [71, 72], cardiovascular diseases [71, 73–77], stroke [77], hyperlipidemia [71], diabetes [74], inflammation [74, 78–81], allergies [74, 78, 79], immune system disorders [72, 82], mutagenesis [74, 83], and cataracts [72] as well as for neurological disorders [72,73, 84], particularly those related to aging, such as cognitive, motoric, and mood decline [85]. Flavonoids have been recognized to exert a variety of biological activities (including estrogenic, antimicrobial, antiviral, and analgesic) [78, 79] and to have hepatoprotective, cytostatic, and apoptotic properties [79]. Some of these protective effects have been confirmed by epidemiological studies [75, 76, 83, 86]. All inflammatory processes include oxygen-activating processes that produce reactive oxygen species, and free radical scavengers or quenchers of activated states warrant metabolic control within certain limits. Cardiovascular disease is related to inflammation and, consequently, is amenable to intervention via molecules with anti-inflammatory effects [67]. With regard to the immune system, flavonoids may preserve T cell–mediated immunity [82]. Flavonoids in the human diet may reduce the risk of various cancers, including hormone-dependent breast and prostate cancers [79], intestinal neoplasia [83], and skin cancer [87].

In vitro flavonoids can bind electrophils, inactivate oxygen radicals, prevent lipid peroxidation, and inhibit DNA oxidation. In cell cultures, they increase the rate of apoptosis, and inhibit bot cell proliferation, and angiogenesis [83], but a direct extrapolation to humans cannot be made on the basis of these data. Flavonoids are present in the diet as glycosylated, esterified, or polymerized derivatives, and human intervention studies have provided evidence that flavonoids are partly absorbed. Due to the low assimilation rate and the high concentrations present, significant flavonoid intake might result in direct effects within the gastrointestinal tract, such as binding of prooxidant iron; scavenging of reactive nitrogen, chlorine, and oxygen species; and perhaps inhibition of cyclooxygenases and lipoxygenases [6].

A growing body of *in vivo* studies is beginning to provide insight into the biological mechanisms of flavonoid action [77]. The nature of polyphenol conjugates *in vivo* has been identified, showing that the biological fate of flavonoids, including their dietary forms, is highly complex and dependent on a large number of processes [88]. The forms reaching the blood and tissues are, in general, neither aglycons (except for green tea catechins) nor the same as the dietary source. As a consequence, the polyphenol conjugates are likely to possess different biological properties and distribution patterns within tissues and cells than polyphenol aglycons. On the other hand, polyphenol concentrations tested should be of the same order as the maximum plasma concentrations achieved after a polyphenol-rich meal [89]. The biological effects of these polyphenols depend on the extent and way in which the circulating metabolites interact with and associate with cells [73].

Antioxidant properties alone are not sufficient to explain the biological properties of flavonoids. Within the last decade, reports on flavonoid activities have been largely associated with enzyme inhibition and antiproliferative activity, which are dependent on particular structures [75]. Although the action mechanisms are not fully understood, recent studies have clearly shown that the role of flavonoids as modulators of cell signalling may be attributed to their effects as anticancer agents, cardioprotectants, and inhibitors of neurodegeneration [90]. Certain flavonoids, especially flavone derivatives, express their anti-inflammatory activity at least in part by modulation of proinflammatory gene expression [80, 81]. The potential neuroprotective effects of dietary flavonoids and their role in modulating oxidative stress may be related to cell signalling cascades, gene expression, and down-regulation of pathways leading to cell death and neuronal apoptosis [85, 91].

9.4.2 Terpenoids

Some terpenoids are the bioactive compounds of traditional herbal remedies used in the treatment of pain, colds, bronchitis, and gastrointestinal diseases. Terpenoids are present in almost every natural food and have been associated with protection from oxidative stress and chronic diseases [92]. Some exhibit cardioprotective action, such as ginkgolides A and B and bilobalide from *G. biloba* [93]. Other relevant properties have been reported, including antibacterial [94], anti-inflammatory [95], anticarcinogenic [40], antimalarial, antiulcer, antimicrobial, and diuretic activities. Protection against a variety of infectious diseases (viral and bacterial) and acaricidal activity have been reported for monoterpenes [96]. The present commercial importance of terpene-based pharmaceuticals is expected to play a more significant role in human disease treatment in the future [97].

9.4.3 Carotenoids

The major biological functions of carotenoids are related to intercellular gap junction communication, cell differentiation, immunoenhancement, and inhibition of mutagenesis. Some carotenoids (α- and β-carotene, β-cryptoxanthin) are precursors of vitamin A and protect against chemical oxidative damage, several kinds of cancer, and age-related macular degeneration. No convincing evidence exists of their protective action against cardiovascular disease [47, 48, 98–100]. *In vitro* studies evidenced that carotenoids can interact with several reactive species and can act as prooxidants, although no documented evidence to date indicates true prooxidant activity *in vivo* [101]. The maximum antioxidant effectiveness of carotenoids in human cells is related to an optimal dose, because higher doses can be less effective or result in cell damage. The relationship between carotenoid intake and cancer has been evaluated, showing an inverse association for lung, colon, breast, and prostate cancer, although negative effects of supplementations have been found [49] and it is not clear if the association between diet and disease is due to the specific carotenoid, other micronutrients present in the specific diet, or the combined effect of several of these active ingredients. Studies on the mechanism of cancer cell growth inhibition by carotenoids at the protein expression level may involve changes in pathways leading to cell growth or cell death, including hormone and growth factor signaling, regulatory mechanisms of cell cycle progression, cell differentiation, and apoptosis [102].

9.4.4 Vitamin E

The actions of tocopherols and tocotrienols have been extensively studied. Vitamin E protects vitamin A, spares selenium and vitamin C, and is the most effective lipid-soluble antioxidant, which protects unsaturated fatty acids in membranes. Other non-antioxidant functions include enhanced immune response and regulation of platelet aggregation [50, 103]. The effects of Vitamin E have been observed at the level of messenger ribonucleic acid (mRNA) or protein and could be related to regulation of gene transcription, mRNA stability, protein translation, and protein stability. Landvik et al. [103] published a compilation of human epidemiological studies on vitamin E, carotenoids, and cancer risk. This vitamin also protects against coronary heart disease [104], aging, cataracts, UV radiation, air pollution, and lipid peroxidation associated with strenuous exercise. Vitamin E bioavailability and metabolism is influenced by intestinal absorption, plasma lipoprotein transport, and hepatic metabolism [105]. Different distribution of vitamin E isoforms in tissues has been reported, being an essential part of the antioxidant defense systems, particularly in the skin, where tocotrienols are preferentially distributed. Tocotrienols are more effective than tocopherols at inhibiting neuronal cell death. It has been suggested that neither the anticarcinogenic effects of tocotrienols nor the neuroprotection are related to the antioxidant properties of tocopherols and tocotrienols [50].

9.4.5 Antioxidant Properties of SC-CO_2 Extracts

Since SFE is a relatively novel application, studies on the biological properties of these extracts will probably increase in the future. Antimicrobial activity of several extracts has been observed for white grape seed fractions [106], spices [107], and marjoram [108]. Protection from ischemic damage was reported for cocoa hull extracts [109]. Antimutagenic and antineoplastic properties have been claimed for SC-CO_2 extracts of plant and spices [107, 110] as well as antimutagenicity for *Terminalia catappa* leave extracts [111].

9.5 DETERMINATION OF ANTIOXIDANT ACTIVITY

The activity of natural antioxidants, which are mixtures or multifunctional systems acting in complex media, cannot be evaluated satisfactorily by a simple test, and contradictory results using different assays have been reported. A comparative evaluation of antioxidants is difficult because, in foodstuffs and biological systems, the activity depends on the substrate, the medium, the oxidation conditions, interfacial phenomena, and the partitioning properties of the antioxidant between phases. The affinity of antioxidants toward air, oil, water, and interfaces explains why polar antioxidants are more active in bulk oils and nonpolar antioxidants are more active in emulsions, a behavior known as the "polar paradox." The need for approved, standardized methods is especially important for comparing food or nutraceuticals in order to provide quality criteria for regulatory issues and health claims. Evaluation of the antioxidant activity at different levels has been suggested [112], including: *i*) quantification and identification of the active compounds, *ii*) evaluation of the radical scavenging activity with more than one method in different solvents, *iii*) evaluation of protection against lipid oxidation in model systems, and *iv*) studies of relevance for food applications and human studies with markers for oxidative stress.

Many *in vitro* methods are performed in the absence of lipids and the partitioning of antioxidants is not evaluated, or these methods do not predict the ability to inhibit oxidation of foods or in biological systems. More realistic information can be achieved by performing several tests and following some general recommendations: *i*) substrates and oxidation conditions should simulate chemical, physical, and environmental conditions in food or biological systems; *ii*) low levels of oxidation should also be considered; *iii*) both initial and secondary products should be measured; and *iv*) the concentrations of catalyst, antioxidants, and substrates should be carefully established and the compositional data should be known to compare samples [3, 16, 18, 20, 113, 114]. The results are also influenced by the specificity and methods used to analyze the progress of oxidation and by the degree of oxidation chosen as end-point for testing [16, 17, 113, 114]. Accelerated oxidation of oils, fats, oil-water emulsions, and muscle foods are relevant during food processing or domestic use [114]. However, under some testing conditions (temperature, partial pressure of oxygen, metal catalysts and other initiators, light or UV radiation), the oxidation mechanisms may change [18]. Methods of expressing antioxidant activity, summarized by Antolovich et al. [18], include the induction period, percentage inhibition of rates, IC_{50} (concentration to achieve 50% inhibition), and scale readings (absorbance, conductivity).

As free radical generation is directly related to oxidation, various methods have been developed based on the ability to scavenge free radicals [5, 19, 115, 116]. Huang et al. [3] and Prior et al. [20] compared the performance and biological relevance of different methods and established a classification based on the mechanisms: hydrogen atom transfer (HAT, measuring the ability to quench free radicals by hydrogen donation), single electron transfer (SET, measuring the ability to transfer one electron to reduce any compound), or a combination of HAT and SET. Among the methods based on HAT, the most frequently employed are the Oxygen Radical Absorbance Capacity (ORAC) and the Total Peroxyl Radical-Trapping Antioxidant Parameter (TRAP). By changing the oxidant sources of peroxyl radicals, the reaction can differentiate quenching of specific oxidants ($O_2^{\cdot-}$, $HO^{\cdot}$, HOCl, $LO(O)^{\cdot}$, $^{\cdot}OONO$, and $^{\cdot}O_2$). Other HAT tests and their target applications are the Total Oxidant Scavenging Capacity test (TOSC, toward hydroxyl radicals, peroxyl radicals, and peroxynitrite); carotenoids bleaching via autoxidation (toward oxidation induced by light or heat or oxidation induced by peroxyl radicals); and LDL Oxidation initiated by Cu (II) or AAPH (with relevance to oxidative reactions that might occur *in vivo*). Methods based on SET reactions are the Ferric Reducing Antioxidant Power (FRAP) and Copper Reduction Assay (CUPRAC). The most frequently employed methods based on both HAT and SET mechanisms are Trolox Equivalent Antioxidant Capacity (TEAC) and DPPH (2,2-diphenyl-1-picrylhydrazyl radical). Both of them are operationally simple and widely used, although the radical anion $ABTS^{\cdot+}$ used in the first is not found in human biology, and the second has several drawbacks [20]. The Folin-Ciocalteu test, used to quantify phenolic content, also measures the effective oxidation/reduction efficiency of all the antioxidants present in the medium.

Extrapolation of antioxidant mechanisms established in food or model systems to *in vivo* situations is not direct. Bioavailability and metabolism of antioxidants must be addressed to know if these compounds reach target tissues because their biological effects may be affected by a variety of factors, including digestion, absorption, metabolism, and the presence of competitive enzymes and other antioxidants or prooxidants. Although *in vitro* assays do not reflect the cellular physiology, metabolism and *in vivo* assays (with animals or humans) are less suited for initial screening of antioxidants than cell culture models because they are expensive and time-consuming [117]. However, cells in culture behave differently from those *in vivo* due to the "culture shock" and to the oxidative stress caused by the process [118]. Apart from the critical general works on the analytical methods to determine antioxidant activity [3, 16, 17, 18, 20, 112, 113, 116], specific revisions concerning food applications [19, 114, 115] have been published. The available methods to measure free radicals and other reactive (oxygen [ROS]/nitrogen/chlorine) species contributing to the development of several diseases by oxidative damage have been revised [4, 119].

9.6 SUPERCRITICAL-CO_2 EXTRACTION OF ANTIOXIDANTS

Depending on the physical state (solid or liquid) of the phase containing the target compounds, SFE can involve solid-liquid or liquid-liquid mass transfer. Solid-liquid extraction is a heterogenous operation involving the transfer of solutes from the vegetal matrix to a fluid. The extraction rate depends on the external mass transfer, effective solute diffusivity in the solid, solute solubility in the solvent, and solute binding to the solid matrix. Batch extraction and semicontinuous extraction are the most commonly used experimental methods. Extraction by solvent flow through a fixed bed of solid particles allows the recovery of fractions obtained along the extraction period. When a liquid stream has to be processed by SFE, both solubility and interphase mass transfer are relevant. Operation is similar to extraction with conventional solvents, and continuous operation can be carried out in single-stage or multistage contact (cross-flow or countercurrent).

9.6.1 Processing Schemes

Different processing schemes have been proposed for SFE of compounds from natural sources. Figures 9.1a to 9.1d present simplified flow diagrams of the most usual alternatives, including:

1. Single extraction stage and fractional separation in several separators. The extract obtained in a single extraction step can be fractionated by releasing pressure in the separators. This disposition is widely used for processing solids and for analytical purposes [120–122].
2. Stagewise extraction at progressively increased severity. After a first stage at low severity (< 15 MPa, no modifier) to extract nonpolar compounds (essential oil and waxes), further SFE of the solid residue is performed at increased severity (up to 50 MPa, 40% modifier) to extract more polar antioxidants [123, 124]. Stepwise extraction needs more solvent than simple extraction with stagewise fractionation of extracts [12, 125], although the extraction yields can be similar.
3. Combination of conventional solvent and SFE of solid samples. A first SFE stage under low severity conditions can be performed to remove volatile compounds and waxes from the solid substrate [126, 127] before extraction with conventional solvents. A hydrothermal treatment, with environmental and operational advantages derived from the nontoxic character of the solvent, has been used for extracting biologically active compounds from SFE-extracted bamboo [128].
4. SFE of dry extracts or solid residues. Solid-liquid SC-CO_2 extraction can be employed to purify commercial extracts, dried extracts from conventional solvent extraction (CSE), or compounds remaining in the solid residue from CSE. The two first schemes have been proposed for enhancing the antioxidant activity and improving the organoleptic properties (dearomatization) of extracts [129, 130]. Improved benefits have been reported for high-molecular-weight compounds, probably due to their lower concentration and interactions with the matrix [131].

Antioxidants have been also obtained by SFE of liquid feed streams, including oils and distillates [132] and juices [133].

Usually, natural raw materials for SFE show both limited contents of the target compounds and low bulk density, making the utilization of large volume extractors necessary [134]. Because of this, processes involving CSE and further purification of

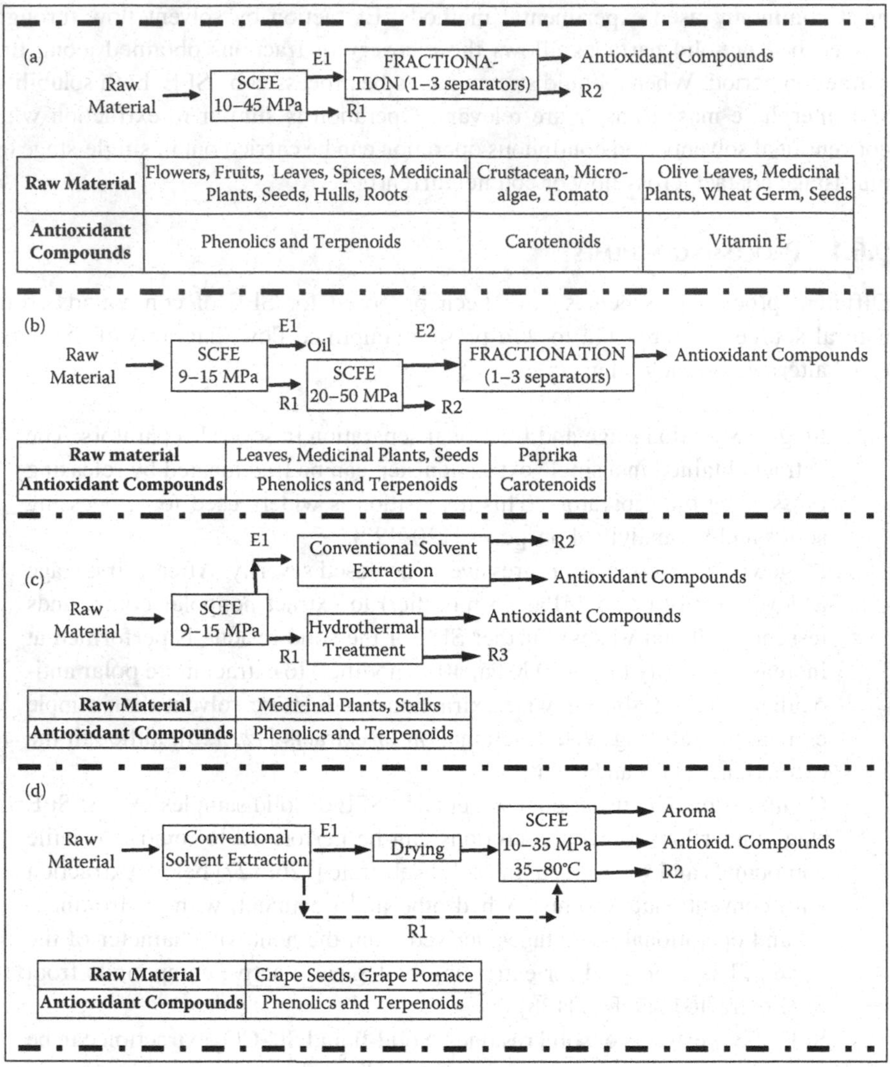

FIGURE 9.1 Processing schemes for extraction of antioxidant compounds involving SFE stages. Nomenclature: E1, E2: extracts; R1, R2, R3: solid residues.

the crude by SFE are comparatively advantageous, as they provide higher yields and/or lower specific CO_2 consumption than direct extraction of the vegetable feedstock.

9.6.2 Effects of the Most Influential Operational Variables

Previous conditioning of the starting material and the experimental conditions employed in extraction and separation influence SFE performance. The nature and properties of vegetable feedstocks or their processing streams (including maturity stage, cultivar, variety, edaphoclimatic conditions) strongly influence the extraction of terpenoids and phenolics [135, 136], carotenoids [137–139], and tocopherols from solid samples. When processing solids, mechanical-thermal conditioning is decisive to facilitate the extraction of intracellular solutes. Reduced particle size favors mass transfer, but too-small particles could limit the performance of fixed beds and grinding may result in losses by volatilization and degradation of active compounds. The major variables influencing SFE of antioxidants (pressure, temperature, solvent flow rate, solvent-to-feed ratio, modifier type, and concentration) should be optimized before operation. Those most influential and specific for SFE in comparison to CSE are further commented.

9.6.2.1 Pressure and Temperature

The solvating power of SFE with CO_2 depends on pressure and temperature. Density gives an estimate of the joint effects of both variables on the solvating power. Besides the operational conditions, the equilibrium solubility of pure compounds is influenced by molecular weight, polarity, and presence of functional groups. When considering extraction from a solid substrate, kinetics and yields also depend on the interaction with the solid matrix.

Effect on the solubility of antioxidant compounds. Equilibrium solubilities are basic information for addressing the design of extraction and separation processes. Solubility data for synthetic antioxidants [140], fat-soluble vitamins [141], and many phenolic compounds have been obtained by different groups and were recently compiled [14, 142]. Solubility data have been reported for pure compounds and their mixtures [143, 144] and for terpenoids from citrus oils [145], as well as for essential oils [9] and their components [8]. Most solubility data refer to a unique solute, and scarce information exists for natural extracts, which are multicomponent mixtures. Since pioneer data on the solubility of tocopherols were published by Chrastil [146], several other studies [147–149] have been reported. Data are also available for mixtures with methyl oleate to simulate the esterified by-product from soybean oil deodorizer distillate [150, 151], for mixtures of this by-product [152, 153] and for crude palm oil [132]. Additional literature concerning pure carotenoids, such as capsaicin [148], β-carotene [148, 154–156], and their mixtures [157], and for natural β-carotene from carrots [156] has been published. Solubility data for β-carotene and tocopherols were compiled by Guglü-Üstündağ and Temelli [137]. Table 9.2 summarizes data from review papers on the solubility of different compounds having antioxidant activity.

Yield and selectivity of antioxidant extraction. Increased pressure results in increased solvent density, allowing higher extraction yields. Increasing pressure beyond a threshold point results in higher fluid viscosity and reduced diffusion coefficients. Pressures over 50 MPa [160] have been reported for the extraction of antioxidants. Operating at high pressure, increased temperatures may decrease the extraction yield due to the reduction in density and the solvent power of the fluid. Operating at pressures close to the critical point, where the density shows higher influence on the solvent power than the vapor pressure, increased temperatures may decrease the extraction yield due to the reduction in density and the solvent power of the fluid. At higher pressures, the increased influence of the solute vapor pressure generally leads to increased solubility. Temperature and pressure show a crossover

TABLE 9.2
Solubility of Selected Antioxidant Compounds in Supercritical CO_2

Compound	P (MPa)	T (K)	Solubility	Ref.
Tocopherols			($y_2 \cdot 10^4$)	
(alpha, delta)	8–35	292–353	—	[141]
	8–35.21	298–353	—	[137]
	9.5–35	303–353	—	[143]
	1–2.52	298–313	2.59–7.31	[12]
Carotenoids			($y_2 \cdot 10^6$)	
(astaxanthin, canthaxanthin, capsanthin,	0.15–50	288–343	—	[141]
β-carotene, lycopene, lutein, zeaxanthin)	2–3.5	313–353	0.09–3.24	[12]
	5–180	288–353	—	[137]
	5–80	288–353	0.019–0.989	[58]
Terpenoids			(mg/g)	
(monoterpene hydrocarbons,	3–11	295–335	—	[59]
sesquiterpene hydrocarbons, oxygenated	8–10	313–333	1.6–CM[a]	[8]
derivatives, aldehydes, ketones)	0.8–13	310–333	—	[143]
Phenolic Compounds			($y_2 \cdot 10^4$)	
(benzoic acid, cinnamic acids, flavonoids)	2–50	308–473	—	[59]
	0.91–2.53	308–318	0.0788–5.61	[12]
	2–40.4	308–473	—	[143]
	79–500	308–373	—	[142]
	0.26–50	308–373	$0.08 \cdot 10^{-4}$–1730	[14]

[a] CM = Complete miscibility

effect whereby higher temperatures improve extraction at high pressures and lower temperatures favor extraction at low pressures. The crossover regions in supercritical fluids, or the point where the slope of solubility *vs.* temperature changes, are also favorable to design separation processes.

The effects of pressure and temperature on the extraction yield of some antioxidants are shown in Figure 9.2 to Figure 9.4. Effects of temperature on texture and color of the extracts have been reported for moso-bamboo extracts [128]. In other cases, slight changes in appearance [176] and significant ones in composition have been reported. This latter effect is due the solvent power (which controls the ability to dissolve different molecules) and to the thermal stability of the solutes. Highly thermal-sensitive compounds require mild extraction conditions (temperatures below 50°C) to avoid alteration. Under these conditions, SFE offers higher yields of active compounds—for example, carnosic acid from rosemary [122], anacardic acid from cashew nut shell [177], hyperforin from *Hypericum perforatum* [178], carnosol from marjoram [179], antioxidants from aloe [165], and matricine from chamomile [180].

9.6.2.2 Modifier

Pure CO_2 under supercritical conditions is a good solvent for lipophilic compounds but is poor for phenolics. Extraction can be enhanced using a modifier able to interact with

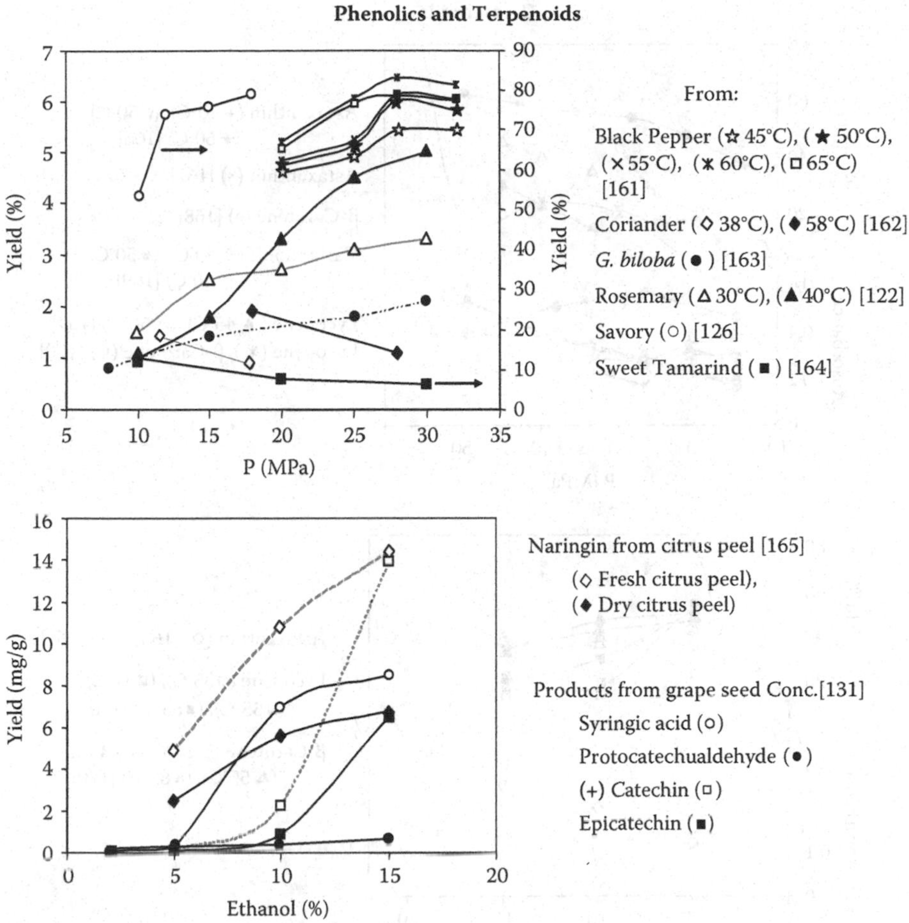

FIGURE 9.2 Effects of pressure, temperature, and ethanol concentration on the extraction yield of phenolics and terpenoids from various sources.

the target compounds, possibly improving yield and selectivity. However, high modifier concentrations may decrease selectivity depending on the size of phenolics [181]. Alcohols are widely used as modifiers, ethanol being the most recommended one on the basis of toxicological and environmental considerations. Ethanol has been employed to increase the solubility of ginsenoids [182], phenols [183], flavonoids [163], terpenoids [184], and carotenoids [166, 170, 185–187]. Methanol has been used for extracting phenolics [188], flavonoids [106, 131, 135, 180, 187], and isoflavones [189] (Figure 9.2). Other modifiers and their target compounds are isopropanol for terpenoids, phenolic ketones, and curcuminoids [190, 191]; propylene glycol for polyphenols [180]; water for phenolic diterpenes and phenolic acids [129]; and acetone; 2,2-dimethoxypropane, chloroform and n-hexane [138, 168, 192] for carotenoids. Vegetable oils have also been proposed for extracting lycopene [139, 193] and caprylic acid [194]. Mixtures of two modifiers have been successfully assayed [107, 190]. Oppositely, ethyl acetate, chloroform [187], and acetic acid [129] were not suitable as modifiers.

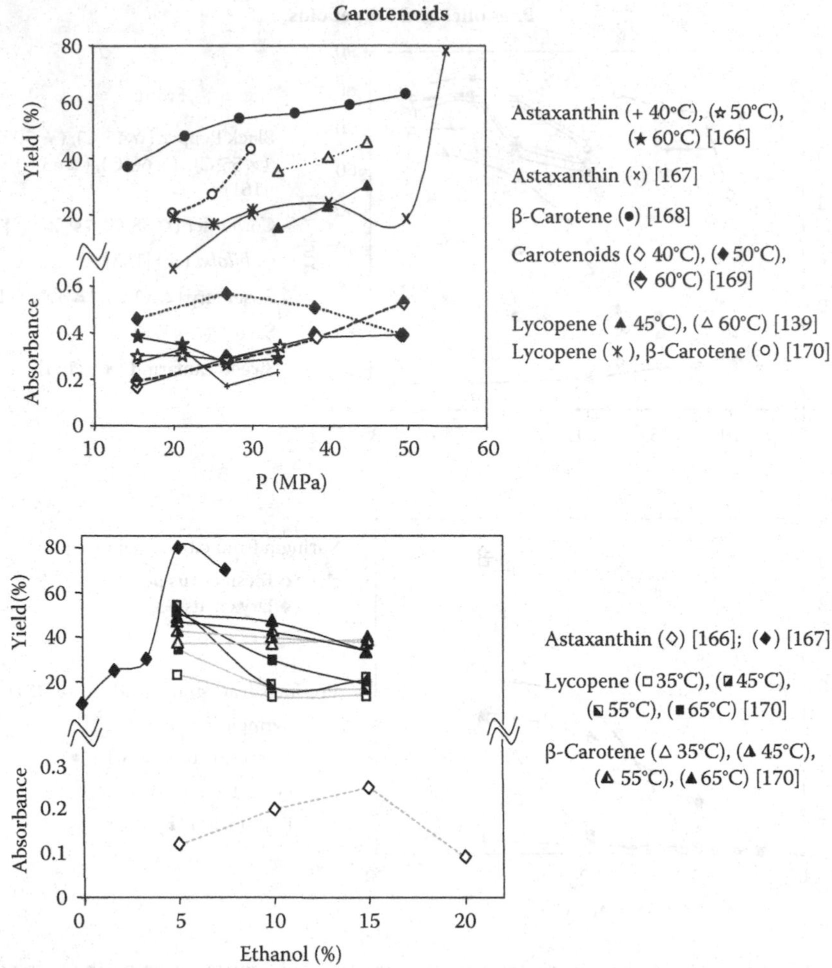

FIGURE 9.3 Effects of pressure, temperature, and ethanol concentration on absorbance and extraction yield of selected carotenoids.

9.6.3 SC-CO_2 Extracts versus Conventional Solvent Extracts

Data concerning the antioxidant activity of SC-CO_2 extracts from solid botanical samples, commercial and crude extracts produced with conventional solvents, and liquid streams are summarized in Table 9.3. Optimization of the operational variables is required for each process, owing to the wide variety of starting materials, target compounds, and conditions employed for extraction and separation. To compare the extracts, both the production conditions and the assay used to quantify the antioxidant activity must be considered. A general comparison between SFE and CSE cannot be established beforehand. Even though conventional, less-selective solvents may allow higher extraction yields [121, 188], the isolated fractions could have unpleasant aromas. Further fractionation by SFE can be used to purify the extract, preserving the antioxidant activity [195]. Similar composition of the extracts obtained using these

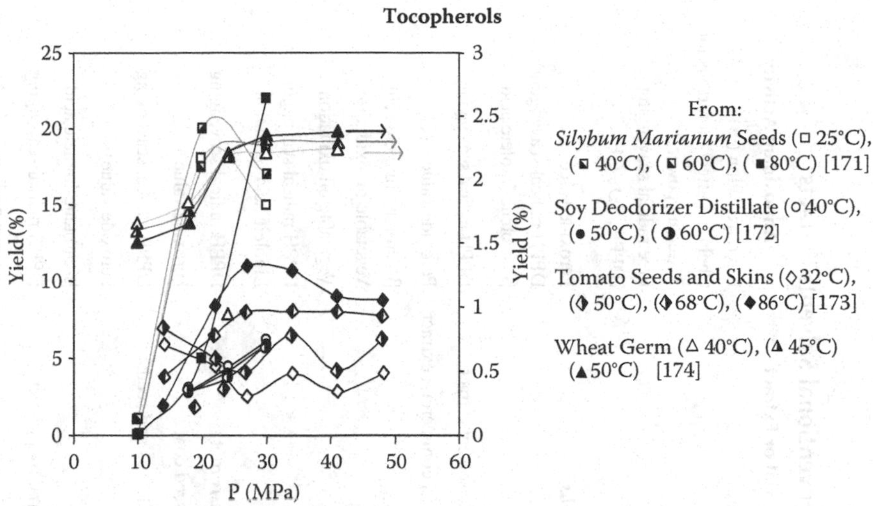

FIGURE 9.4 Effects of pressure, temperature, and ethanol concentration on the extraction yields of tocopherols from various sources.

technologies has been reported for chamomile [180] and marigold [160], whereas different composition was found in extracts from oregano [125] and fennel seeds [176]. Reports indicate that SFE results in higher extraction yields and enhanced selectivity of active compounds than CSE [124, 183, 196]. Superiority of SFE with respect to conventional methods has been reported for eucalyptus leaves [196], black pepper oleoresin [161] and *Lippia alba* stems and leaves [197], owing to the higher concentrations of active compounds. SFE may result in extracts with higher activity when processing substrates with high contents of thermally unstable active compounds and in better odor and color of the isolates [122]. Short processing time and low solvent consumption are additional advantages of SFE [182]. Oppositely, the lower selectivity of conventional solvents could favor the antioxidant activity of extracted fractions showing synergism among components, as observed for turmeric [191], tamarind seed coat [164], marjoram [108], and black sesame [198].

REFERENCES

1. Halliwell, B. and Gutteridge, J.M.C., *Free Radicals in Biology and Medicine*, 3rd ed., Oxford, Oxford University Press, 1989.
2. Halliwell, B., Antioxidant characterization, *Biochem. Pharmacol.*, 49, 1341, 1995.
3. Huang, D., Ou, B. and Prior, R. L., The chemistry behind antioxidant capacity assays, *J. Agric. Food Chem.*, 53, 1841, 2005.
4. Willcox, J.K., Ash, S.L. and Catignani, G.L., Antioxidants and prevention of chronic disease, *Crit. Rev. Food Sci. Nutr.*, 44, 275, 2004.
5. Azzi, A., Davies, K.J.A. and Kelly, F., Free radical biology: Terminology and critical thinking, *FEBS Letters,* 558, 3, 2004.
6. Halliwell, B., Rafter, J. and Jenner, A., Health promotion by flavonoids, tocopherols, tocotrienols, and other phenols: Direct or indirect effects? Antioxidant or not?, *Am. J. Clin. Nutr.*, 81, 268S, 2005.

TABLE 9.3
Antioxidant Activity of Extracts Obtained by SFE of Botanicals, Commercial and Conventional Solvents Extracts

Vegetal Material or Extract	Antioxidant Activity
Aloe barbadensis leaf skin	DPPH radical scavenging
Bupleurum kaoi root ethanolic extracts	Lipid oxidation (%) Free radical scavenging (DPPH, SO·, ·OH)
Camelia sinensis	Lipid oxidation
Coriandrum sativum seeds	DPPH radical scavenging
Curcuma longa	Linoleic acid oxidation β-carotene-linoleic acid
Eucalyptus camaldulensis leaves	Linoleic acid oxidation
Ginkgo biloba leaves methanolic extract	DPPH radical scavenging
Helichrysum italicum flowers	Free radical scavenging (DPPH, SO·) β-carotene-linoleic acid
Humulus lupulus	Linoleic acid oxidation Reducing power
Melissa officinalis	Linoleic acid oxidation
Nigella sativa seeds	β–carotene-linoleic acid model
Origanum vulgare	Peroxide value
Origanum majorana	Vegetable oil oxidation
Orthosiphon spicatus methanolic extract	DPPH radical scavenging
Peumus boldus	ABTS radical scavenging
Propolis ethanol extract	Lipid oxidation (%) Chelating effect on ferrous ions Reducing power Free radical scavenging (DPPH, SO·, ·OH)
Rosmarinus officinalis	Peroxide value DPPH radical scavenging β-carotene-linoleic acid
Rosmarinus officinalis solvent extract	DPPH radical scavenging
Rosmarinus officinalis commercial extract	Peroxide value
Salvia officinalis	β-carotene-linoleic acid Vegetable oil oxidation
Satureja hortensis	Vegetable oil oxidation
Sesamum indicum	DPPH radical scavenging Linoleic acid oxidation
Scutellaria baicalensis root	DPPH radical scavenging
Tamarindus indica seed coat	Peroxide value
Terminalia catappa leaves and seeds	DPPH radical scavenging
Thymus vulgaris	Peroxide value
Thymus vulgaris	β-carotene-linoleic acid
Vitis vinifera industrial by-products	DPPH radical scavenging
Zingiber officinale	β-carotene-linoleic acid

7. Pokorný, J. and Korczak, J., Preparation of natural antioxidant, in *Antioxidants in Food: Practical Applications*, 1st ed., Pokorný, J., Yanishlieva, N. and Gordon, M., Eds., Woodhead Publishing Limited, Abington, Cambridge, England, 2001, pp. 311–330.
8. Reverchon, E., Supercritical fluid extraction and fractionation of essential oils and related products, *J. Supercrit. Fluids*, 10, 1, 1997.
9. del Valle, J.M., de la Fuente, J.C. and Cardarelli, D.A., Contributions to supercritical extraction of vegetable substrates in Latin America, *J. Food Eng.*, 67, 35, 2005.
10. Brunner, G., Supercritical fluids: Technology and application to food processing, *J. Food Eng.*, 67, 21, 2005.
11. Reverchon, E. and De Marco, I., Supercritical fluid extraction and fractionation of natural matter, *J. Supercrit. Fluids*, number 38, p. 146, 2006, available online 27 April 2006.
12. Mukhopadhyay, M., *Natural Extracts Using Supercritical Carbon Dioxide*, CRC Press, Boca Raton, FL, 2000.
13. Herrero, M., Cifuentes, A. and Ibáñez, E., Sub- and supercritical fluid extraction of functional ingredients from different natural sources: Plants, food-by-products, algae and microalgae: A review, *Food Chem.*, 98, 136, 2006.
14. Diaz-Reinoso, B. et al., Supercritical CO_2 extraction and purification of compounds with antioxidant activity, *J. Agric. Food Chem.*, 54, 2441, 2006.
15. Gordon, M.H., The development of oxidative rancidity in foods, in *Antioxidants in Food: Practical Applications*, 1st ed., Pokorný, J., Yanishlieva, N. and Gordon, M., Eds., Woodhead Publishing Limited, Abington, Cambridge, England, 2001.
16. Frankel, E.N. and Meyer, A.S., The problems of using one-dimensional methods to evaluate multifunctional food and biological antioxidants, *J. Sci. Food Agric.*, 80, 1925, 2000.
17. Yanishlieva, N.V., Inhibition of oxidation, in *Antioxidants in Food: Practical Applications*, 1st ed., Pokorný, J., Yanishlieva, N. and Gordon, M., Eds., Woodhead Publishing Limited, Abington, Cambridge, England, 2001.
18. Antolovich, M. et al., Methods for testing antioxidant activity, *Analyst*, 127, 183, 2002.
19. Roginsky, V. and Lissi, E.A., Review of methods to determine chain-breaking antioxidant activity in food, *Food Chem.*, 92, 235, 2005.
20. Prior, R.L., Wu, X. and Schaich, K., Standardized methods for the determination of antioxidant capacity and phenolics in foods and dietary supplements, *J. Agric. Food Chem.*, 53, 4290, 2005.
21. European Parliament and Council Directive No. 95/2/EC, *Official Journal of the European Communities*, 1995.
22. Míková, K., The regulation of antioxidants in food, in *Food Chemical Safety, Vol. 2: Additives*, 1st ed., Watson, D.H., Ed., Woodhead Publishing Limited, Boca Raton, FL, 2002.
23. Ross, J.A. and Kasum, C.M., Dietary flavonoids: Bioavailability, metabolic effects, and safety, *Annual Rev. Nutr.*, 22, 19, 2002.
24. Miller, N.J. and Ruiz-Larrea, M.B., Flavonoids and other plant phenols in the diet: Their significance as antioxidants, *J. Nutr. Env. Med.*, 12, 39, 2002.
25. Soobrattee, M.A. et al., Phenolics as potential antioxidant therapeutic agents: Mechanism and actions, *Mutat. Res.*, 579, 200, 2005.
26. Meyer, A.S. and Frankel, E.N., Antioxidant activity of hydroxycinnamic acids on human low-density lipoprotein oxidation, *Methods Enzymol.*, 335, 256, 2001.
27. Collins, A.R. and Harrington, V., Antioxidants: Not the only reason to eat fruit and vegetables, *Phytochem. Rev.*, 1, 167, 2003.

28. Mansouri, A., Makris, D.P. and Kefalas, P., Determination of hydrogen peroxide scavenging activity of cinnamic and benzoic acids employing a highly sensitive peroxyoxalate chemiluminescence-based assay: Structure–activity relationships, *J. Pharm. Biomed. Anal.*, 39, 22, 2005.
29. Nenadis, N., Zhang, H.Y. and Tsimidou, M.Z., Structure-antioxidant activity relationship of ferulic acid derivatives: Effect of carbon side chain characteristic groups, *J. Agric. Food Chem.*, 51, 1874, 2003.
30. van Acker, S.A. et al., Structural aspects of antioxidant activity of flavonoids, *Free Rad. Biol. Med.*, 20, 331, 1996.
31. Rice-Evans, C.A., Miller, N.J. and Paganga, G., Structure-antioxidant activity relationships of flavonoids and phenolic acids, *Free Rad. Biol. Med.*, 20, 933, 1996.
32. Firuzi, O. et al., Evaluation of the antioxidant activity of flavonoids by "ferric reducing antioxidant power" assay and cyclic voltammetry, *Biochim. Biophys. Acta,* 1721, 174, 2005.
33. Robards, K. et al., Phenolic compounds and their role in oxidative processes in fruits, *Food Chem.*, 66, 401, 1999.
34. Macheix, J.J., Fleuriet, A.M. and Billot, J., *Fruit Phenolics,* CRC Press, Boca Raton, FL, 1990.
35. Heinonen, I.M. and Meyer, A.S., Antioxidants in fruits, berries, and vegetables, *Fruit Vegetable Proc.*, 23, 2002.
36. Proteggente, A.R. et al., The relationship between the phenolic composition and the antioxidant activity of fruits and vegetables, in *Flavonoids in Health and Disease,* 2nd Ed., Rice-Evans, C.A. and Packer, L., Eds., Marcel Dekker, London, 2003, 71.
37. Decker, E. et al., Whole grains as a source of antioxidants, *Cereal Foods World,* 47, 370, 2002.
38. Higdon, J.V. and Frei, B., Tea catechins and polyphenols: Health effects, metabolism, and antioxidant functions, *Crit. Rev. Food Sci. Nutr.*, 43, 89, 2003.
39. Gramza, A. and Korczak, J., Tea constituents (*Camellia sinensis* L.) as antioxidants in lipid systems, *Trends Food Sci. Technol.*, 16, 351, 2005.
40. Grassmann, J., Terpenoids as plant antioxidants, *Vitam. Horm.*, 72, 505, 2005.
41. Escuder, B. et al., Antioxidant capacity of abietanes from *Sphacele salviae, Nat. Prod. Letters,* 16, 277, 2002.
42. Grassmann, J., Hippeli, S. and Elstner, E.F., Plant defense and its benefits for animals and medicine: Role of phenolics and terpenoids in avoiding oxygen stress, *Plant Physiol. Biochem.*, 40, 471, 2002.
43. Milde, J., Elstner, E.F. and Grassmann, J., Synergistic inhibition of low-density lipoprotein oxidation by rutin, γ-terpinene, and ascorbic acid, *Phytomedicine,* 11, 105, 2004.
44. Schieber, A. and Carle, R., Occurrence of carotenoid cis-isomers in food: Technological, analytical, and nutritional implications, *Trends Food Sci. Technol.*, 16, 416, 2005.
45. Di Mascio, P., Kaiser, S. and Sies, H., Lycopene as the most efficient biological carotenoid singlet oxygen quencher, *Arch. Biochem. Biophys.*, 274, 532, 1989.
46. Bohm, V. et al., Trolox Equivalent Antioxidant Capacity of different geometrical isomers of α-carotene, β-carotene, lycopene, and zeaxanthin., *J. Agric. Food Chem.*, 50, 221, 2002.
47. Stahl, W. and Sies, H., Antioxidant effects of carotenoids: Implication in photoprotection in humans, in *Handbook of Antioxidants,* 2nd ed., Cadenas, E. and Packer, L., Eds., Marcel Dekker, New York, 2002, chap. 11.
48. Stahl, W. and Sies, H., Antioxidant activity of carotenoids, *Molecular Aspects of Medicine*, 24(6), 345, 2003.
49. Kiokias, S. and Gordon, M.H., Antioxidant properties of carotenoids *in vitro* and *in vivo, Food Rev. Int.*, 20, 99, 2004.

50. Weber, S.U. and Rimbach, G., Biological activity of tocotrienols, in *Handbook of Antioxidants,* 2nd ed., Cadenas, E. and Packer, L., Eds., Marcel Dekker, New York, 2002, chap. 6.
51. Kitts, D.D., Antioxidant properties of casein-phosphopeptides, *Trends Food Sci. Technol.,* 16, 549, 2005.
52. Chen, L., Remondetto, G.E. and Subirade, M., Food protein-based materials as nutraceutical delivery systems, *Trends Food Sci. Technol.,* 17, 272, 2006.
53. Billaud, C. et al., Maillard reaction products derived from thiol compounds as inhibitors of enzymatic browning of fruits and vegetables: The structure-activity relationship, *Ann. NY Acad. Sci.,* 1043, 876, 2005.
54. Somoza, V., Five years of research on health risks and benefits of Maillard reaction products: An update, *Molec. Nutr. Food Res.,* 49, 663, 2005.
55. Kim, S.K. and Rajapakse, N., Enzymatic production and biological activities of chitosan oligosaccharides (COS): A review, *Carbohydr. Polymers,* 62, 357, 2005.
56. Hall, C., Sources of natural antioxidants: Oilseeds, nuts, cereals, legumes, animal products and microbial sources, in *Antioxidants in Food,* 1st ed., Pokorny, J., Yanishlieva, N. and Gordon, M., Eds., Woodhead Publishing Limited, Abington, Cambridge, England, 2001.
57. Nakagami, T., Tamura, N. and Nakamura, T., Plant phenol compounds as anti-complement agents and therapeutics for complement-associated diseases and health foods containing them, *Jpn. Kokai Tokkyo Koho,* 7, 1995, JP 07223941.
58. Fernández, M.A., Sáenz, M.T. and García, M.D., Anti-inflammatory activity in rats and mice of phenolic acids isolated from *Scrophularia frutescens, J. Pharm. Pharmacol.,* 50, 1183, 1998.
59. Zang, L.Y. et al., Effect of antioxidant protection by p-coumaric acid on low-density lipoprotein cholesterol oxidation, *Am. J. Physiol. Cell Physiol.,* 279, C954, 2000.
60. Cartron, E. et al., Specific antioxidant activity of caffeoyl derivatives and other natural phenolic compounds: LDL protection against oxidation and decrease in the proinflammatory lysophosphatidylcholine production, *J. Nat. Prod.,* 64, 480, 2001.
61. Kwcon, M.II., IIwang, H.J. and Sung, H.C., Identification and antioxidant activity of novel chlorogenic acid derivatives from bamboo (*Phyllostachys edulis*), *J. Agric. Food Chem.,* 49, 4646, 2001.
62. Krishnaswamy, K., Nonnutrients and cancer prevention, ICMR Bull, 31, 2001, available at *http://www.icmr.nic.in/bujan01.pdf*
63. Taniguchi, H. et al., Ferulic acid esters as antioxidants and UV absorbents, *European Pat. Appl.,* EP 681825, 1995.
64. Lee, H.S., Inhibitory activity of *Cinnamomum cassia* bark-derived component against rat lens aldose reductase, *J. Pharm. Pharmaceut. Sci.,* 5, 226, 2002.
65. Aviram, M., Vaya, J. and Fuhrman, B., Licorice root flavonoid antioxidants reduce LDL oxidation and attenuate cardiovascular diseases, *Oxid. Stress Dis.,* 14 (Herbal and Traditional Medicine), Marcel Dekker, Inc., 2004, 595.
66. Aviram, M., Kaplan, M., Rosenblat, M. and Fuhrman, B., Dietary antioxidants and paraoxonases against LDL oxidation and atherosclerosis development, *Handb. Exp. Pharmacol.,* 170 (Atherosclerosis), 2005, 263.
67. Kris-Etherton, P.M. et al., Bioactive compounds in nutrition and health-research methodologies for establishing biological function: The antioxidant and anti-inflammatory effects of flavonoids on atherosclerosis, *Annu. Rev. Nutr.,* 24, 511, 2004.
68. Fuhrman, B. and Aviram, M., Polyphenols and flavonoids protect LDL against atherogenic modifications, *Oxididative Stress Disease,* 8 (Handbook of Antioxidants), 2002, 303–336.
69. Nijveldt, R.J. et al., Flavonoids: A review of probable mechanisms of action and potential applications, *Am. J. Clin. Nutr.,* 74, 418, 2001.

70. Marilyn, E., Dietary flavonoids: Effects on xenobiotic and carcinogen metabolism, *Toxicol. in Vitro*, 20, 187, 2006.
71. Choi, M.S. et al., Cholesterol-lowering properties of citrus flavonoids and polyphenolic compounds and their relevance to antioxidative activity, *Nutr. Sci.*, 6, 31, 2003.
72. Horvathova, K., Vachalkova, A. and Novotny, L., Flavonoids as chemoprotective agents in civilization diseases, *Neoplasma*, 48, 435, 2001.
73. Spencer, J.P.E., Interactions of flavonoids and their metabolites with cell signaling cascades, *Oxididative Stress Disease*, 17 (Nutrigenomics), 2005, 353.
74. Ghosh, D., Anthocyanins and anthocyanin-rich extracts in biology and medicine: Biochemical, cellular, and medicinal properties, *Curr. Top. Nutraceutical Res.*, 3, 113, 2005.
75. Depeint, F. et al., Evidence for consistent patterns between flavonoid structures and cellular activities, *Proc. Nutr. Soc.,* 61, 97, 2002.
76. Erlund, I., Review of the flavonoids quercetin, hesperetin, and naringenin. Dietary sources, bioactivities, bioavailability, and epidemiology, *Nutr. Res.*, 24, 851, 2004.
77. Van Hoorn, D.E.C. et al., Biological activities of flavonoids, *Sci. Med.*, 9, 152, 2003.
78. Das, S. and Rosazza, J.P.N., Microbial and enzymatic transformations of flavonoids, *J. Nat. Prod.*, 69(3), 499, 2006.
79. Hodek, P., Trefil, P. and Stiborova, M., Flavonoids: Potent and versatile biologically active compounds interacting with cytochromes P 450, *Chem.-Biol Interact.*, 139, 1, 2005.
80. Kim, H.P. et al., Anti-inflammatory flavonoids: Modulators of proinflammatory gene expression, *Nat. Prod. Sci.*, 10, 1, 2004.
81. Kim, H.P. et al., Anti-inflammatory plant flavonoids and cellular action mechanisms, *J. Pharmacol. Sci.*, 96, 229, 2004.
82. Strickland, F.M., Boosting the immune system, *Comprehensive Series in Photosciences*, 3(Sun Protection in Man), 2001, 613, 615–636.
83. Hoensch, H.P. and Kirch, W. Potential role of flavonoids in the prevention of intestinal neoplasia: A review of their mode of action and their clinical perspectives, *Int. J. Gastroint. Cancer*, 35, 187, 2005.
84. Dajas, F. et al., Flavonoids and the brain: Evidences and putative mechanisms for a protective capacity, *Curr. Neuropharmacol.,* 3, 193, 2005.
85. Schroeter, H. and Spencer, J.P.E., Flavonoids: Neuroprotective agents? Modulation of oxidative stress-induced MAP kinase signal transduction, *Oxididative Stress Disease,* 9 (Flavonoids in Health and Disease (Second Ed.)), 2003, 233–272.
86. Choi, H.S. et al., Radical-scavenging activities of citrus essential oils and their components: Detection using 1,1-diphenyl-2-picrylhydrazyl, *J. Agric. Food Chem.,* 48, 4156, 2000.
87. Singh, R.P. and Agarwal, R., Flavonoid antioxidant silymarin and skin cancer, *Antioxid. Redox Signaling,* 4, 655, 2002.
88. Walle, T., Absorption and metabolism of flavonoids, *Free Rad. Biol. Med.*, 36, 829, 2004.
89. Kroon, P.A. et al., How should we assess the effects of exposure to dietary polyphenols in vitro?, *Am. J. Clin. Nutr.*, 80, 15, 2004.
90. Schroeter, H. et al., MAPK signaling in neurodegeneration: Influences of flavonoids and of nitric oxide, *Neurobiol. Aging,* 23, 861, 2002.
91. Youdim, K.A. et al., Dietary flavonoids as potential neuroprotectants, *Biol. Chem.*, 383, 503, 2002.
92. Wagner, K.H. and Elmadfa, I., Biological relevance of terpenoids. Overview focusing on mono-, di- and tetraterpenes, *Ann. Nutr. Metab.,* 47, 95, 2003.
93. Pietri, S. et al., Cardioprotective and anti-oxidant effects of the terpenoid constituents of Ginkgo biloba extract (EGb 761), *J. Mol. Cell. Cardiol.,* 29, 733, 1997.

94. Ulubelen, A., Cardioactive and antibacterial terpenoids from some *Salvia* species, *Phytochem.*, 64, 39, 2003.
95. De las Heras, B. et al., Terpenoids: Sources, structure elucidation and therapeutic potential in inflammation, *Curr. Top. Med. Chem.*, 3, 171, 2003.
96. Perrucci, S. et al., Structure/activity relationship of some natural monoterpenes as acaricides against *Psoroptes cuniculi*, *J. Nat. Prod.*, 58, 1261, 1995.
97. Wang, B.J. et al., Antioxidant activity of *Bupleurum kaoi* Liu (Chao et Chuang) fractions fractionated by supercritical CO_2, *Lebensm. Wiss. Technol.*, 38, 281, 2005.
98. Deming, D.M. et al., Carotenoids: Linking chemistry, absorption, and metabolism to potential roles in human health and disease, in *Handbook of Antioxidants,* 2nd ed., Cadenas, E. and Packer, L., Eds., Marcel Dekker, New York, 2002, chap. 10.
99. Granado, F., Olmedilla, B. and Blanco, I., Nutritional and clinical relevance of lutein in human health, *British J. Nutr.*, 90, 487, 2003.
100. Hix, L.M., Lockwood, S.F. and Bertram, J.S., Bioactive carotenoids: Potent antioxidants and regulators of gene expression, *Redox Report*, 9, 181, 2004.
101. Lowe, G.M., Vlismas, K. and Young, A.J., Carotenoids as prooxidants?, *Mol. Aspects Med.*, 24, 363, 2003.
102. Sharoni, Y. et al., Modulation of transcriptional activity by antioxidant carotenoids, *Mol. Aspects Med.*, 24, 371, 2003.
103. Landvik, S.V., Diplock, A.T. and Packer, L., Efficacy of vitamin E in human health and disease, in *Handbook of Antioxidants,* 2nd ed., Cadenas, E. and Packer, L., Eds., Marcel Dekker, New York, 2002, chap. 4.
104. Violi, F., Cangemi, R. and Loffredo, L., Vitamins E and C for prevention of cardiovascular disease, *Curr. Dev. Atheroscler. Res.*, 117, 2006.
105. Traber, M.G., Vitamin E bioavailability, biokinetics and metabolism, *Handbook of Antioxidants,* 2nd ed., Cadenas, E. and Packer, L., Eds., Marcel Dekker, New York, 2002, chap. 5.
106. Palma, M. et al., Fractional extraction of compounds from grape seeds by supercritical fluid extraction and analysis for antimicrobial and agrochemical activities, *J. Agric. Food Chem.*, 47, 5044, 1999.
107. Leal, P.F. et al., Functional properties of spice extracts obtained via supercritical fluid extraction, *J. Agric. Food Chem.*, 51, 2520, 2003.
108. Vági, E. et al., Essential oil composition and antimicrobial activity of *Origanum majorana* L. extracts obtained with ethyl alcohol and supercritical carbon dioxide, *Food Res. Int.*, 38, 51, 2005.
109. Arlorio, M. et al., Antioxidant and biological activity of phenolic pigments from *Theobroma cacao* hulls extracted with supercritical CO_2, *Food Res. Int.*, 38, 1009, 2005.
110. McHugh, M.A. and Krukonis, V.J., *Supercritical Fluid Extraction: Principles and Practice*, Butterworth, Stoneham, MA, 1986.
111. Ko, T.F. et al., Antimutagenicity of supercritical CO_2 extracts of *Terminalia catappa* leaves and cytotoxicity of the extracts to human hepatoma cells, *J. Agric. Food Chem.*, 51, 3564, 2003.
112. Miquel, E.M., Nissen, L.R. and Skibsted, L.H., Antioxidant evaluation protocols: Food quality or health effects. *Eur. Food Res. Technol.*, 219, 561, 2004.
113. Gordon, M.H., Measuring antioxidant activity, in *Antioxidants in Food: Practical Applications*, 1st ed., Pokorný, J., Yanishlieva, N. and Gordon, M., Eds., Woodhead Publishing Limited, Abington, Cambridge, England, 2001.
114. Decker, E.A. et al., Measuring antioxidant reffectiveness in food, *J. Agric. Food Chem.*, 53, 4303, 2005.

115. Halliwell, B., Food-derived antioxidants: How to evaluate their importance in food and *in vivo*, *Handbook of Antioxidants*, 2nd ed., Cadenas, E. and Packer, L., Eds., Marcel Dekker, New York, 2002, chap. 1.
116. Sánchez-Moreno, C., Methods used to evaluate the free radical scavenging activity in foods and biological systems, *Food Sci. Technol. Int.*, 8, 121, 2002.
117. Liu, R.H. and Finley, J., Potential cell culture models for antioxidant research, *J. Agric. Food Chem.*, 53, 4311, 2005.
118. Halliwell, B., Oxidative stress in cell culture: An under-appreciated problem?, *FEBS Letters*, 540, 1, 2003.
119. Halliwell, B. and Whiteman, M., Measuring reactive species and oxidative damage in vivo and cell culture: How should you do it and what do the results mean?, *Br. J. Pharmacol.*, 142, 231, 2004.
120. El-Ghorab, A.H. et al., Antioxidant activity of Egyptian *Eucalyptus camaldulensis var. brevirostris* leaf extracts, *Nahrung*, 47, 41, 2003.
121. Dapkevicius, A. et al., Antioxidant activity of extracts obtained by different isolation procedures from some aromatic herbs grown in Lithuania, *J. Sci. Food Agric.*, 77, 140, 1998.
122. Carvalho, R.N., Supercritical fluid extraction from rosemary (*Rosmarinus officinalis*): Kinetic data, extract's global yield, composition, and antioxidant activity, *J. Supercrit. Fluids*, 35, 197, 2005.
123. Ashraf-Khorassani, M. and Taylor, L.T., Sequential fractionation of grape seeds into oils, polyphenols, and procyanidins via a single system employing CO_2-based fluids, *J. Agric. Food Chem.*, 2440, 2004.
124. Nguyen, U. et al., Process for extracting antioxidants from labiatae herbs, *U.S. Patent*, US 5017397, 1991.
125. Simándi, B. et al., Supercritical carbon dioxide extraction and fractionation of oregano oleoresin, *Food Res. Int.*, 31, 723, 1998.
126. Esquível, M.M., Ribeiro, M.A. and Bernardo-Gil, M.G., Supercritical extraction of savory oil: Study of antioxidant activity and extract characterization, *J. Supercrit. Fluids*, 14, 129, 1999.
127. del Valle, J.M. et al., Recovery of antioxidants from boldo (*Peumus boldus* M.) by conventional and supercritical CO_2 extraction. *Food Res. Int.*, 37, 695, 2004.
128. Quitain, A.T., Katoh, S. and Moriyoshi, T., Isolation of antimicrobials and antioxidants from moso-bamboo (*Phyllostachys heterocycla*) by supercritical CO_2 extraction and subsequent hydrothermal treatment of the residues, *Ind. Eng. Chem. Res.*, 43, 1056, 2004.
129. López-Sebastian, S. et al., Dearomatization of antioxidant rosemary extracts by treatment with supercritical carbon dioxide, *J. Agric. Food Chem.*, 46, 13, 1998.
130. Hadolin, M. et al., Isolation and concentration of natural antioxidants with high-pressure extraction, *Innov. Food Sci. Emerging Technol.*, 5, 2451, 2004.
131. Murga, R. et al., Extraction of natural complex phenols and tannins from grape seeds by using supercritical mixtures of carbon dioxide and alcohol, *J. Agric. Food Chem.*, 48, 3408, 2000.
132. Gast, K., Machado, N. and Brunner, G., Countercurrent extraction of vitamins from crude palm oil, in *State of the Art Book on Supercritical Fluids*, Ainia, Spain, 2004, chap. 20.
133. Señoráns, F.J. et al., Isolation of antioxidant compounds from orange juice by using countercurrent supercritical fluid extraction (CC-SFE), *J. Agric. Food Chem.*, 49, 6039, 2001.
134. Ribeiro, M.A., Bernardo-Gil, M.G. and Esquível, M.M., *Melissa officinalis* L.: Study of antioxidant activity in supercritical residues, *J. Supercrit. Fluids*, 21, 51, 2001.

135. Louli, V., Ragoussis, N. and Magoulas, K., Recovery of phenolic antioxidants from wine industry by-products, *Biores. Technol.*, 92, 201, 2004.
136. Ibáñez, E. et al., Supercritical fluid extraction and fractionation of different pre-processed Rosemary plants, *J. Agric. Food Chem.*, 47, 1400, 1999.
137. Guglü-Üstündağ, O. and Temelli, F., Correlating the solubility behavior of minor lipid components in supercritical carbon dioxide, *J. Supercrit. Fluids,* 31, 235, 2004.
138. Şanal, I.S. et al., Recycling of apricot pomace by supercritical CO_2 extraction, *J. Supercrit. Fluids,* 32, 221, 2004.
139. Vasapollo, G. et al., Innovative supercritical CO_2 extraction of lycopene from tomato in the presence of vegetable oil as co-solvent, *J. Supercrit. Fluids,* 29, 87, 2004.
140. Cortesi, A. et al., Effect of chemical structure on the solubility of antioxidants in supercritical carbon dioxide: Experimental data and correlation, *J. Supercrit. Fluids,* 14, 139, 1999.
141. Johannsen, M. and Brunner, G., Solubilities of the fat-soluble vitamins A, D, E, and K in supercritical carbon dioxide, *J. Chem. Eng. Data*, 42, 106, 1997.
142. Fornari, T. et al., A new development in the application of the group contribution associating equation of state to model solid solubilities of phenolic compounds in SC-CO_2, *Ind. Eng. Chem. Res.,* 44, 8147, 2005.
143. Christov, M. and Dohrn, R., High-pressure fluid phase equilibria: Experimental methods and systems investigated (1994–1999), *Fluid Phase Equilibria,* 202, 153, 2002.
144. Lucien, F. and Foster, N.R., Solubilities of solid mixtures in supercritical carbon dioxide: A review, *J. Supercrit. Fluids,* 17, 111, 2000.
145. Díaz, S., Espinosa, S. and Brignole, E.A., Citrus peel oil deterpenation with supercritical fluids, *J. Supercrit. Fluids,* 35, 49, 2005.
146. Chrastil, J., Solubility of solids and liquids in supercritical CO_2, *J. Phys. Chem.,* 86, 3016, 1982.
147. Chen, C.-C. et al., Vapor-liquid equilibria of carbon dioxide with linoleic acid, α-tocopherol, and triolein at elevated pressures, *Fluid Phase Equilibria,* 175, 107, 2000.
148. Skerget, M. and Knez, Z., Solubility of binary solid mixture β-carotene-capsaicin in dense CO_2, *J. Agric. Food Chem.*, 45, 2066, 1997.
149. Pereira, P.J. et al., High pressure phase equilibrium for α-tocopherol + CO_2, *Fluid Phase Equilibria,* 216, 53, 2004.
150. Fang, T. et al., Phase equilibria for binary systems of methyl oleate-supercritical CO_2 and α-tocopherol-supercritical CO_2, *J. Supercrit. Fluids,* 30, 1, 2004.
151. Fang, T. et al., Phase equilibria for the ternary system methyl oleate + tocopherol + supercritical CO_2, *J. Chem. Eng. Data,* 50, 390, 2005.
152. Stoldt, J., Saure, C. and Brunner, G., Phase equilibria of fat compounds with supercritical carbon dioxide, *Fluid Phase Equilibria,* 116, 399, 1996.
153. Stoldt, J. and Brunner, G., Phase equilibrium measurements in complex systems of fats, fat compounds, and supercritical carbon dioxide, *Fluid Phase Equilibria,* 146, 269, 1998.
154. Cygnarowicz, M.L., Maxwell, R.J. and Seider, W.D., Equilibrium solubilities of β-carotene in supercritical carbon dioxide, *Fluid Phase Equilibria,* 59, 57, 1990.
155. Tuma, D. and Schneider, G.M., Determination of the solubilities of dyestuffs in near- and supercritical fluids by a static method up to 180 MPa., *Fluid Phase Equilibria,* 158–160, 743, 1999.
156. Saldaña, M.D.A. et al., Comparison of the solubility of β-carotene in supercritical CO_2 based on a binary and a multicomponent complex system, *J. Supercrit. Fluids,* 37, 342, 2006.

157. Skerget, M., Knez, Z. and Habulin, M., Solubility of β-carotene and oleic acid in dense CO_2 and data correlation by a density based model, *Fluid Phase Equilibria*, 109, 131, 1995.
158. de la Fuente, J.C., Solubility of carotenoid pigments (lycopene and astaxanthin) in supercritical carbon dioxide, *Fluid Phase Equilibria*, 247, 90, 2006.
159. Dohrn, R. and Brunner, G., High-pressure fluid-phase equilibria: Experimental methods and systems investigated (1988–1993), *Fluid Phase Equilibr.*, 106, 213, 1995.
160. Baumann, D. et al., Supercritical carbon dioxide extraction of marigold at high pressures: Comparison of analytical and pilot-scale extraction, *Phytochem. Anal.*, 15, 226, 2004.
161. Tipsrisukond, N., Fernando, L.N. and Clarke, A.D., Antioxidant effects of essential oil and oleoresin of black pepper from supercritical carbon dioxide extractions in ground pork, *J. Agric. Food Chem.*, 46, 4329, 1998.
162. Yépez, B. et al., Producing antioxidant fractions from hernaceous matrices by supercritical fluid extraction, *Fluid Phase Equilibria*, 194–197, 879, 2002.
163. Yang, C., Xu, Y.R. and Yao, W.X., Extraction of pharmaceutical components from *Ginkgo biloba* leaves using supercritical carbon dioxide, *J. Agric. Food Chem.*, 50, 846, 2002.
164. Luengthanaphol, S. et al., Extraction of antioxidants from sweet Thai tamarind seed coat-preliminary experiments, *J. Food Eng.*, 63, 247, 2004.
165. Giannuzzo, A.N. et al., Supercritical fluid extraction of naringin from the peel of *Citrus paradisi*, *Phytochem. Anal.*, 14, 221, 2003.
166. López, M. et al., Selective extraction of astaxanthin from crustaceans by use of supercritical carbon dioxide, *Talanta*, 64, 726, 2004.
167. Machmudah, S. et al., Extraction of astaxanthin from *Haematococcus pluvialis* using supercritical CO_2 and ethanol as entrainer, *Ind. Eng. Chem. Res.*, 45, 3652, 2006.
168. Jarén-Galán, M., Nienaber, U. and Schwartz, S.J., Paprika (*Capsicum annuum*) oleoresin extraction with supercritical carbon dioxide, *J. Agric. Food Chem.*, 47, 3558, 1999.
169. Montero, O. et al., Supercritical CO_2 extraction of β-carotene from a marine strain of the *Cyanobacterium synechococcus* species, *J. Agric. Food Chem.*, 53, 9701, 2005.
170. Baysal, T. et al., Supercritical CO_2 extraction of beta-carotene and lycopene from tomato paste waste, *J. Agric. Food Chem.*, 48, 5507, 2000.
171. Hadolin, M. et al. High pressure extraction of vitamin E–rich oil from *Silybum marianum*, *Food Chem.*, 74, 355, 2001.
172. Nagesha, G. K., Manohar, B. and Udaya Sankar, K., Enrichment of tocopherols in modified soy deodorizer distillate using supercritical carbon dioxide extraction, *Eur. Food Res. Technol.*, 217, 427, 2003.
173. Rozzi, N.L. et al., Supercritical fluid extraction of lycopene from tomato processing by-products, *J. Agric. Food Chem.*, 50, 2638, 2002.
174. Ge, Y. et al., Extraction of natural vitamin E from wheat germ by supercritical carbon dioxide, *J. Agric. Food Chem.*, 50, 685, 2002.
175. Ge, Y. et al., Optimization of the supercritical fluid extraction of natural vitamin E from wheat germ using response surface methodology, *J. Food Sci.*, 67, 239, 2002.
176. Damjanović, B. et al., Extraction of fennel (*Foeniculum vulgare* Mill.) seeds with supercritical CO_2: Comparison with hydrodistillation, *Food Chem.*, 92, 143, 2005.
177. Smith, R.L., Jr. et al., Separation of cashew (*Anacardium occidentale* L.) nut shell liquid with supercritical carbon dioxide, *Biores. Technol.*, 88, 1, 2003.
178. Seger, C. et al., Characterization of supercritical fluid extracts of St. John's Wort (*Hypericum perforatum* L.) by HPLC-MS and GC-MS, *Eur. J. Pharm. Sci.*, 21, 453, 2004.

179. Vági, E. et al., Phenolic and triterpenoid antioxidants from *Origanum majorana* L. herb and extracts obtained with different solvents, *J. Agric. Food Chem.*, 53, 17, 2005.
180. Hu, Q., Hu, Y. and Xu, J., Free radical-scavenging activity of *Aloe vera* (*Aloe barbadensis* Miller) extracts by supercritical carbon dioxide extraction, *Food Chem.*, 91, 85, 2005.
181. Scalia, S., Giuffreda, L. and Pallado, P., Analytical and preparative supercritical fluid extraction of chamomile flowers and its comparison with conventional methods, *J. Pharm. Biomed. Anal.*, 21, 549, 1999.
182. Wang, H.C., Chen, C.R. and Chang, C.J., Carbon dioxide extraction of ginseng root hair oil and ginsenosides, *Food Chem.*, 72, 505, 2001.
183. Vaher, M. and Koel, M., Separation of polyphenolic compounds extracted from plant matrices using capillary electrophoresis, *J. Chromatogr. A*, 990, 225, 2003.
184. Daukšas, E. et al., Rapid screening of antioxidant activity of sage (*Salvia officinalis* L.) extracts obtained by supercritical carbon dioxide at different extraction conditions, *Nahrung*, 45, 338, 2001.
185. Lim, G.B. et al., Separation of astaxanthin from red yeast *Phaffia rhodozyma* by supercritical carbon dioxide extraction, *Biochem. Eng. J.*, 11, 181, 2002.
186. Suto, K. et al., Determination of magnolol and honokiol in Magnoliae cortex using supercritical fluid chromatography on-line coupled with supercritical fluid extraction by on-column trapping, *J. Chromatogr. A*, 786, 366, 1997.
187. Moraes, M.L.L., Vilegas, J.H.Y. and Lanças, F.M., Supercritical fluid extraction of glycosilated flavonoids from *Pasiflora* leaves, *Phytochem. Anal.*, 8, 257, 1997.
188. Goli, A.H. et al., Antioxidant activity and total phenolic compounds of pistachio (*Pistachia vera*) hull extracts, *Food Chem.*, 92, 521, 2005.
189. Rostagno, M.A., Araujo, J.M.A. and Sandi, D., Supercritical fluid extraction of isoflavones from soybean flour, *Food Chem.*, 78, 111, 2002.
190. Zancan, K.C. et al., Extraction of ginger (*Zingiber officinale Roscoe*) oleoresin with CO_2 and co-solvents: A study of the antioxidant action of the extracts, *J. Supercrit. Fluids*, 24, 57, 2002.
191. Braga, M.E.M., Comparison of yield, composition, and antioxidant activity of Turmeric (*Curcuma longa* L.) extracts obtained using various techniques, *J. Agric. Food Chem.*, 51, 6604, 2003.
192. Cadoni, E. et al., Supercritical CO_2 extraction of lycopene and β-carotene from ripe tomatoes, *Dyes Pigments*, 44, 27, 1999.
193. Sun, M. and Temelli, F., Supercritical carbon dioxide extraction of carotenoids from carrot using canola oil as a continuous co-solvent, *J. Supercrit. Fluids*, 37, 397, 2006.
194. Grigonis, D. et al., Comparison of different extraction techniques for isolation of antioxidants from sweet grass (*Hierochloë odorata*), *J. Supercrit. Fluids*, 33, 223, 2005.
195. Simándi, B. et al., Antioxidant activity of pilot-plant alcoholic and supercritical carbon dioxide extracts of thyme, *Eur. J. Lipid Sci. Technol.*, 103, 355, 2001.
196. Fadel, H. et al., Effect of extraction techniques on the chemical composition and antioxidant activity of *Eucalyptus camaldulensis* var. *brevirostris* leaf oils, *Z. Lebensm. Unters. Forsch.*, 208, 212, 1999.
197. Stashenko, E.E., Jaramillo, B.E. and Martínez, J.R., Comparison of different extraction methods for the analysis of volatile secondary metabolites of *Lippia alba* (Mill.) N.E. brown, grown in Colombia, and evaluation of its in vitro antioxidant activity, *J. Chromatogr. A*, 1025, 93, 2004.
198. Xu, J., Chen, S. and Hu, Q., Antioxidant activity of brown pigment and extracts from black sesame seed (*Sesamum indicum* L.), *Food Chem.*, 91, 79, 2005.

10 Essential Oils Extraction and Fractionation Using Supercritical Fluids

Ernesto Reverchon and Iolanda De Marco

CONTENTS

10.1 Introduction 305
10.2 Solids Processing 307
10.2.1 Selection of the Operating Parameters 308
10.2.2 Examples 312
10.2.2.1 Leaves 312
10.2.2.2 Flowers 314
10.2.2.3 Seeds 314
10.2.2.4 Other Matrices 314
10.2.2.5 Flower Concretes Fractionation 316
10.3 Liquid Feed Processing 318
10.3.1 Selection of the Operating Parameters 319
10.3.2 Examples 320
10.4 Antisolvent Extraction 322
10.4.1 Selection of the Operating Parameters 322
10.4.2 Examples 323
10.4.2.1 Proteins and Aroma Extraction from Tobacco 323
10.5 Mathematical Modelling 324
References 328

10.1 INTRODUCTION

The extraction from natural sources is the most widely studied application of supercritical fluids (SCFs), and several hundred scientific papers on the topic have been published and reviewed [1–9]. Indeed, supercritical fluid extraction (SFE) has immediate advantages over traditional extraction techniques; it is a flexible process due to the possibility of continuous modulation of the solvent power/selectivity of the SCF, and it allows the elimination of polluting organic solvents and of the expensive postprocessing of the extracts for solvent elimination.

Several compounds have been examined as SFE solvents, including hydrocarbons such as hexane, pentane and butane, nitrous oxide, sulfur hexafluoride, and fluorinated hydrocarbons [10]. However, carbon dioxide (CO_2) is the most popular

SFE solvent because it is safe, is readily available, and has a low cost. It allows supercritical operations at relatively low pressures and at near-room temperatures.

The only serious drawback of SFE is that investment costs are higher than those for traditional atmospheric pressure extraction techniques. However, the base process scheme (extraction plus separation) is relatively cheap and very simple to be scaled up to industrial scale.

In this chapter, we focus our analysis on the extraction, isolation, and fractionation of essential oils. Early works on SFE of essential oils frequently used high pressures (> 350 bar), even when supercritical CO_2 (SC-CO_2)–soluble compounds had to be extracted (for example, terpenes, sesquiterpenes, fatty acids). Operating in this manner, the solvent power of the SCF was enhanced, but its selectivity was very low. Since then, the concept of the optimization between solvent power and selectivity has been applied and SFE operating conditions have been chosen to obtain the selective extraction of the compounds of interest, reducing to a minimum the coextraction of undesired compounds [1]. Moreover, for successful extraction, not only must the solubility of the compounds to be extracted be taken into account but also the solubilities of the undesired compounds. Mass transfer resistances due to the structure of the raw material and to the specific location of the compounds to be extracted can also play a relevant role. Microscopic analysis of the natural structure can help in understanding where mass transfer resistances are located. Specific experiments performed varying particle size and supercritical solvent residence time can also be helpful in this sense. The complex interplay between thermodynamics (solubility) and kinetics (mass transfer) has to be understood to properly perform SFE.

Fractional separation of the extracts is another well-known concept that can be useful in improving SFE selectivity. Indeed, in several cases, it is not possible to avoid the coextraction of some compound families that show different solubilities, but there are also different mass transfer resistances in the raw matter. In these cases, one can perform an extraction in successive steps at increasing pressures to obtain the fractional extraction of the soluble compounds contained in the organic matrix. Fractional separation allows fractionation of the SCF extracts, equipping the plant with some separation vessels operating in series at different pressures and temperatures. The scope of this operation is to induce the selective precipitation of different compound families based on different saturation conditions in the SCF.

In several other cases, the feed is a liquid mixture. Therefore, the process to be applied is the continuous liquid extraction in a packed tower. Note that, although the extraction from solids is a discontinuous operation, the packed tower is capable of continuous steady-state operation that allows the processing of large quantities of liquid mixtures in a relatively small apparatus and in a short time.

In some other cases, the material to be treated is a liquid mixture that contains solid compounds dissolved in it. The extraction of these compounds from the liquid solution cannot be performed in a packed tower since the solid matter precipitates on the packings of the bed. In this case, supercritical antisolvent extraction (SAE) can be adopted. The preconditions to apply SAE are similar to the ones characteristic of supercritical antisolvent micronization (SAS): the liquid solvent has to be very soluble in SC-CO_2, whereas the solids have to be insoluble in the SCF. The scope of SAE is not the micronization but the purification of the liquid solution

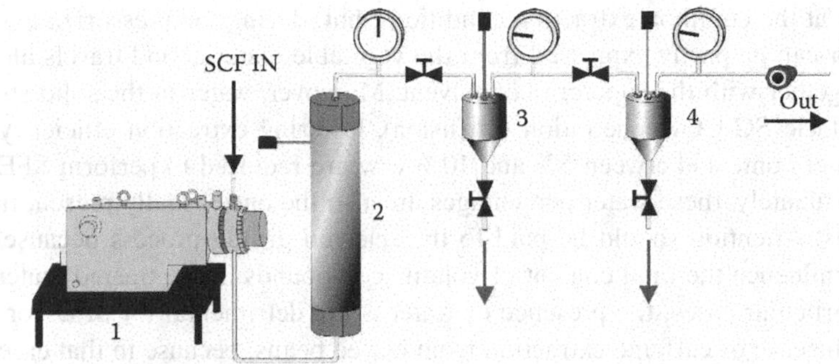

FIGURE 10.1 1) CO_2 pump; 2) extractor; 3) first separator; 4) second separator.

from undesired substances. These conditions can be frequently obtained since many organic solvents are readily soluble in SC-CO_2 even at mild operating conditions and many high-molecular-weight solids show negligible solubilities in SC-CO_2, especially at low CO_2 densities.

Due to the structural complexity and variability (with season, kind, crop, etc.) of the materials to be treated and to the large variety of compounds that can be extracted (different molecular weights, polarities, links with the structure, etc.), these processes are far from being exhaustively studied, although some industrial applications have already been developed. Moreover, an increasing interest has been registered in the extraction of high-added-value essential oils, such as oils that show antioxidant and pharmaceutical properties.

Therefore, in this chapter, we analyze SFE, SAE, and liquid fractionation studies performed on essential oils and related materials and consider the evolution of the extraction processes, products, and materials treated. A critical analysis is performed.

10.2 SOLIDS PROCESSING

Solids processing is the most studied SCF application because the most frequently required separation process is the extraction of one or more compound families from a solid natural matrix. The basic extraction scheme consists of an extraction vessel charged with the raw matter to be extracted. SCF at the exit of the extractor flows through a depressurization valve to a separator in which, due to the lower pressure, the extracts are released from the gaseous medium and collected. As a rule, the starting material is dried and grinded to favor the extraction process and is loaded in a basket located inside the extractor to allow fast charge and discharge of the extraction vessel.

More-sophisticated extraction schemes, such as the one reported in Figure 10.1, contain two or more separators. In this case, it is possible to fractionate the extract in two or more fractions of different compositions by setting opportune temperatures and pressures in the separators [11–28]. Solids preprocessing is also a parameter that can largely influence separation performance. For example, solid drying and particle size optimization, as a rule, have to be taken into account.

Drying of the solid materials is frequently required before extraction because raw vegetable matter can contain up to 90% water. Water is only slightly soluble in

SC-CO_2 at the common extraction conditions; but, during the pressurization of the extractor can be partly expressed from the vegetable material and travels along the plant together with the supercritical solvent. Moreover, water in the solid structure can obstacle SC-CO_2 penetration (diffusion), lowering extraction efficiency. As a rule, water contents between 5% and 10% w/w are required to perform SFE properly. Fortunately, these water percentages are also the ones usually present in dried materials. Attention should be paid to the selected drying process because it can largely influence the final content of volatile compounds in the treated material. In some particular cases, the presence of water is not detrimental for SFE, for example, in the case of caffeine extraction from coffee beans, because in that case water unhooks the caffeine sodium salt from the vegetable matrix.

Other possible variations of the SFE solid processing scheme are multistage extraction and cosolvents addition. Multistep operation involves varying pressure or temperature in each process step [29, 30]. This strategy can be used when the extraction of several compound families from the same matrix that show different solubilities in SC-CO_2 is required. It takes advantage of the fact that SC-CO_2 solvent power can be continuously varied with pressure and temperature. For example, it is possible to perform a first extraction operating at low CO_2 density (e.g., 0.29 g/cm^3, 90 bar, 50°C) followed by a second extraction step at high CO_2 density (e.g., 0.87 g/cm^3, 300 bar, 50°C). The most-soluble compounds (such as the essential oils) are extracted during the first step, whereas the less-soluble compounds (for example, antioxidants and coloring matter) are extracted in the second one [31–34].

A liquid cosolvent can be added to SC-CO_2 to increase its solvent power toward polar molecules. Indeed, SC-CO_2 is a good solvent for lipophilic (nonpolar) compounds, whereas it has a low affinity for polar compounds. Various authors added small quantities of liquid solvents (for example, ethyl alcohol) that are readily solubilized by SC-CO_2 and modify its solvent power [22, 26, 34–69]. This strategy has the drawback that a larger solvent power also implies a lower process selectivity and because, as a rule, the cosolvent is liquid at atmospheric pressure, it will be collected in the separator together with the extracted compounds. Subsequent processing for solvent elimination is required; therefore, one of the advantages of SFE (i.e., solventless operation) is lost.

Another possible process arrangement is the continuous feeding and discharging of the solid to obtain continuous processing of the solid matter [70]. This operation is made possible by adding two solid extruders at the top and bottom of the extractor and can avoid the use of two or more extractors to simulate continuous solid processing. Design and operation of the two extruders is not cheap and simple. A patent exists on this operation mode, but it has not yet been industrially applied.

10.2.1 Selection of the Operating Parameters

Selection of the operating conditions depends on the specific compound or compound family to be extracted; molecular weight and polarity have to be taken into account case by case. However, some general rules can be applied. First of all, processing temperature for thermolabile compounds has to be fixed between 35°C and 60°C (specifically, in the vicinity of the critical point and as low as possible to avoid degradation). The increase of temperature reduces the density of SC-CO_2 (for a fixed pressure), thus reducing the solvent power of the supercritical solvent. However, it also increases the vapor pressure of the compounds to be extracted; thus, the tendency of these compounds to pass in the fluid phase is increased. The most relevant process parameter is the extraction pressure that can be used to tune the selectivity of the SCF. The general rule is this: the higher the pressure, the larger the solvent power and the smaller the extraction selectivity. Frequently, solvent power is described in terms of the SC-CO_2 density at the given operating conditions. CO_2 density can vary from about 0.15 to 1.0 g/cm^3 and is connected to both pressure and temperature. Its variation is strongly nonlinear; therefore, proper selection requires the use of tables of CO_2 properties [71, 72].

The other crucial parameters in SFE are CO_2 flow rate, particle size of the matrix, and duration of the process (extraction time). The proper selection of these parameters has the scope of producing the complete extraction of the desired compounds in the shortest time. They are connected to the thermodynamics (solubility) and kinetics of the extraction process in the specific raw matter (mass transfer resistances). The proper selection depends on the mechanism that controls the process; the slowest one determines the overall process velocity. CO_2 flow rate is a relevant parameter if the process is controlled by external mass transfer resistance or by equilibrium; the amount of supercritical solvent feed to the extraction vessel, in this case, influences the extraction rate. Particle size plays a determining role in extraction processes controlled by internal mass transfer resistances; a smaller mean particle size reduces the length of diffusion of the solvent. However, if particles are too small, they can cause channeling problems inside the extraction bed. Part of the solvent flows through channels formed inside the extraction bed and does not contact the material to be extracted, thus causing a loss of efficiency and yield of the process. As a rule, particles with mean diameters ranging between approximately 0.25 and 2.0 mm are used. The optimum dimension can be chosen case by case considering the water content in the matrix and the quantity of extractable liquid compounds that can produce coalescence among the particles, thus favoring irregular extraction along the extraction bed. Moreover, the production of very small particles by grinding could produce the loss of volatile compounds. Process duration is interconnected with CO_2 flow rate and particle size and has to be properly selected to maximize the yield of the extraction process.

Essential oils are mainly formed by hydrocarbon and oxygenated terpenes and by hydrocarbon and oxygenated sesquiterpenes. They can be extracted from seeds, roots, flowers, herbs, and leaves using the process of hydrodistillation (HD). HD is a very simple process but suffers from many drawbacks: thermal degradation (for example, of *cis*-sabinene hydrate and *cis*-sabinene hydrate acetate in marjoram essential oil), hydrolysis (for example, of linalyl acetate in lavender essential oil), and solubilization in water of some compounds that alter the flavor and fragrance profile of many essential oils extracted by these techniques. In some cases, a comparison between the compositions of the essential oils obtained by SFE and those obtained by HD has been made. For example, in the case of rosemary oil [1], the SFE essential oil contained higher percentages of linalool, verbenone, and isobornyl acetate; their content was almost double that in the hydrodistilled oil. The difference was more evident in terms of the total percentage of oxygenated monoterpenes (which strongly contribute to the fragrance): 73.7% for the SFE oil against 59.4% for the hydrodistilled oil. Organoleptic tests confirmed that the hydrodistilled oil possessed a less intense rosemary aroma.

In some other cases, liquid solvent extraction is also performed; the results are the so-called oleoresins or concretes that are solid or semisolid and contain essential oil compounds together with waxes, fatty acids, coloring matter, antioxidants, and other compounds. Volatile oil or absolute is also sometimes the target of the process that is roughly speaking the volatile fraction of the oleoresin.

Essential oil isolation is an example of extraction plus fractional separation. Indeed, this process can be optimally performed operating at mild pressures (from 90 to 100 bar) and temperatures (from 40°C to 50°C) because, at these process conditions, all the essential oil components are largely soluble in SC-CO_2 [73–76]. For example, at 40°C, linalool is completely miscible with SC-CO_2 at pressures greater than about 85 bar [73], and limonene [74–76], α-pinene [76], and fenchone [76] are completely soluble at about 80 bar.

However, essential oil compounds are at least partly located inside the vegetable structure; therefore, mass transfer resistances have to be considered, too. At the previously discussed operating conditions, essential oil components are extracted together with cuticular waxes (i.e., paraffinic compounds located on the surface of vegetable matter with the scope of controlling its perspiration). Paraffins exhibit a relatively low solubility at these operating conditions [77]. When extraction pressure is increased, their contribution in the extract is more prevalent and other compounds (such as fatty acids) can also be increasingly extracted. Since waxes are on the structure surface, their extraction is controlled by their solubility, whereas essential oil extraction is controlled, at least in part, by internal mass transfer resistances in the vegetable structure. As a result of these interactions, essential oil and waxes are coextracted at all operating conditions. To isolate the essential oil, it is necessary to take advantage of the fact that, at low temperatures (from –5°C to +5°C), waxes are practically insoluble in CO_2, whereas terpenic compounds maintain very large solubilities (they are completely miscible in liquid CO_2). Therefore, it is possible to obtain a fractionation operating, for example, the extraction at 90 bar and 40°C and, then, performing a first separation, for example at 0°C, 90 bar, and a second separation at 15°C, 20 bar. In this manner, the selective precipitation of waxes is obtained in the first separator and no precipitation of the other extracted compounds occurs, whereas, in the second separator, essential oil is recovered. An industrial plant (V = 1200 dm^3) that uses this process arrangement has been constructed and successfully operated since 1996 (Essences, Italy). One must take into account, however, that it is not possible to perform SFE directly at 0°C and 90 bar because the vegetable matter contains many other compound families (antioxidants, colors, etc.) that are soluble at these process conditions and, therefore, a complex mixture of essential oil plus these other compounds is obtained.

Data on essential oils, volatile oils, and oleoresins obtained by SFE are shown in Table 10.1, alphabetically organized by the common name (raw material), the botanical name, and the target components (the extract). In Table 10.1, laboratory, pilot plant, and analytical studies performed using very small extractors are included.

TABLE 10.1
SFE of Oleoresins (OR), Essential Oils (EO), and Volatile Oils (VO)

Raw Material	Botanical Name	Extract	References
	Leaves		
Basil	*Ocimum basilicum*	EO	[73,78]
Eucalyptus	*Eucalyptus globulus* L.	EO	[11]
Laurel	*Laurus nobilis*	EO	[14]
Lemon balm	*Melissa officinalis*	EO	[79]
Lemon bergamot	*Monarda citriodora*	EO	[79]
Lemon eucalyptus	*Eucalyptus citriodora*	EO	[79]
Lemongrass	*Cymbopogon citratus*	EO	[79,80]
Lovage	*Levisticum officinale* Koch.	EO	[22,63,81]
Marjoram	*Origanum majorana* L.	EO	[82,83]
Mint	*Mentha spicata insularis*	EO	[15,78]
Oregano	*Origanum vulgare* L.	EO	[84]
Sage	*Salvia desoleana*	EO	[15,23,63,85]
Spiked thyme	*Thymbra spicata*	EO	[86]
Thyme	*Thyme zygis sylvestris*	EO	[87]
	Flowers		
Chamomile	*Chamomilla recutita* L. R.	EO and OR	[88,89]
Lavender	*Lavandula angustifolia*	EO	[90]
	Seeds		
Aniseed	*Pimpinella anisum* L.	EO	[91]
Fennel	*Foeniculum vulgare* Mill.	EO	[17]
Lovage	*Levisticum officinale* Koch.	EO	[22,63,81]
	Roots		
Celery	*Apium graveolens* L.	EO	[22]
Lovage	*Levisticum officinale* Koch.	EO	[22,63,92]
	Other Matrices		
Bacuri fruit shells	*Platonia insignis* Mart.	EO	[93]
Black pepper fruits	*Piper nigrum* L.	EO	[94,95]
Cashew nut shell	*Anacardium occidentale*	VO	[30,96]
Clove bud	*Eugenia caryophyllata*	EO	[13]
Lemon balm herb	*Melissa officinalis*	EO	[16,21]
Oregano herb	*Origanum vulgare* L.	EO	[61,63,78,97]
Pennyroyal plant	*Mentha pulegium* L.	EO	[50]

continued

TABLE 10.1 (continued)
SFE of Oleoresins (OR), Essential Oils (EO), and Volatile Oils (VO)

Raw Material	Botanical Name	Extract	References
	Other Matrices (continued)		
Red pepper fruits	*Capsicum frutescens* L.	OR	[98]
Star anise	*Illicium anisatum*	EO	[13]
Juniper fruits	*Juniperus communis L.*	VO	[99]
	Concretes		
Jasmine	*Jasminum grandiflorum* L.	VO	[100]
Rose	*Rosa damascena* Mill.	VO	[101]
Tuberose	*Nepeta tuberosa* L.	EO	[12]

In only some cases the operating conditions have optimized to maximize both the yield and the selectivity of the process. Therefore, the yield and the operating conditions indicated by the authors are largely influenced by the final scope of the paper: to isolate the essential oil or to extrapolate its composition from the unfractionated extract ("concrete like"). An analysis on the influence of some process parameters, such as pressure, temperature, extraction time, percentage of cosolvents, and solvent flow rates, is available in some of the papers considered in Table 10.1.

In several cases, the same matrix contains essential oils (or volatile oils) with known biological activity and high-molecular-weight compounds that exhibit nutraceutical or pharmaceutical activity (see Table 10.2 for several examples). A large spectrum of compounds can be inserted in these categories because food additives with nutritional and pharmaceutical properties (nutraceuticals) range from tocopherols to carotenoids to alkaloids to unsaturated fatty acids. Pharmaceutical compounds like Artemisinin (an antimalaria drug), Hyperforin (an antidepressant drug), and sterols can be extracted from various matters. In these papers, a different emphasis is given to these characteristics and the extract is characterized more for its functionality than with respect to essential oil composition.

10.2.2 Examples

As previously stated, essential oils can be extracted from different matrices, including leaves, flowers, and seeds. In this section, we illustrate some examples of extraction.

10.2.2.1 Leaves

Essential oil isolation has been performed as a rule in plants operated with at least two separators in series. An example of SFE from leaves is sage (*Salvia officinalis*) essential oil extraction [85]. The best isolation conditions have been found at 90 bar and 50°C. Sage waxes elimination has been demanded to the first separator, fixing the conditions at 85 bar and –12°C. In the second separator, operating at 17 bar,

TABLE 10.2
Essential Oil–Related and Biologically Active Compounds

Raw Material	Botanical Name	Extract	References
		Leaves	
Aloe vera	*Aloe barbadensis* Miller	α-tocopherol	[45]
Eucalyptus	*Eucalyptus camaldulensis* var. *brevirostris*	Gallic and ellagic acids	[39]
Hawthorn	*Crataegus* sp.	Flavonoids and terpenoids	[48]
Marjoram	*Origanum majorana* L.	Carotenoids and chlorophylls	[19]
Marjoram	*Origanum majorana* L.	Phenolic and triterpenoid antioxidants	[102,103]
Rosemary	*Rosmarinus officinalis* L.	Rosmanol, carnosic acid, and carnosol	[31–33,36, 41,104–107]
Sage	*Salvia officinalis* L.	Carnosolic acid	[23]
Savory	*Satureja hortensis* L.	Oil	[20]
		Flowers	
Chamomile	*Matricaria recutita*	Flavonoids and terpenoids	[48]
Hawthorn	*Crataegus* sp.	Flavonoids and terpenoids	[48]
Marigold	*Calendula officinalis*	Flavonoids and terpenoids	[48]
		Seeds	
Coriander	*Coriandrum sativum*	Tocopherols, flavonoids, and terpenoids	[108]
		Other Matrices	
Anise verbena	*Lippia alba*	Limonene and carvone	[109,110]
Coffee powder	*Coffea arabica*	Aroma	[111]
Ginger rhyzomes	*Zingiber officinale* Roscoe	Gingerols and shogaols	[40,41]
Horsetail plant	*Equisetum giganteum* L.	Oleoresin	[112]
Moso-bamboo plant	*Phyllostachys heterocycla*	Ethoxyquin A, sesquiterpene A, and cyclohexanone A	[46]
Paprika flake	*Capsicum annuum* L.	Carotenoids, tocopherols, and capsaicinoids	[113,114]
Saw palmetto berries	*Serenoa repens*	Fatty acids and β-sitosterol	[56]
Spearmint plant	*Mentha spicata*	Tocopherol	[115]

–6°C, only essential oil is found. The essential oil obtained is wax-free and contains only traces of high-molecular-weight compounds. These high-molecular-weight compounds were identified as two flavones. The main compounds extracted are 1,8-cineole, camphor, and caryophyllene. The asymptotic value of the sage essential oil yield, expressed as weight of extract divided by the weight of the starting material, is 1.35 wt % of the charged material.

10.2.2.2 Flowers

As examples of SFE of essential oils from flowers, we consider lavender [90] and chamomile [89] essential oil extraction and isolation. In the case of lavender flowers [90], the operating conditions, in terms of pressure and temperature, have been fixed as follows: in the extractor, 90 bar and 48°C; in the first separator, 80 bar and –10°C; in the second separator, 25 bar and 0°C. Waxes precipitate in the first separator, and the lavender essential oil precipitates in the second separator and mainly consists of 1,8-cineole, linalool, camphor, 4-terpineol, α-terpineol, and linalyl acetate. The asymptotic value of the lavender essential oil yield is 4.9 wt % of the charged material.

In the case of chamomile flower [89], the operating conditions are similar. The major constituents of this essential oil are oxygenated sesquiterpenes, which represent 78.48% of the oil composition. In Table 10.3, the compositions of the chamomile essential oil and the chamomile cuticular waxes have been reported to give an example of detailed identification of essential oil components. Oxygenated sesquiterpenes contain the most characteristic chamomile essential oil compounds, namely bisabolol oxide B (16.88%), α-bisabolol (0.35%), bisabolone oxide (7.76%), and bisabololoxide A (50.42 %). Matricine (evaluated as chamazulene) represents, in this case, 3.52% and dicycloethers contribute more than 12.97% to the total extract. The yields obtained are 1.18% for essential oil and 0.8% for cuticular waxes.

10.2.2.3 Seeds

In the case of seed oils, at least two SC-CO_2 extractable compound families are contained in the vegetable matrix: essential oil and seed oil; therefore, extraction conditions have to be set to avoid their coextraction. An example is given by fennel essential oil isolation [116]. The first step of the extraction process is performed at 90 bar and 50°C, with the aim of selectively extracting fennel essential oil. Waxes elimination is demanded to the first separator. The extraction and simultaneous isolation of fennel essential oil has been successful. In the first separator, paraffinic waxes are collected with carbon atom numbers between 25 and 37; this wax composition agrees well with that of various other vegetable matter extracted by SC-CO_2 [1]. In the second separator, fennel essential oil is collected. It is mainly formed by estragole (about 80%), anethole, fenchone, and limonene and is not contaminated by waxes or by higher-molecular-weight compounds. An essential oil asymptotic yield of 1.8 wt % of the loaded material has been obtained.

The second step of the extraction process, performed at 40°C and 200 bar, produces the extraction of fennel vegetable oil. Also, in this case, the first separator is used to precipitate coextracted waxes. The white mass collected in the first separator is again formed by paraffins, although their molecular weight is slightly larger (carbon atoms from 25 to 41). Fennel oil is collected in the second separator. However, higher pressures are commonly used in this step of the process to accelerate the extraction of the vegetable oil.

10.2.2.4 Other Matrices

As examples of extraction from different matrices, we consider clove bud and star anise essential oils. In both cases, the best essential oil process conditions are 90 bar

TABLE 10.3
Composition of the Chamomile Essential Oil and of Cuticular Waxes (Adapted from [89])

Compound	Retention Time (min)	Area (%)
	Essential Oil	
6 methyl-5-hepten-2-one	20.02	0.07
Ocimene	25.33	0.11
Linalool	28.26	0.57
Isoborneol	33.07	0.10
Menthol	33.43	< 0.05
4-terpineol	34.02	0.07
α-terpineol	35.00	0.09
n-id. $C_{10}H_{16}O$	38.25	0.11
Nerol	39.58	0.65
Geraniol	41.07	0.24
Menthyl acetate	45.13	0.17
n. id. $C_{12}H_{22}O_2$	48.10	0.17
β-elemene	49.17	< 0.05
β-caryophyllene	51.01	0.13
β-farnesene	53.33	1.53
trans-nerolidol	60.07	0.42
Spathulenol	60.53	0.65
Caryophyllene oxide	63.18	0.17
n. id. $C_{15}H_{26}O$	64.03	0.39
T-cadinol	64.39	0.36
Bisabolol oxide B	65.38	16.88
α-bisabolol	66.14	0.35
Bisabolone oxide	67.09	7.76
Matricine (chamazulene)	69.39	3.52
Bisabolol oxide A	70.53	50.42
n. id. $C_{15}H_{26}O$	71.43	0.34
n. id.	72.09	0.56
n. id. $C_{15}H_{26}O$	74.07	0.18
cis-dicycloether MW 200	77.46	9.64
trans-dicycloether MW 200	78.23	3.33
trans-farnesol	79.02	0.32
cis,trans-farnesol	79.55	0.42
cis-dicycloether MW 214	79.92	< 0.05
trans-dicycloether MW 214	81.18	< 0.05

continued

TABLE 10.3 (continued)
Composition of the Chamomile Essential Oil and of Cuticular Waxes (Adapted from [89])

Compound	Retention Time (min)	Area (%)
Waxes		
Hexadecane	13.12	0.44
Octadecene	17.43	2.17
Docosene	26.82	0.83
Tricosane	29.59	1.64
Tetracosane	32.09	0.31
Pentacosane	34.65	10.50
Hexacosane	36.98	1.52
Methylhexacosane	38.73	0.26
Heptacosane	39.74	17.56
Octacosane	41.78	2.74
Methylheptacosane	43.07	0.85
Nonacosane	43.94	24.12
Triacontane	45.54	2.86
Entriacontane	47.73	19.71
Methyltriacontane	49.24	1.26
Dotriacontane	49.80	1.54
Methylentriacontane	51.39	1.16
Tritriacontane	52.57	9.46
Methyldotriacontane	54.70	1.13

and 50°C for extraction, 90 bar and –10°C in the first separator, and 15 bar and 10°C in the second separator.

The obtained clove bud essential oil comprises mainly eugenol (65.9%), caryophyllene (11.1%), and eugenyl acetate (19.0%). The yield is 20.7% by weight of the material charged in the extractor.

The obtained star anise essential oil contains 94.2% anethole, 1.4% estragole, 1.7% limonene, and 0.3% linalool. The asymptotic yield is 7.3% by weight of the charged material.

10.2.2.5 Flower Concretes Fractionation

When vegetable materials characterized by a very short life have to be processed, as in the case of many flowers, the fragrance production is performed in two steps. The first step consists of solvent extraction, usually by hexane, which yields an intermediate product called "concrete." It is mainly composed of fragrance-related compounds but also contains large quantities of paraffins, fatty acids, fatty acids methyl esters, diterpenic and triterpenic compounds, pigments, and other substances. In the second step, the concrete is postprocessed by steam distillation or solubilization in a large excess of alcohol to obtain a volatile oil containing the fragrance. These conventional postprocessing techniques are subject to some disadvantages, such as thermal degradation, incomplete elimination of nonvolatile compounds, and fractionation of the compounds that form the fragrance. Because flower essential oils have a large commercial value (thousands of dollars per liter), exploring the use of SC-CO_2 extraction as a new way to fractionate the concrete could be beneficial.

In the case of rose concrete [101], a preliminary study showed that the major volatile compounds contained in the starting material were 2-phenylethanol (25.1%), citronellol (4.7%), and 2-phenylethyl acetate (2.7%). However, it also contained many other compounds that do not contribute to rose fragrance formation, like paraffins. Among these, two long chain paraffinic alcohols have been detected. They are commonly called "steroptens" and are characteristic of rose concrete and adversely contribute to rose fragrance.

Rose concrete has been warmed up to 35°C, mixed with 2-mm diameter glass beads to perform SC-CO_2 processing, and then charged into the extractor. The mixing has been performed to obtain a thin layer of concrete around the glass beads. This procedure has been used to maximize the contact surface between the concrete and the supercritical solvent and to avoid channeling. The solution at the exit of the extractor, as for solid extraction, flowed into the two separators operated in series, in order to fractionate the extract. The first separator was set at 80 bar and –16°C, whereas the second separator was set at 15 bar and 0°C to minimize the loss of volatile compounds in the gaseous CO_2 stream at the exit of the apparatus. At the end of the extraction process, glass beads were recovered by washing with warm ethanol.

The optimized SFE is performed at 80 bar and 40°C, and a maximum yield of volatile compounds of 49% by weight of the charged material has been recovered in the second separator using fractional separation. The process was extremely selective: no unwanted compounds were detected. The waxes yield in the first separator was about 2% by weight operating at 80 bar and 40°C but can reach up to 10.4% by weight if the extraction is performed at 120 bar and 40°C.

At the end of the SFE process at 80 bar and 40°C, on the exhausted charge, a further extraction step has been performed operating at 120 bar and 40°C for 100 min. The higher-pressure step yields a further 14.9% by weight of the charge. The extract is still liquid but contains only 1.0% 2-phenylethanol, whereas the percentage of steroptens is 21.3%. The second step was performed to evaluate whether other valuable compounds were still contained in the starting material and were not previously extracted at 80 bar and 40°C.

In the case of tuberose flowers, concrete fractionation is also required because SC-CO_2 extraction performed directly on the tuberose flower is not applicable at an industrial scale since the yield in essential oil from flowers is less than 0.1% by weight. Tuberose concrete fractionation using SC-CO_2 has been performed [12] with the aim of separating volatile oil from the higher-molecular-weight compounds. The extraction process has been coupled again with the fractional separation technique that uses two separation stages operating in series.

Systematic SFE tests on tuberose concrete have been performed in the range of 80 to 100 bar, operating at a temperature of 40°C and analyzing the product collected in the two separators by gas chromatography–mass spectrometry. By operating at 80 bar, the researchers obtained the maximum content of fragrance compounds in the extract collected in the second separator.

Since the different compound families that constitute the tuberose volatile oil (hydrocarbon terpenes, oxygenated terpenes, and benzene derivatives) show different extraction rates, the compositions of the tuberose oil changes during the extraction process. The extract recovered after 20 minutes of extraction contains a high percentage (> 83%) of oxygenated compounds (monoterpenes and benzene derivatives). The volatile fraction recovered in the extraction time interval between 360 and 480 minutes consists of a lower percentage (< 79%) of oxygenated compounds with an increment of the percentage of the trans-methylisoeugenol and of the eugenyl acetate, whereas the most volatile compounds, such as methyl benzoate and methyl salicylate, are reduced.

The fraction recovered in the extraction time interval between 690 and 750 minutes contains very low quantities of aroma compounds but still contains large quantities of trans-methylisoeugenol, eugenyl acetate, and lactones.

By dividing the tuberose oil compounds into three families, hydrocarbon compounds (monoterpenes and sesquiterpenes), oxygenated compounds with 10 or less carbon atoms (monoterpenes and benzene derivatives), and oxygenated compounds with 15 or more carbon atoms (sesquiterpenes and lactones), the contribution of each compound family can be calculated as the sum of the area contribution of all compounds belonging to that family. Tuberose oil contains a low percentage of hydrocarbon compounds (monoterpenes and sesquiterpenes) and their percentage decrease by increasing the extraction time. The percentage of oxygenated compounds with 10 or less carbon atoms also decreases during the extraction, whereas the percentage of oxygenated compounds with 15 or more carbon atoms increases, especially at extraction times longer than 450 minutes. This means that different solubilities and perhaps mass transfer resistances characterize the various compound families during the extraction process and the extraction time plays a relevant role in the final composition of tuberose oil. Moreover, by interrupting the extraction of the oil at different times, it is possible to fractionate the fragrance and to obtain an extract in which top-notes or bottom-notes prevail.

It is also possible to divide the tuberose oil compounds in two groups: the fragrance compounds (oxygenated compounds) and the nonfragrance compounds (hydrocarbon compounds). The yield curves show an exponential trend against the extraction time; the hydrocarbon compounds curve gets flat after the first 300 minutes of extraction, whereas the yield curve of fragrance compounds asymptotizes only when complete extraction is performed (after 750 minutes).

As confirmed by the two examples, SFE is generally applicable to flower concretes and can be very selective. Thus, it can be used to recover in a relatively cheap, single-step SFE operation essential oils widely used in the perfume industry. A better product is also obtained with SFE than with traditional techniques.

10.3 LIQUID FEED PROCESSING

The fractionation of liquid mixtures into two or more fractions is another relevant process. In a typical apparatus, two pumps deliver the liquid solution and SC-CO_2 to

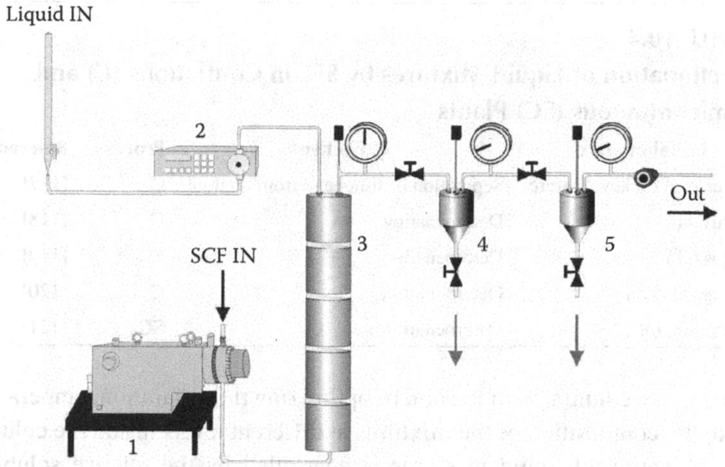

FIGURE 10.2 1) CO_2 pump; 2) liquid pump; 3) packed tower; 4) first separator; 5) second separator.

the packed column. The packing is an inert material characterized by a large specific surface whose scope is to favor the contact between the liquid and the SCF. SC-CO_2 generally flows along the column from the bottom to the top, whereas the liquid solution is usually added to the top. However, it is also possible to feed the liquid at an intermediate position along the column and to add a recycle of part of the fluid phase exiting at the top. A scheme of the apparatus is shown in Figure 10.2.

10.3.1 Selection of the Operating Parameters

Selection of the operating parameters is based on the different solubilities of the liquids to be separated in SC-CO_2. The ideal case is obtained when only the compounds to be extracted are soluble in SC-CO_2 and all the other liquid components are completely insoluble. However, this case is rare and a limited solubility of the other liquid compounds forming the mixture has to be taken into account. For this reason, pressure and temperature of the process have to be accurately chosen to select the conditions at which the maximum difference in solubility exists among the compounds to be extracted and all the other compounds in the mixture. Also in this case, CO_2 density is frequently used as a criterion to find the conditions of maximum selectivity. The difference in density between the liquid and SC-CO_2 is another parameter to be taken into account; to allow the countercurrent operation, SCF density has to be lower than the density of the liquid mixture.

The traditional operation of packed columns requires that liquid flow rate be larger than the minimum amount that assures the complete wetting of the packing. The feed ratio is also selected to avoid the massive entrainment of the liquid in the fluid phase (flooding). These conditions have to be respected also when a SCF is used as the fluid processing medium. The classical calculation in terms of the number of theoretical equilibrium stages required for separation can also be applied. A possible variation of this processing scheme can consist of the adoption of a temperature

TABLE 10.4
Fractionation of Liquid Mixtures by SFE in Continuous (C) and Semicontinuous (SC) Plants

Initial mixture	Objective	Process	References
Citrus peel oil key mixture	Separation of limonene from linalool	C	[117]
Citrus oil	Deterpenation	C	[118]
Citrus oil	Deterpenation	SC	[119]
Orange peel oil	Deterpenation	C	[120]
Oreganum oil	Deterpenation	SC	[121]

profile along the column, with the aim of optimizing the separation temperature with respect to the composition of the mixtures at different levels inside the column.

The extraction of liquid mixtures is controlled by the relative solubilities in SC-CO_2 of the various compounds forming the mixture—that is, the thermodynamic limitation of the process. Mass transfer between the two phases represents the kinetic limitation. The distance from the equilibrium condition is the driving force for the separation along the column.

10.3.2 Examples

The fractionation of liquid mixtures by SFE has been proposed for various applications, including some essential oils, with the aim of improving their fragrance and eliminating hydrocarbon terpenic compounds that can rapidly decompose and therefore can shorten the shelf life of the product. The problem does not have a simple solution. Selective elimination of hydrocarbon terpenes (deterpenation) is required in citrus peel oils because these compounds contribute to a small extent to the citrus peel fragrance and are rapidly oxidized by air and can undergo structural rearrangements. These essential oils are mainly formed by hydrocarbon and oxygenated terpenes, but also contain small quantities of sesquiterpenes and high-molecular-weight compounds like coumarins, psoralens, and waxes. Hydrocarbon terpene percentage can range from 60% to 99%. In Table 10.4, some examples of fractionations of liquid mixtures are reported.

The fractionation of a peel oil with SC-CO_2 was studied using a mixture of four key compounds [117]. The composition of this mixture was determined by the following considerations. Limonene is a hydrocarbon terpene and it is the predominant compound in all peel oils, with concentrations between 30% and 80%. Therefore, it can be chosen as the most abundant key compound, with a concentration of 60% by weight. Linalool is one of the most representative among the oxygenated compounds and, therefore, its concentration was set at 20% by weight, representing the second most abundant component in the key mixture. γ-Terpinene has a molecular weight similar to limonene but is a hydrocarbon terpene that has a volatility very near to oxygenated compounds. Therefore, its presence in the mixture (10%) allows evaluation of the effectiveness of separation for those compounds that are more difficult to separate than limonene. Linalyl acetate (10%) is representative of oxygenated compounds with a molecular weight higher than linalool.

Systematic experiments [117] have been performed on a 2-m column operated in countercurrent and equivalent to about 2.5 theoretical equilibrium stages. The operating pressure and temperature ranged between 75 and 90 bar and between 40°C and 80°C, respectively. Solvent-to-feed ratios of 60, 80, and 120 were used. The effects of different feed insertion points and column packings were also tested. Experimental results indicated that fractionation could be successfully obtained between 75 and 80 bar and between 50°C and 80°C. In general, an increase in solubility corresponds to a decrease in selectivity and, thus, optimization of the separation is required. Experiments also indicated that temperature helps separation and, furthermore, increases the recovery of oxygenated compounds. The upper limit to the operating temperature is given, however, by the thermal stability of the product. The total and partial refluxes of the extract at the column top show a definitely positive effect on the separation. However, only when the operation in a two-column series was performed (i.e., using a length of column equivalent to about 5 equivalent stages), researchers observed the elimination in the top product of linalyl acetate and the reduction of linalool content to about 0.8%.

Liquid feed can be alternatively fractionated by adsorption/desorption of the liquid mixture on an adsorbent. For example, SC-CO_2 has been used to selectively desorb bergamot peel oil components from silica gel [122]. The maximum desorption selectivity has been obtained operating at 40°C, in two successive pressure steps. The first step, performed at 75 bar, produces the selective desorption of hydrocarbon terpenes; the second one, performed at 200 bar, assures the fast desorption of all the oxygenated compounds.

In another work [123], thirteen terpenes forming a mixture characteristic of peel essential oils were selectively adsorbed on silica gel using SC-CO_2. These compounds are α-pinene, β-pinene, myrcene, limonene, γ-terpinene, β-caryophyllen, citronellyl acetate, geranyl acetate, linalyl acetate + geraniol, linalool, citronellal, and citral. They can be grouped in four families of pseudocomponents: hydrocarbon terpenes (like limonene), a sesquiterpene (β-caryophyllene), terpene acetates (like geraniol), and oxygenated compounds (like linalool). The experiments were performed at different pressures (130 to 210 bar), temperatures (37°C to 57°C), and concentrations (0.9 to 7.6 $g/kg_{solvent}$). The different pseudocomponents show a different adsorption behavior that can be justified looking at the interactions with the active sites of the adsorbent. Hydrocarbon terpenes and sesquiterpenes can be adsorbed on CH_3-groups of silica gel and are rapidly shifted by the more polar compounds (displacement). Oxygenated compounds can also bond with OH-groups (silanol). Compounds with higher molecular weights or higher polarities can move (displace) the other species from the adsorption site in which they were located. The displacement of the less strongly adsorbed compounds is more evident at lower concentrations and allows the selective recovery of the various fractions at the exit of the adsorption/desorption column.

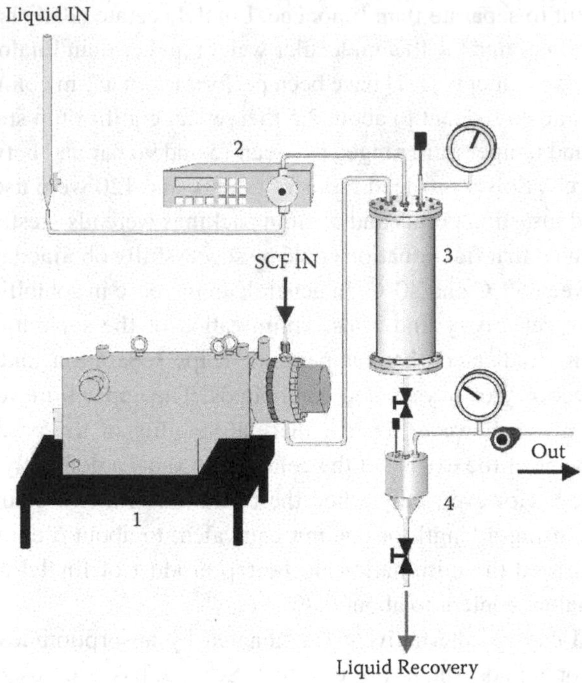

FIGURE 10.3 1) CO_2 pump; 2) liquid pump; 3) precipitator; 4) separator.

10.4 ANTISOLVENT EXTRACTION

The recovery of solid compounds from a liquid mixture requires different process approaches since the fixed bed extractor does not adapt to process liquid mixtures. In addition, the packed tower cannot be used in these cases because the solid compounds precipitate on the internal packings.

The Supercritical antisolvent extraction (SAE) process is conceptually very similar to Supercritical antisolvent micronization (SAS), but the scope of the process is the recovery of one or more solid compounds from a liquid mixture. It consists of the continuous flow of SC-CO_2 and of the liquid mixture in a pressurized precipitation vessel. If the process conditions have been properly selected, the liquid is rapidly dissolved in the SCF, whereas the solid precipitates at the bottom of the precipitation vessel. Therefore, in a possible representation of the process, two pumps deliver the liquid solution and the SCF, respectively. The precipitation vessel is used to collect the solid and a vessel located downstream the precipitator and operated at lower pressure (for example, 30 bar and 25°C) is used to recover the liquid. A schematic of the apparatus is shown in Figure 10.3.

10.4.1 Selection of the Operating Parameters

The first step of this process is the formation of a spray of the liquid solution. The intent of this operation is to produce a very large liquid surface due to the formation of small liquid droplets to strongly enhance the rate of solubilization of the liquid phase in the supercritical medium. For the same reason, the process is performed at operating conditions at which the liquid solvent is completely soluble in SC-CO_2. The knowledge of solubility data on the liquid solvents and of the solids in SC-CO_2 is mandatory for the proper selection of process temperature and pressure.

In the case of SAE, the interactions between thermodynamics constraints and mass transfer mechanisms also control the process performance. The enhanced mass transfer that characterizes SCFs is again a distinctive advantage of their use as extraction media, together with the fast and complete separation by simple depressurization between the supercritical solvent and the liquid.

A limitation of this process is the possible formation of a ternary mixture liquid/solid/SC-CO_2. Indeed, the presence of the liquid can induce an increase of the solubility of the solid compounds in SC-CO_2. In this case, the liquid can act as a cosolvent from the point of view of solid solubilization. When this phenomenon occurs, the part of the solid retained in the fluid phase obviously does not precipitate and is lost in the liquid recovered in the separation vessel. The limit case is the complete solubilization of the solid in the fluid phase that produces the process failure.

10.4.2 Examples

Until now, SAE has been used in a limited number of processes (for example, the recovery of essential oil from a mixture essential oil + triglyceride oil [124, 125]) but it has a large potential for future applications, some of which are discussed here.

10.4.2.1 Proteins and Aroma Extraction from Tobacco

A process has been developed for the recovery of tobacco proteins, amino acids, and aroma using ethyl alcohol [126]. A fractionation process is required to separate protein-related (solid) compounds from tobacco aroma, obtaining the simultaneous elimination of the liquid solvent. The ethanolic extract is prepared using an ethanol solution 1% potassium hydroxide using a solution containing a tobacco blend of approximately 10 to 1 (v/w), for 3 hours at 40°C. Potassium hydroxide is added to the solution for its stabilization.

Ethanol is readily soluble in SC-CO_2; therefore, the couple solvent-antisolvent should not deserve problems in SAE processing. Proteins and their derivative amino acids should be virtually insoluble in SC-CO_2 due to their composition and their molecular weights. Flavoring compounds can also be readily extracted. Therefore, based on these considerations, a fractionation of the ethanolic extract is, in principle, possible if processing conditions are selected to induce proteic compound precipitation and the transfer in the fluid phase of the liquid solvent together with flavoring components.

Experiments have been performed at different pressures and solid concentrations. Good results have been obtained at 150 bar, 40°C, and 100 mg/mL. The ethanolic extract has been efficiently fractionated, and the yield of precipitated material was about 40% (w/w of extract).

10.5 MATHEMATICAL MODELLING

Mathematical modelling gives the possibility to generalize experimental results and, if successful, to obtain indications about systems different from those studied (simulation). Moreover, it is useful in the development of scaling-up procedures from laboratory to pilot and industrial scales. For these reasons, several attempts at mathematical modelling of SFE have been presented in the literature [1, 127, 128] and some of these models are related to SFE of essential oils.

A model should not be a mere mathematical instrument, but it should reflect the physical knowledge of the solid structure and the experimental observations. Therefore, mathematical models that have no physical correspondence to the materials and the process studied are of limited validity, although they can be used to fit some experimental data.

Three different approaches have been proposed for the mathematical modelling of SFE: (1) empirical [129, 130], (2) based on heat and mass transfer analogy [131, 132], and (3) differential mass balances integration [127, 133, 134]. The most proper analysis is obtained from the integration of the differential mass balances: time-dependent concentration profiles are obtained for fluid and solid phases.

In facing mathematical modelling of SFE, several general aspects have to be taken into account:

1. *Solid material structure:* Knowledge of the botanical aspects or scanning electron microscope (SEM) analysis of the material is necessary to visualize the structure. For example, seeds are essentially formed by specialized structures that operate as small recipients containing the oil. Their shape and structure change seed by seed, but the general organization is always the same.
2. *Location of the compounds to be extracted:* The distribution of the solute within the solid substrate may be very different. The extractable substances may be free on the surface of the solid material or inside the structure of the material itself. Essential oil can be located near the leaf surface in glandular trichomes or in vacuoles (intracellular structures located well inside the leaf) [29, 135].
3. *Interactions of solutes with the solid matrix:* Depending on the interactions between the compounds and the solid structure, different equilibria may be involved. Indeed, if the material has no interactions with the matrix, equilibrium solubility has to be taken into account. The material can be adsorbed on the outer surface or inside the solid structure; in this case, a partitioning equilibrium between solid and fluid phase occurs.
4. *Broken-intact cell structures:* Part of the compounds to be extracted may be near the surface of the structure due to cell breaking during grinding. Moreover, membranes modifications may occur due to drying, freeing part of the soluble material.
5. *Shape of particles:* Particles may be spherical, platelike, or other shapes as a result of the original shape of the material (for example, leaves) and of the grinding process. Their shape can influence the diffusion path of the supercritical solvent [29, 127].

TABLE 10.5
Mathematical Modelling of SFE of Essential Oils

Raw Material	Extract	Type of Model	References
Basil leaves	Essential oil	DMBE	[136]
Caraway seeds	Essential oil	DMBE	[136]
Clove bud	Essential oil	DMBE	[137]
Fennel seeds	Essential oil	DMBE	[116, 138]
Ginger rhizomes	Oleoresin	DMBE	[139]
Jalapeno pepper flakes	Oleoresin	DMBE	[140]
Lavender flower	Essential oil	DMBE (shrinking-core)	[141]
Marigold	Oleoresin	Various models proposed	[142]
Marjoram	Essential oil	DMBE	[136]
Orange flower concrete	Volatile oil	DMBE	[143]
Oregano bracts	Essential oil	HMT (single-plate model)	[29]
Pennyroyal	Essential oil	DMBE	[135]
Pepper, black	Essential oil	DMBE	[144]
Rosemary leaves	Essential oil	DMBE	[136]

From the point of view of the extraction mechanisms, other considerations are necessary. The equilibrium may exist if:

1. the material is largely available
2. it is distributed on or near the surface
3. the kind of equilibrium depends on the interactions (if any) with the solid structure

Mass transfer resistances, in general, may be of two types: external or internal (and, in this case, various possibilities have to be considered). To take into account mass transfer resistances, differential mass balances are applied.

Essential oils for which mathematical modelling of SFE has been attempted are reported in Table 10.5. We have also indicated if the model is based on empirical kinetic equations, on the analogy between heat and mass transfer (HMT), or on differential mass balances equations (DMBE) along the extraction bed or on a single particle.

A broken-intact cell model of the SFE of essential oils can be based on the following hypotheses:

1. The behavior of all compounds extracted is similar and can be described by a single pseudocomponent with respect to the mass transfer phenomena.
2. Concentration gradients in the fluid phase develop at larger scales than the particle size (i.e., concentration variations in the fluid phase have a characteristic length scale larger than the diameter of particles).
3. The solvent flow rate, with superficial velocity u, is uniformly distributed in all the sections of the extractor.

4. The volume fraction of the fluid, ε, is not affected by the reduction of the solid mass during extraction.
5. The solute in the solid is present in two separate phases. One phase includes the solute contained inside the internal structure of the particles. It fills a fraction φ_t of the overall volume occupied by the seed particles. The other phase is made of the solute freely available on the particle surface. The concentration here is always the same and, according to the hypotheses, it is equal to the pure solute density ρ_0 (the solute is freely available on the surface and, therefore, its concentration is constant).
6. The fraction of the volume filled by the free solute before extraction is $\varphi_f = 1 - \varphi_t$.
7. The fraction of the volume occupied by the free solute during the extraction is $\psi\varphi_f$, where is $\psi \leq 1$.
8. A linear equilibrium relationship applies between phases.

According to the above hypotheses, the mass balance on the solute in the extractor is:

$$\varepsilon \cdot \rho_f \cdot D_L \cdot \frac{\partial^2 C}{\partial z^2} + \varepsilon \cdot \rho_f \cdot \frac{\partial C}{\partial t} + \rho_f \cdot u \cdot \frac{\partial C}{\partial z} + (1-\varepsilon)\phi_f \cdot \rho_s \cdot \frac{\partial P}{\partial t}$$

$$+ (1-\varepsilon)\phi_f \cdot \rho_0 \cdot \frac{\partial \psi}{\partial t} = 0 \quad (10.1)$$

where D_L is the axial dispersion; u the superficial velocity; ρ_f the fluid density, which is supposedly not affected by the presence of the solute; ρ_s the bulk density of the nonsoluble solid that is the mass of nonsoluble solids in the vegetable materials per unit of filled particle volume, that is the total volume of the particle minus the volume of broken cells.

The general mass balance on the phase of the free solute alone is:

$$\rho_0 \frac{\partial \psi}{\partial t} = -\frac{k_f a(\rho_0 - K_\psi C)}{(1-\varepsilon)\phi_f} \quad \text{until } \psi > 0, \quad (10.2)$$

otherwise,

$$\frac{\partial \psi}{\partial t} = 0 \quad (10.3)$$

where a is the specific surface of the particles and k_f is the external mass transfer resistance.

The mass balance on tied solute is:

$$\frac{\partial P}{\partial t} = -\frac{k_i \cdot a(P - K_p C)}{(1-\varepsilon)\phi_t} \quad (10.4)$$

where k_i is the internal mass transfer resistance.

This system of equations has a unique solution when the initial conditions (i.c.) on C, P, and ψ and the boundary condition (b.c.) on C are given:

$$\text{(i.c.)}\quad \text{At } t=0:\quad C=C_0;\quad P=P_0;\quad \psi=\psi_0;\quad \text{for each } z \tag{10.5}$$

$$\text{(b.c.)}\quad \text{At } z=0:\quad \frac{u}{\varepsilon}\cdot C - D_L\cdot\frac{\partial C}{\partial z}=0\quad \text{for each } t \tag{10.6}$$

$$\text{(b.c.)}\quad \text{At } z=L:\quad \frac{\partial C}{\partial z}=0\quad \text{for each } t. \tag{10.7}$$

The set of differential equations can be numerically integrated using a finite difference method.

Reverchon et al. [145–148] used SEM analysis to confirm the presence of broken cells on particle surfaces. The concept of broken and intact cells was combined with equilibrium relationships for either free solute [149, 150] or solute interacting with matrix [133, 135]. Both types of equilibrium were also assumed to occur simultaneously by various authors, the free solute in broken cells and the interacting solute in intact cells [135, 137, 146, 147, 151].

Sovovà [128] proposed that a general model approach can be applied to seed oil and essential oil extraction. The model is based on the division of the process into two extraction periods: the first one governed by phase equilibrium and the second one by internal diffusion in particles, taking into account the concept of broken and intact cells to explain the sudden reduction of the extraction rate after the first extraction step. This effect is particularly evident in the case of seed oil extraction. The new feature of the model is the description of the first extraction period considering different types of phase equilibria: independent on matrix (solubility equilibrium), adsorbed on matrix (partition between the two phases), and different flow patterns, mainly dispersion. The model has been verified on data sets from literature related to seeds (almond) and essential oils (orange peels, pennyroyal). This model presents a limit in the case of essential oil extraction when the extractable material is located only inside the matrix; the concept of broken (in the surface) and intact cells (inside the particle) is no longer applicable and the first part of extraction controlled by equilibrium does not apply.

Gaspar et al. [29] modeled the extraction of oregano essential oil. The model is based on the prevalent geometry of particles: those obtained from leaves tend to maintain a plate-like geometry. Mass balances on the particle have been proposed.

Adsorption-desorption processes can also be treated as extraction processes: adsorption-desorption isotherms being the equilibrium curves due to interactions of the solutes between the solid matrix and the fluid phase. Differential mass balances in this case can also describe the extraction process. This approach has been used by Reverchon [152] to model the selective desorption from silica gel of two key compounds of essential oils: limonene (representative of the hydrocarbon terpenes fraction) and linalool (representative of the oxygenated terpenes fraction). The model has also been extended to the fractional desorption of bergamot peel oil [122], describing a two-step desorption process, with the first step performed at 40°C, 75 bar to desorb hydrocarbon terpenes and the second step at 40°C, 200 bar to desorb the oxygenated compounds. Later, Reverchon et al. [123] modeled the selective adsorption on silica gel of a complex terpenic mixture formed by 13 components. The mixture was divided into four families considered as four pseudo–key components. The integration of differential mass balances gave account of the competition among the different compounds for the occupation of the adsorption sites and of displacement effects observed at the exit of the adsorption bed.

Mathematical modelling has also been performed in this case [143, 153]; the volatile oil has been considered as a mixture of four compound families (pseudo-components) extracted from an active layer of concrete put on a spherical inert core. Successful modelling of terpenes oxygenated terpenes and oxygenated sesquiterpenes extraction was obtained.

Mathematical modelling of countercurrent packed column has been studied by only a few authors [154–158] and only in some cases with reference to natural matter fractionation [154, 157]. The most interesting work is the one proposed by Ruivo et al. [154], which performed the dynamic modelling and simulation of a packed column. They used experimental data from a model binary mixture formed by squalene and methyl oleate, fractionated using SC-CO_2. The model was formed by a set of partial differential equations that correspond to the differential mass balances on the packed column and algebraic equations that describe the mass transfer, the hydrodynamics of the two-phase flow through the packings and the ternary thermodynamic equilibrium for the studied system. The column was considered at constant temperature. A fairly good agreement was obtained between measured and predicted composition profiles of the outlet streams over time.

REFERENCES

1. Reverchon, E., Supercritical fluid extraction and fractionation of essential oils and related products, *J. Supercrit. Fluids*, 10, 1, 1997.
2. Meireles, M.A.A., Supercritical extraction from solid: Process design data (2001–2003), *Current Opin. Solid State Mat. Sci.*, 7, 321, 2001.
3. Lang, Q. and Wai, C.M., Supercritical fluid extraction in herbal and natural product studies—A practical review, *Talanta*, 53, 771, 2001.
4. Turner, C., Eskilsson, C.S. and Bjorklund, E., Review: Collection in analytical-scale supercritical fluid extraction, *J. Chromat. A*, 947, 1, 2002.
5. Brunner, G., Supercritical fluids: Technology and application to food processing, *J. Food Eng.*, 67, 21, 2005.
6. Stahl, E., Quirin, K.W. and Gerard, D., *Verdichtete Gase zur Extraction und Raffination*, Springer, Berlin, 1997.
7. Moyler, D.A., Extraction of flavours and fragrances with compressed CO_2, in *Extraction of Natural Products Using Near-Critical Solvents*, King, M.B. and Bott, T.R., Eds., Blackie, Glasgow, 1993.
8. Kerrola, K., Literature review: Isolation of essential oils and flavour compounds by dense carbon dioxide, *Food Rev. Int.*, 11, 547, 1995.
9. Rosa, P.T.V. and Meireles, M.A.A., Supercritical technology in Brazil: Systems investigated (1994–2003), *J. Supercrit. Fluids*, 34, 109, 2005.

10. Smith, R.M., Supercritical fluids in separation science—The dreams, the reality, and the future, *J. Chromatogr. A*, 856, 83, 1999.
11. Della Porta, G. et al., Isolation of eucalyptus oil by supercritical fluid extraction, *Flavour Fragr. J.*, 14, 214, 1999.
12. Reverchon, E. and Della Porta, G., Tuberose concrete fractionation by supercritical carbon dioxide, *J. Agric. Food Chem.*, 45, 1356, 1997.
13. Della Porta, G. et al., Isolation of clove bud and star anise essential oil by supercritical CO_2 extraction, *Lebensm. -Wiss. U. -Technol.*, 31, 454, 1998.
14. Caredda, A. et al., Supercritical carbon dioxide extraction and characterization of *Laurus nobilis* essential oil, *J. Agric. Food Chem.*, 50, 1492, 2002.
15. Marongiu, B. et al., Extraction and isolation of *Salvia desoleana* and *Mentha spicata* subsp. *insularis* essential oils by supercritical CO_2, *Flavour Fragr. J.*, 16, 384, 2001.
16. Marongiu, B. et al., Antioxidant activity of supercritical extract of *Melissa officinalis* subsp. *officinalis* and *Melissa officinalis* subsp. *inodora*, *Phytother. Res.*, 18, 789, 2004.
17. Simandi, B. et al., Supercritical carbon dioxide extraction and fractionation of fennel oil, *J. Agric. Food Chem.*, 47, 1635, 1999.
18. Simandi, B. et al., Pilot-scale extraction and fractional separation of onion oleoresin using supercritical carbon dioxide, *J. Food Eng.*, 46, 183, 2000.
19. Vagi, E. et al., Recovery of pigments from *Origanum majorana* L. by extraction with supercritical carbon dioxide, *J. Agric. Food Chem.*, 50, 2297, 2002.
20. Esquivel, M.M., Ribeiro, M.A. and Bernardo-Gil, M.G., Supercritical extraction of savory oil: Study of antioxidant activity and extract characterization, *J. Supercrit. Fluids*, 14, 129, 1999.
21. Ribeiro, M.A., Bernardo-Gil, M.G. and Esquivel, M.M., *Melissa officinalis* L.: Study of antioxidant activity in supercritical residues, *J. Supercrit. Fluids*, 21, 51, 2001.
22. Dauksas, E. et al., Effect of fast CO_2 pressure changes on the yield of lovage (*Levisticum officinale* Koch.) and celery (*Apium graveolens* L.) extracts, *J. Supercrit. Fluids*, 22, 20, 2002.
23. Dauksas, E. et al., Rapid screening of antioxidant activity of sage (*Salvia officinalis* L.) extracts obtained by supercritical carbon dioxide at different extraction conditions, *Nahrung/Food*, 45, 338, 2001.
24. Wang, B.-J., Lien, Y.-H. and Yu, Z.-R., Supercritical fluid extractive fractionation—Study of the antioxidant activities of propolis, *Food Chem.*, 86, 237, 2004.
25. Wang, B.-J. et al., Hepatoprotective and antioxidant effects of *Bupleurum kaoi* Liu (Chao et Chuang) extract and its fractions fractionated using supercritical CO_2 on CCl_4–induced liver damage, *Food Chem. Toxicol.*, 42, 609, 2004.
26. Vasapollo, G. et al., Innovative supercritical CO_2 extraction of lycopene from tomato in the presence of vegetable oil as co-solvent, *J. Supercrit. Fluids*, 29, 87–96, 2004.
27. Kiriamiti, H.K. et al., Supercritical carbon dioxide processing of pyrethrum oleoresin and pale, *J. Agric. Food Chem.*, 51, 880, 2003.
28. Kiriamiti, H.K. et al., Pyrethrin extraction from pyrethrum flowers using carbon dioxide, *J. Supercrit. Fluids*, 26, 193, 2003.
29. Gaspar, F. et al., Modelling the extraction of essential oils with compressed carbon dioxide, *J. Supercrit. Fluids*, 25, 247, 2003.
30. Smith, R.L., Jr., et al., Separation of cashew (*Anacardium occidentale* L.) nut shell liquid with supercritical carbon dioxide, *Biores. Technol.*, 88, 1, 2003.
31. Ibanez, E. et al., Supercritical fluid extraction and fractionation of different preprocessed rosemary plants, *J. Agric. Food Chem.*, 47, 1400, 1999.
32. Senorans, F.J. et al., Liquid chromatographic-mass spectrometric analysis of supercritical fluid extracts of rosemary plants, *J. Chromatogr. A*, 870, 491, 2000.

33. Ramirez, P. et al., Separation of rosemary antioxidant compounds by supercritical fluid chromatography on coated packed capillary columns, *J. Chromatogr. A*, 1057, 241, 2004.
34. Grigonis, D. et al., Comparison of different extraction techniques for isolation of antioxidants from sweet grass (*Hierochloe odorata*), *J. Supercrit. Fluids*, 33, 223, 2005.
35. Dauksas, E., Venskutonis, P.R. and Sivik, B., Supercritical fluid extraction of borage (*Borago officinalis* L.) seeds with pure CO_2 and its mixture with caprylic acid methyl ester, *J. Supercrit. Fluids*, 22, 211, 2002.
36. Bicchi, C., Binello, A. and Rubiolo, P., Determination of phenolic diterpene antioxidants in rosemary (*Rosmarinus officinalis* L.) with different methods of extraction and analysis, *Phytochem. Anal.*, 11, 236, 2000.
37. Le Floch, F. et al., Supercritical fluid extraction of phenol compounds from olive leaves, *Talanta*, 46, 1123, 1998.
38. del Valle, J.M. et al., Extraction of boldo (*Peumus boldus* M.) leaves with supercritical CO_2 and hot pressurized water, *Food Res. Intern.*, 38, 203, 2005.
39. El-Ghorab, A.H. et al., Antioxidant activity of Egyptian *Eucaliptus camaldulensis* var. brevirostris leaf extracts, *Nahrung/Food*, 47, 41, 2003.
40. Zancan, K.C. et al., Extraction of ginger (*Zingiber officinale* Roscoe) oleoresin with CO_2 and co-solvents: A study of the antioxidant action of the extracts, *J. Supercrit. Fluids*, 24, 57, 2002.
41. Leal, P.F. et al., Functional properties of spice extracts obtained via supercritical fluid extraction, *J. Agric. Food Chem.*, 51, 2520, 2003.
42. Braga, M.E.M. et al., Comparison of yield, composition, and antioxidant activity of turmeric (*Curcuma longa* L.) extracts using various techniques, *J. Agric. Food Chem.*, 51, 6604, 2003.
43. Yoda, S.K. et al., Supercritical fluid extraction from *Stevia rebaudiana* Bertoni using CO_2 and CO_2 + water: Extraction kinetics and identification of extracted components, *J. Food Eng.*, 57, 125, 2003.
44. Rostagno, M.A., Araujo, J.M.A. and Sandi, D., Supercritical fluid extraction of isoflavones from soybean flour, *Food Chem.*, 78, 111, 2002.
45. Hu, Q., Hu, Y. and Xu, J., Free radical-scavenging activity of Aloe vera (*Aloe barbadensis* Miller) extracts by supercritical carbon dioxide extraction, *Food Chem.*, 91, 85, 2005.
46. Quitain, A.T., Katoh, S. and Moriyoshi, T., Isolation of antimicrobials and antioxidants from moso-bamboo (*Phyllostachys heterocycla*) by supercritical CO_2 extraction and subsequent hydrothermal treatment of the residues, *Ind. Eng. Chem. Res.*, 43, 1056, 2004.
47. Saldana, M.D.A. et al., Extraction of methylxanthines from guaranà seeds, matè leaves, and cocoa beans using supercritical carbon dioxide and ethanol, *J. Agric. Food Chem.*, 50, 4820, 2002.
48. Hamburger, M., Baumann, D. and Adler, S., Supercritical carbon dioxide extraction of selected medicinal plants—Effects of high pressure and added ethanol on yield of extracted substances, *Phytochem. Anal.*, 15, 46, 2004.
49. Klimes, J., Sochor, J. and Kriz, J., A study of the conditions of the supercritical fluid extraction in the analysis of selected anti-inflammatory drugs in plasma, *Il Farmaco*, 57, 117, 2002.
50. Aghel, N. et al., Supercritical carbon dioxide extraction of *Mentha pulegium* L. essential oil, *Talanta*, 62, 407, 2004.
51. Cui, Y. and Ang, C.Y.W., Supercritical fluid extraction and high-performance liquid chromatographic determination of phloroglucinols in St. John's wort (*Hypericum perforatum* L.), *J. Agric. Food Chem.*, 50, 2755, 2002.

52. Cao, X.-L. et al., Supercritical fluid extraction of catechins from *Cratoxylum prumifolium* Dyer and subsequent purification by high-speed counter-current chromatography, *J. Chromatogr. A*, 898, 75, 2000.
53. Cao, X. and Ito, Y., Supercritical fluid extraction of grape seed oil and subsequent separation of free fatty acids by high-speed counter-current chromatography, *J. Chromatogr. A*, 1021, 117, 2003.
54. Yang, C., Xu, Y.-R. and Yao, W.-X., Extraction of pharmaceutical components from *Ginkgo biloba* leaves using supercritical carbon dioxide, *J. Agric. Food Chem.*, 50, 846, 2002.
55. Shu, X.-S., Gao, Z.-H. and Yang, X.-L., Supercritical fluid extraction of sapogenins from tubers of *Smilax china*, *Fitoterapia*, 75, 656, 2004.
56. Catchpole, O.J. et al., Supercritical extraction of herbs I: Saw palmetto, St. John's wort, kava root, and echinacea, *J. Supercrit. Fluids*, 22, 129, 2002.
57. Sovovà, H. et al., Near-critical extraction of pigments and oleoresin from stinging nettle leaves, *J. Supercrit. Fluids*, 30, 213, 2004.
58. Lim, G.-B. et al., Separation of astaxanthin from red yeast *Phaffia rhodozyma* by supercritical carbon dioxide extraction, *Biochem. Eng. J.*, 11, 181, 2002.
59. Quan, C. et al., Supercritical fluid extraction and clean-up of organochlorine pesticides in ginseng, *J. Supercrit. Fluids*, 31, 149, 2004.
60. Ling, Y.-C., Teng, H.-C. and Cartwright, C., Supercritical fluid extraction and clean-up of organochlorine pesticides in Chinese herbal medicine, *J. Chromatogr. A*, 835, 145, 1999.
61. Leeke, G., Gaspar, F. and Santos, R., Influence of water on the extraction of essential oils from a model herb using supercritical carbon dioxide, *Ind. Eng. Chem. Res.*, 41, 2033, 2002.
62. Illés, V. et al., Extraction of hiprose fruit by supercritical CO_2 and propane, *J. Supercrit. Fluids*, 10, 209, 1997.
63. Menaker, A. et al., Identification and characterization of supercritical fluid extracts from herbs, *C. R. Chimie*, 7, 629, 2004.
64. Chiu, K.-L. et al., Supercritical fluids extraction of Ginkgo ginkgolides and flavonoids, *J. Supercrit. Fluids*, 24, 77, 2002.
65. King, J.W., Supercritical fluid extraction of *Vernonia galamensis* seeds, *Ind. Crops and Products*, 14, 241, 2001.
66. Sanal, I.S. et al., Determination of optimum conditions for SC-(CO_2 + ethanol) extraction of β-carotene from apricot pomace using response surface methodology, *J. Supercrit. Fluids*, 34, 331, 2005.
67. da Cruz Francisco, J. et al., Application of supercritical carbon dioxide for the extraction of alkylresorcinols from rye bran, *J. Supercrit. Fluids*, 35, 220, 2005.
68. Teberikler, L., Koseoglu, S.S. and Akgerman, A., Deoling of crude lecithin using supercritical carbon dioxide in the presence of co-solvents, *J. Food Sci.*, 66, 850, 2001.
69. Teberikler, L., Koseoglu, S.S. and Akgerman, A., Selective extraction of phosphatilylcholine from lecithin by supercritical carbon dioxide/ethanol mixture, *JAOCS*, 78, 115, 2001.
70. Eggers, R., Voges, S.K. and Jaeger, Ph.T., Solid bed properties in supercritical processing, in I. Kikic, M. Perrut, Eds., *Proceedings of the 9th Meeting on Supercritical Fluids*, Trieste (Italy) 2004, E11 (pdf).
71. http://webbook.nist.gov/chemistry/fluid.
72. Span, R. and Wagner, W., A new equation of state for carbon dioxide covering the fluid region from the triple-point temperature to 1100 K at pressures up to 800 MPa, *J. Phys. Chem. Ref. Data*, 25, 1509, 1996.
73. Raeissi, S. and Peters, C.J., Bubble-point pressures of the binary system carbon dioxide + linalool, *J. Supercrit. Fluids*, 20, 221, 2001.

74. Iwai, Y. et al., High-pressure vapor-liquid equilibria for carbon dioxide + limonene, *J. Chem. Eng. Data*, 41, 951, 1996.
75. Marteau, P., Obriot, J. and Tufeu, R., Experimental determination of vapor-liquid equilibria of CO_2 + limonene and CO_2 + citral mixtures, *J. Supercrit. Fluids*, 8, 1995, 20.
76. Akgün, M., Akgün, N.A. and Dinçer, S., Phase behaviour of essential oil components in supercritical carbon dioxide, *J. Supercrit. Fluids*, 15, 117, 1999.
77. Reverchon, E., Russo, P. and Stassi, A., Solubilities of solid octacosane and triacontane in supercritical carbon dioxide, *J. Chem. Eng. Data*, 38, 458, 1993.
78. Dýaz-Maroto, M.C., Perez-Coello, M.S. and Cabezudo, M.D., Supercritical carbon dioxide extraction of volatiles from spices: Comparison with simultaneous distillation-extraction, *J. Chromatogr. A*, 947, 23, 2002.
79. Rozzi, N.L. et al., Supercritical fluid extraction of essential oil components from lemon-scented botanicals, *Lebensm. –Wiss. U. –Technol.*, 35, 319, 2002.
80. Carlson, L.H.C. et al., Extraction of lemongrass essential oil with dense carbon dioxide, *J. Supercrit. Fluids*, 21, 33, 2001.
81. Dauksas, E., Venskutonis, P.R. and Sivik, B., Supercritical CO_2 extraction of the main constituents of lovage (*Levisticum officinale* Koch.) essential oil in model systems and overground botanical parts of the plant, *J. Supercrit. Fluids,* 15, 51, 1999.
82. Reverchon, E., Fractional separation of SCF extracts from marjoram leaves: Mass transfer and optimization, *J. Supercrit. Fluids*, 5, 256, 1992.
83. Rodrigues, M.R.A. et al., The effects of temperature and pressure on the characteristics of the extracts from high-pressure CO_2 extraction of *Majorana hortensis* Moench, *J. Agric. Food Chem.*, 51, 453, 2003.
84. Simandi, B. et al., Supercritical carbon dioxide extraction and fractionation of oregano oleoresin, *Food Res. Intern.*, 31, 723, 1998.
85. Reverchon, E., Taddeo, R. and Della Porta, G., Extraction of sage essential oil by supercritical CO_2: Influence of some process parameters, *J. Supercrit. Fluids*, 8, 302, 1995.
86. Sonsuzer, S., Sahin, S. and Yilmaz, L., Optimization of supercritical CO_2 extraction of *Thymbra spicata* oil, *J. Supercrit. Fluids*, 30, 189, 2004.
87. Moldao-Martins, M. et al., Supercritical CO_2 extraction of *Thymus zygis* L. subsp. *sylvestris* aroma, *J. Supercrit. Fluids*, 18, 25, 2000.
88. Povh, N.P., Marques, M.O.M. and Meireles, M.A.A., Supercritical CO_2 extraction of essential oil and oleoresin from chamomile (*Chamomilla recutita* [L.] Rauschert), *J. Supercrit. Fluids*, 21, 245, 2001.
89. Reverchon, E. and Senatore, F., Supercritical carbon dioxide extraction of chamomile essential oil and its analysis by gas chromatography-mass spectrometry, *J. Agric. Food Chem.*, 42, 154, 1994.
90. Reverchon, E., Della Porta, G. and Senatore, F., Supercritical CO_2 extraction and fractionation of lavender essential oil and waxes, *J. Agric. Food Chem.*, 43, 1654, 1995.
91. Rodrigues, V.M. et al., Supercritical extraction of essential oil from aniseed (*Pimpinella anisum* L.) using CO_2: Solubility, kinetics, and composition data, *J. Agric. Food Chem.*, 51, 1518, 2003.
92. Dauksas, E. and Venskutonis, P.R., Extraction of lovage (*Levisticum officinale* Koch.) roots by carbon dioxide. 1. Effect of CO_2 parameters on the yield of the extract, *J. Agric. Food Chem.*, 46, 4347, 1998.
93. Monteiro, A.R. et al., Extraction of the soluble material from the shells of the bacuri fruit (*Platonia insignis* Mart) with pressurized CO_2 and other solvents, *J. Supercrit. Fluids*, 11, 91, 1997.
94. Ferreira, S.R.S. et al., Supercritical fluid extraction of black pepper (*Piper nigrun* L.) essential oil, *J. Supercrit. Fluids*, 14, 235, 1999.

95. Tipsrisukond, N., Fernando, L.N. and Clarke, A.D., Antioxidant effects of essential oil and oleoresin of black pepper from supercritical carbon dioxide extractions in ground pork, *J. Agric. Food Chem.*, 46, 4329, 1998.
96. Patel, R.N., Bandyopadhyay, S. and Ganesh, A., Extraction of cashew (*Anacardium occidentale*) nut shell liquid using supercritical carbon dioxide, *Biores. Techn.*, 97, 847, 2006.
97. Rodrigues, M.R.A. et al., Chemical composition and extraction yield of the extract of *Origanum vulgare* obtained from sub- and supercritical CO_2, *J. Agric. Food Chem.*, 52, 3042, 2004.
98. Duarte, C. et al., Supercritical fluid extraction of red pepper (*Capsicum frutescens* L.), *J. Supercrit. Fluids*, 30, 155, 2004.
99. Barjaktarovic, B., Sovilj, M. and Knez, Z., Chemical composition of *Juniperus communis* L. fruits supercritical CO_2 extracts: Dependence on pressure and extraction time, *J. Agric. Food Chem.*, 53, 2630, 2005.
100. Reverchon, E., Della Porta, G. and Gorgoglione, D., Supercritical CO_2 fractionation of jasmine concrete, *J. Supercrit. Fluids*, 8, 60, 1995.
101. Reverchon, E. and Della Porta, G., Rose concrete fractionation by supercritical CO_2, *J. Supercrit. Fluids*, 9, 199, 1996.
102. Vagi, E. et al., Phenolic and triterpenoid antioxidants from *Origanum majorana* L. herb and extracts obtained with different solvents, *J. Agric. Food Chem.*, 53, 17, 2005.
103. El-Ghoran, A.H., Mansour, A.F. and El-massry, K.F., Effect of extraction methods on the chemical composition and antioxidant activity of Egyptian marjoram (*Majorana hortensis* Moench), *Flavour Fragr. J.*, 19, 54, 2004.
104. Lopez-Sebastian, S. et al., Dearomatization of antioxidant rosemary extracts by treatment with supercritical carbon dioxide, *J. Agric. Food Chem.*, 46, 13, 1998.
105. Ibanez, E. et al., Combined use of supercritical fluid extraction, micellar electrokinetic chromatography, and reverse phase high performance liquid chromatography for the analysis of antioxidants from rosemary (*Rosmarinus officinalis* L.), *J. Agric. Food Chem.*, 48, 4060, 2000.
106. Tena, M.T. et al., Supercritical fluid extraction of natural antioxidants from rosemary: Comparison with liquid solvent sonication, *Anal. Chem.*, 69, 521, 1997.
107. Hadolin, M. et al., Isolation and concentration of natural antioxidants with high-pressure extraction, *Inn. Food Science Emerg. Technol.*, 5, 245, 2004.
108. Yepez, B. et al., Producing antioxidant fractions from herbaceous matrices by supercritical fluid extraction, *Fluid Phase Equil.*, 194–197, 879, 2002.
109. Braga, M.E.M. et al., Supercritical fluid extraction from *Lippia alba*: Global yields, kinetic data, and extract chemical composition, *J. Supercrit. Fluids*, 34, 149, 2005.
110. Stashenko, E.E., Jaramillo, B.E. and Martinez, J.R., Comparison of different extraction methods for the analysis of volatile secondary metabolites of *Lippia alba* (Mill.) N.E. Brown, grown in Colombia, and evaluation of its in vitro antioxidant activity, *J. Chromatogr. A*, 1025, 93, 2004.
111. Ramos, E. et al., Obtention of a brewed coffee aroma extract by an optimized supercritical CO_2-based process, *J. Agric. Food Chem.*, 46, 4011, 1998.
112. Michielin, E.M.Z. et al., Composition profile of horsetail (*Equisetum giganteum* L.) oleoresin: Comparing SFE and organic solvents extraction, *J. Supercrit. Fluids*, 33, 131, 2005.
113. Daood, H.G. et al., Extraction of pungent spice paprika by supercritical carbon dioxide and subcritical propane, *J. Supercrit. Fluids*, 23, 143, 2002.
114. Uquiche, E., del Valle, J.M. and Ortiz, J., Supercritical carbon dioxide extraction of red pepper (*Capsicum annuum* L.) oleoresin, *J. Food Eng.*, 65, 55, 2004.
115. Gomez-Coronado, D.J.M. et al., Tocopherol measurement in edible products of vegetable origin, *J. Chromatogr. A*, 1054, 227, 2004.

116. Reverchon, E. et al., Supercritical fractional extraction of fennel seed oil and essential oil: Experiments and mathematical modelling, *Ind. Eng. Chem. Res.*, 38, 3069, 1999.
117. Reverchon, E., Marciano, A. and Poletto, M., Fractionation of a peel oil key mixture by supercritical CO_2 in a continuous tower, *Ind. Eng. Chem. Res.*, 36, 4940, 1997.
118. Budich, M. et al., Countercurrent deterpenation of citrus oils with supercritical CO_2, *J. Supercrit. Fluids*, 14, 105, 1999.
119. Sato, M., Goto, M. and Hirose, T., Fractional extraction with supercritical carbon dioxide for the removal of terpenes from citrus oil, *Ind. Eng. Chem. Res.*, 34, 3941, 1995.
120. Stahl, E., Quirin, K.W. and Gerard, D., *Verdichtete Gase zur Extraction und Raffination,* Springer, Berlin, 1987.
121. Kubat, H., Akman, U. and Hortacsu, O., Semi-batch packed-column deterpenation of origanum oil by dense carbon dioxide, *Chem. Eng. Proc.*, 40, 19, 2001.
122. Reverchon, E. and Iacuzio, G., Supercritical desorption of bergamot peel oil from silica gel—Experiments and mathematical modelling, *Chem. Eng. Sci.*, 52, 3553, 1997.
123. Reverchon, E., Lamberti, G. and Subra, P., Modelling and simulation of the supercritical adsorption of complex terpene mixtures, *Chem. Eng. Sci.*, 53, 3537, 1998.
124. Catchpole, O.J. and Bergmann, C., Continuous gas anti-solvent fractionation of natural products, in *Fifth Meeting on Supercritical Fluids,* Vol. 1, Perrut, M. and Subra, P., Eds., Nice, France, 1998, 257.
125. Catchpole, O.J., Hochmann, S. and Anderson, S.R.J., Gas antisolvent fractionation of natural products, in *Process Technology Proceedings 12 (High Pressure Engineering),* R. von Rohr, Ch. Trepp, Eds., Elsevier, Amsterdam, 1996, 309.
126. Scrugli, S. et al., A preliminary study on the application of supercritical antisolvent technique to the fractionation of tobacco extracts, in *Eighth Meeting on Supercritical Fluids*, Besnard, M. and Cansell, F., Eds., Bordeaux, France, 2002, 901.
127. Reverchon, E., Mathematical modelling of supercritical extraction of sage oil, *AIChE J.*, 42, 1765, 1996.
128. Sovová, H., Mathematical model for supercritical fluid extraction of natural products and extraction curve evaluation, *J. Supercrit. Fluids*, 33, 35, 2005.
129. Kandiah, M. and Spiro, M., Extraction of ginger rhizomes: Kinetic studies with supercritical carbon dioxide, *Int. J. Food Sci. Technol.*, 25, 328, 1990.
130. Nguyen, K., Barton, P. and Spencer, J., Supercritical carbon dioxide of vanilla, *J. Supercrit. Fluids*, 4, 40, 1991.
131. Bartle, K.D. et al., A model for dynamic extraction using a supercritical fluid, *J. Supercrit. Fluids*, 3, 143, 1990.
132. Reverchon, E., Donsì, G. and Sesti Osséo, L., Modeling of supercritical fluid extraction from herbaceous matrices, *Ind. Eng. Chem. Res.*, 32, 2721, 1993.
133. Sovovà, H. et al., Supercritical carbon dioxide extraction of caraway essential oil, *Chem. Eng. Sci.*, 49, 2499, 1994.
134. Roy, B.C., Goto, M. and Hirose, T., Extraction of ginger oil with supercritical carbon dioxide: Experiments and modeling, *Ind. Eng. Chem. Res.*, 35, 607, 1996.
135. Reis-Vasco, E.M.C. et al., Mathematical modelling and simulation of pennyroyal essential oil supercritical extraction, *Chem. Eng. Sci.*, 55, 2917, 2000.
136. Goodarznia, I. and Eikani, M.H., Supercritical carbon dioxide extraction of essential oils: modelling and simulation, *Chem. Eng. Sci.*, 53, 1387, 1998.
137. Reverchon, E. and Marrone, C., Supercritical extraction of clove bud essential oil: Isolation and mathematical modelling, *Chem. Eng. Sci.*, 52, 3421, 1997.
138. Reverchon, E. et al., Supercritical fractional extraction of fennel seed oil and essential oil: Experiments and mathematical modelling, *Ind. Eng. Chem. Res.*, 38, 3069, 1999.
139. Martinez, J. et al., Multicomponent model to describe extraction of ginger oleoresin with supercritical carbon dioxide, *Ind. Eng. Chem. Res.*, 42, 1057, 2003.

140. del Valle, J.M., Jimenez, M. and de la Fuente, J.C., Extraction kinetics of pre-pelletized jalapeno peppers with supercritical CO_2, *J. Supercrit. Fluids*, 25, 33, 2003.
141. Akgun, M., Akgun, N.A. and Dincer, S., Extraction and modelling of lavender flower essential oil using supercritical carbon dioxide, *Ind. Eng. Chem. Res.*, 39, 473, 2000.
142. Campos, L.M.A.S. et al., Experimental data and modelling the supercritical fluid extraction of marigold (*Calendula officinalis*) oleoresin, *J. Supercrit. Fluids*, 34, 163, 2005.
143. Reverchon, E., Della Porta, G. and Lamberti, G., Modelling of orange flower concrete fractionation by supercritical CO_2, *J. Supercrit. Fluids*, 14, 115, 1999.
144. Ferreira, S.R.S. and Meireles, M.A.M., Modeling the supercritical fluid extraction of black pepper (*Piper nigrum* L.) essential oil, *J. Food Eng.*, 54, 263, 2002.
145. Reverchon, E. and Marrone, C., Modeling and simulation of the supercritical CO_2 extraction of vegetable oils, *J. Supercrit. Fluids*, 19, 161, 2001.
146. Marrone, C. et al., Almond oil extraction by supercritical CO_2: Experiments and modelling, *Chem. Eng. Sci.*, 53, 3711, 1998.
147. Reverchon, E., Kaziunas, A. and Marrone, C., Supercritical CO_2 extraction of hiprose seed oil: Experiments and mathematical modelling, *Chem. Eng. Sci.*, 55, 2195, 2000.
148. Reverchon, E. et al., Supercritical fractional extraction of fennel seed oil and essential oil: Experiments and mathematical modelling, *Ind. Eng. Chem. Res.*, 38, 3069, 1999.
149. Sovová, H., Rate of the vegetable oil extraction with supercritical CO_2. I. Modelling of extraction curves, *Chem. Eng. Sci.*, 49, 409, 1994.
150. Štástová, J. et al., Rate of the vegetable oil extraction with supercritical CO_2. III. Extraction from sea buckthorn, *Chem. Eng. Sci.*, 51, 4347, 1996.
151. Reis-Vasco, E.M.C. et al., Mathematical modelling and simulation of pennyroyal essential oil supercritical extraction, *Chem. Eng. Sci.*, 55, 2917, 2000.
152. Reverchon, E., Supercritical desorption of limonene and linalool from silica gel: Experiments and modelling, *Chem. Eng. Sci.*, 52, 1019, 1997.
153. Reverchon, E. and Poletto, M., Mathematical modelling of supercritical CO_2 fractionation of flower concretes, *Chem. Eng. Sci.*, 51, 3741, 1996.
154. Ruivo, R. et al., Dynamic model of a countercurrent packed column operating at high-pressure conditions, *J. Supercrit. Fluids*, 32, 183, 2004.
155. Ramchandran, B. et al., Dynamic simulation of a supercritical fluid extraction process, *Ind. Eng. Chem. Res.*, 31, 281, 1992.
156. Camy, S. and Condoret, J.-S., Dynamic modelling of a fractionation process for a liquid mixture using supercritical carbon dioxide, *Chem. Eng. Proc.*, 40, 499, 2001.
157. Benvenuti, F. and Gironi, F., Dynamic simulation of a semicontinuous extraction process using supercritical solvent, in *Proceedings of the Fifth Italian Conference on Supercritical Fluids and their Applications*, A. Bertucco, Ed., 1999, 281.
158. Cesari, G. et al., A computer-program for the dynamic simulation of a semi-batch supercritical fluid extraction process, *Comput. Chem. Eng.*, 13, 1175, 1989.

11 Processing of Spices Using Supercritical Fluids

Mamata Mukhopadhyay

CONTENTS

11.1 Introduction 337
11.2 Importance of Spices 338
11.3 Beneficial Attributes of Spices 339
11.4 Bioactive Ingredients in Spices 341
11.5 Saleable Spice Products 343
11.6 Conventional Extraction Methods 346
11.7 Supercritical Carbon Dioxide as the Extractant 347
11.8 Commercial SCFE Process 349
11.9 Comparison of Spice Extracts by Conventional and SCFE Processes 351
11.10 Process Analysis of SCFE from Selected Spices 354
11.10.1 Celery Seed (*Apium graveolen*) 354
11.10.2 Red Chili 356
11.10.3 Paprika 357
11.10.4 Ginger 358
11.10.5 Nutmeg 358
11.10.6 Black Pepper 359
11.10.7 Vanilla 359
11.10.8 Cardamom 359
11.10.9 Fennel, Caraway, and Coriander 360
11.9.10 Garlic 361
11.10.11 Cinnamon 362
11.11 Correlation for Spice Oil Solubility in SC-CO_2 362
11.12 Conclusions 364
References 365

11.1 INTRODUCTION

In recent years, increasing demand for superior quality and safety of foods and medicines, as well as concern for environmental pollution during their commercial production, have triggered stringent regulations on the toxin levels in foods and medicines as well as on the discharge of pollutants to the environment. In addition, there has been increasing consumer preference for natural substances. All these factors have given strong impetus to development of cost-effective new technologies, such as the one for eco-friendly extraction from natural substances employing green and safe solvents. In recent years, supercritical fluid extraction (SCFE) has emerged as a highly promising environmentally benign technology for production of natural extracts, such as flavors, fragrances, spice oils, and oleoresins; natural antioxidants; natural colors; nutraceuticals; and biologically active principles. The state of a substance is called *supercritical* when both temperature and pressure exceed their critical point values. A supercritical fluid (SCF) combines the twin beneficial properties, namely high density (which imparts high solvent power) and high compressibility (which permits high selectivity due to large variability of solvent power by small changes in temperature and pressure). In addition, it offers very attractive extraction characteristics, owing to its favorable diffusivity, viscosity, surface tension, and other thermo-physical properties.

Since the 1980s, several potential applications of SCFE techniques have been reported. So far, the most popular SCF has been carbon dioxide (CO_2), owing to its easy availability, low cost, nonflammability, nontoxicity, and a spectrum of solvent properties in a single substance. Its critical temperature is 31.1°C and its critical pressure is 73.8 bar. Dense or supercritical carbon dioxide (SC-CO_2) could very well be the most commonly used solvent in this century due to its wide-ranging applications. Its near-ambient critical temperature makes it ideally suitable for processing of thermally labile natural substances. It is generally regarded as safe (GRAS), and it yields microbial-inactivated, contaminant-free, tailor-made extracts of superior organoleptic profile and longer shelf life, with high potency of active ingredients. The SCFE technique ensures high consistency and reliability in the quality and safety of the bioactive heat-sensitive botanical products, as it does not alter the delicate balance of bioactivity of natural molecules. All of these advantages are almost impossible in conventional processes. Therefore, SCFE technology using SC-CO_2 as the solvent is an ideal alternative to the conventional techniques for extraction of bioactive ingredients from spices. This chapter presents some of the principles and methods of SCFE technology for processing of spices for use in food products, medicines, and dietary supplements.

11.2 IMPORTANCE OF SPICES

By definition, a spice is an "aromatic, pungent vegetable substance used to flower food," as originated from the Latin name "species aromatacea." An herb is defined as a "plant without woody tissue that withers and dies after flowering." Spices and herbs both function as flavoring agents for food and, accordingly, the U.S. Food and Drug Association includes spices and herbs together in a class of aromatic vegetables that impart flavor and seasoning to food rather than nutritional value. Thus, *spice* implies a tropical herbal plant or some part of it that is used in cooking and in condiments as well as in candies, cosmetics, fragrances, and medications in order to provide aroma, flavor, and color, along with stimulating pungency and taste [1].

In ancient ages, spices were employed for embalming, preserving foods, and masking bad odors. The Ebers Papyrus, written in Egypt in about 1500 B.C., even mentions some common spices, such as coriander, cumin, fenugreek, and mint, and describes how these spices were used in foods and medicines. In Medieval and Renaissance times, Greeks and Romans used to spend vast fortunes on trade with Arabia, which was then the center of the spice trade. Exotic spices used to be exhibited as a symbol of wealth and power, so much so that there were many long explorations in search of sources of spices. The extraordinary voyages made for these spices resulted in the discovery of the New World and, in turn, demonstrated that the globe could be circumnavigated by sea. The strong motivation to control spice sources lured the British to India, the Portuguese to Brazil, the Spanish to Central and South America and the Philippines, the French to Africa, and the Dutch to Indonesia. However, countries exploring these new regions had to deal with the natives to establish monopolistic control of power over the spice-growing regions and major spice trade routes. Over this long historical journey, many of the exotic spices of earlier attraction, such as nutmeg and saffron, lost their pride of place, while less valued spices having some medicinal values, such as garlic, peppers, and other common herbs, have now become increasingly popular [2].

11.3 BENEFICIAL ATTRIBUTES OF SPICES

In present days, most people aspire for good health and there is a growing fascination in the use of natural health care products. It is interesting to note that more than 80% of the world's population believes that "prevention is better than cure," for which they prefer botanical products having health promoting, disease preventing, medicinal properties. Hardly any spices have no medicinal effects [2]. The most commonly used spices are well proven to be medicinal; for example, black pepper, cayenne, cinnamon, garlic, ginger, licorice, onion, and chives all contain a variety of biologically active compounds [3] that are GRAS. Table 11.1 lists some spices and their synergistic therapeutic benefits based on the number of bioactive components present for a specific biological action. It is now believed that the naturally occurring synergistic effect of the total extract renders better effectiveness for a specific biological action than the isolated active ingredient [2].

Spices can be used to promote health, cure disorders, and prevent diseases, from cancer and diabetes to liver and heart problems to obesity. Diverse agro-climatic zones prevalent in some geographical locations are responsible for producing biodiverse flora and fauna. These zones led to the development of the ancient medicinal sciences, like Ayurveda, Sidda, and Unani, based on the regional natural resources available in India and China. Natural flavors and fragrances obtained from spices and herbs are often used to relieve stress by aromatherapy. Spices and herbs are also used for gastrointestinal therapies, aphrodisiacs, and nonspecific tonics. The more pungent ones are counterirritants and can be used for pain relief and anti-inflammatory effects. Many have antibacterial or antifungal properties. Spices are claimed to prevent cancers due to their strong antioxidant properties, though most of the more potent medical benefits have not been validated owing to the difficulty of identifying the relevant bioactive compounds.

For years, researchers have recognized that garlic reduces hypertension, cholesterol, respiratory and urinary tract infections, and digestive and liver disorders. It also cures diphtheria, hepatitis, ringworm, typhoid, and bronchitis and inhibits pathogenic bacteria, amoebae, fungi, and yeast, even at levels of 10 ppm. It is both an antioxidant and an antiseptic. It is thus popularly believed that "a garlic

TABLE 11.1
Number of Bioactive Components Responsible for Specific Therapeutic Benefits of Common Spices

Spice (Part)	Therapeutic Benefits (Correlated to Number of Bioactive Compounds)*
All spice (Fruit)	P(8), AC(5), AI(3), AB(2), AU(2), AA(2), H(3), AD(2), AAM(2) AG(2), AM(2)
Black mustard	P(8), AC(10), AS(3), FC(3), AB(6), AO(4), AV(4), AG(3), L(2)
Black pepper (Fruit)	P(31), AC(21), AS(16), AI(8), FC(10), AB(14), AO(4), AV(6), ST(7), AA(4), AN(8), H(8), AG(5)
Cardamon	P(2), AC(8), AI(2), HG(2), AM(2), HP(3)
Cassia (Bark)	P(10), AC(7), AS(5), ADB(3), AO(3), AU(4), AV(3), AA(3), AD(3), AM(3), I(4)
Cinnamon (Bark)	P(29), AC(14), AS(10), AI(7), FC(10), AB(11), AU(5), AV(6), ST(8), AA(4), H(8), AG(4), ADB(3)
Clove (Bud)	P(11), AC(10), AS(4), AI(6), FC(5), AO(3), AU(5), ST(3), AN(2), AA(3), AG(2)
Coriander (Fruit)	P(40), AC(27), AS(9), AI(8), FC(11), AB(20), AO(7), AV(12), ST(8), H(7), AG(6), AAM(5)
Cumin (Fruit)	P(27), AC(11), AS(5), AI(7), FC(6), AB(11), AO(5), AU(5), AV(7), ST(6), H(6), AG(3), AAM(3)
Garlic (Bulb)	P(23), AC(21), AS(6), FC(8), AB(13), AO(9), AU(6), AV(5), ST(5), AA(9), AD(5), HG(6), HP(5)
Ginger (Rhizome)	P(43), AC(25), AS(11), FC(18), AB(17), AO(6), AU(13), AV(6) ST(11), H(7), HP(8)
Licorice (Root)	P(45), AC(26), AS(23), AI(12), FC(21), AB(20), AO(10), AU(6), AV(8), ST(6), AN(9), AA(5), E(8)
Nutmeg (Seed)	P(32), AC(15), AS(11), FC(14), AB(15), AU(4), AV(4), ST(6), AN(5), AA(6), E(4), H(6), AG(3)
Poppy	P(5), AO(3), AU(6), HPT(4), AD(5), AM(4)
Sesame	ADB(4), P(7), AC(17), AB(5), AO(7), AD(7)
Turmeric (Rhizome)	P(15), AC(9), AS(4), AI(5), FC(7), AB(8), AO(3), AU(6), AV(3) AN(3), H(4), AG(3), I(4)
Vanilla (Fruit)	P(20), AC(7), AS(9), FC(9), AB(7), AO(7), AU(3), AV(3), AN(5) E(4)

* *P:* Pesticidal; *AC:* Anticancerous; *AS:* Antiseptic; *AI:* Anti-inflammatory; *FC:* Fungicidal; *AB:* Antibacterial; *E:* Expectorant; *H:* Herbicidal; *AG:* Analgesic; *AD:* Antidepressant; *HG:* Hypoglycermic; *AM:* Antimigraine

TABLE 11.1 (continued)
Number of Bioactive Components Responsible for Specific Therapeutic Benefits of Common Spices

AO: Antioxidant	*ADB:* Antidiabetic
AU: Antiulcerous	*AAM:* Antiasthmatic
AV: Antiviral	*L:* Laxative
ST: Sedative	*I:* Immunostimulant
AN: Anaesthetic	*HP:* Hepatoprotective
AA: Antiaggregant	

Source: Mukhopadhyay, M., *Natural Extracts Using Supercritical Carbon Dioxide*, CRC Press, Boca Raton, FL, 2000. With permission.

clove a day keeps the doctor away." Ginger is often used to cure colds, cough, asthma, tuberculosis, joint pain, high cholesterol, low blood pressure, and even motion sickness. It is a heart stimulant, bactericide, and antidepressant. Onion is known to be a blood cleanser, weight regulator, and antidiabetic. It also prevents colds and infections. Clove and cinnamon are known painkillers, antifungals, and antidiabetics and can be used for sparing insulin. They generally cure indigestion, nausea, and hyperacidity; inhibit tuberculosis; relieve fever, insomnia, and allergies; and even lower blood pressure. Cumin cures piles, hoarseness of voice, dyspepsia, jaundice, insomnia, colds, and fever. It also cures hook-worm infection. Turmeric is well known for its anti-inflammatory, antiseptic, and anticarcinogenic properties. It cures arthritis, respiratory tract infection, skin allergies, bronchial asthma, and viral hepatitis. Essence of jojoba is considered a unique product of nature as it is almost identical to natural skin oil and is used for herbal skin care. Moisturizers and beauty oils are also made from natural oils (e.g., olive, almond, wheat grass, and aloe vera). Tea tree oil is an unique herbal oil that is an antibacterial, antifungal, anti-infective, and antiseptic. It is obtained from needle-like leaves of a small herb, a native of Australia that is now extensively grown in the U.S.

11.4 BIOACTIVE INGREDIENTS IN SPICES

Spices may be classified according to their therapeutic benefits or bioactive ingredients, in addition to their aroma, taste, color, and consistency imparted to foods and medicines. Table 11.2 lists groups of active ingredients present in spices and their therapeutic values. The volatile fraction of a spice is known as its *essential oil* and is responsible for the essence or flavor of the spice. Essential oils are found in various parts of a plant (e.g., sandalwood, clove bud, cinnamon bark, orange peel, rose and jasmine flowers). The essential oil constituents may be classified into four major groups: monoterpenes, diterpenes, sesquiterpenes, and oxygenated compounds. The compounds belonging to the last group, namely esters, ketones, alcohols, and ethers, are very specific to the species or the genus of the spice plant. Even though these compounds are present in very small quantities, they are the substances responsible for the characteristic flavor of the spice, the absence of which sometimes changes the aroma completely.

TABLE 11.2
Classification of Bioactive Constituents in Spices

Group	Example	Spice: Active Ingredient	Therapeutic Value
Alkaloids	Bitter amines	Chili: Capsaicin	Counter-irritant for pain
Bioflavonoids	Phenolic pigments	Rosemary: Luteolin	Antioxidant
Essential oils	Mixtures of volatiles	Clove: Various	Aphrodisiac, perfume
Glycosides	Carbohydrate derivatives	Garlic: Alliin	Expectorant
Phenylpropanoids	Cinnamic acid derivatives	Cinnamon: Eugenol	Topical anesthetic
Resins	Terpene oxidants	Myrrh: Resin acids	Antibacterial
Saponins	Soapy hemolysants	Licorice: Glycyrrhizin	Anti-inflammatory
Sterols	Steroid precursors	Sesame: Linoleic acid	Antioxidant
Tannins	Polyphenolics	Tea: Catechin	Antioxidant
Terpenes	Isoprene derivatives	Ginger: Zingiberene	Antinauseant
Carotenoids	Carotenes	Paprika, red chili: β-carotene	Antioxidant, color
Anthocyanins	Curcumin derivatives	Turmeric: Kokum	Natural color

The relatively nonvolatile fraction of a spice extract is viscous and resinous, and so it is called *oleoresin.* It is responsible for the taste or pungency of the spice. This comprises nonvolatile constituents or large-molecular-weight compounds, such as fatty acids, resins, paraffin waxes, and alkaloids. Most of the ingredients that are responsible for the medicinal attributes of a spice are present in this fraction. For example, the compound responsible for the medicinal value in black pepper is piperine, an alkaloid having bitter taste that is present in the oleoresin. It is not present in the essential oil and does not contribute to the aroma of black pepper. Among the bioactive compounds present in oleoresin, another important group deters the formation and propagation of free radicals and is called *antioxidants.* They prevent diseases caused by oxidative damage (e.g., aging, cataracts, coronary heart disease, cancer, memory loss, Alzheimer's disease, and kidney-failure). These natural bioactive chemical compounds are often commercially called *nutraceuticals,* in line with the term *pharmaceuticals,* and are also termed *phytochemicals* if they are derived from leaves, roots, stems, seeds, or fruits. The phytochemicals present in functional foods include phenolics and polyphenolics, and some are commercially used as natural antioxidants. Vitamins A, C, and E and flavonoids are some natural antioxidants that are added in small concentrations to foods as supplements for preservation. The provitamin A activity in spices is due to the carotenoids present in spices and plays an important role as an antioxidant. Spices and herbs that possess antioxidant properties include clove, turmeric, allspice, rosemary, mace, sage, oregano, thyme, nutmeg, ginger, cassia, cinnamon, savory, black and white pepper, aniseed, and basil. In the food industry, these natural antioxidants are used along with synthetic antioxidants, such as butylated hydroxy anisole, butylated hydroxy toluene, tertiary butyl hydro quinone, and propyl gallate.

In addition to the antioxidants, some spices (e.g., turmeric, paprika, saffron, mustard, and black pepper) have *natural color* as an important ingredient, which is used for food color. Occurrence of the component curcumin in turmeric (*Curcuma longa*) has made it beneficial not only as a natural color to food processing industries but also as an antioxidant, anti-inflammatory, antimutagenic, and antivenom agent for human health. A large number of bioactive components are present in the essential oil obtained from turmeric, such as, curcumin, ar-turmeric, and turmerone. Some of the bioactive compounds present in other commonly used domestic spices are listed in Table 11.3.

11.5 SALEABLE SPICE PRODUCTS

Spices are not necessarily sold as pure spices in whole or ground form but are preferred in the form of blends and formulations for the ease of usage. In many cases, additives are added to improve the quality or shelf life. Many pure spices sold in the form of powder undergo caking. Accordingly, anticaking additives are often added to maintain the spice dry and free flowing. For example, a silica gel (sodium silicate) is often added as an anticaking agent. Calcium stearate, magnesium stearate, and potassium stearate are also used as effective anticaking agents. Some spices are sold as blends of spices, such as curry powder. The ingredients and their compositions in a spice blend may change with the food applications, as in the case of chili powder.

Alternatively, spice extracts can replace spice powder in food and flavor formulations. Spice extracts can provide the true essence of spice in the form of the volatile essential oil, the taste components in the form of the nonvolatile resinous fraction, and the food colors in the form of the pigments. The formulations with flavors, spice oils, and oleoresin are very much an art rather than a standard technology and vary depending on the buyer's preferences for food habits and health care products. Liquid spice flavors are added to the edible gum, powdered starch, or cellulose substance and care is taken so that the blend can retain its powdery nature by the addition of anticaking agents. Furthermore, flavor dehydrates, such as dehydrated chicken, meat, and cheese powders, are added to the blend. Each spice formulation is devised so that it is effective due to its own characteristics, and each formula has to be within the limits specified by the regulatory boards. Spice blending equipment, such as blenders and filters, should be cleaned on a regular basis to ensure that there is no contamination with residual spice blend from the previous operation.

Liquid spice extracts, such as essential oils and oleoresins, provide food technologists with many advantages, as food manufacturers can select the specific flavor profiles with much greater precision than if they were to simply use blends of whole spices. In addition, hygienic concerns as well as transportation costs are greatly reduced if the oils and oleoresins are extracted close to the areas where the spices are grown. Spice oils and oleoresins primarily find uses in processed meat, fish and vegetables; soups, sauces, chutneys and dressings; cheeses and other dairy products; baked foods, confectionery, snacks, and beverages. The demand for spice oils and oleoresins is increasing globally day by day, due to the increasing demand for spicy fast-food snacks to be introduced into the market and with an eye toward developing a characteristic taste in the snacks for the future. Spice oils and oleoresins

TABLE 11.3
Biologically Active Constituents in Common Spices

Spice	Plant Part	Bio-Active Constituents (ppm)
Garlic (*Allium sativum*)	Bulb	Ajoene, Allicin, Alliin, Allistatin-I, Allistatin-II, Arginine (6000–15000), Ascorbic Acid (100–800), Choline, Citral, Diallyl disulfide, Geraniol, Glutamic acid (8050–19320) Linalool, Niacin (4–17), Scorodin-A, Tryptophan (660–1584)
Ginger (*Zingiber officinale*)	Root	Acetaldehyde, Ascorbic Acid (0–310), Asparagine Borneol (55–1100), Bornylacetate (2–50), Camphene (25–550) Chavicol, 1-8 Cineole (30–650) Citral (0–13500), Dehydrogingerdione, Geraniol (2– 50), Gingerdiones, Gingerols (18200), Hexahydrocurcumin, Limonene, Linalool(30–650), Methionine (670–735), Myrcene (2–50), Pinene (5–200), Selinene (35–700), Shogaols (1800), Zingerone, Zingibain, Tryptophan (630–690).
Clove (*Syzygium aromaticum*)	Bud	Anethole, Benzaldehyde, Carvone, Caryo-phylene (7400–8160), Chavicol (465–510), Cinnamaldehyde, Elagic-Acid, Eugenol (108000–120000), Eugenol acetate (36000–40000), Furfural, Gallic acid, Kaemferol, Linalool (1), Methyl Eugenol (310–340)
Cassia and cinnamon (*Cinnamonum, cassia & verum*)	Bark	Benzaldehyde (25–100), Camphene, Camphor, Caryophyllene (135–1315), 1,8Cineole (165–1800), Cinnamaldehyde (6000–30000), Cuminaldehyde (5–100), p-cymene (55–445), Eugenol (220–3520), Farnesol (3–10), Furfural (3–10), Limonene (45–180), Linalool (230–950), MethylEugenol, Myrcene (5–20), Niacin (8), Pinene (20–235), Piperitone (7–25), Safrole, Terpineol (1–260)
Cumin (*Cuminum cyminum*)	Fruit	Anisaldehyde (835), Ascorbic acid (0–75), Bornyl acetate (35), delta-3-carene (270), beta-carotene (5), Carveol (435), Caryophyllene (140–320), 1,8 Cineol(40–135), Copaene (30), p-Cymene(810–12600), Farnesol (830), Limonene (60–695), Linalool (30–315), Methyl Chavicol (30), Myrcene (35–120), Niacin (45), Pinene (10–6600), Piperitone (170), Terpinene (25–11800), Terpinen-4-0l (30), Terpineol (30–275)
Coriander (*Coriander sativum*)	Fruit	Anethole (1–2), Ascorbic acid (180–6290), Borneol (2–50), Camphor (100–1300), Carvone (20–25), Caryophyllene (1–8), 1-8 Cineole, p cymene (70–725), Ferrulicacid (460–1360), Geraniol (30–440), Limonene (30–1230), Linalool (4060–16900), Pinene (50–13750), Terpineol (30–40), Vanillic acid (220–960)
Cardamon (*Elettaria cardamomum*)	Fruit	Borneol (30–8000), Camphene (10–30), Camphor (5–20), 1,8 Cineole (525–56000), Citronellal, Citronellol (10–40), p-cymene (130–28000), Geraniol (45–140), Limonene (595–9480), Linalool (1285–8000), Myrcene (335–3000), Nerol 10–30), Nerylacetate, Pinene (70–3000), Terpinen-4-ol (250–23200), Terpinene (20–140)

Turmeric (*Curcuma domestica*)	Rhizome	Ascorbic Acid (0–290), Bisdesmethoxy curcurmin (60–27000), Borneol (15–350), Camphor (100–720), 1,8,Cineole (30–720) Cinnamic acid, curumin (10–38500), p-cymene, Niacin (5–60), p-Tolmethyl-carbinol (500–1750), Turmerone (1800– 43200)
Black pepper (*Piper nigrum*)	Fruit	Ascorbic acid (10), Benzoic Acid, Borneol Camphor, Carvacrol, Carveol, Caryophyllene, 1,8 Cineole, Cinnamic acid, Citral, Citronellal, p-cymene, Eugenol, Limonene, Linalool, Myrcene, Myristicin, Pinene, Piperidine, Piperine, Safrole, Terpinen-4-ol
Black mustard (*Brassica nigra*)	Seed/Leaf	Allyl isothiocyanate (6510–11760, seed), Arginine (1810–26657, LF), Ascorbic acid (235–4000, LF), β-carotene (30–475, LF), Erucic acid (770–11340, LF), Methionine (230–3390, LF), Niacine (3–48, LF), Tryptophan (270–3975, LF)
Saffron (*Crocus sativus*)	Flower	β-carotene, 1-8 Cineole, Crocetin, Crocin (20000), Delphinidin, Hentria Contane, Kaemferol, Lycopene, Myricetin, Naphthalene, Pinene, Quercetin
Licorice (*Glycyrrhiz glabra*)	Root	Acetic acid (2), Anethole (1), Betaine, Choline, O-Cresol, Estragole, Eugenol (1), Ferulic acid, Glycyrhizic Acid (100000–240000), Guaiacol, Kaemferol, Linalool, Mannitol, Niacin (70)
Mace and nutmeg (*Myristica fragans*)	Seed	Borneol (4200–25600), 1,8-Cineole (440–3500), p-cymene (120–960), Elemicin (20–3500), Eugenol (40–320), Furfural (15000), Geraniol, Limonene (720–5760), Linalool, Methyl Eugenol (20–900), Myrcene (740–5920), Myristicin (800–12800), Pinene (3000–4000), Safrole (120–2720), Terpinen-4-ol (600–4800), Terpineol (120–9600)
Thyme (*Thymus vulgaris*)	Plant	Borneol (15–1460), Bornyl acetate (15–540), Caffeic Acid, Camphene (15–270), delta-3- Carnene (510), Beta-carotene (20–25), Carvarol (15–18720), Chlorogenic acid, 1,8 Cineole (80–4590), p-cymene (145–20800), Geraniol (0–10660), Limonene (15–5200), Linalool (180–17420), Methionine (1370–1980), Myrcene (35–675), Niacin (50), Pinene (15–1600), Rosmarinic Acid (5000–6000), Terpinen-4-ol (70–8320), Terpineol (35–6500), Thymol (15–24000), Tryptophan (1860–2000), Ursolic acid (15000–18800)
Red Pepper/Chilli (*Capsicum corrals*)	Fruit	Arginine (400–8000), Ascorbic Acid (350–20000), Asparagine, Betaine, Capsaicin (100–2200), beta-carotene (0–460), Chlorogenic acid, Hesperidin, Methionine (100–1900), Niacin (5–170), Oxalic acid, Tryptophan (100–2000)
Vanilla (*Vanilla plantifolia*)	Fruit	Acetaldehyde, Acetic acid, Anisaldehyde, Benzaldehyde, Benzoic Acid, Creosol, Eugenol, Furfural, Guaiacol, Vanillic acid, Vanillin (13000–30000), Vanillyl Alcohol.

Source: Mukhopadhyay, M., *Natural Extracts Using Supercritical Carbon Dioxide*, CRC Press, Boca Raton, FL, 2000. With permission.

are particularly suitable for such snacks because they can be used very conveniently (without having to handle in bulk raw spices such as ginger, garlic, chili, onion, cardamom, and cinnamon). For example, approximately 7000 kg of onions are needed to produce 1 kg of highly concentrated onion oil.

Most spice extracts can be supplied in both oil- and water-soluble forms. As a result, spice oleoresins and essential oils are now shipped in dispersions of edible oil or other liquids. Furthermore, dispersions can be standardized with other ingredients, such as mono-, di-, or triglycerides or polysorbates. Another technique for easier spice application is to make a liquid emulsion of spice oleoresins, essential oils, and a starch or spray dry the essential oil to a powder. In this case, the spray-drying process has to make sure that other products are not spray-dried on the same equipment to avoid contamination. Oleoresins, spice oils, substrates, and diluent all should individually meet reliable standardization.

The blend formulations for spice seasonings use flavor enhancers and other flavor ingredients, such as monosodium glutamate, sodium erythorbate (used in deli meats), dextrose, maltodextrin, and hydrolyzed vegetable proteins. Many of these flavorings and ingredients are corn or soy based. Other processed ingredients are derived from products that go through a multistage conversion process of enzymolysis, fermentation, and regeneration until the final product is achieved. The quality and safety of the formulation are assured before marketing [4]. More information may be gathered from the references provided [5,6].

11.6 CONVENTIONAL EXTRACTION METHODS

Figure 11.1 outlines the various alternative steps involved in the conventional methods for production of spice extracts. The spice aroma or essential oil is traditionally produced by steam distillation (SD) of the ground spice or SD of the extracts obtained by solvent extraction (SE) or aqueous alkaline extraction (AE) of the ground spice. A variety of solvents, such as alcohols, acetone, and hexane, can be used for extraction of spices. However, removal of these organic solvents leaves some residual solvent behind, which requires the desolventization at elevated temperatures. This can cause chemical modifications of the oleoresins.

Table 11.4 lists a few examples of spice extracts that are produced commercially with their percentage yields of essential oils and oleoresins from ground spice, as reported by Marion et al. [7]. Variations in yield and quality of spice extracts may occur due to variations in the origin and harvesting time of spices.

The yield and quality of extracts also depend on preprocessing operations, such as grinding, the technique of extraction, and the nature of the solvent which, in turn, are decided based on the desired specification of the end product in terms of its aroma, flavor, and solubility. Each extract plays a specific role in the formulation and its selection is the key to the product development, as per the specific requirement of value addition.

Conventional extraction methods, such as SE, AE, and direct or indirect SD (i.e., hydrodistillation [HD]), are not selective. As a result, the extracts often contain color (e.g., chlorophyll) or some other undesirable components. Therefore, further purification is imperative using a number of techniques, such as color

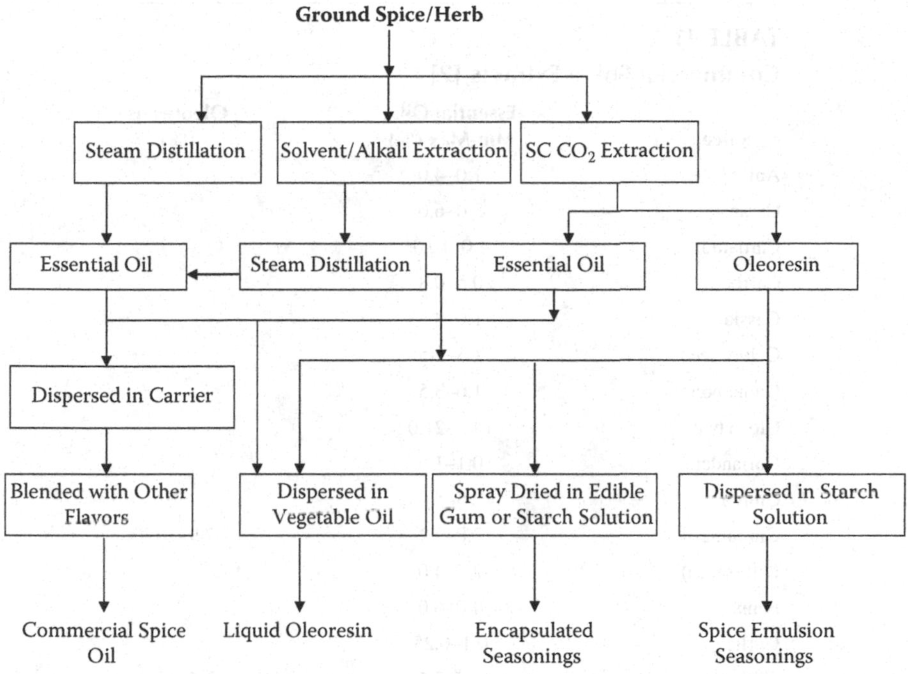

FIGURE 11.1 Various alternative steps for spice extraction. (From Mukhopadhyay, M., *Natural Extracts Using Supercritical Carbon Dioxide*, CRC Press, Boca Raton, FL, 2000. With permission.)

adsorption by activated charcoal, drying using silica gel, chromatographic separation, vacuum fractionation, or molecular distillation. Owing to the ban on the usage of the chlorinated solvents, the most commonly used solvent for SE today is hexane. For most spice oils and oleoresins in the international market, the residual hexane content in products has to be reduced to less than 25 ppm. This limit is expected to go down further. Hence, the SE process may be phased out in the near future. The food industry needs to combat strict regulations and comply with measures for safety, reliability, and standardization of natural products to be consumed as nutrients and food additives. This may be achieved by adopting SCFE techniques, as SC-CO_2 can recover the active ingredients in natural form without degradation or contamination. Over the last two decades, SCFE has emerged as a superior alternative to conventional processes such as SD and SE in the food, pharmaceutical, and cosmetics industries.

11.7 SUPERCRITICAL CARBON DIOXIDE AS THE EXTRACTANT

Several spice extracts—such as those from basil, black pepper, cardamom, chili, cinnamon, clove, cumin, fennel, fenugreek, ginger, garlic, nutmeg, paprika, savory, turmeric, and vanilla—are now commercially produced using SC-CO_2, as it is currently the most desirable SCF solvent for extraction of natural products. The solubility of the extract in SC-CO_2 increases with pressure or density of SC-CO_2 and decreases with

TABLE 11.4
Commercial Spice Extracts [2]

Spice	Essential Oils, Min-Max (%)	Oleoresins (%)
Anise	1.0–4.0	
Caraway	3.0–6.0	
Cardamon	4.0–10.0	
Carrot	0.5–0.8	
Cassia	1.0–3.8	
Celery seed	1.5–2.5	
Cinnamon	1.6–3.5	
Clove bud	14.0–21.0	
Coriander	0.1–1.0	
Cumin	2.5–5.0	
Curcuma	2.0–7.2	7.9–10.4
Dill (seeds)	2.5–4.0	
Fennel	4.0–6.0	
Garlic	0.1–0.25	
Ginger	0.3–3.5	3.5–10.3
Marjoram	0.2–0.3	—
Mace	8.0–13	22.0–32.0
Nutmeg	2.6–12	18.0–37.0
Pepper	1.0–3.5	5.0–15.0
Pimento berry	3.3–4.5	6.0
Saffron	0.5–1.0	
Savory	0.5–1.2	14.0–16.0
Vanilla	29.9–47.0	

Source: Mukhopadhyay, M., *Natural Extracts Using Supercritical Carbon Dioxide*, CRC Press, Boca Raton, FL, 2000. With permission.

temperature up to a limiting pressure (termed *cross-over pressure*), beyond which the solubility increases with both pressure and temperature. This phenomenon is utilized for recovering and recycling CO_2 after extraction by simply lowering the pressure, increasing the temperature, or both in the separators. There is no solvent residue in the extract as CO_2 is in a gaseous state at the ambient condition.

A broad range of selectivity and extractability can be achieved using SC-CO_2 just by manipulating the operating conditions, such as pressure and temperature, thereby targeting the specific compounds of interest. Because SCFE is highly selective, the concentration of the desired active compound in the total extract is higher and the yield of the desired active compound is closer to the total yield. There is rarely any need for additional processing steps for SCF extracts, whereas organic solvent–extracted oleoresins include undesirable resins that precipitate and make the solution cloudy, requiring an additional step of filtration. Even steam-distilled oil forms an immiscible layer due to the presence of monoterpene hydrocarbons, to a large extent, which inhibit the solubility of oil in soft drinks and beverages. The remarkable selectivity of SC-CO_2 over organic solvents facilitates the recovery of spice extracts with desirable constituents and superior blending characteristics. Studies on SCFE with SC-CO_2 and SE with alcohol indicate that, although the overall yield obtained using alcohol as the solvent is higher due to coextraction of undesirable components (which yields an accordingly higher quantity of total extract), the percentage of the desired active compound in that extract is lower [2].

In addition to selective extraction and the absence of organic solvent residues, SCFE offers another unique advantage; namely, simultaneous fractionation of different compounds is possible using the same solvent, SC-CO_2. For example, the active components in black pepper can be extracted with SC-CO_2 and separated into two fractions by changing pressure and temperature; the first fraction is enriched in oleoresin and the second fraction in essential oil. Accordingly, SC-CO_2 can, in a single process of SCFE, selectively extract the oleoresin and essential oil fractions (as opposed to processing of spices by SE, AE, or SD) and then separate them by sequential depressurization. Furthermore, most raffinate (the material left over after extraction) is uncontaminated and has a high market value due to the content of fiber and protein, which remain insoluble in SC-CO_2.

However, in view of the fact that SC-CO_2 is essentially nonpolar, it is unsuitable for extracting water-soluble constituents. This seeming disadvantage may be easily overcome by adding a food-grade polar cosolvent (typically in very small quantities, say 3 to 5 mole %) to SC-CO_2. The binary homogeneous mixture is then capable of extracting water-soluble or high-molecular-weight compounds. The best candidates for such cosolvents, especially for foods and nutraceuticals, are ethanol, ethyl acetate, and in some cases water. The remarkable value-addition that the SC-CO_2 extracts offer as natural concentrates, in addition to their advantages from the standpoint of environment and health, has generated a great deal of commercial interests for using SC-CO_2 as the extractant in the food industry.

11.8 COMMERCIAL SCFE PROCESS

For solid feeds, SCFE is usually a semi-batch process in which CO_2 flows in a continuous mode, whereas the feed is charged in the extractor basket in batches. However, for better viability on the commercial scale, the process is made semicontinuous using multiple extraction and separation vessels, as schematically described in the flow diagram shown in Figure 11.2. Extraction and separation of the extract are often carried out in stages, by maintaining different conditions of pressure and temperature in the extractors and separators. This allows easy fractionation of the extract for enrichment of the specific active components, which are subsequently fractionated in each of the separators. It is thus possible to produce a variety of products using the same hardware by merely changing pressure, temperature, and cosolvent concentration and make a plant for multiple products [2]. The commercial SCFE process works in a closed loop with constant circulation of CO_2 in the system, with a typical

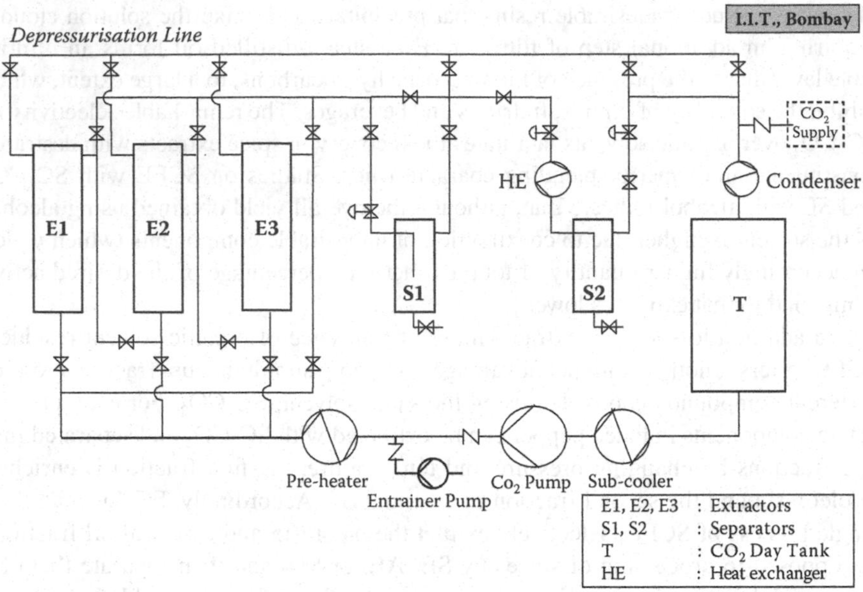

FIGURE 11.2 Process flow sheet of SCFE of spices.

batch time of 2 to 4 hours. Typical operating conditions for SCFE are in the range of 100 to 500 bar and 40°C to 80°C.

The feed for the SCFE process needs to have low moisture level (less than 10%) and be in the form of ground powder (100 to 300 mesh). However, preprocessing should be done in such a way that there are minimal losses of essential oils and active ingredients and negligible thermal and chemical degradation due to the rise in temperature or exposure to atmosphere, respectively. Accordingly, drying could be carried out in a fluidized bed drier in an inert environment, such as in flowing nitrogen or carbon dioxide. Similarly, grinding could be achieved in a liquid CO_2 or dry-ice, precooled grinder while controlling the humidity of the incoming air to avoid condensation of moisture in the feed.

It is generally believed that SCFE is capital-intensive, due to the requirements of the process to be operated at high pressures with very precise process control. There is also a general concern that SCFE technology is energy-intensive. However, it is interesting to note that the energy needed to attain a supercritical state ($P > 73.8$ bar, $T > 31.1$°C) is more than compensated for by the negligible energy required for solvent recovery from the extract by a simple step of depressurization. As a result, the overall energy consumption of SCFE using CO_2 is lower than that for traditional SD or SE, due to steam generation in SD and due to solvent evaporation and blowing off of steam for removal of residual solvent from the residues in SE. For example, the removal of residual solvent from an extract by SE requires about 8 kWh of energy per kilogram of plant extract [8], whereas extraction with SC-CO_2 requires one-tenth of this energy. Also, the solvent loss in the batch SE process is up to one-third of the feed solvent (though it is somewhat lower [10% to 15%] in the continuous SE process), whereas the loss of CO_2 in the SCFE process is negligible because CO_2 can be easily regenerated and recycled. The solvent requirement in the batch process is 10 to 20 times that of the feed charged and that in the continuous SE process is much lower, namely about 3 to 4 times the feed, whereas an hourly circulation rate of SC-CO_2 is in the range of 16 to 24 times the amount of feed charged.

The relatively higher investment required in the SCFE process is well balanced by other benefits of SCFE, such as low solvent (CO_2) cost, lower batch times, higher concentrations of active desirable components in the extract, and no additional purification- and pollution-abatement-related costs. SCFE also generates practically no effluent. In addition, the extracted residue (cake) does not undergo any degradation or contamination, unlike in SE and SD. Residue from the SCFE process has a market value as it retains all the useful ingredients, such as edible proteins and fibers. This can be sold as a high value by-product to yield additional revenue.

The normal SCFE process simultaneously and separately yields both liquid and solid products, starting with the same feed of spices in a single step, unlike SD and SE. In SD, the steam volatile essential oil, which is a liquid product, is distilled out, whereas in SE, the liquid (essential oil) and solid (oleoresin) products are obtained together. Subsequently, SD or SE with another solvent is employed to recover the essential oil from the mixed product. Liquid spice products are more stable, have a more reproducible quality than their conventional forms, and contain the characteristic aroma, taste, and odor. During their utilization, a smaller quantity is required for obtaining the same effect. The standardization of these new liquid spice products implies controlling their composition. A detailed feasibility study shows that even at the existing price (of extracts from SE and SD) of oil and oleoresins, the investment in SCFE is profitable, which justifies it to be the preferred route from a long-term perspective. The instrumentation and control system necessary for the SCFE process is designed to provide accurate control of the parameters, ensuring high consistency, reliability, and standardization of the final product.

11.9 COMPARISON OF SPICE EXTRACTS BY CONVENTIONAL AND SCFE PROCESSES

Although recovery of essential oils from spices by SD and AE has been practiced for centuries, the oils produced by these processes may contain artifacts formed during the processing, in addition to the fact that the recovery of the oils is quite low, as some of the components are not steam volatile. When SE involves the use of organic solvents to extract essential oils from ground spices, the quality of the extract is decided by the presence of residual solvent, artifacts formed due to thermal degradation during the recovery of the solvents, or by coextraction of undesirable components due to polarity of the solvent. A polar solvent is likely to extract most polar components from spices, some of which may even be undesirable. Extraction of spices with SC-CO_2 or subcritical liq-CO_2 is most favorable for commercial production of essential oils and oleoresins, as CO_2 is a natural solvent and is ideally suitable for thermally labile natural products. The oleoresins are extracted at relatively high pressures, whereas the essential oils are recovered at relatively low pressures, which can form a clear solution when added directly to soft drinks. But simultaneous extraction of oleoresins and essential oils at a very high pressure in the range of 250 to 350 bar,

TABLE 11.5
Yields by SD and SC-CO_2 Extraction with a Cosolvent

	Steam Distillation	SC-CO_2 Extraction			
Spice	Yield (%)	Cosolvent	Extractor P/T (bar/°C)	Separator P/T (bar/°C)	Yield (%)
Allspice	2.5	Ethanol	300/40	55/37	5.3
Basil	0.5	Ethanol	200/40	56/15	1.3
Cardamom	4.0	Methyl acetate	150/60	50/9	5.8
Coriander	0.6	Ethanol	300/40	54/13	1.3
Ginger	1.1	Ethanol	300/40	52/11	4.6
Juniper berry	1.5	Hexane	300/60	52/11	7.2
Marjoram	2.06	Ethanol	250/40	50/35	1.7
Oregano	3.0	Ethanol	150/40	55/14	5.4
Rosemary	1.44	Ethanol	250/60	53/12	7.5
AU	1.34	Hexane	250/60	53/12	7.5
Sage	1.1	Methyl acetate	200/40	53/12	4.3
Thyme	1.85	Hexane	150/46	50/9	2.1

Source: Mukhopadhyay, M., *Natural Extracts Using Supercritical Carbon Dioxide*, CRC Press, Boca Raton, FL, 2000. With permission.

followed by stage-wise selective fractionation at supercritical and subcritical conditions, is preferred for commercial production. The operating conditions of pressure, temperature, and cosolvent are appropriately selected in order to obtain a specific product profile. The cosolvent is often selected on the consideration that it can be left behind in the extracted product or used for making the formulation with allowance made for dilution level. The cosolvent selected is a food-grade GRAS organic solvent, such as ethanol, ethyl acetate, acetic acid, or water. The advantage of SCFE with cosolvent-mixed SC-CO_2 over SE is that the former is selective and retains all other advantages of the SCFE process because the cosolvent added to SC-CO_2 is in very small amount (3 to 5 mole %). Most of the cosolvent after the SCFE process escapes with CO_2 in the separator, ensuring marginal content of residual solvent in the final extract. The extraction yields obtained by Calame and Steiner [9] using SD and SC-CO_2 are given in Table 11.5.

It is now an established fact that SC-CO_2 extraction at optimized conditions yields much more active ingredients than SE or SD. But other factors, such as particle size, preprocessing methods (e.g., drying and grinding), time of extraction and storage after harvesting, and even geographical origin of the raw spice, responsible for the recovery of the extract are not included in Table 11.5. According to the experience of the author, higher yields may be obtained from some of these spices with SC-CO_2, even without a cosolvent, as can be seen later in Table 11.7.

Not only the yields of the extracts but also their organoleptic (sensory) characteristics may be different for extracts obtained by different methods. Accordingly, the criteria for selection of the best process condition are based on the desired

TABLE 11.6
Spice Constituents (Area %) by Various Methods

Constituents	Distillation (%)	L CO_2 (%)	SC-CO_2 (%)	Hexane (%)
Ginger Extract (by GC)				
α-Curcumene	10.0		3.7	2.3
α-Zingiberene	44.0		19.6	12.1
β-Zingiberene	8.0		3.4	2.0
β-Bisabolene	8.3		3.7	2.4
β-Sesquiphellandrene	17.8		7.9	4.9
Zingerone	0.8		0.7	0.3
Ginger Extract (by HPLC)				
6-gingerol	0.2		16.4	0.9
8-gingerol	0.3		3.1	0.7
10-gingerol	—		3.8	0.8
6-shogaol	0.3		2.8	6.3
8-shogaol	—		—	1.6
Cumin Extract[2] (by GC)				
α-pinene	—		1.1	—
(Ethyl Ether)-pinene	13.0		21.0	
p-cymene	13.0		9.4	
γ-terpinene	24.8		20.0	
Cuminaldehyde	16.0	20.3[19]	21.0	11.4
Cymol	33.4[20]	26.7[19]	15.2	13.5
Clove Extract (by GC)				
(Ethyl Ether)				
Eugenol	76.4	77.1	71.8*	73.3
Eugenol acetate	5.6	4.9	11.1*	4.6
β-caryophyllene	5.8	8.5	9.3*	10.4

* Pilot plant experiment at IIT Bombay

Source: Mukhopadhyay, M., *Natural Extracts Using Supercritical Carbon Dioxide*, CRC Press, Boca Raton, FL, 2000. With permission.

quality. In any case, CO_2 extracts have more top notes, more back notes, no off notes, no degradation, more shelf life, and better aroma and blending characteristics than steam distilled and hexane extracts, as can be seen in Table 11.6. In general, the extract produced by SE contains all the ingredients that are soluble in the organic solvent, including the volatile oils and resins. Some triglycerides (lipids) present in spices are coextracted and act as nature's own fixative resulting easy and proper blending.

The yield of essential oil by SD of cumin (2.5%) is less than that by SC-CO_2 extraction (3.5%) at 120 bar and 40°C [10]. A comparison of the composition of the clove extracts obtained by liquid CO_2, SE with ethyl ether, and SD indicates that liquid CO_2 and ethyl ether extracts are similar, though liquid CO_2 extract has the characteristics of both essential oil and oleoresin [11]. Liquid and SC-CO_2 extracts are always transparent and contain more active components that are closer to those in the fresh or natural spice, due to low operating temperature and inert environment. SC-CO_2 extraction followed by fractional separation was carried out for a variety of spices using a 10-L extractor capacity pilot plant at Indian Institute of Technology (IIT), Bombay. The composition of the active ingredients of the extracts separated in the two separators, as analyzed by either gas chromatography (GC) or gas chromatography-mass spectrometry (GC-MS), are indicated in Table 11.7 (except for pepper, where piperine was quantified by ultraviolet [UV] method). The yields and the compositions of the SC-CO_2 extracts are compared with those of hexane-extracted products from the same spices (Table 11.7). SC-CO_2 extraction yields are better for clove, cumin, and black pepper. In most cases, the concentrations of the active ingredients were higher in the SC-CO_2 extracted product [2].

11.10 PROCESS ANALYSIS OF SCFE FROM SELECTED SPICES

Spice extracts are usually a complex mixture of volatile essential oils, waxes, triglycerides, and resinous and other miscellaneous materials, with the composition of the constituents contributing to aroma, flavor, and pungency selected depending on the specific application. Accordingly, for customized applications, SC-CO_2 extraction of spices requires fractional separation of selected groups of constituents. This can be achieved in two ways: by stage-wise extraction followed by depressurization of the extract-laden SC-CO_2 or by single-stage extraction at a very high pressure followed by stage-wise depressurization for fractional separation. In the former method, the volatile oil is first extracted at relatively milder conditions and, subsequently, the nonvolatile oleoresins are extracted at relatively more-severe conditions. In the latter method, the finely ground spice is more or less completely extracted at a relatively more-severe condition to recover both oleoresins and volatile oil simultaneously and efficiently so that the time of extraction is greatly reduced. The extract-laden SC-CO_2 is subsequently depressurized in two or three separators at predetermined conditions so that specific products are selectively fractionated and collected. The second method offers significant advantages, as the quality of the product is improved and the batch time for extraction is reduced, resulting in higher production capacity and cost effectiveness of the SC-CO_2 extraction process. Simultaneous fractionation at precisely selected conditions allows production of customized quality fractions and elimination of undesirable contaminants from them. The specific advantages of this SC-CO_2 extraction and fractionation process are mentioned in the following subsections with respect to a few common spices.

11.10.1 CELERY SEED (*APIUM GRAVEOLEN*)

A comparison of chemical composition of celery seed oil by HD and SC-CO_2 extraction [12] at 100 bar and 40°C indicated that the HD oil contained mostly monoterpenes, whereas the SC-CO_2 extracted oil contained mostly phthalides (Table 11.8).

TABLE 11.7
Yields and Concentrations of Active Ingredients in Extracts with SC-CO_2 and Hexane

Spices (Active Ingredient)	SC-CO_2 Extraction (200 bar, 40°C) (by wt.)			Solvent (Hexane) Extraction (by wt.)	
	Yield (%)	% Ess. Oil	% Oleoresin	Yield (%)	% Extract
Clove	23.8			16.8	
Eugenol		71.8	—		70.7
Eugenol acetate		11.1	—		11.3
(by GC, 10% FF AP)					
Cumin	21.0			12.2	
Cymol		15.2	—		13.5
Cuminaldehyde		15.3	—		11.4
(by GC-MS, DB5)					
Coriander	3.6			20.0	
Linalyl acetate		7.8	—		5.8
D-linalool		13.0	—		—
(by GC-MS, DB-5)					
Ginger	4.6			4.9	
Zingiberene		26.7	1.6		31.6
Gingerol		5.65	10.1		5.4
(by GC-MS, DB-5)					
Cinnamon	3.0			5.11	
Cinnamicaldehyde		77.5			45.0
(by GC, SPB-1)					
Pepper	4.6			5.0	
Piperine		—	53.0		46.4
(by UV method)					
Ajwain	4.5			5.18	
Thymol		63.6	—		24.6
(by GC OV-101)					

Source: Mukhopadhyay, M., *Natural Extracts Using Supercritical Carbon Dioxide*, CRC Press, Boca Raton, FL, 2000. With permission.

The SC-CO_2 extract contained some additional components, such as fatty acids, which were not present in the HD oil. Monoterpenes constituted 57.6% of the HD oil, whereas the SC-CO_2 oil contained 56.8% phthalides [12]. The low level of phthalides (15.2%) in the HD oil was attributed to their high boiling points and low volatility in steam. More than 10 hr of HD was necessary for their complete recovery. Phthalides are cyclic esters or lactones with outstanding odor characteristics of celery. The odor of the SC-CO_2-extracted oil is more intense and less terpenic. Therefore, the SC-CO_2-extracted oil is preferred to the HD oil to impart the celery flavor. With

TABLE 11.8
Composition (%) of Celery Seed Essential Oil by $SC\text{-}CO_2$ Extraction and Hydrodistillation

Class of Compounds	$SC\text{-}CO_2$ Extraction	HD
Monoterpenes	16.1	57.6
Oxygenated monoterpenes	0.2	0.6
Sesquiterpenes	19.7	23.3
Phthalides	56.8	15.2
Others	4.8	0.3

Source: Mukhopadhyay, M., *Natural Extracts Using Supercritical Carbon Dioxide*, CRC Press, Boca Raton, FL, 2000. With permission.

TABLE 11.9
Composition of Essential Oils from Celery Seeds and Leaves by $SC\text{-}CO_2$

Component	Celery Seeds (100 bar, 40°C)	Chinese Celery Seed (100 bar, 40°C)	Celery Leaves (90 bar, 40°C)
Limonene	3.7	14.9	33.4
β-Selinene	33.8	17.6	3.0
α-Selinene	5.3	1.8	0.5
Butyl phthalide	19.8	5.5	2.8
Sedanenclide	—	22.4	—
Bedanolid	—	28.8	—
Germacrone	21.0	—	45.4

Source: Mukhopadhyay, M., *Natural Extracts Using Supercritical Carbon Dioxide*, CRC Press, Boca Raton, FL, 2000. With permission.

$SC\text{-}CO_2$ extraction carried out from celery seeds at a relatively moderate pressure of 100 bar and 40°C, the yield of essential oil was merely 2.03% of the charged material [13]. The yield of essential oil from celery leaves by $SC\text{-}CO_2$ extraction at 90 bar and 40°C was even lower (0.04%). The compositions of the essential oils from celery seeds and celery leaves by $SC\text{-}CO_2$ extraction were found to be significantly different, as can be seen in Table 11.9. The celery seed extracts contained more paraffin and fatty acid methyl esters than the celery leaf extract.

11.10.2 Red Chili

$SC\text{-}CO_2$ extraction of red chili is carried out in the pressure range of 300 to 500 bar and 80°C to 100°C, with simultaneous fractionation of the extracts into light and heavy fractions. The light fraction contains most of the capsaicin (i.e., the compound responsible for the hotness of the spice), in addition to the essential oil, whereas the heavy fraction contains triglycerides and the color compounds, in addition to

TABLE 11.10(A)
Composition in Light and Heavy Fractions of Chili Extract

Products	% Capsaicin	% Dihydro-Capsaicin	Total % Capsaicinoid
Raw material	0.21	0.14	0.39
Light fraction	8.10	4.05	13.50
Commercial product	1.83	1.52	3.93
Heavy fraction	0.57	0.31	0.95
Ratio L/H	14.2	13.1	14.2

TABLE 11.10(B)
SCFE from Fresh Chili with Successive Increase in Temperature and Pressure with 5% Acetic Acid (by wt.) of Feed Charged

Pressure (bar)	Temperature (°C)	Time (min)	Yield (% wt. of feed)	Capsaicinoid (% of extract)
100	40–50	60	0.065	47.0
130	50–60	45	0.040	57.5
150	60–70	45	0.010	87.4
175	70–75	45	0.003	98.8
250	75	45	0.106	0.8

Source: Mukhopadhyay, M., *Natural Extracts Using Supercritical Carbon Dioxide*, CRC Press, Boca Raton, FL, 2000. With permission.

a small quantity of capsaicin [14], as shown in Table 11.10a. The distribution of capsaicin in the two fractions is adjusted by selecting the conditions in the separators for fractionation. SCFE of a more pungent variety was performed [28] over a lower temperature range (35°C to 70°C) and a wider pressure range (100 to 550 bar). It was shown that the capsaicinoid (capsaicin, dihydrocapsaicin and the like) could be raised to 75% in the 0.2% to 0.3% oleoresin if extraction was carried out from the dried material and up to 99% from fresh material. It was observed the capsaicinoid content increased with time of extraction and addition of a cosolvent, such as 5% acetic acid, and it increased further by successively increasing the temperature, as presented in Table 11.10b.

11.10.3 Paprika

Paprika is useful in industry for its natural color. For SC-CO_2 extraction of paprika, most of the color compounds are collected in the heavy fraction, while aroma is collected in the light fraction. Research indicates [14] that the color value of SC-CO_2-extracted product could reach as high as 7200 ASTA, whereas a normal commercial product is characterized to have a color value in the range of 1000 to 2000 ASTA.

TABLE 11.11(A)
Light and Heavy Fractions of $SC\text{-}CO_2$-Extracted Ginger Oleoresin

Product	% 6-G	% 8-G + 6-S	% 10-G + 8-S	(8G+6S) % Total	Total % Extract
Raw material	0.87	0.14	0.27	0.11	1.28
Heavy fraction	13.95	2.58	4.37	0.12	20.90
Commercial product	2.81	5.83	1.19	0.52	11.12
Light fraction	1.43	0.61	0.36	0.25	2.40
Ratio H/L	9.8	4.2	12.1	0.5	8.7

G : Gingerol; S : Shogaol

TABLE 11.11(B)
Compositions (%) Light and Heavy Fractions of $SC\text{-}CO_2$-Extracted Ginger Essential Oil

Product	Raw Material	Heavy Fraction	Light Fraction	Ratio L/H
Essential oil (ml/100 g)	2.0	4.4	98.8	22.5
β-pinene	2.5	0.5	2.6	5.2
Camphene	7.0	1.6	7.3	4.6
Cineole	8.4	2.3	8.6	3.7
Limonene	1.2	0.3	1.2	4.0
Zingiberene	21.8	17.4	22.7	1.3
Bisabolene	8.7	8.6	8.8	1.0
Sesquiphellandrene	11.9	12.7	11.9	0.9

Source: Mukhopadhyay, M., *Natural Extracts Using Supercritical Carbon Dioxide*, CRC Press, Boca Raton, FL, 2000. With permission.

11.10.4 GINGER

Ginger extract using $SC\text{-}CO_2$ is fractionated into two fractions: the essential oil–enriched light fraction and the oleoresin-enriched heavy fraction, the compositions of which are compared in Table 11.11a and Table 11.11b. Gingerols (G) and shogaols (S) are the compounds responsible for the pungency of ginger, and they are mostly collected in the heavy fraction of the $SC\text{-}CO_2$ extract, as can be seen in Table 11.11a. Shogaols, being the oxidation products of gingerols, are present in very less quantities in the $SC\text{-}CO_2$-extracted fractions. A product of desired specification can be formulated by combining the two fractions in a suitable proportion [14].

11.10.5 NUTMEG

$SC\text{-}CO_2$ extraction and fractionation of nutmeg can yield good quality nutmeg butter as the heavy fraction with very little volatile oil and nutmeg oil as the light fraction, in which the undesirable hallucinatory compound myristicin is present in negligible concentration [14]. It is possible to use SC-CO_2 to produce nutmeg oil devoid of this compound. This is an important advantage of the SCFE process, as the presence of this compound in nutmeg oil is banned in some countries.

11.10.6 Black Pepper

When black pepper is extracted and fractionated into two fractions using SC-CO_2, the light fraction may be completely free from piperine, the active ingredient of pepper, whereas the heavy fraction may be enriched with up to 60% piperine [14]. Besides the concentration of the specific component, all SC-CO_2-fractionated products are of superior quality. The rate of extraction at 500 bar is almost double the rate at 300 bar and 60°C. The production capacity of the fractions may be enhanced four times at 500 bar using a cascade of four extractors. Thus, the operating cost of extraction can be reduced to one-fourth of that obtained by the traditional SCFE plant. The current commercial practice is to follow this technique to improve the efficiency and cost effectiveness of SC-CO_2 extraction of major spices.

11.10.7 Vanilla

Natural vanilla fragrance is extracted from cured vanilla beans. Green vanilla beans are cured to bring about hydrolysis of the glucosides present in the beans to generate vanillin and other flavor and fragrance components. The curing process changes the green vanilla beans into dark, brownish, soft beans. The current commercial extraction method uses aqueous alcohol of 35 to 40 vol.% in concentration at a temperature as high as 87°C in a number of steps, making the extract thermally degraded. SC-CO_2 extraction of cryogenically ground, dried beans resulted 10.6% yield of oleoresin at 110 bar and 36°C, which is even higher than 5.3–8.4% yields by alcohol extraction [15]. The vanilla oleoresin contained as high as 16% to 36% vanillin by SC-CO_2 extraction, which amounted to 74% to 97% recovery of the total vanillin content, respectively. Other flavor and fragrance constituents in the natural vanilla extract are p-hydroxy benzaldehyde, vanillic acid, and p-hydroxybenzoic acid. The quality of the extract is, however, characterized by its vanillin content. The compositions of the natural vanilla extracts by traditional alcohol extraction and SC-CO_2 extraction are compared in Table 11.12. The highest purity vanillin could be obtained by SC-CO_2 extraction of water-presoaked beans, though the yield was only 3%. On the other hand, cryo-grinding apparently releases more compounds and, accordingly, the yield was also high (10.6%). The purity of the alcohol extract as well as percent recovery of vanillin (61%) is lower than that of the SC-CO_2 extract. Even the color of the extract, which is yellow compared to the dark brown color of the alcoholic extract, is superior in the case of SC-CO_2 extraction.

11.10.8 Cardamom

SC-CO_2 extraction of cardamom requires much higher pressure (100 bar) than subcritical propane, which requires as low as 20 bar to yield the same amount of

TABLE 11.12
Comparison of Compositions of Vanilla Extracts

Solvent (Beans)	$SC-CO_2$ (120 bar, 33°C) (Dry)	(Ground)	(Water soaked)	Ethanol + H_2O (Water soaked)
p-hydroxybenzoic acid (area %)	0.2	0.1	0.1	1.1
Vanillic acid (area %)	0.1	1.3	0.1	1.1
p-hydroxy benzaldehyde (area %)	0.6	1.9	0.9	2.7
Vanillin (mass %)	21.0	16.1	36.3	20.0[a]
Unknown	0.0	2.4	0.0	8.0

[a]: Water-free basis.

Source: Mukhopadhyay, M., *Natural Extracts Using Supercritical Carbon Dioxide*, CRC Press, Boca Raton, FL, 2000. With permission.

TABLE 11.13
Yield of Cardamom Oil and Pigment by $SC-CO_2$ and Propane

Process Conditions	Yield % (g/g oil)	β-Carotene (g/g oil)	Chlorophyll (g/g oil)	Pheophytin
$SC-CO_2$ (80 bar, 25°C)	5.65	0.8	0.65	—
$SC-CO_2$ (100 bar, 35°C)	5.45	2.1	0.30	—
$SC-CO_2$ (200 bar, 35°C)	5.95	3.9	0.36	0.33
$SC-CO_2$ (300 bar, 35°C)	6.65	5.8	4.53	2.36
CO_2 + ethanol (100 bar, 25°C)	5.28	1.64	9.65	2.10
Ethanol	—	0.80	11.95	2.60
Propane (50 bar, 25°C)	7.24	18.6	10.80	4.80
Propane (20 bar, 25°C)	6.85	16.2	3.40	2.10

Source: Mukhopadhyay, M., *Natural Extracts Using Supercritical Carbon Dioxide*, CRC Press, Boca Raton, FL, 2000. With permission.

essential oil. Addition of ethanol to $SC-CO_2$ does not greatly increase the yield, but increases the coextraction of pigments, as can be seen in Table 11.13. Reduction in pressure of $SC-CO_2$ usually reduces the contents of β-carotene, chlorophyll, and pheophytin in the extract.

The amount of pigment extracted is significantly more when subcritical propane is used as the extractant. However, better recovery of aroma (Table 11.14) is possible with $SC-CO_2$ at 100 bar and 35°C, as reported by Illes et al. [16].

11.10.9 Fennel, Caraway, and Coriander

Recovery of active components from fennel, caraway, and coriander by different methods of extraction is compared in Table 11.15. It is clear that $SC-CO_2$ extracts are richer in active components, owing to better selectivity of the extractant [17].

TABLE 11.14
Peak Area (× 10^3) of Aroma Constituents of Cardamom Oil by SCF

	β-Pinene	Cineole	Linalool	α-Terpinol	Borneole
CO_2 (80 bar, 25°C)	16.1	295	34.8	47.8	356
CO_2 (100 bar, 35°C)	27.6	450	73.5	91.2	579
CO_2 (300 bar, 35°C)	17.4	341	32.7	46.4	340
Propane (20 bar, 25°C)	15.5	286	25.6	36.9	304
Propane (50 bar, 25°C)	26.9	386	72.1	82.7	521
CO_2 + ethanol (100 bar, 25°C)	6.5	198	5.8	8.9	112

Source: Mukhopadhyay, M., *Natural Extracts Using Supercritical Carbon Dioxide*, CRC Press, Boca Raton, FL, 2000. With permission.

TABLE 11.15
Recovery of Active Components from Fennel, Caraway, and Coriander by Various Methods

		SC-CO_2					
Active	P (bar)	80	100	200–300	Ultrasound		
Component	T (°C)	28	30	35	water	Hexane	Steam
			Fennel				
Fenchon	10.7	13.1	9.2	21.9	16.3	0.3	
Estragol	1.6	0.5	1.5	6.6	3.1	1.7	
Trans anethole	68.2	50.8	72.5	70	70	77.6	
			Caraway				
Limonene			33.5	32.0	33.3	30.1	
D-carvone			56.9	54.0	54.3	50.2	
			Coriander				
Linalool	20–30	15	80–85	67	80	79	

Source: Mukhopadhyay, M., *Natural Extracts Using Supercritical Carbon Dioxide*, CRC Press, Boca Raton, FL, 2000. With permission.

11.9.10 Garlic

SC-CO_2 extraction of valuable ingredients from garlic is comparable to that by hexane [18]. The major components of garlic oil are diallyl disulfide (30%), diallyl trisulfide (30%), and diallyl sulfide (15%). Alliin, a major garlic active ingredient, is known to degrade to allicin by an enzymatic reaction, and other garlic components are also susceptible to oxidation with temperature. A comparison of high-performance liquid chromatography (HPLC) and GC analysis of extracts obtained by SE with a variety of solvents with varying polarity with that by SC-CO_2 indicated that the former contained

TABLE 11.16
Compositions of Cinnamon Leaf Oil and Cinnamon Bark Oil

	% Leaf Oil[a]	% Bark Oil		
Component	Steam Distillate	Steam Distillate	$SC\text{-}CO_2$ Extract (200 bar, 60°C)	$SC\text{-}CO_2$ + Ethanol Extract (200 bar, 60°C)
Eugenol	85–95	3.3	2.0	2.8
Caryophyllene	6	Traces	2.1	1.6
Cinnamicaldehyde	38	7.88	1.98	6.8
Isoeugenol	21.9	1.2	0.4	
Linalool	20.1	0.9	0.8	
Cinnamyl acetate	23.6	5.1	1.8	
o-methroxy cinnamic aldehyde	1.3	1.6	2.6	

[a]: From reference (Wright, 1994)

Source: Mukhopadhyay, M., *Natural Extracts Using Supercritical Carbon Dioxide*, CRC Press, Boca Raton, FL, 2000. With permission.

more components. This is attributed to degradation of the components in SE. Clinical tests also indicated that the $SC\text{-}CO_2$ garlic extract has more potent bioactivity, close to that of raw garlic [18].

11.10.11 Cinnamon

Two types of essential oil namely, leaf and bark oil, are produced from two different parts of a cinnamon tree. Cinnamon leaf oil is mainly produced in Sri Lanka. It is also produced in India and Seychelles. Most cinnamon oil is produced from leaves. Bark oil amounts to only 15% of total production. Leaves yield 1% oil. However, the root bark yields 3% oil [19]. The comparison of the compositions of extracts from Srilankan cinnamon bark and leaves is given in Table 11.16. SD of the cinnamon bark yields 1.4% oil, whereas $SC\text{-}CO_2$ extraction at 200 bar and 60°C results 1.5% yield. However, addition of ethanol as a cosolvent increases the yield to 2.6% [9]. The leaf oil rich in eugenol makes it a substitute for clove oil and may be used for conversion to vanillin. Bark oil is more valuable than the leaf oil, although both find wide uses in flavoring and pharmaceutical industries.

11.11 CORRELATION FOR SPICE OIL SOLUBILITY IN $SC\text{-}CO_2$

Solubility of spice oils in $SC\text{-}CO_2$ is an important process parameter needed for design and scale-up of the commercial SCFE plant. Solubility depicts the maximum possible solvent capacity of $SC\text{-}CO_2$ at a given temperature, pressure, or cosolvent concentration in $SC\text{-}CO_2$, though the actual loading or dissolution of the solute is much less than this solubility in the presence of the solid substrate. However, the neat solubility (without the presence of the substrate) behavior of spice oil can suffice for selection of the process conditions for the most efficient performance of the SCFE process. Because experimental measurement of solubility is tedious and

TABLE 11.17
Black Pepper Oil Solubility in SC-CO_2

T (°C)	P (bar)	Density of SC-CO_2 (g/cm³)	Oil Solubility (g/cm³ CO_2)
30	150	0.8478	0.0755
40	150	0.7812	0.0728
50	150	0.7010	0.06015
30	200	0.8909	0.1006
40	200	0.8404	0.08774
50	200	0.7851	0.7812
30	300	0.9486	0.13698
40	300	0.9106	0.1243
55	300	0.8712	0.1093

Source: Silva, D.C.M.N. et al., Correlating solubility values of black pepper oil in supercritical CO_2 using empirical models, *in Proceedings of the Sixth International Symposium on Supercritical Fluids*, Nice, France, 2003, Tome 1, 279. With permission.

time consuming at different conditions, a reliable correlation can serve the purpose, as it can be utilized for estimation of solubility spice oil in SC-CO_2. Chrastil [21] related the equilibrium solubility of a solute in SC-CO_2 by a linear relationship in terms of its density as:

$$\ln y^* = k \ln \rho + \frac{a}{T} + b \tag{11.1}$$

where, y^* is the solute solubility (g/L), T (K) is the temperature, ρ is the density of SC-CO_2 (g/L), and a, b, and k are adjustable constants that can be evaluated from the limited experimental data.

De Valle and Aguilera [22] modified the Chrastil's correlation by adding one more regressable constant to widen its validity for the temperature range from 20°C to 80°C and for pressures varying from 150 to 280 bar as:

$$\ln y^* = k \ln \rho + \frac{a}{T} + \frac{b}{T^2} + C \tag{11.2}$$

Silva et al. [23] correlated the experimental data (as reported in Table 11.17) in terms of the density of SC-CO_2 as reported by Angus [24]. The constants in the correlations are presented in Table 11.18. Ferreira et al. [25] reported SCFE of black pepper essential oil from which the solubility data were generated and were correlated in terms of vapor pressures (P^s), considering oil as a pseudo-pure component:

$$y^* = \frac{P^S}{P} \exp\left[A + B\rho\right] \tag{11.3}$$

TABLE 11.18
Parameters for the Solubility Correlations of Black Pepper Oil

Correlation	a	b	c	k
Chrastil	−14807.9	26.123	—	3.84
De Valle & Aguilera	70207.45	−13500128.15	−107.409	3.84

Source: Silva, D.C.M.N. et al., Correlating solubility values of black pepper oil in supercritical CO_2 using empirical models, *in Proceedings of the Sixth International Symposium on Supercritical Fluids*, Nice, France, 2003, Tome 1, 279. With permission.

where A and B are empirical constants evaluated from the experimental data. However, this requires vapor pressure of spice oil as a function of temperature.

Essential oil obtained from turmeric (*Curcuma longa*) contains a large number of components, such as curcumin, ar-turmeric (42%), turmerone (12%), and the rest (< 5%). Considering it to be a single component by researchers [26] the solubility of turmeric oil in SC-CO_2 was modeled using the steady-state extraction data at the initial period as well as using Naik's correlation [27] as:

$$Y = \frac{Y_\infty t}{B+t} \tag{11.4}$$

where Y = extraction yield (kg extract/kg curcumin) × 100
t = CO_2 mass (kg CO_2/kg curcumin)
Y_∞ = extraction yield at equilibrium
B = CO_2 mass needed to reach the half of Y_∞.

A comparison of the predicted solubilities with the corresponding experimental data is given in Table 11.19. It may be noted here that the fraction [0.5 Y_∞/B] is similar to the slope of the extraction curve at the initial stage (i.e., when the extraction of spice oil is controlled by its solubility). Both methods result similar agreement with the experimental data and may be considered for ascertaining the solubility behavior.

11.12 CONCLUSIONS

SCFE of spices is considered a superior alternative to the conventional techniques of SD, SE, and ASE for simultaneous production of essential oils and oleoresins in a single step. SCFE ensures high consistency and reliability in the quality and safety of the bioactive natural molecules. SC-CO_2 is GRAS and yields contaminant-free, tailor-made extracts of superior organoleptic profile, with high potency of active ingredients without any residual organic solvent and artifacts. The extracts are very close to that in nature in smell and taste and have longer shelf lives due to coextraction of antioxidants and better blending characteristics due to coextraction of triglycerides. SCFE is known to be commercially viable for high-value, low-volume extracts, and if multiple products are obtained operating the same plant at the respective optimized process conditions.

TABLE 11.19
Solubility of Essential Oils of *Curcuma longa*

T (°C)	P (bar)	Solubility (g/100 g CO_2) By Naik's Model	Solubility (g/100 g CO_2) From Extraction Data
30	100	0.39	0.67
	150	0.82	1.00
	200	0.87	1.12
	250	0.95	1.20
	280	1.59	—
40	100	0.17	0.20
	150	0.58	0.63
	200	1.24	1.34
	250	1.67	1.51
	280	1.88	1.74
50	100	0.19	0.31
	155	0.77	0.89
	200	1.51	1.53
	250	2.54	1.96
	280	2.80	2.16
35	280	1.59	1.54

Source: Blasco, M. et al., SCFE of *Curuma longa*: Solubility of essential oil, *in Proceedings of the Sixth International Symposium on Supercritical Fluids*, Nice, France, 2003, Tome 1, 279. With permission.

REFERENCES

1. Darling, M. Louis, Biomedical Library, UCLA, *Spices: Exotic Flavors & Medicines*, Available at http://unitproj.library.ucla.edu/biomed/spice/
2. Mukhopadhyay, M., *Natural Extracts Using Supercritical Carbon Dioxide*, CRC Press, Boca Raton, FL, 2000.
3. Duke, J.A., Biologically active compounds in important spices, in *Spices, Herbs, and Edible Fungi*, Charalambous, G., Ed., Elsevier Science Publishers, Netherlands, 225–250, 1994.
4. Rosen, R.T., *Ta'am Tov B'Tuv Ta'am: A Flavorful Blend of Kashrus and Spices*, Available at http://www.kashrut.com/articles/spices/ September 12, 2006.
5. Tainter, D.R. and Grenis, A.T., *Spices and Seasonings: A Food Technology Handbook*, Second Edition, Culinary and Hospitality Industry Publications Service, 1997.
6. Raghavan Uhl, S., *Handbook of Spices, Seasonings, and Flavorings*, Culinary and Hospitality Industry Publications Service, 1996.
7. Marion, J.P., Audrin, A., Maignial, L. and Brevard, H., Spices and their extracts: Utilization, selection, quality control, and new developments, in *Spices, Herbs, and Edible Fungi*, Charalambous, G., Ed., 71–95, 1994.
8. Pellerin, P., Comparing extraction by traditional solvents with supercritical extraction from an economic point and environmental standpoint, in *Proceedings of the Sixth International Symposium on SCFs*, France, 2003, Tome 1, 13.

9. Calame, J.P. and Steiner, R., Supercritical extraction of flavors, in *Theory and Practice of Supercritical Fluid Technology*, Hirata, M. and Ishikawa, T., Eds., Tokyo Metropolitan Univ., 275–318, 1987.
10. Gangadhara Rao, V.S.G. and Mukhopadhyay, M., Selective extraction of spice oil constituents by supercritical carbon dioxide, *Proceedings of the Annual Convention of Indian Institute of Chemical Engineers,* Baroda, India, 1988.
11. Meireles, M.A.A. and Nikolov, Z.L., Extraction and fractionation of essential oils with liquid CO_2, in *Spices, Herbs, and Edible Fungi*, Charalambous, G., Ed., Elsevier Science Publishers, Netherlands, 171–199, 1994.
12. Zhang, J. et al., Volatile compounds of a SCF extract of Chinese celery seed, in *Proceedings of the Fourth International Symposium on Supercritical Fluids*, Sendai, Japan, 1994, 235–237.
13. Della Porta, G., Reverchon, E. and Ambrousi, A., Pilot plant for isolation of celery and parsley essential oil by SC-CO_2, in *Proceedings of the Fifth Meeting of Supercritical Fluids,* Nice, France, 1998, Tome 2, 613–618.
14. Nguyen, U.Y., Anstee, M. and Evans, D.A., Extraction and fractionation of spices using SCF CO_2, in *Proceedings of the Fifth Meeting of Supercritical Fluids,* Nice, France, 1998, Tome 2, 523–528.
15. Nguyen, K., Barton, P. and Spencer, J.S., Supercritical carbon dioxide extraction of vanilla, *J. Supercrit. Fluids*, 4, 40–46, 1991.
16. Illes, V., Daood, H., Karsai, E. and Szalai, O., Oil extraction from cardamom crop by sub and supercritical carbon dioxide and propane, *in Proceedings of the Fifth Meeting of Supercritical Fluids,* Nice, France, 1998, Tome 2, 533–538.
17. Then, M., Daood, H., Illes, V. and Bertalan, L., Investigation of biologically active compounds in plant oils extracted by different extraction methods, in *Proceedings of the Fifth Meeting of Supercritical Fluids,* Nice, France, 1998, Tome 2, 555–560.
18. Nawrot, N. and Wenclawiak, B., Supercritical fluid extraction of garlic followed by chromatography, in *Proceedings of the Second International Symposium on Supercritical Fluids*, Boston, 1991, 451–455.
19. Mahindru, S.N., *Indian plant perfumes,* Metropolitan, New Delhi, India, 1992.
20. Gangadhara Rao, V.S.G., *Studies on Supercritical Extraction of Spices,* Ph.D. Dissertation, Indian Institute of Technology, Bombay, 1990.
21. Chrastil, J., Solubility of solids and liquids in supercritical gases, *J. Phys. Chem*, 86, 3016–3021, 1982.
22. De Valle, J.M. and Aguilera, J.M., An improved equation for predicting the solubility of vegetable oils in supercritical CO_2, *Ind. Eng. Chem. Res.,* 27(8), 1551–1553, 1988.
23. Silva, D.C.M.N., Ferreira, S.R.S. and Meireles, M.A., Correlating solubility values of black pepper oil in supercritical CO_2 using empirical models, *in Proceedings of the Sixth International Symposium on Supercritical Fluids*, Nice, France, 2003, Tome 1, 279.
24. Angus, S., Ramstrong, B. and De Reuck, K.M., *International Thermodynamic Tables of the Fluid State*: *Carbon Dioxide,* New York, Pergamon Press, 3, 1976.
25. Ferreira, S.R.S. et al., SCFE of black pepper essential oil, *J. Supercrit. Fluids,* 14(3), 235–245, 1999.
26. Blasco, M. et al., SCFE of *Curuma longa*: Solubility of essential oil, in *Proceedings of the Sixth International Symposium on Supercritical Fluids*, Nice, France, 2003, Tome 1, 341.
27. Naik, S.N., Lentz, H. and Maheshawari, R.C., Extraction of perfumes and flavour from plant materials with liquid carbon dioxide at liquid–vapour equilibrium conditions, *Fluid Phase Equilibria*, 49, 115–126, 1989.
28. McGaw, D.R., Holder, R., Commissiong, E. and Maxwell, A., Extraction of volatile and fixed oil products from hot pepper, in *Proceedings of the Sixth International Symposium on Supercritical Fluids*, Nice, France, 2003, Tome 1, 111.

12 Preparation and Processing of Micro- and Nano-Scale Materials by Supercritical Fluid Technology

Eckhard Weidner and Marcus Petermann

CONTENTS

12.1 Introduction 367
12.2 Particle Generation by High-Pressure Spray Processes 368
12.2.1 Rapid Expansion of a Supercritical Solution 370
12.2.2 Antisolvent Processes 372
12.2.3 Spraying of Gas Saturated Liquids 373
12.2.4 Economics 375
12.3 Composites with High-Pressure Spray Processes 376
12.3.1 Spray Agglomeration with a High-Pressure Spray Process 376
12.3.2 Liquid-Filled Composites with a High-Pressure Spray Technology 378
12.4 Processing of Nutraceuticals with Supercritical Fluid Technology 380
12.5 Conclusions 384
References 384

12.1 INTRODUCTION

The generation of nano- and micro-particles and the formation of particulate composites have become more and more important in many industrial areas. In food, pharmaceutical, material, and life science industries, existing and new products in new application forms with tailor-made properties are being developed faster and faster. To form particulate products, different well-established processes are available. Powders can be obtained by crystallization, grinding, or spray drying processes. However, all these techniques have drawbacks, especially if sensitive substances such as nutraceuticals or bioactive systems have to be processed. In classical crystallization techniques, solvents—in many cases organic solvents—have to be used as auxiliary media. Resultant residues of these solvents may be found in the products and have to be removed by time-consuming and expensive technologies. Similar

FIGURE 12.1 Morphology of grinded particles.

drawbacks appear in most spray drying processes. In addition, the high temperatures necessary to evaporate solvents may cause product degradation. Grinding processes, which are typically solvent free, can only be applied for brittle substances. To achieve sufficient brittleness, deep-freeze conditions are sometimes required. But even if milling is possible, only particles with sharp edges are available (Figure 12.1).

The increasing demand for new product properties and the drawbacks of existing processes are causing a steady search for new technological possibilities for the formation of particulate systems. Some promising techniques include using supercritical fluids (SCFs) to generate nano- and micro-scaled particle systems with well-defined morphologies and, therefore, product behavior. In addition to being used for pure particle formation, these techniques are being used more and more to generate composites consisting of two or more substances, even those in different states of aggregate (liquid/solid). This allows manufacturing of high-quality products offering tailor-made properties, such as controlled release of active substances [1–5].

12.2 PARTICLE GENERATION BY HIGH-PRESSURE SPRAY PROCESSES

Generating particles from pure substances or composites by high-pressure technologies requires unit operations similar to those used for classical low-pressure processes. Those unit operations comprise, for instance, melting, dissolving, mixing, spraying, separating, and pumping. Performing such unit operations under high pressures and in the presence of SCFs requires specific adaptations in plant design and processing. Due to extensive R&D work and industrial experience, those adaptations of machines and apparatuses are meanwhile known quite well. In spite of the fact that major technical problems are solved for high-pressure applications, it is undisputed that those processes are more challenging from the technical and economical points of view than are low-pressure processes. New possibilities to create value-added products that are not accessible with classical technologies are a strong driving force for the industrial use of high-pressure processes. A series of such processes and process modifications that allows generating new product forms has been developed in the past years. An extensive number of acronyms are used to characterize those processes. Those acronyms will not be listed here, but the processes are grouped according to thermophysical principles they use.

In rapid expansion of supercritical solution (RESS) processes, the substance to be powderized is dissolved in the SCF. In the particles from gas-saturated solutions (PGSS) processes, the subcritical fluid or SCF is admixed, dispersed, and dissolved in the substance to be powderized. In both types of processes, the particles are generated by expansion in capillaries or nozzles [6–9]. In the so-called "antisolvent processes," such as the gas anti-solvent (GAS) process, the product to be powderized is first dissolved in a classical organic solvent. Afterward, this solution is admixed with a SCF. This causes a decrease of fluid density and leads to reduced solvent power of the organic solvent. Resultant particles are precipitated within the mixture of organic solvent and SCF [10–13]. Compared to traditional crystallization processes, in which organic solvents are evaporated at high temperatures or in a vacuum, GAS processes work at low temperatures and achieve supersaturation much faster. Therefore, GAS processes are advantageous compared to traditional crystal lization processes, even if the residual solvent problem has to be solved.

All processes have in common that the substance to be powderized has to be brought into a liquid or dispersed form. This is achieved by melting the substance, by dissolving the product in classical solvents, or by dispersing the product in a liquid. In many cases, this step is performed at ambient or slightly elevated pressures. Then the product is compressed by continuously operated plunger pumps, gear pumps, or extruders. The type of dosing system used depends mainly on the properties of the product (e.g., melting point and viscosity*). Dosing of substances with high viscosities at high pressures requires special technical solutions, some of which have been elaborated in the past years [14]. Many research activities demonstrated that high pressure, in connection with the unique properties of SCFs, opens the chance to generate powders from highly viscous liquids, which cannot be sprayed by classical technologies such as spray drying. Sprayability is achieved by a considerable reduction of viscosity and surface tension if such highly viscous liquids are effectively admixed with a SCF that is sufficiently soluble [15, 16]. Admixing is achieved via stirrers, dispersers, impinging jets, static mixers, or membranes that are operated under high pressure.

Improvement of sprayability is not the only advantage of SCFs. After a SCF is admixed, the physical properties of the substances are changed dramatically [17–19]. The most important effects are reduction of viscosity and melting point depression due to dissolved gas. Both effects allow handling substances near or even below their melting point under ambient pressure. Due to cooling of the SCF that occurs during expansion, the temperatures are even (much) lower after the particles have been formed via expansion. As the heat of solidification is removed by direct heat transfer from the particles to the coexpanded gas (in the case of RESS and PGSS), solidification occurs much faster (some 10 milliseconds) [20–22] and the temperature stress on the particles is lower than that during classical air drying processes, whereby the

* In discontinuous processes, products are dosed either as liquid or solid into an autoclave, where additionally compressed gas is added and admixed. The elevated pressure in the autoclave is used to transport the products into the next process steps.

solvent is evaporated by heat transfer between liquid droplets and the surrounding hot air or gas [23]. Furthermore, unlike many traditional spray drying processes, the supercritical technologies are intrinsically free from contacting the products with air or oxygen. So these technologies are suitable for sensitive substances [24]. Another positive effect is that as long as the contact between the generated powders and air is carefully avoided, dust explosions may not occur.

An important—and sometimes underestimated—step of all high-pressure processes is the collection of the particles. Using spray processes like RESS, PGSS, and concentrated powder form (CPF), the particles have to be separated from a gas stream. Depending on particle size and particle concentration in the gas stream, different separation techniques might be suitable. Coarse fractions can be separated just by settling the particles in spray towers or by enlarging the separation forces in cyclones. For finer particles and lower particle concentration filters, membranes or sinter plates can be used to collect the manufactured product. For the antisolvent processes, in which particles are formed by precipitation in a liquid, a solid-liquid separation must be used. This separation is achieved by settling or by different filter systems.

After the particles are separated from the SCF or the solvent, postprocessing might be necessary. In the case of antisolvent precipitation, solvent-wet particles are obtained, which have to be dried. This could be done, for example, by flushing with heated gas or by additional SCF extraction steps [25]. Depending on the application of the product, a whole range of further posttreatment steps might be considered (coating, sieving, agglomeration, sifting, size fractionation, dispersion, and so on), which could either be applied alone or used in combination with the upstream particle generation process. Some of these combinations have already been studied [26–28].

An often-discussed issue for all processes is gas recycling. If liquid solvents are in the system, solvent removal from the gas is required in order to avoid enrichment in the recycle gas. In solvent-free processes, the particles have to be removed carefully before recompressing the gas in order to avoid plugging of the recycling system. Both purification methods can technically be applied, but each is connected with additional costs for equipment and operation. On a case-by-case basis, one must consider whether recycling of the gas is feasible and reasonable according to economic and environmental aspects. Recycling might be a disadvantage of the use of SCF for powder generation, as the pressure differences between preexpansion and postexpansion require high energies for recompressing the gas. Therefore, reducing the gas demand for particle generation as far as possible is recommended. Some processes (e.g., the PGSS- or CPF-method) allow generating 1 kg of powder with 0.1 to 1 kg of gas. If a cheap SCF, such as carbon dioxide (CO_2), is used, gas recycling for small- and medium-sized plants might be more expensive than using fresh gas.

12.2.1 Rapid Expansion of a Supercritical Solution

One of the oldest processes that uses the special properties of compressed gases is the so-called RESS process [29–37]. A flow scheme for this process is presented in Figure 12.2. The substance to be powderized is stored in an extraction vessel.

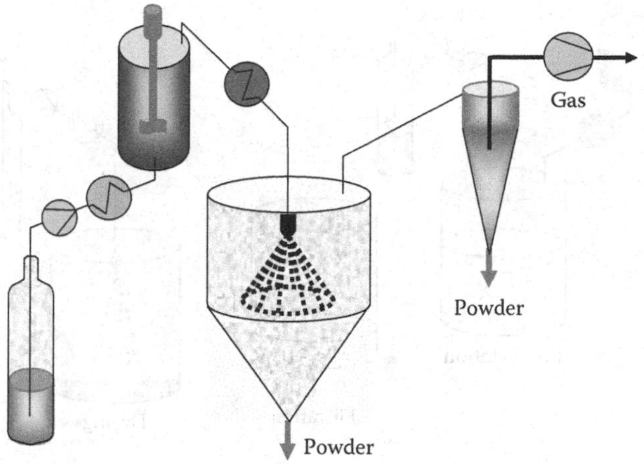

FIGURE 12.2 RESS process scheme.

Compressed gas (in most cases CO_2) is led through the vessel. Under high pressures, sometimes up to 800 bar, the product is (partly) soluble in the gas. The so-formed solution is dosed through a heat exchanger and finally is expanded via a nozzle. Caused by the rapid depressurization, the dissolution power of the gas is reduced, supersaturation occurs, and a precipitation of fine particles is induced. Particle formation can be influenced by the pressure during extraction, the concentration of the dissolved substance, the temperature before depressurization, the geometry of the nozzle, and conditions in the spray chamber after the expansion. Very fine powders in the range of 0.1 to 10 μm with narrow particle distributions are obtained by properly adjusting the process parameters.

The RESS process is characterized by a rather simple setup in laboratory or small production scale, but it is limited by the poor solubility of many substances in CO_2. Sometimes more than 100 kg of CO_2 would be necessary to manufacture 1 kg of the particulate product. Subsequently, the particles have to be separated from very highly diluted gas streams, which is a procedural challenge.

Therefore, the RESS process offers a high potential for high value-added products such as pharmaceuticals and cosmetics. Investigations with the model substance griseofulvin showed that, compared with products micronized with classical processes, an accelerated dissolving behavior can be achieved with the RESS product. In addition, researchers observed better absorption behavior in an in-vitro test system [38]. In all nano-scaled processes, the posttreatment of the particles after particle formation is a challenging task. Nano-particles tend to agglomerate, and redispersing such systems is very difficult or sometimes nearly impossible. To stabilize RESS particles in their nano-scale, researchers proposed to collect them in liquid-containing emulsifiers [39]. Newly published papers showed that, using this technique, a long-term stability of nanosuspensions is obtained with particles smaller than 100 nm and concentrations of up to 11 g/dm³ [40]. Other researchers sprayed ternary mixtures consisting of a SCF, the active substance, and a polymer; they obtained resultant powders with encapsulated active substances in different concentrations [41–43].

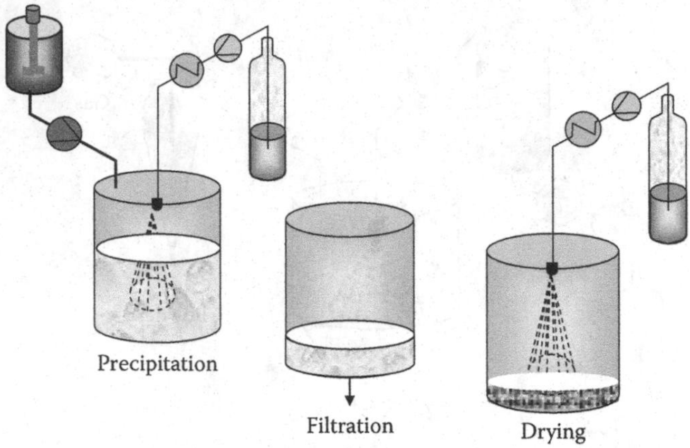

FIGURE 12.3 Antisolvent-process (GAS) process scheme.

12.2.2 Antisolvent Processes

Another important group of high-pressure processes are the so-called "antisolvent processes." Similar to the RESS process, these techniques are suitable for manufacturing powders of pure substances and composites from nano to micro scale. A lot of different process modifications are described [44–51], for example:

- GAS: gas antisolvent
- SAS: supercritical fluid antisolvent
- PCA: precipitation with a compressed antisolvent
- ASES: aerosol solvent extraction system
- SEDS: solution-enhanced dispersion by supercritical fluids.

All the processes behind these different acronyms make use of an effect well known in classical crystallization techniques. By adding a third component (antisolvent) to a solution, the solubility of the dissolved component is reduced and finally the substance precipitates. In high-pressure technology, the antisolvent is a supercritical or near-critical fluid. The use of compressed gases instead of other antisolvents opens the way to new particle morphologies and new composites [52, 53].

The first three processes (GAS, SAS, and PCA) are typically operated discontinuously. A simplified process scheme is shown in Figure 12.3. The substance or mixture to be powderized is dissolved in a liquid (mostly organic) solvent. To precipitate particles, the solution has to be contacted with a SCF. This is achieved in different ways:

- If the liquid solution is provided in a vessel and afterward the SCF is dosed into that vessel, the process is called the *GAS process*.
- In the SAS and PCA processes, the SCF is provided in a high-pressure vessel and the solution is sprayed into that supercritical solution.
- In the SEDS and ASES processes, both fluids (the solution and the SCF) are mixed in nozzles and sprayed into autoclaves.

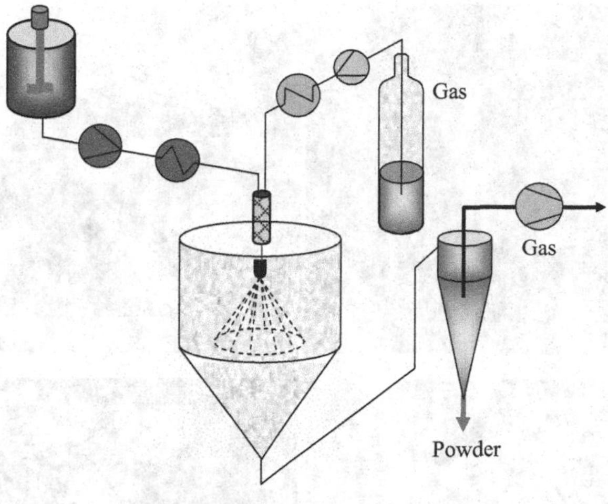

FIGURE 12.4 PGSS process scheme.

In all of these processes, admixing of the gas causes a precipitation of the dissolved product in the liquid solution. In the next process step, the so-formed particles have to be separated from the liquid. In laboratory scale, filters integrated into the high-pressure vessels typically perform this step. In a third process step, the collected particles have to be dried. In an elegant way, this is achieved by adding fresh SCF to the pressure vessel. The liquid is extracted from the precipitated powder by the gas. One advantage of this processing is the temperature required in the supercritical drying step, which is moderate compared with the temperature required for conventional drying procedures.

The antisolvent processes have been successfully tested with many different products. Beside the particle generation from explosives like β-HMX and nitroguanidin, polymers (polyacrylnitril, polycaprolacton) and other organic substances (hydroquinone and phenanthrene) have been powderized. As in the RESS process, the main focus of research is addressing pharmaceuticals like ascorbic acid, insulin, and paracetamol.

12.2.3 Spraying of Gas Saturated Liquids

The RESS process and the antisolvent processes are typically carried out in discontinuous or semicontinuous mode. The PGSS process may rather easily be operated in a continuous mode and, therefore, is also suitable for products manufactured in larger quantities [54–57].

Figure 12.4 illustrates a principle flow scheme of the PGSS process. To spray a gas-saturated liquid, the SCF has to be admixed with the product to be powderized under elevated pressures. Typically, the product has to be melted or liquefied by adding a solvent in advance at low-pressure conditions. Subsequently this fluid is pumped via high-pressure pumps to a mixing device (mostly static mixers), where the SCF is admixed. Under high-pressure conditions the SCF is partly soluble in the melt, dispersion or solution. The solubility causes a reduction of viscosity and of

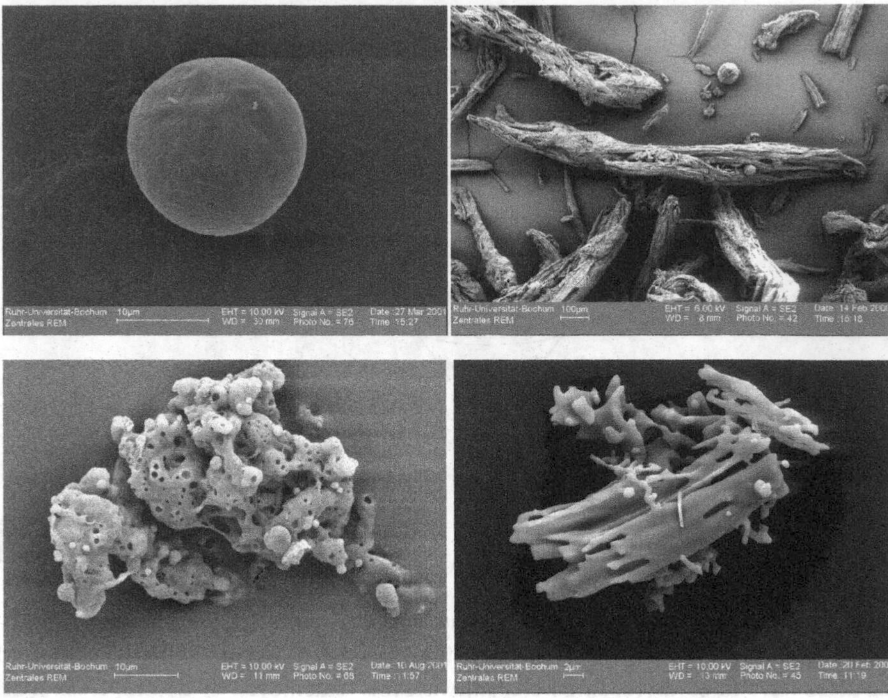

FIGURE 12.5 Particle morphologies of PGSS particles.

interfacial tension. Both effects result in improved sprayability, even for substances that normally cannot be atomized by spray processes. Afterward, the gas-enriched fluids are depressurized via a nozzle into a spray tower. Normally, the spray tower is operated at ambient pressure. Due to the volume increase of the expanding gas, the product is disintegrated into fine droplets. Simultaneously, the gas cools down immediately during expansion. Although the temperature in spray tower can in principle be adjusted by additional heating or cooling, most applications do not need this temperature control. The temperature in the spray tower is mostly set by the preexpansion conditions in static mixer. Typical temperatures in the spray tower lie in the range of –20°C to 100°C. If the resultant temperature is low enough, the liquid/melt reaches the solidification point and the droplets freeze. Particle size and particle size distribution of the obtained powders can be adjusted by changing the SCF, the pressure in the mixing device, the temperature before expansion, and the geometry of the nozzle.

As an example, different morphologies and particle sizes of powders are illustrated in Figure 12.5. The technique can be used for the powderization of many different systems. In addition to organic substances such as citric acid and polyethylene glycol (PEG), certain pharmaceuticals (e.g., nifedipine and tobramycin) were successfully micronized. Moreover, composites and even reactive systems like powder coatings can be handled with this technique [58].

One advantage of this process, compared with other high-pressure techniques, is the low to moderate consumption of SCF. Typically, 0.5 to 5 kg of SCF are necessary

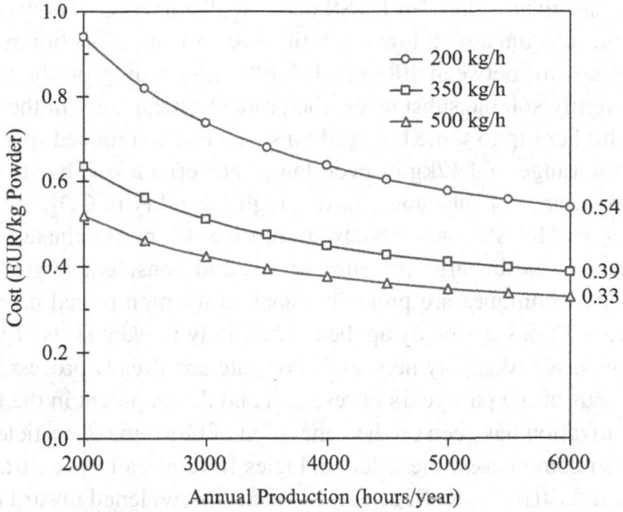

FIGURE 12.6 Production cost for PGSS process (gas consumption 2 kg/kg product). Diagram courtesy of Natex, Austria.

to produce 1 kg of powder. The low gas demand together with the relatively simple construction of the PGSS plants allows the production of huge quantities of products. The design of the process and the control of the product properties depends on the thermodynamic and fluid dynamic behavior, for example, the solubility of the compressed gases in the product to be powderized, the viscosity, and flow behavior of the gas-containing melts.

12.2.4 Economics

Figure 12.6 presents estimated costs for industrial PGSS (non-Good Manufacturing Practice [GMP]) production facilities. The diagram gives the total costs for 1 kg of product, including costs for investment, personnel, energy, and gas consumption [57, 59]. Depending on the hourly capacity and the annual production hours, costs range from thirty cents to 1€ (about $0.75) per kilogram of powder. The estimated costs are mainly due to personnel (40%) and carbon dioxide consumption (40%). Compared with the costs of classical micronization techniques, such as milling or spray drying, the costs are on the same order of magnitude.

Cost studies have been published for other high-pressure processes, like RESS and GAS. Rantakylä analyzed the antisolvent process SAS [60]. Estimated manufacturing costs for a new GMP plant are around 50 to 300 €/kg (38 to 230 $/kg) product without a feedstock price. This is for a 4000 to 8000 kg/year production rate and 5 to 10 wt% feed concentration of the starting material in an organic solvent. An effective way to decrease the manufacturing costs is to increase the raw material concentration in solvent. Weber et al. [61] provide data for a non-GMP PCA process. For an initial solvent concentration of 10 wt% and a production of 11.25 kg/hr (corresponding to 87 MT/year), the costs per kilogram of powder are around 8 € ($6). If the capacity is doubled, the specific costs are reduced to approximately 5 €/kg ($3.80/kg).

Türk [62] has given values for RESS plants with a fixed CO_2 flow of 2.35 MT/hr. In the case of a substance with low solubility, the annual production is 1.78 MT and the specific costs are between 100 and 140 €/kg (depending on the time of depreciation). For highly soluble substances, the production capacity in the same plant is considerably higher (up to some hundred tons), leading to reduced specific costs that might reach the range of 1 €/kg or even lower. Nevertheless it has to be noted that only a limited number of substances have a high solubility in CO_2.

The costs for RESS and antisolvent processes are dominated by relatively high investment costs for large pressure vessels and considerable gas consumption. Therefore, these techniques are preferably applied for high-priced products, such as pharmaceuticals. PGSS is already applied industrially in plant sizes of some hundred kg/hr. Fats, fat derivatives, polymers, and chocolate are already processed in industry.

A main focus of the past years of research and development in the field of supercritical micronization has been on the generation of tailor-made particles from single components. In some cases, these technologies have already been transferred successfully into industrial scale. Recently, the focus has widened toward the formation of composites. Technologies with SCFs offer an increased number of possibilities to generate composites with new functionalities. The following section highlights two examples how SCF can be used to produce such products.

12.3 COMPOSITES WITH HIGH-PRESSURE SPRAY PROCESSES

For commercial success, it has become more and more attractive to design tailor-made particle systems that allow, for example, the controlled release of active agents or offer durable protection of sensitive ingredients. Classical processes like spray drying, crystallization, and in-situ polymerization processes are in principle able to produce such composites. In spray drying, the high temperature level limits the technique to insensible substances. In crystallization processes, complete encapsulation is hard to achieve and the particle shape is difficult to control. For polymerization processes, only a few material combinations are suitable. In this field, a few techniques using SCFs are established and the results are very promising. These SCF technologies allow the generation of powders with properties that are difficult or even impossible to achieve by classical methods.

12.3.1 Spray Agglomeration with a High-Pressure Spray Process

The processes described above lead to reduced particle size of the raw material. SCF processes are not limited to particle size reduction. Different shape-forming methods have been established in the last few years (e.g., coating, agglomeration, impregnation, and dispersion processes) [63, 64].

The CPF technique is a spray agglomeration technique that allows the production of liquid-loaded composites with loadings of up to 90 wt%. The agglomerates obtained have a high mechanical stability and good flow behavior. Figure 12.7 illustrates the flow scheme of the CPF process.

The liquid to be powderized is dosed with a high-pressure pump from the storage vessel to a static mixer. Here, a second stream of SCF, mostly CO_2, is added.

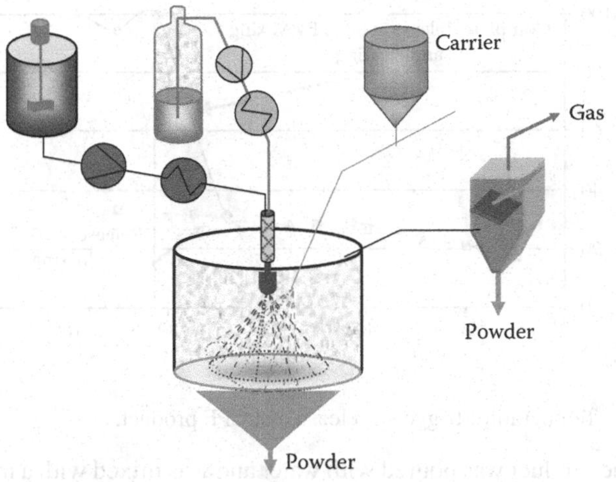

FIGURE 12.7 CPF process scheme.

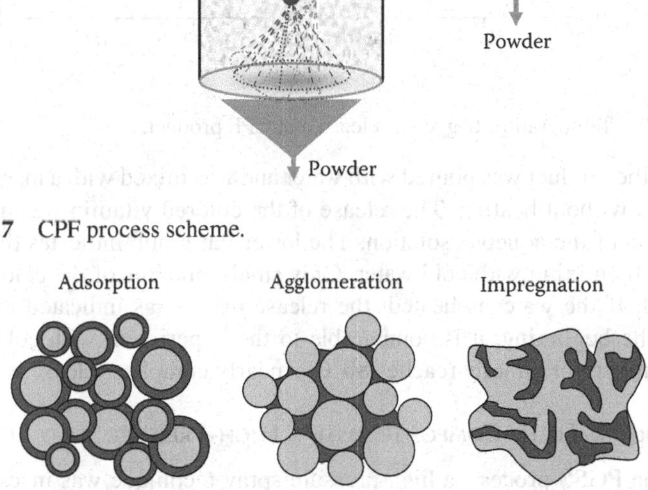

FIGURE 12.8 Binding mechanism of CPF products.

Under pressures of up to 200 bar, the gas and liquid are mixed and subsequently depressurized via a nozzle to atmospheric pressure. The volume increase of the gas leads to the formation of very fine droplets. Temperatures in the spray tower are controlled by the preexpansion conditions in the static mixer and are typically in the range of –20°C to 60°C. Therefore, this technique is especially suitable for processing of temperature-sensitive or volatile substances. By adding a solid carrier with a pneumatic conveying system into the spray tower, the liquid droplets are bound. Solid, free-flowing agglomerates are formed that can have a maximum of 90 wt% liquid content. The liquid is bound by adsorption to the surface of the carrier and by capillary forces between the single particles in the agglomerates or even in porous structures of the single particles (Figure 12.8). The formation of such agglomerates was tested with many different substances (e.g., natural extracts of basil, pepper, lemon oil, α-tocopherol, whiskey). Silicic acid, celluloses, and starches were used as carriers. In all of these experiments, the first task was to get free-flowing powders that can be easily handled in postprocessing. In addition to the flowability of the products, the release of the bound liquid is of importance. By varying the carrier material, products with defined release behaviors can be achieved for the food, pharmaceutical, and cosmetic industries.

Figure 12.9 illustrates the controlled release of a CPF product [65]. For this product, vitamin B_2 was sprayed on a potato starch using the CPF technique.

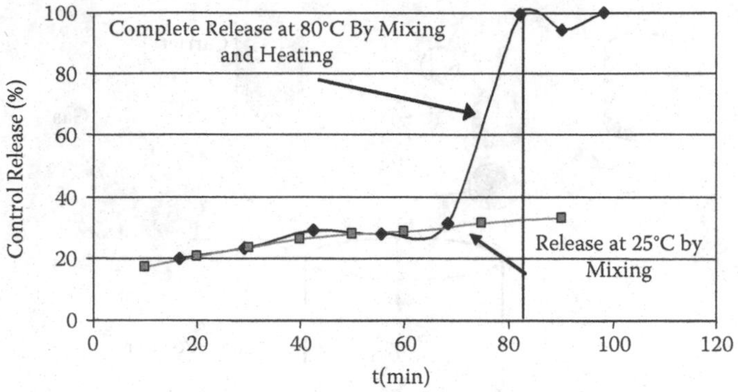

FIGURE 12.9 Temperature-triggered release of a CPF product.

Afterward, the product was poured with water and was mixed with a magnetic stirrer bar with and without heating. The release of the colored vitamin was measured by the extinction of the aqueous solution. The lower flat graph indicates the release of the vitamin by mixing with cold water. Only small amounts of the colored vitamin are released. If the water is heated, the release occurs (as indicated by the upper curve). At the beginning, it is comparable to the experiment with cold water, but when the temperature finally reaches 80°C, a nearly complete release is obvious.

12.3.2 Liquid-Filled Composites with a High-Pressure Spray Technology

Based on the PGSS process, a high-pressure spray technique was investigated that allows to manufacture composites consisting of a core material, which could be a liquid or a solid dispersed in a liquid, and a shell material that must be solid under storage conditions [66–68]. A principle flow scheme of this process is illustrated in Figure 12.10. Both components have to be provided in a pumpable form. High-pressure pumps are used to feed the shell materials and the liquid core materials to the static mixer, where the components are dispersed. In addition, a SCF is pumped into the mixer. Depending on the system and the mixer size, a more or less stable dispersion is obtained. Subsequently, this dispersion is depressurized into a spray tower. The shell material solidifies due to temperature reduction of the expanding gas. The core material is encapsulated in the shell. Figure 12.11 shows in principle the morphologies that can be obtained on the one side with a liquid dispersed in a melt and on the other side with a dispersion of two immiscible melts. One advantage of this technique is that the dispersion or emulsions formed in the static mixer can be either stable or unstable. The residence time after mixing is extremely short (some milliseconds to seconds) so that a phase split does not occur before expansion. After the expansion, solidification of the shell material happens instantly; the dispersion is stabilized by solidification.

As an example for the manufactured composites, three scanning electron microscope (SEM) pictures of a solid wax/liquid PEG composites are shown in Figure 12.12. The lighter gray regions consist of wax; the dark regions in the pictures show the bound liquid PEG. In the range of 50 to 60 wt.-%, a change from closed to open composites is observed.

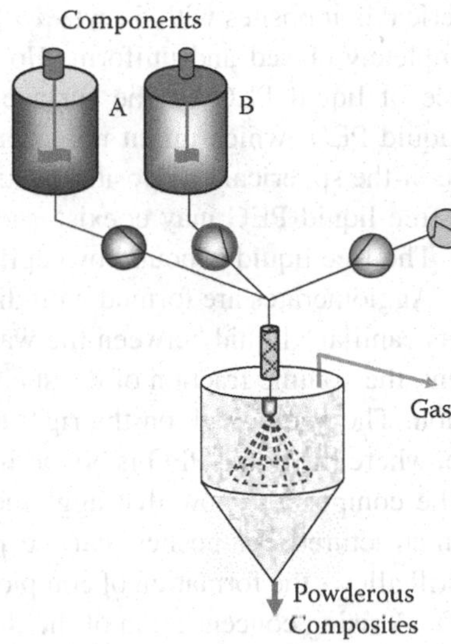

FIGURE 12.10 Process scheme of composite process.

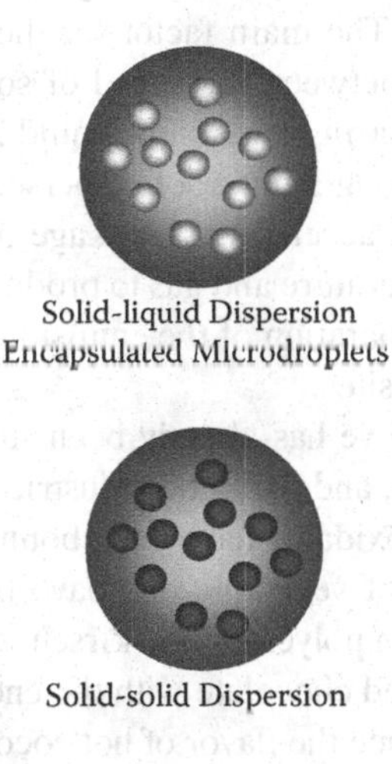

FIGURE 12.11 Morphologies of composites.

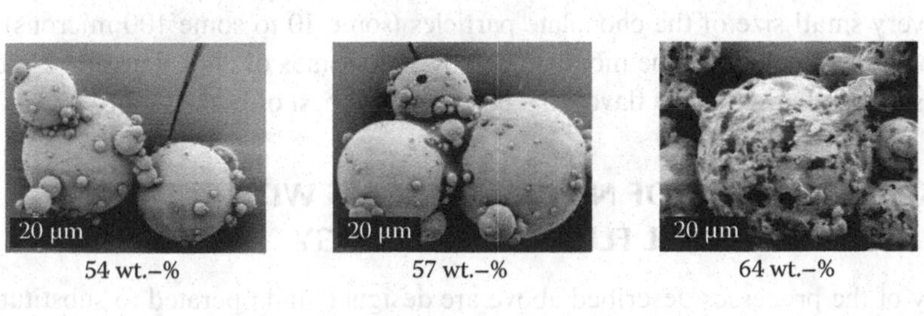

FIGURE 12.12 SEM pictures of composite particles (wax/PEG system).

The surface of the spherical composites with 54 wt.-% of PEG on the left-hand side of Figure 12.12 is completely closed and uniform. No darker regions, which would indicate the presence of liquid PEG on the surface, can be detected. By increasing the amount of liquid PEG, which might not sufficiently be admixed in the static mixer, the surface of the spherical composites is still closed and uniform. Nevertheless, in that case, free liquid PEG may coexist with droplets that consist of wax and dispersed PEG. The free liquid is bound by capillary forces in between the solidified wax particles. Agglomerates are formed with dispersed PEG encapsulated in the wax and PEG as capillary liquid between the wax particles. By further increase of the liquid content, the volume fraction of the shell material is too low to allow complete encapsulation. The photograph on the right-hand side shows a particle with an open structure, where the liquid PEG is bound in pores of the wax.

The morphologies of the composites show that agglomerates, single particles as well as closed and open-structured composites, can be produced. The highest concentration of PEG that still allows the formation of completely closed composites was approximately 60 wt.-%. A rising concentration of the liquid favors the generation of open composites. This can be understood by focusing on the basics of particle formation. To form a composite, the liquid has to be admixed to the shell material as the dispersed phase of an emulsion. Subsequently, the mixture is sprayed and solidified using an expanding gas. The main factor for the generation of closed or open composites is the difference between the speed of solidification and phase separation of the emulsion. An increasing amount of liquid leads to a rising drop diameter of the dispersed phase or a rising number of dispersed droplets. A rising number of dispersed droplets leads to an accelerated breakage of the emulsion. With constant process parameters (i.e., temperature and gas to product ratio), the solidification time will be comparable but the separation of the emulsion is much faster. This results in the generation of open composites.

The process described above has already been successfully applied to different products in the chemical, food, and cosmetic industries. Water has been encapsulated in fat; liquid aromas and antioxidants have been bound in a fat matrix to reduce the losses during storage, different vegetable oils have been encapsulated in PEGS, a paraffin wax has been bound in polyester, and kirsch was encapsulated in a chocolate matrix [69, 70]. The micronized chocolate with the encapsulated kirsch aroma could be used, for example, to enhance the flavor of hot cocoa or to bring additional aroma into any chocolate product. The release of the aroma for these differs from products on the market where liquors are encapsulated in macroscopic structures, like pralines. The very small size of the chocolate particles (some 10 to some 100 microns) leads to immediate melting in the mouth. Thereby, the aromas of chocolate and kirsch are released together to form a flavor that combines the best of both (Figure 12.13).

12.4 PROCESSING OF NUTRACEUTICALS WITH SUPERCRITICAL FLUID TECHNOLOGY

Many of the processes described above are designed and operated to substitute one classical process task (e.g., milling). To gain larger benefits, different process tasks

FIGURE 12.13 Chocolate–kirsch composite.

can be combined and solved in one SCF-assisted process [71, 72]. Table 12.1 gives an overview of some nutraceuticals processed with SCFs.

SCF technology is typically a low-temperature technique and is normally carried out in completely inert atmospheres. Therefore, these techniques are especially suitable for thermo- and oxygen-sensitive substances. Green tea and especially polyphenol extracts from green tea leaves are widely used in nutraceutical applications. Polyphenols, substances known to stabilize oil and fat products, are antioxidants that have been discussed for cancer prevention and for dental caries prevention, to name just two positive effects. To isolate these polyphenols from green tea leaves, a water extraction is made. After filtration, this aqueous extract is dried with classical spray drying techniques. In spray drying, high temperatures are necessary to evaporate water. In addition, most spray dryers work with heated air and therefore the polyphenols may suffer from thermal and oxidative stress during processing. Resultant green tea products may contain lower concentrations of antioxidants than can be achieved with gentler processing.

One possibility for obtaining solid green tea products without degradation of the antioxidants is to use SCF technology. Therefore, a process that combines the drying step of aqueous green tea extracts with particle formation was designed. The flow scheme of this process is presented in Figure 12.14. The green tea extract used for the drying and pulverization experiments was obtained by an extraction performed at 60°C, by mixing 1 kg of extract in 10 kg of deionized water for 15 minutes. This extract is dosed to a vessel by a high-pressure pump through a static mixer. Here, preheated CO_2 is added under elevated pressures. The residence time in the mixer is extremely short (< 1 sec); therefore, it is possible to raise the temperature above 100°C, even sometimes as high as 180°C, without degradation of the product. Subsequently, the mixture is depressurized into a spray tower and thus quenched

TABLE 12.1
Nutraceuticals Processed with Supercritical Fluids

Process	Substance	Reference
RESS	Benzoic acid	[76]
	Ibuprofen	[77, 78]
	Aspirin	[79]
	Caffeine	[76]
	Griseofulvin	[38, 80]
	Lidocaine	[81]
GAS	Mefenamic acid	[82]
	Copper-Inomethacin	[78, 83]
	Insulin	[84–86]
	Paracetamol and ascorbic acid	[47]
	β-carotene	[87]
SAS	Amoxycilin	[25, 88]
	Dextran, cholesterol	[89–91]
	Inulin	[92]
	Lecithin	[93]
	Organic pigments	[94]
PGSS	Felodipine	[55]
	Glucose	[79]
	Albuterol sulphate	[79, 95]
	Cromolyn sodium	[79, 95]
	Glucose oxidase	[96]
CPF	Flavour extracts	[24, 97]
	Emulsions	[98]

immediately to low temperatures. Fine droplets with large surfaces are formed. The water is extracted already in the static mixer at high temperatures or is taken up by the dry CO_2 after expansion and, finally, solid green tea extract is precipitated in the spray tower. The obtained green tea powders are shown in Figure 12.15. The evaporated water can be withdrawn with the expanded CO_2.

Table 12.2 shows the polyphenol concentrations and water content of the different products. The raw material (tea leaves) has had a water content of 2.97 weight %. The ground leaves were extracted with water (leaves:water/1:10 [g/g]), and the water extract was dried with a low-temperature vacuum evaporation (40°C) and with the SCF process. The main process parameters of the supercritical drying process are shown in Table 12.3. The residual water content in the vacuum process after 6 hours was determined to 8.82 weight %. In SCF processing, 5.09 weight % was obtained. The polyphenol concentration after water extraction could be increased for all three types of polyphenols compared with the raw material. The PGSS drying step shows the same or slightly higher concentrations of polyphenols than the water extract dried

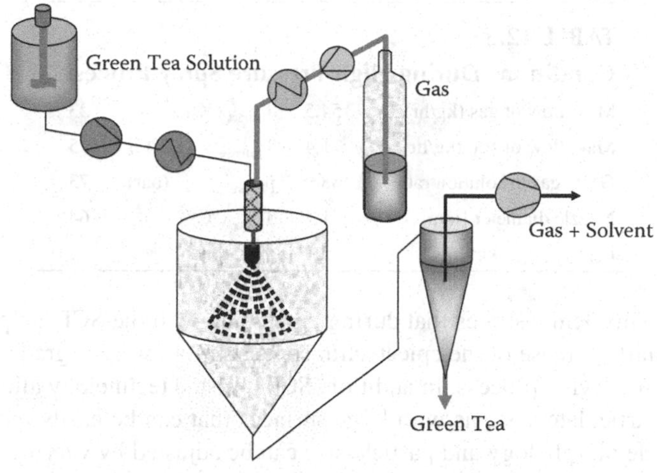

FIGURE 12.14 Green tea processing with supercritical fluids.

FIGURE 12.15 Picture and SEM of powderous green tea extract.

TABLE 12.2
Polyphenol Content at Different Process Steps

	Water Content (%)	Polyphenols (g/100 g dry raw material)		
Process Step	Residue (%)	Epicatechin (EC)	Epigallocatechin-Gallate (EGCG)	Epicatechin-Gallate (ECG)
Raw material	2.97	0.97	3.92	1.41
Water extract (1:10) vacuum dried for analyses	8.82	2.31	4.07	1.50
Supercritical fluid dried	5.09	2.16	4.90	1.70

TABLE 12.3
Conditions During High-Pressure Spray Process

Mass flow of gas (kg/hr)	54.5	$T_{extract}$ (°C)	23
Mass flow of sol. (kg/hr)	1.4	$T_{before\ expansion}$ (°C)	125
GSR (gas to solution ratio)	38.9	$p_{before\ expansion}$ (bar)	73
Nozzle diameter [mm]	1.4	T_{tower} (°C)	64
$m_{sprayed\ solution}$ (kg)	5.1	p_{tower} (bar)	1.0

in vacuum. This demonstrates that during processing with the SCF, no polyphenols (except a small decrease of the epicatechin concentration) were degraded compared with a vacuum drying process. In addition, SCF-assisted technology allows the production of particulate systems with large surfaces that can be easily redissolved in water. Particle morphology and particle size can be adjusted by varying the process parameters [73–75]. In vacuum drying, a bulky product is obtained that has to be grinded to get fine particles.

12.5 CONCLUSIONS

In the last 15 to 20 years, numerous processes for particle generation using SCFs have been proposed and applied for substances from the food, polymer, pharmaceutical, life science, and nutraceutical industries. The main focus of these applications has been on the micronization of pure substances. The thermodynamic and fluid-dynamic properties of certain single-component model systems (e.g., PEGs, triglycerides, naphthalene) in the presence of compressed gases, mostly CO_2, have been studied intensively. This fundamental research has led to an improved understanding of the processes for particle generation. As a result, high-pressure technology becomes more and more established in industry. Meanwhile, high-pressure plants with capacities of some grams per hour to some hundred kilograms per hour can be designed and built by several specialized plant constructors. Cost analysis shows that these techniques can be competitive to classical micronization techniques. SCFs not only open new possibilities for processing substances that are difficult to handle (e.g., substances with low melting points, high viscosities, or sticky surfaces) but also allow generation of composites with customized properties. These new chances motivate research for improved understanding of the processes and the development of new products.

REFERENCES

1. Reverchon, E., Micro- and nano-particles produced by supercritical fluids assisted techniques: Present status and perspectives, *Chem. Eng. Trans.*, 2, 1, 2002.
2. Jung, J. and Perrut, M., Particle design using supercritical fluids: Literature and patent survey, *J. Supercrit. Fluids*, 20, 179, 2001.
3. Knez, Z. and Weidner, E., Precipitation of solids with dense gases, in *High Pressure Process Technology: Fundamentals and Application*, Bertucco, A. and Vetter, G., Eds., Elsevier, 2002.

4. Weidner, E., Powderous composites by high pressure spray processes, in *Proceedings of the Sixth International Symposium on Supercritical Fluids,* Versailles, France, 2003, 3, 1483.
5. Gamse, T., Schwinghammer, S. and Marr, R., Erzeugung feinster partikel durch einsatz von überkritischen fluiden, *Chem. Ing. Techn.,* 77, 669, 2005.
6. Türk, M. et al., Stabilization of pharmaceutical substances by the rapid expansion of supercritical solutions (RESS), in *Proceedings of the Sixth International Symposium on Supercritical Fluids,* Versailles, France, 2003, 3, 1747.
7. Reverchon, E., Supercritical-assisted atomization to produce micro- and/or nano-particles of controlled size and distribution, *Ind. Eng. Chem. Res.,* 41, 2405, 2002.
8. Reverchon, E. and Della Porta, G., Micronization of antibiotics by supercritical assisted atomization, *J. Supercrit. Fluids,* 26, 243, 2003.
9. Shariati, A. and Peters, C.J., Measurements and modeling of the phase behavior of ternary systems of interest for the GAS process: I. The system carbon dioxide+1-propanol+salicylic acid, *J. Supercrit. Fluids,* 23, 195, 2002.
10. Yeo, S., Kim, M. and Lee, J., Recrystallisation of sulfathiazole and chlorpropamide using the supercritical antisolvent process, *J. Supercrit. Fluids,* 25, 143, 2003.
11. Ventosa, N., Sala, S. and Veciana, J., DELOS process: A crystallisation technique using compressed fluids: 1. Comparison to the GAS crystallization methode, *J. Supercrit. Fluids,* 26, 33, 2003.
12. Sun, X. et al., The characteristics of coherent structures in the rapid expansion flow of the supercritical carbon dioxide, *J. Supercrit. Fluids,* 24, 231, 2002.
13. Hanna, M. and York, P., Patent WO 95/01221, 1994.
14. Weidner, E. et al., Manufacture of particles from high viscous melts using supercritical fluids, reviewed. Proc. "High Pressure in Venice", ISBN 88-900775-1-4, 2, 735, 2002.
15. Peter, S. et al., Interfacial tension in binary systems containing a dense gas, in *Supercritical Fluids: Fundamentals for Application,* Kiran, E. and Levelt-Sengers, J.M.H., Eds., Boston, Kluwer Academic Publishers, 731, 1994. Proceedings of the NATO Advanced Study Institute on Supercritical Fluid—Fundamentals for Application, Kemer, Antalya, Turkey, July 18–31, 1993; NATO Dordrecht, Kluwer NATO ASI Series: Series E, 273.
16. Jakob, H., Peter, S. and Weidner, E., Die Viskositaet koexistierender Phasen bei der ueberkritischen Fluidextraktion, *Chem. Ing. Techn.,* 59, 37, 1987.
17. Jaeger, P., Eggers, R. and Baumgartl, H., Interfacial properties of high viscous liquids in a supercritical carbon dioxide atmosphere, *J. Supercrit. Fluids,* 24, 203, 2002.
18. Gulari, E. et al., Rheology of molten polystyrene with dissolved supercritical and near-critical gases, *J. Polym. Sci.,* 37, 2771, 1999.
19. De Simone, J.M. et al., Viscosity effects on the thermal decomposition of bis(perfluoro-2N-propoxypropionyl) peroxide in dense carbon dioxide and fluorinated solvents, *J. Am. Chem. Soc.,* 123, 30, 7199, 2001.
20. Petermann, M., *Herstellung von Pulverlacken durch Verspruehung gashaltiger Schmelzen,* Ph.D. Thesis, Erlangen, 1999.
21. Kayrak, D., Akman, U. and Hortacsu, Ö., Micronization of Ibuprofen by RESS, *J. Supercrit. Fluids,* 26, 17, 2003.
22. Kappler, P. et al., Size and morphology of particles generated by spraying polymer-melts with carbon dioxide, in *Proceedings of the Sixth International Symposium on Supercritical Fluids,* Versailles, France, 2003, 1891.
23. Sievers, R.E. et al., Micronization of water-soluble or alcohol-soluble pharmaceuticals and model compounds with a low-temperature bubble dryer, *J. Supercrit. Fluids,* 26, 9, 2003.
24. Grüner, S., Otto, F. and Weinreich, B., CPF-technology: A new cryogenic spraying process for pulverization of liquid, in *Proceedings of the Sixth International Symposium on Supercritical Fluids,* Versailles, France, 2003, 1935.

25. Reverchon, E. et al., Pilot scale micronization of amoxicillin by supercritical antisolvent precipitation, *J. Supercrit. Fluids,* 26, 1, 2003.
26. Best, W. et al., German Patent 2 943 267, *BASF AG*, 1979.
27. Subramaniam, B. et al., Patent WO 97/31691.
28. Krause, H., Niehaus, M. and Teipel, U., German Patent DE 19711393, 1998.
29. Türk, M., *Erzeugung von organischen Nanopartikeln mit ueberkritischen Fluiden,* Habilitationsschrift, University Karlsruhe, 2001.
30. Türk, M. et al., Micronization of pharmaceutical substances by rapid expansion of supercritical solutions (RESS): Experiments and modelling, *Part. Syst. Charact.*, 19, 327, 2002.
31. Reverchon, E. and Donsi, G., Salycylic acid solubilization in supercritical CO_2 and its micronisation by RESS, *J. Supercrit. Fluids,* 6, 241, 1993.
32. Türk, M. et al., Micronisation of pharmaceutical substances by rapid expansion of supercritical solutions (RESS): A promising method to improve the bioavailability of poorly soluble pharmaceutical agents, *J. Supercrit. Fluids,* 22, 75, 2002.
33. Helfgen, B., Türk, M. and Schaber, K., Hydrodynamic and aerosol modelling of the rapid expansion of supercritical solutions (RESS), *J. Supercrit. Fluids,* 26, 225, 2003.
34. Diefenbacher, A. and Türk, M., Phase equilibria of organic solid solutes and supercritical fluids with respect to the RESS process, *J. Supercrit. Fluids,* 22, 175, 2002.
35. Orlovic, A. and Skala, D., Materials processing, using supercritical fluids, *Hem. Ind.*, 59, 213, 2005.
36. Sun, Y.P. et al., Polymeric nanoparticles from rapid expansion of supercritical fluid solution, *Chem. Eur. J.*, 11, 1366, 2005.
37. Moribe, K., Micronisation of phenylbutazone by rapid expansion of supercritical CO_2 solution, *Chem. Pharm. Bull.*, 53, 1025, 2005.
38. Türk, M. et al., Micronization of pharmaceutical substances by the rapid expansion of supercritical solutions (RESS): A promising method to improve bioavailability of poorly soluble pharmaceutical agents, *J. Supercrit. Fluids,* 22, 75, 2002.
39. Sane, A. and Thies, M.C., The formation of fluorinated tetraphenylporphyrin nanoparticles via rapid expansion process: RESS vs RESOLV, *J. Phys. Chem. B*, 109, 19688, 2005.
40. Türk, M., Herstellung organischer nanopartikeln und deren stabilisierung in wässrigen lösungen (RESSAS), *Chem. Ing. Tech.*, 75, 792, 2003.
41. Türk, M. and Wahl, M., Utilization of supercritical fluid technology for the preparation of innovative carriers loaded with nanoparticular drugs, *2004 International Congress for Particle Technology* (*PARTEC*). Ed. S.E. Pratsinis, Nueremberg, 2004.
42. Türk, M., Untersuchungen zum Coating von submikronen Partikeln mit dem CORESS-Verfahren, *Chem. Ing. Techn.,* 76, 3, 2004.
43. Thakur, R. and Gupta, R.B., Formation of phenytoin nanoparticles using rapid expansion of supercritical solution with solid cosolvent (RESS-SC) process, *Int. J. Pharm.*, 308, 190, 2006.
44. Kümmel, R., Weiß, C. and Bertling, J., Smart materials: New ways to microstructured particles of environmental concern, *Adv. Environ. Mat.,* II, 175, 2001.
45. Weber, A. and Kümmel, R., Nanoparticles by precipitation in supercritical fluids – possibilities and limitations, in *Proceedings of the Seventh Meeting on Supercritical Fluids,* Antibes, France, 2001, 49.
46. Weber, A. et al., A production plant for gas antisolvent crystallization, in *Proceedings of the Fifth Meeting on Supercritical Fluids,* Nice, France, 1998, 281.
47. Weber, A., Tschernjaew, J. and Kümmel, R., Coprecipitation with compressed antisolvents for the manufacture of microcomposites, in *Proceedings of the Fifth Meeting on Supercritical Fluids,* I, 1999, 243.
48. Krause, H., Niehaus, M. and Teipel, U., German Patent 19711393, 1998.

49. Yeo, S. et al., Supercritical anti-solvent process for a series of substituted para-linked aromatic polyamides: phase equilibria and morphology, *Macromolecules*, 26, 6207, 1993.
50. Berends, E.M. et al., Crystallization of phenanthrene from toluene with carbon dioxide by the GAS process, *AIChE J.*, 42, 431, 1996.
51. Perez de Diego, Y. et al., Opening new operating windows for polymer and protein micronisation using the PCA process, *J. Supercrit. Fluids,* 36, 216, 2006.
52. Thiering, R., Dehghani, F. and Foster, N.R., Current issues relating to anti-solvent micronisation techniques and their extension to industrial scales, *J. Supercrit. Fluids,* 21, 159, 2001.
53. Weiß, C. et al., Modeling mass transfer and crystallization in disperse systems, *Chem. Eng. Technol.*, 23, 485, 2000.
54. Weidner, E., Powder generation by high pressure spray processes, in *Proceedings of the International Meeting on High Pressure Chemmical Engineering*, Karlsruhe, Wissenschaftliche Berichte FZKA 6271, 1999, 217.
55. Knez, Z., Micronisation of pharmaceuticals using supercritical fluids, in *Proceedings of the Seventh Meeting on Supercritical Fluids*, Antibes, France, 1, 21, 2000.
56. Weidner, E., Petermann, M. and Blatter, K., Die herstellung von pulverlacken durch versprühen gashaltiger schmelzen, *Chem. Ing. Techn.*, 72, 743, 2000.
57. Münüklü, P., *Particle Formation of Ductile Materials using the PGSS Technology with Supercritical Carbon Dioxide,* Ph.D. Thesis, University Delft, Netherlands, 2005.
58. Weidner, E., Petermann, M. and Knez, Z., Multifunctional composites by high pressure spray processes, in *Current Opinion in Solid State and Material Science*, Netherlands, Elsevier, 385, 2004.
59. Weidner, E., Powderous composites by high pressure spray processes, in *Proceedings of the Sixth International Symposium on Supercritical Fluids*, Versailles, France, 1483, 2003.
60. Rantakylä, M., *Particle Production by Supercritical Antisolvent Processing Techniques,* PhD Thesis, Helsinki University of Technology, Helsinki, 2004.
61. Weber, A. et al., A production plant for gas antisolvent crystallization, in *Proceedings of the Fifth Meeting on Supercritical Fluids*, Nice, France, 1998, 3, 281.
62. Tuerk, M., Erzeugung von organischen Nanopartikeln mit überkritischen Fluiden, Habilitationsschrift, Karlsruhe, 2001.
63. Grüner, S., *Entwicklung eines Hochdrucksprühverfahrens zur Herstellung hochkonzentrierter, flüssigkeitsbeladener Pulver,* Ph.D. Thesis, University Erlangen, Germany, 1999.
64. Grüner, S. et al., CPF-Process for powder generation from liquids, in *Proceedings of the Fifth Conference on Supercritical Fluids and Their Applications*, Garda, Italy, 1999, 471.
65. Grüner, S. and Otto, F., Adalbert Raps Forschungszentrum, Unpublished Data, Weihenstephan, 2005.
66. Brandin, G. et al., Tailor-made composites by high pressure spray processes, in *Proceedings of the Fifth World Congress on Particle Technology*, Orlando, FL, 2006.
67. Weidner, E. et al., Fluid-filled micro particles using PGSS technology, in *Proceedings of the Third International Meeting on High Pressure Chemical Engineering*, Erlangen, Germany, 2006.
68. Brandin, G., *Herstellung pulverförmiger Komposite mittels Hochdrucksprühverfahren,* Ph.D. Thesis, Ruhr-University, Bochum, 2005.
69. Brandin, G. et al., Fluid-filled micro particles by the use of the PGSS technology, *Advanced Powder Technology*, accepted for publication, 2006.
70. Weidner, E. et al., Production of powdery phase change materials using the PGSS process, *J. Supercrit. Fluids,* not yet published, 2006.

71. Weidner, E. et al., Manufacture of particles from high viscous melts using supercritical fluids, *rev. Proc. "High Pressure in Venice"*, ISBN 88-900775-1-4, 2, 735, 2002.
72. Petermann, M. et al., Micronization of polymer solutions by PGSS-drying, in *Proceedings of the Eighth Conference on Supercritical Fluids and Their Applications*, Ischia, Italy, 2006.
73. Weidner, E., Final report: Product engineering with neutraceuticals with superior quality, PRONUTRA, EU-Project, GRD1-1999-10212, 2003.
74. Meterc, D. et al., Extraction of natural substances and drying with PGSS process, *Slovenian Chemical Days 2006*, Maribor, 2006.
75. Meterc, D., Ph.D Thesis, Ruhr-University Bochum, in preparation.
76. Reverchon, E., Della Porta, G. and Falivene, M.G., Process parameters controlling the supercritical anti-solvent micronization of some antibiotics, in *Proceedings of the Sixth Meeting on Supercritical Fluids, Chemistry and Materials,* Nottingham, UK, 1999.
77. Chroenchaitrakool, M. et al., Micronization by RESS to enhance the dissolution rates of poorly water soluble pharmaceuticals, in *Proceedings of the Fifth International Symposium on Supercritical Fluids*, Atlanta, GA, 2000.
78. Foster, N.R. et al., Application of dense gas techniques for the production of fine particles, *AAPS PharmSci*, 5, 2, 105–111, 2003.
79. Manning, M.C. et al., U.S. Patent 5 770 559, 1998.
80. Martin, H.-J. et al., Nanoscale particles for pharmaceutical purpose by rapid expansion of supercritical solutions (RESS). Part II: Characterization of the product and use, in *Proceedings of the Seventh Meeting on Supercritical Fluids*, Antibes, France, 2000, 1, 53.
81. Frank, S.G. and Ye, C., Small particle formation and dissolution rate enhancement of relatively insoluble drugs using rapid expansion of supercritical solutions (RESS) processing, in *Proceedings of the Fifth International Symposium on Supercritical Fluids (CD-ROM)*, 2000.
82. Foster, N.R. et al., Processing pharmaceuticals using dense gas technology, *Proceedings of the Fifth International Symposium on Supercritical Fluids (CD-ROM)*, 2000.
83. Warwick, B. et al., Micronization of copper-indomethacin using gas anti-solvent processes, *Ind. Eng. Chem. Res.*, 41, 8, 1993–2004, 2002.
84. Thiering, R., Dehghani, F. and Foster, N.R., Micronization of model proteins using compressed carbon dioxide, in *Proceedings of the Fifth International Symposium on Supercritical Fluids*, Atlanta, 2000.
85. Thiering, R. et al., The influence of operating conditions on the dense gas precipitation of model proteins, *J. Chem. Technol. Biotechnol.*, 75, 29–41, 2000.
86. Thiering, R. et al., Solvent effects on the controlled dense gas precipitation of model proteins, *J. Chem. Technol. Biotechnol.*, 75, 42–53, 2000.
87. Cocero, M.J., Ferrero, S. and Vicente, S., GAS crystallization of β-carotene from ethyl acetate solutions using CO_2 as antisolvent, in *Proceedings of the Fifth International Symposium on Supercritical Fluids*, Atlanta, 2000.
88. Reverchon, E. et al., in *Proceedings of the Sixth Conference on Supercritical Fluids and Their Applications*, Maiori, Italy, 2001, 301.
89. Subra, P. and Vega, A., in *Proceedings of the 15th International Congress on Chemical and Process Engineering*, CHISA 2002, Prague, Czech Republic, 2002.
90. Pellikaan, H.C. and Wubbolts, F.E., Nozzel construction for particle formation using supercritical antisolvent precipitation, *in Proceedings of the Sixth International Symposium on Supercritical Fluids*, Versailles, France, 2003, 1765.
91. Reverchon, E. et al., Supercritical fluid antisolvent micronisation of some biopolymers, *J. Supercrit. Fluids*, 18, 239, 2000.

92. Jung, J., Clavier, J.Y. and Perrut, M., Gram to kilogram scale-up of supercritical anti-solvent process, in *Proceedings of the Sixth International Symposium on Supercritical Fluids*, Versailles, France, 2003, 1683.
93. Magnan, C. et al., Soy lectin micronisation with a compressed fluid antisolvent—influence of process parameters, *J. Supercrit. Fluids*, 19, 69, 2000.
94. Nagahama, K. and Kamoshita, C., in *Proceedings of the Fourth International Symposiom on High Pressure Process Technology and Chemical Engineering*, Venice, Italy, 2002.
95. Sloan, R. et al., Supercritical fluid processing: Preparation of stable protein particles, in *Proceedings of the Fifth Meeting on Supercritical Fluids*, Nice, France, 1998, 1, 301.
96. Reverchon, E., Supercritical anti-solvent precipitation: Its application to microparticle generation and products fractionation, in *Proceedings of the Fifth Meeting on Supercritical Fluids*, Nice, France, 1998, 1, 221.
97. Petermann, M. et al., CPF—Concentrated powder form—A high pressure spray agglomeration technique, in *Proceedings of Spray Drying 01 Conference,* Dortmund, Germany, 2001, 143.
98. Wehowski, M. and Weidner, E., Water-containing agglomerates by high pressure spraying according to the concentrated powder form (CPF) process, *CIT*, 77, 3, 274–278, 2005.

Index

α-Carotene, sources of, 281
α-Linolenic acid, 57, 78
α-Tocopherol, 113, 115
β-Carotene
    Chrastil parameters of from fish oil, 160
    fruit and vegetable oil extraction and, 86
    overview of, 56, 196–197
    physical properties of, 53
    separation of isomers of, 196–198
    sources of, 281
β-Cryptoxanthin, 281

A

Absorption, 26
Acetone, 34
Acorns, 63, 70
Activity coefficients, 13
Adlay seed oil, 223–224
Adsorption
    chromatographic separations and, 156
    concentrated powder form process and, 377
    extraction process and, 26, 327–328
    liquid feed extraction and, 321
    procyanidin extraction and, 233
    solute separation and, 218
Aerosol solvent extraction systems (ASES), 372–373
Agglomeration, 371, 376–378, 380
Aging, free radical theory of, 276–277
Aglycons, 280
Aguaribay, 245
Ajwain, 355
Algae, 192. *See also* Microalgae
Alkaloids, 342
Almonds, 52, 57, 59, 62, 63, 66, 69–71
Amaranthus grain, 80, 81, 85
Ammonia, 3
Andreadoxa, 249
Anethole, 261, 314
Angelica sinensis, 228–230
Anise, 261
Annatto, 245
Anthocyanadins, 280
Anthocyanins, 342
Anticaking agents, 343
Antimicrobial activity, 285
Antioxidants
    A. sinensis, *L. chuanxiong* hort and, 228
    carotenoids as, 281, 284
    conventional solvent extraction of, 292–293
    determination of activity of, 285–286
    effect of pressure and temperature on extraction of, 289–292
    lycopene as, 56
    overview of, 275–276
    overview of natural, 277–282
    phenolics as, 280, 282–283
    SC-CO_2 extraction of, 286–292
    spices and, 342
    terpenoids as, 280–281, 283
    tocols and, 59–60
    types and regulation of, 276–277
    vitamin E as, 281–282, 284
Antisolvent extraction. *See* Supercritical antisolvent extraction
Antisolvent processes, 369, 372–375. *See also Specific processes*
Apricots, 73, 79
Aqueous alkaline extraction, 346
Arachidonic acid, 57
Arnica, 249
Aroeira, 249
Aromatic compounds, 11, 245–247
Arruda de serra, 249
Artemisia, 249
Artemisinin, 312
Arteriosclerosis, 142
Arthritis, 142–143
Arthrospira (Spirulina) spp., 205–209
ASES. *See* Aerosol solvent extraction systems
Astaxanthin, 193, 195, 198–201
ATBC study, 56
Atherosclerosis, 56, 60, 228
Atomization, 33–34
Avocado, 249
Ayurveda, 338

B

Baccharia, 249
Bacuri, 244, 248
Bamboo piper, 245
Basil, 245, 249
Batch reactors, 20
Bergamot peel oil, 328
BHA. *See* Butylated hydroxyanisole
BHC. *See* Hexachlorocyclohexanes
BHT. *See* Butylated hydroxytoluene
Binaries behavior, 15–16
Binding, 377
Binodal curves, 19

Bioethanol, 15
Bioflavonoids, 342
Biofuels, 126, 192
Black mustard, 345
Black oreo, 147, 148
Black pepper
  bioactive compounds from, 244, 245
  biologically active constituents of, 345
  extraction of bioactive compounds from, 359
  solubility of oils in SC-CO_2 and, 363, 364
Black wattle, 249
Blends, 346
Blowout discs, 41
Boldo, 249
Borage, 73, 79
Botrycoccus braunii, 191–193
Brazil, 244
Brazilian ginseng, 249, 261
Breakage, 324
Broken-intact cells, 324, 325–326
Buriti fruit, 87, 248
Bushy lippia, 245
Butylated hydroxyanisole (BHA), 276, 277
Butylated hydroxytoluene (BHT), 276, 277

C

Cacao, 249
Caking, 343
Calcium stearate, 343
Campesterol, 58
Camphor, 313
Cancers
  adlay seed oil and, 223–224
  carotenoids and, 284
  lycopene and, 56–57
  omega-3 fatty acids and, 142–144
  phenolics and, 282, 283
  sitosterol and, 59
  squalene and, 58
  terpenoids and, 283
  tocols and, 60
  Vitamin A and, 146
  vitamin E and, 284
Candida antarctica lipase, 158
Canthaxanthin, 193, 195, 200, 281
Cap automation mechanisms, 29–30
Caprylic acid methyl ester, 78
Capsaicin, 252, 342, 356–357
Capsules, 52
Caraway, 360–361
Cardamon, 344, 359–360, 361
Cardiovascular disease
  carotenoids and, 284
  phenolics and, 282
  polyunsaturated fatty acids and, 57
  terpenoids and, 283
  tocols and, 60
  vitamin E and, 284
CARET study, 56
Carnahan-Starling equation, 7
Carotenes, 56, 193. *See also* β-Carotene
Carotenoids
  algae and, 193–195, 198–201, 206–209
  as antioxidants, 280, 281
  biological properties of, 284
  cosolvents and, 89
  fruit and vegetable oil extraction and, 86
  rice germ extraction and, 84
  solubility of in SC-CO_2, 290
  in specialty oils, 56–57
  spices and, 342
Carrier materials, extraction process and, 26
Carrots, 87, 90
Caryophyllene, 313, 316
Cascading extraction vessels, 29
Cashews, 250
Cassia, 344
Catalysts, 105–106
Catechins, 280
Celery seed extraction, 354–356
Cell cycle, 284
Cellular location, 324
Cellulosic structure, 252–254
Cereal oils, 80–84
Chamazulene, 314
Chamomile
  bioactive compounds from, 245
  composition of essential oil from, 315–316
  essential oil extraction from, 314
  overall extraction curve for, 259
Charge time, 29–30
Cherry, 58, 72, 73, 79, 80
Chile, 244
Chilean hops, 250
Chili extracts, 356–357
Chili peppers, 346
Chlorella vulgaris, 192, 193–196, 202
Chocolate-kirsch composites, 380, 381
Cholesterol, 58, 59, 145, 338
Chrastil correlation, 159, 160, 363–364
Chromatography
  gas chromatograph with electrical conductivity detector analysis, 233–234
  high-speed countercurrent, 224
  overview of, 156
  supercritical fluid, 36–37, 173–175
Chromobacterium viscosum lipase, 157
Cinnamon, 341, 344, 355, 362
Citronella, 245
Cleaning-in-place, 35
Cloudberries, 87
Clove bud oils, 225–228, 244, 245, 314–316

Cloves
biologically active constituents of, 344
cost of manufacturing and, 261
extraction of by various methods, 353
therapeutic benefits of, 341
yields and concentrations of active ingredients from, 355
CMC-Na, 222–223
CO_2 (SC)
advantages of processing with, 52, 158, 276, 338
cereal oil extraction in, 80–84
cosolvents, TCM processing and, 220–221, 222
fruit and vegetable oil extraction in, 84–90
liquid-liquid immiscibility and, 11
nut oil extraction in, 62–72
processing of TCM and, 220
seed oil extraction in, 72–80
solvent properties of, 3
specialty oil extraction in, 61–62
TCM processing and, 217–219
Coca, 250
Cod, 143, 153
Codex Alimentarius, 277
Coffee, 26, 246
Cold processing, specialty oils and, 52
Collection, high-pressure spray processes and, 370
Color
algae and, 193
fruit and vegetable oil extraction and, 90
nut oil extraction and, 71
paprika extraction and, 357
seed oil extraction and, 78, 80
spices and, 343
Composites, 376–380
Compounds, 28–30, 31–34
Concentrated powder form (CPF) process, 370, 376–378, 382
Concretes, 310, 312, 316–318
Conjugated double bonds, 56
Constant extraction rate periods, 257–260
Control systems, 41–42
Conventional solvent extraction
of antioxidants, 292–293
cereal oil extraction and, 84
fruit and vegetable oil extraction and, 90
nut oil extraction and, 72
seed oil extraction and, 80
specialty oils and, 52
Copaiba, 250
Copper Reduction Assay (CUPRAC), 286
Coriander, 246, 344, 355, 360–361
Coronaridine, 261
Cosolvents
A. sinensis, *L. chuanxiong* hort and, 229–230
algal extraction and, 200–201
BHC extraction from radix ginseng and, 235–236
cereal oil extraction and, 83
essential oil extraction and, 308
fruit and vegetable oil extraction and, 86, 89
heat treatment for removal of, 91
nut oil extraction and, 69, 71
polarity and, 6, 349
processing of TCM and, 220–221, 222
procyanidin extraction and, 231–232
SC-CO_2 extraction and, 166–168
seed oil extraction and, 77–78
solubility and, 123
vegetable oils as, 89
vitamin E and, 352
Cost estimates
cellulosic structure and, 252–254
industrial process implementation and, 44–48
for PGSS, 375–376
for selected Latin American plants, 260–261
selection of parameters for, 254–260
SFE for Latin American plants and, 243, 254–262
spice oil extraction and, 350–351
Cost of manufacturing (COM). *See* Cost estimates
Countercurrent extraction columns, 18, 31–32, 36, 328
Couplings. *See* Drive couplings
CPF (concentrated powder form) process, 370, 376–378, 382
Critical curves, 115
Critical point, 2
Critical pressure, 115
Critical properties, 3, 5, 30
Crossover effects, 30, 348
Croton, 246
Cryoprotection, 58
Crystallization, 151–152, 152–156, 372–373
Cumin, 341, 344, 353, 355
CUPRAC. *See* Copper Reduction Assay
Cupuassu, 248
Curcuma longa, 365
Curcumin, 343
Cuticular waxes, 310, 315–316

D

Decaffeination, 26
Degradation
fish oil extraction and, 148–149, 151
green tea leaves and, 381
hydrodistillation and, 309
lipid oxidation and, 276–277
molecular distillation of tocopherols and, 105
Degree of extraction, 71
Degree of saturation, 152–156
Dehydration, 11–15

Dense-gas fractionation, 17–18
Density
 fog phenomenon and, 126
 nut oil extraction and, 71
 overview of, 4
 phase equilibrium and, 2–3
 solvent power and, 1
Deodorizer distillate (DOD), 104, 108. *See also* Tocopherols
Depressurization rate, 27–29, 67–68
Deterpenation, 320
DHA. *See* Docosahexaenoic acid
Diacyl glycerol ethers, 144–146, 176–178
Diallyl sulfides, 361–362
Diffusion coefficients, 1, 4
Diffusion-controlled rate periods, 257–260
Dilophus ligulatus, 192
Diphtheria, 338
Discharge time, 29–30
Dispersal, 369
Dissolution, 369
Distillation
 celery seed essential oil and, 356
 essential oils and, 309
 of fish oils, 149–151
 of menhaden oil, 150
 problems with tocopherol concentration using, 104–105
 SFE and for TCM processing, 223–224
 spice constituent extraction and, 353
 spice extraction and, 346–347
Distribution coefficients, 113, 115
Diterpenes, 341
Docosahexaenoic acid (DHA), 57, 142–144, 168–169
Docosapentanoic acid (DPA), 57
DOD. *See* deodorizer distillate. *See also* Tocopherols
DPA. *See* Docosapentanoic acid
DPPH radicals, 286
Drive couplings, 32, 33
DSS, 222–223
Dunaliella spp., 191, 192, 196–198
Dynamic axial columns, 36–37

E

Ebers Papyrus, 338
Ecdysterone, 261
Echium, 73, 79
Economics. *See* Cost estimates
EDTA. *See* Ethylenediamenetetraacetic acid
Eicosanoids, 57
Eicosapentanoic acid (EPA)
 algae and, 201–202, 202–205
 extraction of from fish oils, 142–144
 overview of, 57
 SC-CO_2 extraction of, 168–169
 structure of, 144
Emulsions, 346
Encapsulation, 380
Entrainment, 105, 201, 209
Enzymatic transformation, 156–158, 175–176
Ephedrine, 222
Equilibration time, 66, 76
Equilibria, 151–152
Equilibrium calculations, 8–9
Equipment, 34–35, 219–220
Erva baleeira, 246
Essential oils
 antisolvent extraction and, 322–323
 celery seed extraction and, 356
 examples of, 320–322
 extraction of from flowers, 314, 316–318
 extraction of from leaves, 312–313
 extraction of from seeds, 314
 flower concretes fractionation and, 316–318
 ginger extraction and, 358
 liquid feed extraction and, 318–319
 mathematical modeling of extraction of, 324–328
 operating parameter selection for, 319–340
 overview of, 305–307, 343, 346
 solids processing and, 307–312
 sources of, 311
 spices and, 341, 342
Esterification, 105–106, 156–157
Estragole, 314
Ethane, 3, 11
Ethanol. *See also* Cosolvents
 cereal oil extraction and, 83
 as cosolvent, 91
 fruit and vegetable oil extraction and, 89
 nut oil extraction and, 69
 processing of TCM and, 220–221
 seed oil extraction and, 77–78
 squalene extraction and, 168
 urea inclusion complexation and, 155–156
Ethylenediamenetetraacetic acid (EDTA), 277
Eucalyptus, 246
Eugenia carophyllata, 225–228
Eugenol, 225–227, 316
Evening primrose, 73, 79
Expansion. *See* Rapid expansion of supercritical solution (RESS) process
Explosives, 373
Extract materials, 26
Extraction. *See also* Conventional solvent extraction
 of compounds from liquid feed, 31–34
 of compounds from solid matrix, 28–30
 control systems for, 41–42
 equipment design and, 34–35
 heat exchangers for, 38–39

industrial process implementation for, 42–48
overview of, 48–49
of oxychemicals, 11–15
piping, valves and, 39–41
process development and, 218
process overview, 26–28
processing parameters for solids extraction, 30–31, 308–309
pumps and compressors for, 37–38
from vegetable matrices, 18–19
vessels for, 34, 35–37
Extraction time, 68–69, 77, 83
Extraction vessels, 29, 34, 35–37

F

Falling extraction rate periods, 257–260
FAME. *See* Fatty acid methyl esters
Fanshensu, 225
Fatty acid methyl esters (FAME), 105–106, 106–107, 126–136
Fatty acids
cereal oil extraction and, 83–84, 85
Chrastil parameters of from fish oil, 160
of fish oils, 143
fruit and vegetable oil extraction and, 89–90
nut oil extraction and, 70
seed oil extraction and, 78, 79
Feed materials, 26
Fenchone, 314
Fennel, 246, 261, 314, 360–361
Fermenters, algae and, 191
Ferric Reducing Antioxidant Power (FRAP), 286
Ferulic acid, 228–230
Fish oils
chromatographic separations of, 156
distillation of, 149–151
enzymatic transformation of, 156–158
low-temperature crystallization of, 151–152
omega-3 fatty acids of, 142–144
overview of, 141–142
overview of separation and fractionation technologies for, 147–149
phase equilibria of in SC-CO_2, 158–168
polyunsaturated fatty acids and, 57, 168–176
squalene and diacyl glycerol ethers of, 144–146, 176–178
urea crystallization of, 152–156
vitamin A and, 146, 178–181
wax esters of, 146–147, 181
Fixed costs, 260–261
Flammability, 18–19
Flavanones, 280
Flavones, 280, 313
Flavonoids, 280
Flavonols, 280
Flax, 74, 79
Flow control valving, 39–40
Flow rates and directions
cereal oil extraction and, 83
fruit and vegetable oil extraction and, 86, 89
nut oil extraction and, 68
seed oil extraction and, 77
solids extraction processing parameters and, 31
Flowers, 251, 313, 314, 316–318
Fog phenomenon, 126
Fractional extraction processes, 26–27, 150–151
Fractionation
of essential oil extracts, 306, 310
FAME removal during tocopherol concentration and, 126–136
of oils, 17–18
of rose concrete, 317–318
spice extraction and, 354
Fragrances, 316–318, 358, 359, 361
FRAP. *See* Ferric Reducing Antioxidant Power
Free fatty acid esters, 163
Free fatty acids, 105–106
Free radicals, 276–277, 285–286
French paradox, 282
Fruit oils, 84–90
Fugacity, 4–6
Fugacity coefficients, 6

G

Gallates, 276, 277–278
Gamma-linoleic acid (GLA), 57 (all-*cis*-6,9,12-octadecatrienoic acid), 205–206
Garlic, 338–340, 344, 361–362
GAS (gas antisolvent) process, 15, 372–374, 382
Gas Chromatograph with Electrical Conductivity Detector (GC-ECD) analysis, 233–234
Gas recycling, 370
Gas salting out effect, 4
Gas saturated liquids (PGSS), 369, 373–375, 382
Gases, physical properties of, 4
Gas-liquid alternating circulation system, 107–109
General expenses, 260–261
Genetic engineering, 191
Ginger
bioactive compounds from, 244, 246, 344
cost of manufacturing and, 261
extraction of bioactive compounds from, 358
extraction of by various methods, 353
overall extraction curve for, 259, 260
return on investment and, 47, 48
therapeutic benefits of, 341
yields and concentrations of active ingredients from, 355
Ginseng, 233–236, 249, 261
GLA. *See* Gamma-linoleic acid
Glycerides, 105–106

Glycosides, 342
Good Manufacturing Practice (GMP) compliance, 47–48, 375
Grapefruit, 250
Grapes, 74, 79, 230–233, 250
Green pepper basil, 246
Green tea leaves, 381–384
Green-lipped mussel oil, 181
Grinding, 76, 86, 368
Group contribution equation of state (GC-EOS) model, 7
Guaco, 250
Guarana, 250

H

Haematococus pluvialis, 192, 198–201
Halibut, 143
HAT. *See* Hydrogen atom transfer
Hazard and Operatability (HAZOP) studies, 38
Hazelnuts, 63, 66, 69, 70
HD. *See* Hydrodistillation
Heat exchangers, 38–39
Heat treatment, 91
Helmholtz residual energy, 7
Hemicellulose, 69
Herring, 143
Hexachlorocyclohexanes (BHC), 233–236
Hexane, 80, 347, 353, 355
Hibiscus, 74, 79
High-critical temperature (high Tc) fluids, 3
High-pressure spray processes, 368–376, 376–380, 384
High-speed countercurrent chromatography (HSCCC), 224
Hiprose, 74, 87
Hoki liver oil, 172
Horsetail (giant), 246
HSCCC. *See* High-speed countercurrent chromatography
Hybrid hibiscus, 74, 79
Hydrocarbons, 55, 160, 192–193
Hydrodistillation (HD), 309, 346–347, 356
Hydrogen atom transfer (HAT), 286
Hydrolysis, 151, 156–157, 309
Hyperforin, 312
Hypertension, 228, 338
Hypnea charoides, 192, 201–202

I

Impregnation, 377
Industrial process implementation, 42–48
Inflammation, 57, 282
Interactions, 324
Interfacial tension, 4
Isochrisis galbana, 192
Isoflavones, 280
Isofugacity criterion, 4–5
Isolation valving, 39–40
Isomerization, 151
Isoprenoids. *See* Terpenoids
Isopropanol, 15

J

Jackfruit, 250
Jalapeno peppers, 250
Jojoba, 248, 341

K

Kanglaite Injection, 223–224
Khoa, 246
Kinetics, 257–258
Kirsch, 380, 381
Koenen and Gaube diagrams, 12

L

Labor, 47
Latin American plants
  cost estimates and, 243, 254–262
  examples of SFE from, 244–252
  overview of SFE of bioactive compounds from, 243–244
  SFE process for, 252–254
Lavender, 314
Leaves, 312–313
Lecithin, 33–34
Lemon verbena, 247
Lemongrass, 247
Licorice, 346
Limonene, 314, 316, 320, 327
Linalool, 314, 321, 327, 361
Lingusticum chuanxiong hort, 228–230
Linoleic acid, 55, 57
Linolenic acid, 69, 78, 90, 201–202, 230–233
Lipases, 156–158
Lipids, 18–19, 190, 248, 276–277, 281–282
Lippia sidoides, 247
Liquid feeds, 31–34, 306
Liquid-filled composites, 378–380
Liquid-liquid equilibrium, 7
Liquid-liquid immiscibility, 9–11
Liquids, 4
Lobenzarit preparation, 16
Low-critical temperature (low Tc) fluids, 3
Low-temperature crystallization, 151–152
Lunaria, 79
Lutein, 53, 84, 86, 200

Lycopene, 53, 56–57, 86, 89, 281
Lyprinol, 181

M

Mace, 346
Macela, 247
Mackerel, 143
Macroalgae, 190
Macroporous resin adsorption technology, 233
Magnesium stearate, 343
Mangos, 250
Manufacturing costs. *See* Cost estimates
Marigolds, 247, 250
Marine macroalgae, 190
Mass transfer models, 255, 325–327
Mastranto, 247
Matricine, 314
MD extraction, 224
Mechanical mixing, 32–33
ME-DOD. *See* Methyl esterified DOD
Melting, 369
Menhaden, 143, 150
Mesityl oxide, 34
Methanol, 3, 155–156
Methanolysis, 105–106
Methyl esterified DOD (ME-DOD). *See also* Tocopherols
  composition of, 130
  FAME removal from, 126–129
  phase behavior of, 106–124, 124–126
  pressure and, 131–134
  pretreatment and, 105–106, 129–131
Methyl oleate
  distribution coefficients of, 113, 115
  FAME removal and, 106–107, 110–111
  gas-liquid interface and, 117
  separation factor of in SC-CO_2 fractionation, 122–123
Microalgae
  Botrycoccus braunii, 191–193
  Chlorella vulgaris, 193–196
  Dunaliella spp., 196–198
  Haematococus pluvialis, 198–201
  Hypnea charoides, 201–202
  Nannochloropsis spp., 202–205
  overview of, 189–191, 209, 244
  Spirulina spp., 205–209
Microemulsion. *See* Surfactants
Micronization, 13, 15–17. *See also* Supercritical antisolvent micronization
Micro-particles, 367–368. *See also* High-pressure spray processes
Migration rates, 156
Milk thistle, 74
Minerals, 69
Miscibility, 9–11, 115
Mixtures, 43
Modeling, 7, 255, 324–327
Modified Huron-Vidal 2 (HHV2) model, 7
Modifiers, 290–291. *See also* Cosolvents
Moisture content
  cereal oil extraction yield and, 80
  fruit and vegetable oil extraction and, 86
  nut oil extraction and, 66
  seed oil extraction and, 76
  solids extraction processing parameters and, 31
  spice extraction and, 350
Molecular distillation, 104–105, 150
Mongolia mushrooms, 225
Monoterpenes, 341, 355
Morphology, 31
Mortierella sp., 192
mRNA, 284
Mullet, 143
Multicomponent fluids, 8
Multiplunger pumps, 37–38
Multistage extraction, 308
Munch, 72, 74, 79
Myristicin, 359

N

Nannochloropsis spp., 192, 202–205
Nano-particles, 367–368, 371. *See also* High-pressure spray processes
Nanosuspensions, 371
Natural products. *See* Traditional Chinese medicines and natural products
Near-critical region, 2, 7, 8–9
Nebuilizing, 32
Neem, 79
n-Hexane, 3, 109, 119, 168, 291
Nonclassical supercritical effects, 13–14
Non-Random Two Liquids (NRTL) model, 7
Nut oils, 62–72
Nutmeg, 346, 358–359
Nutraceuticals, 342, 380–384

O

Oats, 81, 85
Ochronomas danica, 192
Odor, 208. *See also* Fragrances
OEC. *See* Overall extraction curves
Oils, 17–19
Oleic acids, 69, 78
Oleoresins
  cost of manufacturing of, 261
  defined, 244, 342
  ginger extraction and, 358
  from Latin American plants, 245–247, 252
  liquid solvent extraction and, 310

overview of, 343, 346
sources of, 311
vanilla extraction and, 359
Olives, 87, 248
Omega-3 fatty acids, 142–144
Onions, 341
Operating costs. *See* Cost estimates
Optimization, 9
ORAC. *See* Oxygen Radical Absorbance Capacity
Orange (sweet), 247
Orange roughy, 147, 148
Oregano, 247, 327
Organic acids, 277
Organochlorine pesticide, 233–236
Osteoarthritis, 60
Overall extraction curves (OEC), 256, 257–260
Oxidation. *See also* Antioxidants
CO_2 processing and, 52
distillation of fish oils and, 151
fish oil extraction and, 149
nut oils and, 62, 72
seed oil extraction and, 80
squalene and, 145
Oxychemicals, 11–15
Oxygen Radical Absorbance Capacity (ORAC), 286
Oxygenated compounds, 341

P

Palmarosa, 247
Palmitic acid, 69, 90
Palms, 248
Paprika powder, 248, 357–358
Paraffins, 11, 310
Paragual, 244
Particle shape, 37, 324–325, 368, 379
Particle size
A. sinensis, *L. chuanxiong* hort and, 228–229
cereal oil extraction yield and, 80
clove bud oils and, 226
fruit and vegetable oil extraction and, 86
nut oil extraction and, 62, 66
seed oil extraction and, 76
solids extraction processing parameters and, 31
Particulates, 367–368, 368–376
Passion flower, 251
Passion fruit, 248
PCA. *See* Precipitation with compressed antisolvent
Peanuts, 64, 66
Pecans, 64, 66, 69, 70
Pectins, 69
Peel oils, 320
PEG, 380
Pejibaye, 248
Pepper, 355
Percolation method, 230
Peroxidation, 281–282
Pesticides, 233–236
PGSS (gas saturated liquids), 369, 373–375, 382
Phaffia rodozyma, 192
Phase equilibria
chromatographic separations and, 156
of fish oils in SC-CO_2, 158–168
for Latin American bioactive compounds, 253
of methyl oleate-DOD, 124–126
multiple, 6–8
overview of, 9–11
phase equilibrium analyzers and, 43
solid solubilities and, 4–6
tocopherol concentration and, 110–113, 117–119
Phase equilibrium analyzers, 43
Phase equilibrium diagrams, 2
Phase equilibrium engineering, 8–11, 20
Phenolics
as antioxidants, 280
biological properties of, 282–283
effect of pressure and temperature on yield of, 291
as primary antioxidants, 277
solubility of in SC-CO_2, 290
Phenylpropanoids, 342
Photobioreactors, 191
Photosynthetic capacity, 190–191
Phthalides, 355
Physical properties, overview of, 4
Phytochemicals, defined, 342
Phytoplankton, 189
Phytyl chains, 59
Pigments, 56, 193, 202–205, 281, 360
Pilayella littorallis, 192
Pink trumpet tree, 251
Piperine, 359
Piping, 39–41
Piprioca, 247
Pistachios, 64, 69
Pitanga, 251
Plant polyphenols, 230–233
Polarity, 6, 217, 229, 349
Polyethylene Terephthalate (PET) films, 252
Polymerization, 151
Polyphenols, 381–384
Polysaccharides, 225
Polyunsaturated fatty acids
algae and, 201–202
extraction of from fish oils, 168–176
of fish oils, 142
in specialty oils, 57–58
vitamin E and, 60
Potassium stearate, 343
Pravastatin, 58

Precipitation with compressed antisolvent (PCA), 372–373
Preservation, antioxidants and, 276
Pressure. *See also* Depressurization rate
CO_2 extraction and, 218, 289–290
FAME removal during tocopherol concentration and, 131–134
fruit and vegetable oil extraction and, 86
nut oil extraction and, 66–68
phase equilibrium and, 2
seed oil extraction and, 76–77
selection of cost estimate parameters and, 254–257
solvent recycle and, 45, 46
tocopherol SC-CO_2 concentration and, 117–119
Pressure reduction, 218
Pressure vs. temperature diagrams, overview of, 8
Pressured fractional distillation, 223–224
Pretreatments, 105–106, 116–118, 129
Preventative antioxidants, 277
Primary antioxidants, 277
Primrose, 73, 79
Process design
extraction from vegetable matrices and, 18–19
fractionation of oils and, 17–18
overview of, 26–28
oxychemical extraction, dehydration and, 11–15
particle micronization and, 15–17
supercritical reactions and, 19–20
Processing parameters, 30–31
Procyanidins, 230–233
Propane, 3, 13–14, 18–19, 359–360
Prostaglandins, 57
Protocatechualdehyde, 225
Provitamin A, 342. *See also* β-Carotene
Pseudomonas sp., 157, 158
Psoralen, 224
Pumpkins, 75, 79
Pumps and compressors, 37–38
Pupunha, 248

Q

Quinones, 277

R

Radix ginseng, 233–236
Raffinate, 31–32
Rapeseed, 248
Rapid expansion of supercritical solution (RESS) process
cost studies of, 375–376
nutraceuticals produced by, 382
overview of, 369, 370–371
solubility and, 15
spraying of gas saturated liquids and, 373–375
Recycling, 9, 44–47, 174, 370
Red chili extraction, 346, 356–357
Redundancy, control systems and, 41
Regulation of antioxidants, 277–278
Relative volatility, 18
Residence time, 31
Residual oils, 48
Resins, 342. *See also* Oleoresins
Resistance temperature detector (RTD) sensors, 41
Respiratory tract, 338
RESS. *See* Rapid expansion of supercritical solution process
Retinol, 146
Retrograde behavior, 5
Revenue estimates, 47
Rice bran, 81–82, 84, 85, 248
Roasting, 62
Rose concrete, 317
Rose hips, 75, 79, 248
Rosemary, 247, 261

S

SAE. *See* Supercritical antisolvent extraction
Safety, 41, 233–236
Saffron, 244, 346
Sage, 312–313
Salmon, 143
Saponification values, 107
Saponins, 342
Saprolegnia parasitica, 192, 195–196
Sardine oil, 171
SAS. *See* Supercritical antisolvent micronization
Saturation degree of, 152–156
SC-CO_2. *See* CO_2-SC
Scenedesmus obliquus, 191, 192
SDS, 222–223
Sea buckthorn, 75
Sebum, 144
Secondary antioxidants, 277
SEDS. *See* Solution-enhanced dispersion by supercritical fluids
Seed oils
celery seed extraction and, 354–356
characteristics of products extracted from, 78–80
cosolvents and, 77–78
essential oil extraction and, 313, 314
extraction time and, 77
flow rate and direction and, 77
modeling extraction of, 327
moisture, equilibration time and, 76
overview of, 72–76
particle size and, 76

temperature, pressure and, 76–77
Selectivity, 3–4, 26–27
Sensors, 41
Sensory properties, 280, 352–353
Separability, 3
Separation, 26, 218
Sequential depressurization, 27–29
Sesame, 75, 79
Sesquiterpenes, 314, 321, 341
SET. *See* Single electron transfer
SFC. *See* Supercritical fluid chromatography
Sharks, 145
Shea nut oil, 72
Shell-and-tube heat exchangers, 39, 40
Shogaols, 358
Short-path distillation, 104–105, 150
SHS, 222–223
Sidda, 338
Silanol, 321
Silybum marianum, 72, 77
Single electron transfer (SET), 286
Sitosterol, 58
Sitosterolemia, 58–59
Skeletonema costatum, 192
Small spined oreo, 147, 148
Solid matrices, 28–30
Solids, 30–31
Solubility
of β-carotene isomers, 197–198
Chrastil correlation and, 159, 160
CO_2 extraction and, 61, 289–290
cosolvents and, 123
crystallization separations and, 151–152
of fatty acids from fish oils, 159–162
hydrodistillation and, 309–310
for Latin American bioactive compounds, 253, 256–257
of methyl oleate and α-tocopherol in SC-CO_2, 113, 114
nut oil extraction and, 67
phase equilibrium analyzers and, 43
solids extraction processing parameters and, 30
spice oils and, 362–364, 365
Solution-enhanced dispersion by supercritical fluids (SEDS), 372–373
Solvent loading, 29
Solvent power, 3, 15, 26
Solvent recycle, 9, 44–47, 174
Solvent-feed ratios, 30–31
Solvents. *See also* Cosolvents
chlorinated, 347
crystallization and, 367
density-dependent nature of, 1
high-pressure spray processes and, 369
reaction with piping surfaces and, 39
Span-80, 223
Specialty oils. *See also Specific oils*
bioactives in, 52–55
carotenoids in, 56–57
extraction of, 61–62
overview of, 52, 90–91
polyunsaturated fatty acids in, 57–58
squalene in, 58
sterols in, 58–59
tocols in, 59–61
Sperm whales, 147, 148
Spice extracts, 343, 346, 348
Spices
antioxidants from, 277
beneficial aspects of, 339–341
bioactive compounds from, 341–343
black pepper extraction and, 359
cardamom extraction and, 359–360
celery seed extraction and, 354–356
cinnamon extraction and, 362
commercial SCFE process for, 349–351
conventional extraction of, 346–347, 351–354
defined, 338
fennel, caraway, coriander extraction and, 360–361
garlic extraction and, 361–362
ginger extraction and, 358
importance of, 338–339
nutmeg extraction and, 358–359
overview of, 337–338, 364–365
paprika extraction and, 357–358
red chili extraction and, 356–357
saleable products from, 343–346
SC-CO_2 extraction of, 347–349, 351–354
solubility of oils in SC-CO_2 and, 362–364
specific therapeutic benefits of, 340–341
vanilla extraction and, 359
Spiny Dogfish, 143
Spirulina maxima, 205–207, 244
Spirulina platensis, 207–209
Spirulina sp., 192
Splinefitting, 258–260
Spray agglomeration, 376–378
Squalene
amaranth oil and, 83, 84
from fish oils, 144–146, 176–178
oil fractionation and, 17
physical properties of, 55
SC-CO_2 extraction of, 165–168, 176–178
in specialty oils, 58
Standardization, 285
Steam distillation, 346, 352–353
Sterols, 54–55, 58–59, 342
Steroptens, 317
Stevia, 244, 247, 251
Stigmasterol, 58
Stripping, 18
Structure, 324

Sucrose, 244
Sugars, 280
Supercritical reactions, 19–20
Supercritical antisolvent extraction (SAE), 306–307, 322–323
Supercritical antisolvent micronization (SAS), 306–307, 322, 372–373, 375–376, 382
Supercritical fluid antisolvent. *See* Supercritical antisolvent micronization
Supercritical fluid chromatography (SFC), 36–37, 173–175
Supercritical fluids, defined, 217, 338
Supercritical region, 2
Surface tension, 1
Surfactants, 221–223
Symposia, 219

T

Tabernaemontana, 251, 261
TAG form of polyunsaturated fatty acids, 157
Tannins, 342
Tanshinone, 225
TBHQ. *See* Tertiary butyl hydroquinone
TEAC. *See* Trolox equivalent antioxidant capacity
Tecanalysis software, 262
Temperature
  clove bud oils and, 227
  CO_2 extraction and, 218, 289–290
  distillation of fish oils and, 151
  enzymatic transformation of fatty acids and, 156–157
  fruit and vegetable oil extraction and, 86
  nut oil extraction and, 66–68
  phase equilibrium and, 2
  seed oil extraction and, 76–77
  selection of cost estimate parameters and, 254–257
  tocopherol SC-CO_2 concentration and, 117–119
Terpenes, 280, 320–321, 342
Terpenoids (isoprenoids), 144, 280–281, 283, 290, 291
Tertiary butyl hydroquinone (TBHQ), 276
Texture, 71, 290
Thar Technologies, 32, 33
Thermal conductivities, 4
Thermowell isolation, 41
Thrombosis, 142
Thyme, 346
Tobacco, 323
Tocols, 53–54, 59–61
Tocopherols
  binary phase equilibria of, 107–113
  conventional extraction and, 72
  distribution coefficients of, 113–118
  effect of pressure and temperature on yield of, 293
  equilibrium lines for, 123–124
  fundamental research on concentration of, 106–107
  molecular distillation and, 104–105
  molecular structure of, 104
  nut oil extraction and, 69
  oil fractionation and, 17
  overview of, 104, 136–138
  phase behavior of ME-DOD system and, 124–126
  pretreatment before concentration of, 105–106
  propane solvents and, 72
  regulation of, 277
  rice bran oil and, 84
  separation factor of with methyl oleate, 122–123
  separation of with SC-CO_2 fractionation, 126–136
  solubilities of, 113, 290
  ternary phase equilibria of, 118–122
  wheat germ extraction and, 83
Tocotrienols, 284
Tomatoes, 75, 79, 88, 90
Torulaspora delbrueckii, 192
Total Peroxyl Radical-Trapping Antioxidant Parameter (TRAP), 286
Traditional Chinese medicines and natural products
  of edible and medicinal ingredients from grape seeds, 230–233
  equipment made in China for, 219–220
  essential oil from clove buds and, 225–228
  of medical ingredients from *A. sinensis* and *L.chuanxiong* hort and, 228–230
  of organochloride pesticide from ginseng, 233–236
  overview of, 216–217, 236–237
  overview of active compounds extracted from, 221, 222
  SFE and enhanced separation methods and, 223–224
  SFE and ultrasound-enhanced extraction and, 223
  SFE with CO_2 in presence of solvent and, 220–221
  SFE with CO_2 in presence of surfactant and, 221–223
  SFE with pure supercritical CO_2 and, 220
  supercritical fluid processing of, 217–219
  use of combinations of extraction methods and, 224–225
Traditional processing methods. *See* Conventional solvent extraction
TRAP. *See* Total Peroxyl Radical-Trapping Antioxidant Parameter
Tree tea oil, 341
Triacylglycerols, 52, 160

Triglycerides, 69
Trilinolein, 89–90
Triple point, 2
Trolox Equivalent Antioxidant Capacity (TEAC), 286
Trout, 143
Tuberose volatile oil, 318
Tucuman, 248
Tuna oil, 172, 175
Turbines, 46–47
Turmeric
  bioactive compounds from, 244, 251, 343, 345
  solubility of essential oils from, 364
  therapeutic benefits of, 341
Tween-80, 223
Type V phase behavior, 9–10

U

Ucuuba, 248
Ultrasound, 223
Ultraviolet radiation, 145
Unani, 338
Urea crystallization, 152–156
Urea inclusion complexation, 155–156
Urea-fatty acid ratio, 152–154

V

Valves, 39–41
Vanilla, 346, 359, 360
Vanillin, 359
Vapor pressure, 149–150
Vapor-liquid equilibrium (VLE), 7
Vegetable matrices, 18–19
Vegetable oils, 17, 84–90
Vessels, 29, 34, 35–37
Vetivergrass, 247
Vinca, 251
Viscosity, 1, 4, 32, 134–136
Vitamin A, 146, 160, 178–181
Vitamin E, 277, 281–282, 284. *See also* Tocopherols
VLE. *See* Vapor-liquid equilibrium
Voacangine, 261
Volatile oils
  cost of manufacturing of, 261
  from Latin American plants, 245–247, 252
  liquid solvent extraction and, 310
  sources of, 311–312
  spices and, 341
Volatility, 18, 67
Volume, 2, 71

W

W3 fatty acids, 201–202, 223
Walnuts, 65, 70
Water, 3, 13–14
Wax ester oils, 146–147, 181
Waxes, 146–147, 181, 310, 315–316
Wertheim's statistical association fluid theory, 7
Wheat germ, 82, 85
Wheat plumule, 82
Workflows, 42–43

X

Xanthophylls, 56, 193
Xylopia aromatica, 247

Z

Zeaxanthin, 84, 281

Printed in the United States
by Baker & Taylor Publisher Services